CHROMATOGRAPHY AND CAPILLARY ELECTROPHORESIS IN FOOD ANALYSIS

RSC Food Analysis Monographs

Series Editor: P.S. Belton, *The Institute of Food Research, Norwich, UK.*

The aim of this series is to provide guidance and advice to the practising food analyst. It is intended to be a series of day-to-day guides for the laboratory worker, rather than library books for occasional reference. The series will form a comprehensive set of monographs providing the current state of the art on food analysis.

Dietary Fibre Analysis
by David A.T. Southgate, *Formerly of the AFRC Institute of Food Research, Norwich, UK.*

Quality in the Food Analysis Laboratory
by Roger Wood, *Joint Food Safety and Standards Group, MAFF, Norwich, UK*, Anders Nilsson, *National Food Administration, Uppsala, Sweden,* and Harriet Wallin, *VTT Biotechnology and Food Research, Espoo, Finland*

Chromatography and Capillary Electrophoresis in Food Analysis
by Hilmer Sørensen, Susanne Sørensen and Charlotte Bjergegaard, *Royal Veterinary and Agricultural University, Frederiksberg, Denmark*, and Søren Michaelsen, *Novo Nordisk A/S, Denmark*

How to obtain future titles on publication

A standing order plan is available for this series. A standing order will bring delivery of each new volume immediately upon publication. For further information, please write to:

Turpin Distribution Services Ltd.
Blackhorse Road
Letchworth
Herts. SG6 1HN

Telephone: Letchworth (01462) 672555

Chromatography and Capillary Electrophoresis in Food Analysis

Hilmer Sørensen, Susanne Sørensen, Charlotte Bjergegaard
Royal Veterinary and Agricultural University, Frederiksberg, Denmark

Søren Michaelsen
Novo Nordisk A/S, Denmark

ISBN 0-85404-561-9

A catalogue record for this book is available from the British Library.

Published by The Royal Society of Chemistry,
Thomas Graham House, Science Park, Milton Road, Cambridge CB4 0WF, UK

For further information see our web site at www.rsc.org

Typeset by Paston Prepress Ltd, Beccles, Suffolk, NR34 9QG
Printed by MPG Books Ltd, Bodmin Cornwall UK

Preface

Chromatography (Greek; Chromato = χρῶμα = colour, graphy = γρᾶφω = description) is an ancient word now used in another broader sense for techniques of separation of structurally different analytes between two phases.

Chromatography has evolved from liquid–liquid partition chromatography techniques originally developed in natural product chemistry at the beginning of this century.[1–8] These techniques have provided the basis for various modifications, including gas chromatography (GC) introduced about 50 years ago and the liquid chromatography (LC) separation principles often referred to as: liquid–liquid-, liquid–solid-, adsorption-, partition-, ion-exchange-, ion-pair-, size-exclusion-, solid phase extraction (SPE)-, normal- and reversed phase-, chemically bonded phase-, dynamically modified phase-, and chiral separation chromatography. Liquid chromatography (LC), with its new developments in column chromatography, comprises all of these techniques and is used as high performance liquid chromatography (HPLC) and fast polymer liquid chromatography (FPLC), in addition to techniques such as paper chromatography (PC) and thin layer chromatography (TLC). The techniques based on GC[9] are considered to be outside the scope of this book, which is limited to the analyses of amphiphilic and hydrophilic natural products including proteins with focus on enzymes and antibodies, whereas techniques concerning nucleic acid and gene technology are excluded.

Electrophoretic techniques are considered as an important part of the methods described, and these methods are not limited to the traditional techniques using a solid support for the buffer phase or solutions, as is the case for slab gel electrophoresis and high voltage electrophoresis (HVE). Focus is placed on new techniques utilizing the idea behind electrophoresis in free solution, originally known from Tiselius methods,[10] as well as isotachophoresis described by Hjerten,[11] and developed as capillary electrophoresis (CE) by Jorgenson and Lukacs.[12] Efficient techniques known as high-performance capillary electrophoresis (HPCE)[13] have been developed, including important methods based on pseudophases which were developed about ten years ago as micellar electrokinetic capillary chromatography (MECC) by Terabe *et al.*[14] HPCE and especially MECC have the potential for efficient chromatographic separations which are not limited to charged analytes but include advanced utilization of the above mentioned LC principles in separations involving low molecular weight (low-M_r) and high molecular weight (high-M_r) charged and uncharged biomolecules.[15,16] In HPCE–MECC, the electro-osmotic flow (EOF) forces the analytes to move, and the direction of EOF depends on the selected buffer and micellar system. The choice of system is made according to the structure and properties of analytes to be separated[13,16,17] utilizing the influence of appropriate modifiers and buffer components to give the required analyte partition between buffer and micellar phases in MECC.

New developments in chromatography and capillary electrophoresis that are suitable to methods in food analysis, natural product chemistry and biochemistry constitute an area where the most efficient solutions are obtained by combined use of knowledge of LC as well as the structure and properties of biomolecules, their function, and transport in living organisms. These biochemistry subjects and transport processes have many similarities with advanced ideas behind new analytical techniques. The highly efficient, specific and exciting processes in living cells have been developed as an evolution process from the beginning of life on Earth. Compared to only one century devoted so far to the development of our known chromatography and electrophoresis techniques, it is not surprising that we still have much to learn. In living cells and organisms the transport, 'chromatography', occurs under gentle conditions in hydrophilic media, on the borderline with lipophilic media. The processes comprise hydrophilic, lipophilic, amphiphilic molecules and micelles, in channels, capillaries, membranes ('size-exclusion') and with binding (association–dissociation) of biomolecules to each other in a specific way including 'chiral separations'.

Working with biomolecules and efficient methods of food analysis requires dealing with native biomolecules. We have to avoid or be aware of the problems connected with possible artefact formation during the sequences of steps included in the methods. This demands a strategy in order to evaluate the structure and properties of available chromatographic media and techniques in relation to the structure and properties of biomolecules or analytes. The optimal procedure for a required group separation and purification prior to analysis depends also on structure and properties of the components which dominate quantitatively and impurities that are likely to interfere in the material or food matrix containing the analytes. Plant materials are thus different from animal materials owing to dietary fibres and phenolics specific for plant materials. These plant constituents are able to bind various types of other biomolecules and in some respect they can function as 'chromatographic phases' disturbing extraction, separation and analysis of the analytes. Several fast and simple SPE and combined column chromatographic techniques, as well as specially developed supercritical fluid techniques (SFT), supercritical fluid extraction (SFE), supercritical fluid chromatography (SFC), and fast LC methods known as 'fast' or 'flash' chromatography (FC), are described as the methods of choice for group separations.

The present text *Chromatography and Capillary Electrophoresis in Food Analysis* comprises a description of chromatographic and electrophoretic principles and procedures for the analysis of various low-M_r and high-M_r hydrophilic and amphiphilic biomolecules. The descriptions are arranged in 14 chapters, covering the main subjects on LC and electrophoresis, especially HPCE, together with an Appendix.

Each of the 14 chapters and the Appendix are preceded by a summary of the background knowledge together with concepts required, and selected applied examples of the strategy and methods employed for analysis of specific biomolecules are described.

The developments now obtained in LC, CE and SFT have resulted in new opportunities for relatively simple, fast, efficient and cheap methods of analysis for the individual compounds in the various groups of high- and low-M_r compounds occurring in feed, food and natural products. This trend seems to continue, and it is obviously strongly needed in connection with the increasing number of new technologies and as tool in control of feed and food quality, in studies of the bioavailability of nutrients, antinutrients, xenobiotics and in determination of authenticity and adulteration of feed and food.

Success in these areas of research and in analytical–biochemical work depends very much on the ability of analysts to utilize the required and appropriate group separation techniques prior to quantitative analytical determinations. Furthermore, some basic knowledge of chemotaxonomy is a great advantage. Chemotaxonomy will thus be of the utmost importance in relation to the design of an optimal strategy for the analytical procedures, and for evaluating which groups of compounds that may be present and relevant to consider. The majority of low-M_r natural products are of plant origin, and many of these compounds are specific for the plant or plant part producing them. Betalains are specific for the order Centrosperma; occurrence of glucosinolates is limited to plants of the order Capparales and a few other plants; different kinds of protein-type proteinase inhibitors have been identified as constituents of plants belonging to *Leguminosae*, *Brassicaceae*, *Solanaceae*, *Graminaceae*, and this is also the case for various non-protein amino acids, alkaloids, heteroaromatics, and other types of natural product. Chemotaxonomic knowledge and appropriate LC methods can also be an important tool in revealing authenticity and adulteration of food and feed. The trend toward more efficient and specific methods of analysis for the determination of the great number of individual compounds in the various groups of natural products, food and feed additives is thus needed for many reasons. It is therefore our hope that the methods and techniques now described will be of value and will help to solve the various analytical problems in biochemistry, natural product chemistry, feed and food analysis, where the starting points are the real and natural matrix systems containing often complicated mixtures of high- and low-M_r compounds and ions.

References

[1] D.T. Day, *Proc. Am. Phil. Soc.*, 1897, **36**, 112; *Science*, 1903, **17**, 1007.
[2] M. Tswett, *Trav. Soc. Nat. Varsovie*, 1903, **14**.
[3] M. Tswett, *Ber. Dtsch. Botan. Ges.*, 1906, **24**, 316; *ibid.* 384.
[4] M. Steiger and T. Reichstein, *Helv. Chem. Acta*, 1938, **31**, 546.
[5] A.J.P. Martin and R.L.M. Synge, *Biochem. J.*, 1941, **35**, 1358.
[6] R. Consden, A.H. Gordon and A.J.P. Martin, *Biochem. J.*, 1944, **38**, 224.
[7] A. Tiselius, *Science*, 1941, **94**, 145.

[8] L. Zechmeister, *Progress in Chromatography, 1938–1947*, Chapman and Hall, London, 1950.
[9] R.L.Grob, *Modern Practice of Gas Chromatography*, Wiley-Interscience, New York, 1985.
[10] A. Tiselius, *Trans. Faraday Soc.*, 1937, **33**, 524.
[11] S. Hjerten, *Chrom. Rev.*, 1967, **9**, 122.
[12] J.W. Jorgenson and K.D. Lukacs, *Anal. Chem.*, 1981, **53**, 1298.
[13] S.F.Y. Li, *Capillary Electrophoresis, Principles, Practice and Applications*, Journal of Chromatography Library, Elsevier, Amsterdam, 1992.
[14] S. Terabe, K. Otsuka and T. Ando, *Anal. Chem.*, 1985, **57**, 834.
[15] S. Terabe, K. Otsuka, K. Ichikawa, A. Tsuchiya and T. Ando, *Anal. Chem.*, 1984, **56**, 111.
[16] S. Michaelsen and H. Sørensen, *Pol. J. Food Nutr. Sci.*, 1994, **3/44**, 5.
[17] B. Chankvetadze, *Capillary Electrophoresis in Chiral Analysis*, John Wiley & Sons, London, 1997.

Contents

Chapter 1 General Aspects of Experimental Biochemistry **1**

1 Introduction 1
2 Quantities, Symbols, and Units 1
3 Basic Mathematics and Statistics 2
4 Laboratory Safety 7
5 Recording and Reporting Laboratory Results 8
6 Selected General and Specific Literature 11

Chapter 2 Buffers and Micelles **12**

1 Introduction 12
2 Protolytic Activity, Acids, and Bases 12
3 Activity Coefficients and Ionic Strength 14
4 Electrolytes and Ampholytes 16
5 Buffer Solutions and Buffer Capacity 18
6 Preparation of Buffers 23
7 Selection of Buffers 26
8 Organic Solvents as Modifiers of Aqueous Solutions 28
9 Amphipathic Molecules and Micelles in Aqueous Solution 30
10 Selected General and Specific Literature 34

Chapter 3 Binding, Association, Dissociation, and Kinetics **35**

1 Introduction 35
2 Binding Forces – Position of Equilibrium 36
3 Specificity of Proteins/Enzymes Towards Ligands and Antinutrients 38
4 Analysis of Binding Experiments 38
5 Selected General and Specific Literature 67

Chapter 4 Extraction of Native Low-M_r and High-M_r Biomolecules **69**

1 Introduction 69
2 Extraction of Low-M_r Biomolecules 70
3 Amphipathic Membrane Constituents 74

4 Dialysis 85
5 Ultrafiltration 88
6 Centrifugation 89
7 Selected General and Specific Literature 91

Chapter 5 Spectroscopy and Detection Methods 92

1 Introduction 92
2 Atomic Spectroscopy 94
3 UV–Vis Spectroscopy 94
4 Colorimetry 111
5 Fluorimetry 111
6 Diode-array Detection 114
7 Infrared Spectroscopy (IR) 114
8 Nuclear Magnetic Resonance (NMR) 114
9 Mass Spectrometry 117
10 Selected General and Specific Literature 119

Chapter 6 Liquid Chromatography 120

1 Introduction 120
2 Separation of Analytes between the Two Phases in LC 121
3 Surface Tension, Osmotic Pressure, Viscosity, Frictional Coefficient, and Diffusion 122
4 Basic Concepts in Liquid Chromatography 130
5 Paper and Thin-layer Chromatography 136
6 Permeation- or Size-exclusion Chromatography 142
7 Affinity Chromatography 144
8 Hydrophobic Interaction Chromatography 148
9 Solid Phase Extractions and Reversed Phase Chromatography 149
10 Selected General and Specific Literature 150

Chapter 7 Ion-exchange Chromatography 152

1 Introduction 152
2 Ion-exchanger Materials 153
3 Choice of Ion-exchanger 155
4 Regeneration of Ion-exchangers 159
5 Packing of Ion-exchange Columns and Sample Application 159
6 Group Separation of Low-M_r Compounds 161
7 Column Chromatographic Separation of Low-M_r Compounds 163
8 Column Chromatographic Separation of High-M_r Compounds 164
9 Flash Chromatography 166

10 Chromatofocusing 166
11 Selected General and Specific Literature 167

Chapter 8 High-performance Liquid Chromatography and Fast Polymer Liquid Chromatography 168

1 Introduction 168
2 Instrumentation 169
3 Columns and Column Materials 170
4 Liquid Chromatography Principles including Ion-pair Reversed-phase Chromatography and Chiral Chromatography 171
5 Applications of HPLC/FPLC 176
6 Selected General and Specific Literature 177

Chapter 9 Electrophoresis 178

1 Introduction 178
2 Electro-osmosis, Stern Layer, and Zeta Potential 180
3 Zone Electrophoresis 182
4 Moving Boundary Electrophoresis – Isotachophoresis 188
5 Specialized Electrophoretic Techniques 190
6 Detection and Recovery 204
7 Selected General and Specific Literature 207

Chapter 10 High-performance Capillary Electrophoresis 208

1 Introduction 208
2 Basic Principles 208
3 Instrumentation 221
4 Selected Modes of HPCE Operation and Examples of Applications 229
5 Parameters of Analysis Affecting Separation in HPCE 251
6 Development of a HPCE Method of Analysis 266
7 Conclusion 276
8 Selected General and Specific Literature 276

Chapter 11 Analytical Determination of Low-M_r Compounds 278

1 Introduction 278
2 Inorganic Ions and Elemental Analyses 279
3 Low-M_r Organic Compounds with Positive Net Charge, Alkaloids and Biogenic Amines 280
4 Ampholytes, Amino Acids, Peptides, and Heteroaromatics 291
5 Carboxylates, Phosphates, Sulfonates, and Sulfates 298
6 Low-M_r Compounds without Protolytically Active Groups. Carbohydrates, Glycosides, Esters 304
7 Special Groups of Bioactive Compounds 308

8 Conclusion 311
9 Selected General and Specific Literature 313

Chapter 12 Protein Purification and Analysis **315**

1 Introduction 315
2 Proteins in Food Matrices 315
3 Handling of Protein Fractions 319
4 Strategy for Protein Purification 320
5 Sample Preparation 321
6 Group Separation 323
7 High Resolution Techniques 326
8 Methods of Protein Analyses 328
9 Selected Examples of Protein Purification Procedures 340
10 Selected General and Specific Literature 372

Chapter 13 Immunochemical Techniques **374**

1 Introduction 374
2 Antibodies 374
3 Immunochemical Methods of Analyses 379
4 Immunoaffinity Chromatography 391
5 Selected General and Specific Literature 393

Chapter 14 Analysis of Dietary Fibre **394**

1 Introduction 394
2 Composition of DF 394
3 Gravimetric Methods of Analyses 397
4 Chromatographic Methods of Analyses 399
5 Further Characterization of DF 408
6 Conclusion 411
7 Selected General and Specific Literature 412

Appendix Supercritical Fluid Extraction (SFE) and Supercritical Fluid Chromatography (SFC) **413**

1 Introduction 413
2 Theory 414
3 SFE 417
4 SFC 419
5 Selected General and Specific Literature 429

Subject Index **430**

Abbreviations Used in the Text

ε	Molar extinction coefficient
ε^*	Permittivity
ε_o	Permittivity in vacuum
ε_p	Dielectric constant in solution
ε_r	Dielectric constant (relative permittivity)
μ_e	Electrophoretic mobility
μ_{eap}	Apparent electrophoretic mobility
μ_{eo}	Electro-osmotic mobility
A	Absorbance
ABEE	4-Aminobenzoic acid ethyl ester
ABTS	2,2′-Azinodi(ethylbenzthiazoline)sulfonate
ACES	*N*-(2-Acetamido)-2-aminoethanesulfonic acid
AcO^-	Acetate
ADA	*N*-(2-Acetamido)-iminodiacetic acid
ADF	Acid detergent fibre
ANS	1-Anilinonaphthalene-8-sulfonate
AOAC	Association of Official Analytical Chemists
BAPA	N_α-Benzoyl-L-arginine-4-nitroanilide
BBI	Bowman–Birk inhibitor
BCA	Bicinchoninin acid
BCIP	Bromochloroindolylphosphate
Bicine	*N*,*N*-Bis(hydroxyethyl)glycine
bis	*N*,*N*′-Methylene-bis-acrylamide
BSA	Bovine serum albumin
C(%)	Cross-linking (%)
c.m.c.	Critical micellar concentration
CAE	Capillary affinity electrophoresis
CAPS	3-(Cyclohexylamino)-propanesulfonic acid
CBQCA	3-(4-Carboxybenzoyl)-2-quinolinecarboxaldehyde
CDTA	Cyclohexanediaminetetraacetic acid
CEC	Capillary electro-osmotic chromatography
CFZE	Capillary free zone electrophoresis
CGE	Capillary gel electrophoresis
CHAPS	3-[3-(Chloroamidopropyl)dimethylammonio]-1-propanesulfonate

CHAPSO	3-[3-(Chloroamidopropyl)dimethylammonio]-2-hydroxy-1-propanesulfone
CHES	3-(Cyclohexylamino)-ethanesulfonic acid
CI	Chemical ionization
CIEF	Capillary isoelectric focusing
CITP	Capillary isotachophoresis
CM	Carboxymethyl
COSY	Correlated spectroscopy
CTAB	Cetyltrimethylammonium bromide
CTAC	Cetyltrimethylammonium chloride
CZE	Capillary zone electrophoresis
DAB	Diaminobenzidine
DEAE	Diethylaminoethyl
DF	Dietary fibre
DM	Dry matter
DMSO	Dimethyl sulfoxide
Dns	Dansyl
DTAB	Dodecyltrimethylammonium bromide
DTAC	Dodecyltrimethylammonium chloride
$E^{1\%}_{1\,cm}$	Extinction of 1% solution in 1 cm light path
E	Electric field
E	Extinction
EDTA	Ethylenediaminetetraacetic acid
EI	Electron impact
ELISA	Enzyme-linked immunosorbent assay
EOF	Electro-osmotic flow
EtOH	Ethanol
F	Faraday
F^*	Force; electric field
F'	Friction
FAB	Fast atom bombardment
Fab	Fragment; antigen binding
FC	Flash chromatography/Fast chromatography
Fc	Fragment; constant
FPLC	Fast polymer liquid chromatography
FSCE	Free solution capillary electrophoresis
FT	Fourier transform
FZCE	Free zone capillary electrophoresis
GC	Gas chromatography
GLC	Gas–liquid chromatography
GLP	Good laboratory practice
GLUPHEPA	α-Glutaryl-L-phenylalanine-4-nitroanilide
HDL	High density lipoprotein
HEPES	*N*-2-Hydroxyethylpiperazine-*N'*-ethanesulfonic acid
HEPPS	*N*-2-Hydroxyethylpiperazinepropanesulfonic acid
HETP	Height equivalent to a theoretical plate

HIC	Hydrophobic interaction chromatography
HICE	Hydrophobic interaction capillary electrophoresis
HOAc	Acetic acid
HPCE	High-performance capillary electrophoresis
HPLC	High-performance liquid chromatography
HVE	High voltage paper electrophoresis
I	Intensity of transmitted radiation
IDF	Insoluble dietary fibre
IEF	Isoelectric focusing
Ig	Immunoglobulin (IgG, IgA, *etc.*)
I_0	Intensity of incident radiation
I-P RPC	Ion-pair reversed phase chromatography
IR	Infrared
IsoPrime	Preparative isoelectric membrane electrophoresis
IsoPrime^{TM}	Preparative Isoelectric Membrane Electrophoresis
K_M	Metal binding constant
K_m	Michaelis–Menten constant
KSTI	Kunitz soybean trypsin inhibitor
L-BAPA	N_α-Benzoyl-L-arginine-4-nitroanilide
LC	Liquid chromatography
LDL	Low density lipoprotein
LE	Leading electrolyte
M	Molarity
mAb	Monoclonal antibody
MCE	Microemulsion capillary electrophoresis
MD	Mean deviation
MECC	Micellar electrokinetic capillary chromatography
MEKC	Micellar electrokinetic chromatography
MeOH	Methanol
MES	2-(*N*-Morpholino)-ethanesulfonic acid
MHz	Megahertz
MOPS	3-(*N*-Morpholino)-propanesulfonic acid
MOPSO	3-(*N*-Morpholino)-2-hydroxypropanesulfonic acid
M_r	Molecular weight
MS	Mass spectrometry
MT	Migration time
N	Newton
N	Normality
N	Number of theoretical plates
NA	Normalized peak area
NBT	Nitrobluetetrazolium
NC	Nitrocellulose paper
NDF	Neutral detergent fibre
NMR	Nuclear magnetic resonance
N_0	Avogadro's constant
NOE	Nuclear Overhauser effect

NOESY	Nuclear Overhauser effect spectroscopy
NSP	Non-starch polysaccharides
OPA	*o*-Phthalaldehyde
OPD	*o*-Phenylenediamine
PAB	*p*-Aminobenzyl
pAb	Polyclonal antibody
PAGE	Polyacrylamide gel electrophoresis
PAL	Phenylalanine ammonia lyase
PAS	Periodic acid–Schiff
PBE	Polybuffer exchanger
PC	Paper chromatography
PEG	Poly(ethylene glycol)
PHA	Phthalate
$\mathrm{p}I = \mathrm{p}H_{\mathrm{i}}$	Isoelectric point
PIPES	Piperazine-*N*,*N*′-bis(2-ethanesulfonic acid)
PMA	Pyromellitic acid
PMSF	Phenylmethylsulfonyl flouride
PPI	Pea protein-type proteinase inhibitor
PTH	Phenylthiohydantoin
PVDF	Poly(vinylidene diflouride)
QAE	Quaternary aminoethyl
QMA	Quaternary methylamine
R	Gas constant
R	Hydrodynamic radius
RCF	Relative centrifugal field
RI	Refractive index
RMT	Relative migration time
RNA	Relative normalized peak area
RPC	Reversed phase chromatography
RPPI	Rapeseed protein-type proteinase inhibitor
RS	Resistant starch
R_s	Resolution
RSD	Relative standard deviation
RZ	Reinheits zahl
SCS	Sodium cetyl sulfate
SDF	Soluble dietary fibre
SDS	Sodium dodecyl sulfate
SDS–PAGE	Sodium dodecyl sulfate–polyacrylamide gel electrophoresis
SE	Sulfoethyl
SFE	Supercritical fluid extraction
SI units	Système Internationale d'Unités
SOS	Sodium octadecyl sulphate
SP	Sulfopropyl
SPE	Solid phase extraction
STS	Sodium tetradecyl sulfate
T	Transmittance

T(%)	Total acrylamide (%)
TCA	Trichloroacetic acid
TDF	Total dietary fibre
TE	Terminating electrolyte
TEMED	*N*,*N*,*N'*,*N'*-Tetramethylethylenediamine
TES	*N*-Tris(hydroxymethyl)methyl-2-aminoethanesulfonic acid
TLC	Thin-layer chromatography
TLE	Thin layer electrophoresis
TMA	Trimellitic acid
TMB	3,3′,5,5′-Tetramethylbenzidine
TOCSY	Totally correlated spectroscopy
Tricine	Tris(hydroxymethyl)glycine
Tris	Tris(hydroxymethyl)methylamine
TTAB	Tetradecyltrimethylammonium bromide
TTAC	Tetradecyltrimethylammonium chloride
U	Electric mobility
U	Enzyme unit

CHAPTER 1

General Aspects of Experimental Biochemistry

1 Introduction

Experimental work demands a knowledge of some basic areas such as quantities, symbols, and units commonly used in biochemistry, as well as of laboratory safety rules and general working practices, including the registration and evaluation of results, the search for literature, *etc*. Whereas quantities, symbols, and units are beyond dispute, variations of course exist in how things are done at different laboratories/institutions. Some general aspects can, however, be stated from which one hopefully may get inspiration.

2 Quantities, Symbols, and Units

The following tables and lists are not complete, but are intended as guidelines to frequently used quantities, symbols, and units. A more comprehensive overview can be found in Ref. 1.

A physical quantity is the product of a numerical value (a pure number) and a unit. The base units may be given in the centimetre – gram – seconds (cgs) system or, more often, in SI units (Système International d'Unités; The International System of Units) (Table 1.1).

Units with special names and symbols are derived from these basic units. Examples of commonly used units are listed in Table 1.2 together with some units found in older literature but which are also of current interest.

Some values of selected fundamental constants are:

Faraday (F)	9.6485309×10^{4} C mol^{-1}
Electron volt (eV)	96.6 kJ mol^{-1} = 23.1 kcal mol^{-1}
Electronic charge (e^-)	$F/N_0 = 4.8 \times 10^{-10}$ electrostatic units (esu) $= 14.4 \times 10^{-8}$ V $= 1.60 \times 10^{-19}$ C, 1 mol e^- $= 6.023 \times 10^{23} e^-$
Avogadro constant (N_0)	6.0221367×10^{23} mol^{-1}
Planck constant (h)	$6.6260755 \times 10^{-34}$ J s $= 1.58 \times 10^{-34}$ cal s $= 6.62 \times 10^{-27}$ erg s

Table 1.1 *The seven basic units in the SI Unit systems*

Quantity	*SI unit*	*Symbol*
Length	metre	m
Mass	kilogram	kg
Time	second	s
Electric current	ampere	A
Thermodynamic temperature	Kelvin	K
Luminous intensity	candela	cd
Amount of substance	mole	mol

Absolute temperature of the 'ice' point, 0 °C ($T_{0\,°C}$)	273.16 K
Gas constant (R)	$R = \frac{PV}{nT}$ (P = pressure = 0, and T = temperature = 0 °C) = 8.314510 J K^{-1} mol^{-1} = 1.987 cal K^{-1} mol^{-1} = 0.082 l atm K^{-1} mol^{-1}
Pressure–volume product (PV) for 1 mol of a gas at 0 °C and zero pressure	22.414 L atm mol^{-1} = 22414 cm^3 atm mol^{-1}
Boltzmann constant (k_B)	$R/N_0 = 1.380658 \times 10^{-23}$ J K^{-1}
Acceleration due to gravity (g)	9.80665 m s^{-2}
Speed of light in vacuum (c_0)	2.99792×10^8 m s^{-1}

An enzyme unit (U) is defined as the amount of enzyme catalysing the conversion of 1 μmol of substrate into product in 1 min under defined conditions (often at 25 or 30 °C and at optimal assay conditions, including pH). The SI unit for enzyme activity is the katal (kat), which is the amount of enzyme converting 1 mol s^{-1} of substrate into product under optimal/defined conditions; 1 katal = 6×10^7 U; 1 U = 1.67×10^{-8} kat.

Units are often expressed as multiples or sub-multiples by the use of prefixes (Table 1.3). Knowledge of the Greek alphabet may be useful in various connections (Table 1.4). A Periodic Table of the elements is shown in Figure 1.1.

3 Basic Mathematics and Statistics

This section reviews some fundamental concepts within mathematics and statistics relevant for the evaluation of analytical results. For a more comprehensive discussion, the reader is referred to textbooks on these subjects.

Table 1.2 *Selected units*

Quantity	*Unit*	*Symbol for Unit*	*Definition*
Length	ångström	Å	10^{-1} nm = 10^{-10} m
Mass	gram	g	10^{-3} kg
Time	min	min	60 s
	hour	h	3600 s
	day	d	86400 s
Area	metre squared	m^2	
	centimetre squared	cm^2	$cm^2 = 10^{-4}\ m^2$
Volume	metre cubed	m^3	
	decimetre cubed	dm^3	$dm^3 = 10^{-3}\ m^3$ = litre (L)
	centimetre cubed	cm^3	$cm^3 = 10^{-6}\ m^3$ = ml
Partial specific volume		v	$m^3\ kg^{-1} = cm^3\ g^{-1}$
Concentration*	molarity	M	mol dm^{-3}
	normality	N	equivalents dm^{-3}
Percent		%	$g \cdot 10^2\ g^{-1}$
Per thousand		‰	$g \cdot 10^3\ g^{-1}$
Part per million		ppm	$g \cdot 10^6\ g^{-1} = mg\ kg^{-1}$
Specific gravity			$g\ ml^{-1} = g\ cm^{-3}$
Force	newton	N	$J\ m^{-1} = kg\ m\ s^{-2}$
	dyne	dyn	10^{-5} N
Pressure	pascal	Pa	$N\ m^{-2} = kg\ m^{-1}\ s^{-2}$ = 10 dyn cm^{-2}
	bar	bar	105 Pa
	standard atmosphere	atm	101325 Pa = 760 mmHg = 1013 mbar
Power	watt	W	$J\ s^{-1} = kg\ m^2\ s^{-3}$
Energy	joule	J	$kg\ m^2\ s^{-2}$ = N m
	calorie	cal	4.184 J
	erg	erg	10^{-7} J
Temperature	kelvin, Celsius	K, °C	0 °C = 273.16 K
Specific heat			$J\ kg^{-1}\ K^{-1}$
Enthalpy		H	J
Entropy		S	$J\ K^{-1}$
Thermal conductivity		κ, k, λ	$W\ m^{-1}\ K^{-1}$
Viscosity	poise	P	$g\ cm^{-1}\ s^{-1} = 10^{-1}\ N\ s\ m^{-2}$
Diffusion coefficient		D	$m^2\ s^{-1}$
Electric current		I	A
Electric potential	volt	ϕ, V	$V = J\ A^{-1}\ s^{-1} = J\ C^{-1} = kg\ m^2\ s^{-3}\ A^{-1}$
Electric resistance	ohm	Ω	$V\ A^{-1} = kg\ m^2\ s^{-3}\ A^{-2}$
Electric conductance	siemens	S	$\Omega^{-1} = kg^{-1}\ m^{-2}\ s^3\ A^2$
Electric charge	coulomb	C	A s
Electric mobility		U	$cm^2\ s^{-1}\ V^{-1}$

*Basically, concentrations can be expressed in terms of mass or moles. Mass concentration is based on a weight divided by volume, the commonly used unit being grams per litre (g dm^{-3}) which is the same as milligrams per millilitre (mg cm^{-3}). The molar concentration (M) is defined as the number of moles per litre (mol dm^{-3}), often expressed as millimolar concentrations (mM = mmol dm^{-3}). Calculation of mole number is based on knowledge to the molar weight (g mol^{-1}) of the actual compound:

$$\text{Number of moles} = \frac{\text{Mass (g)}}{\text{Molar weight (g mole)}}$$

Most often, the molarity and needed volume of a solution is given, leaving the mass to be found: Mass (g) = Molar weight (g mole) × Molarity (mol/dm^{-3}) × Volume (dm^{-3}).

Table 1.3 *Prefixes and symbols used for multiples or sub-multiples of units*

Multiple	*Prefix*	*Symbol*	*Sub-multiple*	*Prefix*	*Symbol*
10	deca	da	10^{-1}	deci	d
10^2	hecto	h	10^{-2}	centi	c
10^3	kilo	k	10^{-3}	milli	m
10^6	mega	M	10^{-6}	micro	μ
10^9	giga	G	10^{-9}	nano	n
10^{12}	tera	T	10^{-12}	pico	p
10^{15}	peta	P	10^{-15}	femto	f
10^{18}	exa	E	10^{-18}	atto	a

Table 1.4 *The Greek alphabet*

Alpha	A	α	Nu	N	ν
Beta	B	β	Xi	Ξ	ξ
Gamma	Γ	γ	Omicron	O	o
Delta	Δ	δ	Pi	Π	π
Epsilon	E	ε	Rho	P	ρ
Zeta	Z	ζ	Sigma	Σ	σ
Eta	H	η	Tau	T	τ
Theta	Θ	θ	Ypsilon	Y	υ
Iota	I	ι	Phi	Φ	ϕ
Kappa	K	κ	Chi	X	χ
Lambda	Λ	λ	Psi	Ψ	ψ
Mu	M	μ	Omega	Ω	ω

Significant figures. The concept of significant figures is most simply illustrated by the use of examples, *e.g.* 0.0073. This number has two significant figures as the leading zeroes are not regarded as significant, except as a way of showing the magnitude of the number. The number 7.3×10^{-3} also has two significant figures. However, zeroes after the first figure higher than zero are included, *e.g.* 0.00730 has three significant figures, as it is neither 0.00731 nor 0.00732, *etc.* An example of a number higher than 1 is 24500, which has five significant figures, whereas 2.45×10^4 has 3 significant figures.

Analyses generally lead to results with a limited number of significant figures compared to, for example, mathematically defined constants such as $\pi = 3.1415 \ldots$ or $e = 2.7182 \ldots$. The uncertainty of results may be random or systematic. Results of analyses should only be given with the number of reliable figures, often 2–4. This is, however, only for the final results. During calculations, rounded figures should not be used.

Mean value. The mean value $\bar{x}$ of n data ($x_1, x_2, x_3, \ldots, x_n$): $\bar{x} = \frac{1}{n}\sum x_i$.

I 1	II 2	3	4	5	6	Transition Metals 7	8	9	10	11	12	III 13	IV 14	V 15	VI 16	VII 17	VIII 18
1 H 1.0079																	2 He 4.0026
3 Li 6.941	4 Be 9.0122											5 B 10.811	6 C 12.011	7 N 14.007	8 O 15.999	9 F 18.998	10 Ne 20.180
11 Na 22.990	12 Mg 24.305											13 Al 26.982	14 Si 28.086	15 P 30.974	16 S 32.066	17 Cl 35.453	18 Ar 39.948
19 K 39.098	20 Ca 40.078	21 Sc 44.956	22 Ti 47.88	23 V 50.942	24 Cr 51.996	25 Mn 54.938	26 Fe 55.847	27 Co 58.933	28 Ni 58.69	29 Cu 63.546	30 Zn 65.39	31 Ga 69.723	32 Ge 72.61	33 As 74.922	34 Se 78.96	35 Br 79.904	36 Kr 83.80
37 Rb 85.468	38 Sr 87.62	39 Y 88.906	40 Zr 91.224	41 Nb 92.906	42 Mo 95.94	43 Tc [98]	44 Ru 101.07	45 Rh 102.91	46 Pd 105.42	47 Ag 107.87	48 Cd 112.41	49 In 114.82	50 Sn 118.71	51 Sb 121.75	52 Te 127.60	53 I 126.90	54 Xe 131.29
55 Cs 132.91	56 Ba 137.33	57 *La 138.91	72 Hf 178.49	73 Ta 180.95	74 W 183.85	75 Re 186.21	76 Os 190.2	77 Ir 192.22	78 Pt 195.08	79 Au 196.97	80 Hg 200.59	81 Tl 204.38	82 Pb 207.2	83 Bi 208.98	84 Po [209]	85 At [210]	86 Rn [222]
87 Fr [223]	88 Ra 226.03	89 **Ac 227.03	104 Unq [261]	105 Unp [262]	106 Unh [263]												

Atomic weights are based on ^{12}C = 12 and conform to the 1987 IUPAC report values rounded to 5 significant digits. Numbers in [ ] indicate the most stable isotope.

* Lanthanides	58 Ce 140.12	59 Pr 140.91	60 Nd 144.24	61 Pm [145]	62 Sm 150.36	63 Eu 151.97	64 Gd 157.25	65 Tb 158.93	66 Dy 162.50	67 Ho 164.93	68 Er 167.26	69 Tm 168.93	70 Yb 173.04	71 Lu 174.97
** Actinides	90 Th 232.04	91 Pa 231.04	92 U 238.03	93 Np 237.05	94 Pu [244]	95 Am [243]	96 Cm [247]	97 Bk [247]	98 Cf [251]	99 Es [252]	100 Fm [257]	101 Md [258]	102 No [259]	103 Lr [262]

Figure 1.1 *Periodic table of the elements*

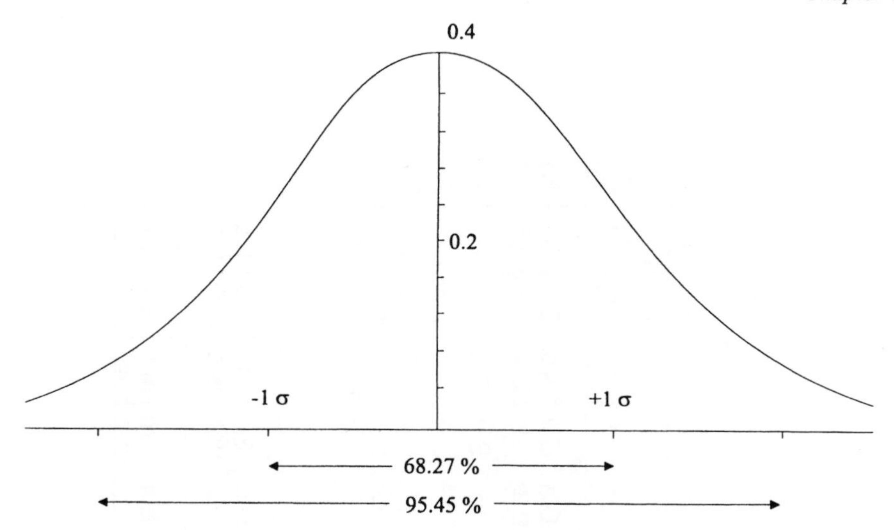

Figure 1.2 *The normal distribution*

Standard deviation. The standard deviation (σ) is the average square root of the squares of the deviations from the mean value: $\sigma = \sqrt{\frac{1}{n}\sum(x_i - \bar{x})^2}$. With a restricted number of observations, $\sigma_{(n-1)} = \sqrt{\frac{1}{n-1}\sum(x_i - \bar{x})^2}$ should be used.

Standard deviation of the mean value. The standard deviation of the mean value or relative standard deviation (RSD) is defined as $\sigma/\bar{x}$.

Normal distribution. The normal distribution applies to measurements of continuous values. Normal distribution diagrams have 68.27% of all values within a plus or minus one standard deviation of the mean (Figure 1.2), 95.45% within $\pm 2\sigma$, and 99.73% within $\pm 3\sigma$.

Mean deviation. The mean deviation (MD) is defined as $\text{MD} = \frac{1}{n}\sum \mid x_i - \bar{x} \mid$. The mean deviation between two results can be given in the same way: $\text{MD} = \frac{1}{2}\sum \mid x_1 - x_2 \mid$.

Repeatability. The repeatability is the standard deviation for a set of analyses performed in the same laboratory with the same apparatus and by the same analyst within a limited period. The data must have a normal distribution.

Reproducibility. The reproducibility is the standard deviation for a set of analyses performed in different laboratories, and/or with different apparatus, and/or by different analysts, and/or over different periods. The data must have a normal distribution.

Linear regression. The statistical procedure used to find the line of best fit is based on the principle of least squares. The equation describing the best line is $y = b + \alpha x$ where

$$\alpha = \frac{\sum(x_i - \bar{x})(y_i - \bar{y})}{\sum(x_i - \bar{x})}$$

α is the slope of the best straight line and can be calculated from any two points (x_1, y_1 and x_2, y_2) on the line [$\alpha = (y_2 - y_1)/(x_2 - x_1)$]; b is the intercept of the line with the y-axis. These figures (α and b) are easily found by the use of most calculation software for PCs or with calculators. This is also true for the correlation coefficient r; $r = \pm 1.0$ describes a perfectly linear positive/negative correlation, whereas $r = 0$ means that no linear correlation exists between the x- and y-values. A value of $r \geqslant 0.998$ is normally considered satisfactory for most biochemical analyses.

Hyperbola and power functions. The equation for two parameters following a hyperbolic curvature is $y = k_1 + k_2 x^{-1}$. If the data follow a power function, the equation is $y = kx^{\bar{n}}$. Data following such functions are often found in analytical and experimental biochemistry. These equations can be transformed into linear functions: single/double reciprocal and logarithmic functions, respectively (Section 3.4), which then give a better basis for evaluation of the experimental results.

4 Laboratory Safety

Before dealing with analytical work, one should recognize the importance of safe laboratory practice. Most often, a laboratory is a working place for a group of people, meaning that *your* knowledge of basic safety rules matters for more than yourself. Some very general laboratory safety rules are listed below. Most laboratories have additional rules and specialized laboratories, *e.g.* laboratories working with micro-organisms, may require further precautions.

- Do not smoke in the laboratory.
- Do not eat or drink in the laboratory.
- Never wear contact lenses in the laboratory.
- Safety glasses should be used when necessary.
- Naked flames should not be used outside the fume cupboard. Always check that no inflammable liquids or other inflammable materials are present in the fume cupboard.
- Be aware of the location of fire exits, emergency showers, and the use and whereabouts of fire-fighting equipment.
- Be aware of the location of eye-wash bottles.
- Never heat plugged flasks and other closed glassware.
- Never pipette by mouth.
- Be careful when attaching or detaching a rubber tube to glassware. Use, for example, a pair of scissors. Broken glass is often involved in laboratory accidents.

- Read and follow specific safety instructions from manufacturers on hazardous chemicals.
- Get a basic knowledge of national legislation as well as company or university 'rules' on safety and environmental aspects of laboratory rules.
- Always use a fume cupboard when working with hazardous chemicals.
- Waste chemicals at working places, *e.g.* in fume cupboards, in or around balances, centrifuges, *etc.*, should be thoroughly cleaned up.
- Always mark samples, chemicals, test tubes, *etc.* with name, date and type of content.

Whenever an experiment is to be started, one has to check for possible hazards in the working procedure and of the chemicals involved. Guidance from experienced people is valuable, but new statements may have arisen and these should be considered. Hazardous chemicals will be labelled, and you should become familiar with the symbols used (Figure 1.3). Proper disposal of waste chemicals and solutions depends on particular rules. Be aware of them.

If an accident occurs, there are some basic rules which should be followed:
Corrosion. Corrosive chemicals on skin and in eyes; the affected area should be bathed immediately in water; use the eye-wash bottle or the emergency shower thoroughly for a long period.
Combustion. Burnt skin should be cooled immediately with ice or cold water. The cooling should be continued for a long period.
Fire. Minor fires can be extinguished by the use of wet cloths or fire-fighting equipment. Fire in hair or clothes should be put out with a fire blanket or emergency shower.

5 Recording and Reporting Laboratory Results

Recommendable laboratory routines include good recording habits. It may be very difficult to remember even the most obvious things done just a few weeks ago, and analytical work should not be based on 'I think I did . . .' but preferably on 'It says here that I did . . .'. The form of recording may of course vary depending on the actual analysis work, but some basic rules are recommendable in most cases. These may be supplemented whenever needed. The notes recorded should at least include:

- Your name, address, and phone number in case you lose your notebook/scribbling block.
- Date and time for the laboratory work.
- A description of the experiments including time schedules, exact working procedure, amounts of various materials used, concentrations, dilution's, notes concerning the apparatus involved, *etc.*
- The results obtained and immediate comments on them. Observations on colour, smell, heat/air development, *etc.* may be valuable supplements to recorded figures.
- Calculations, rough graphs, *etc.*

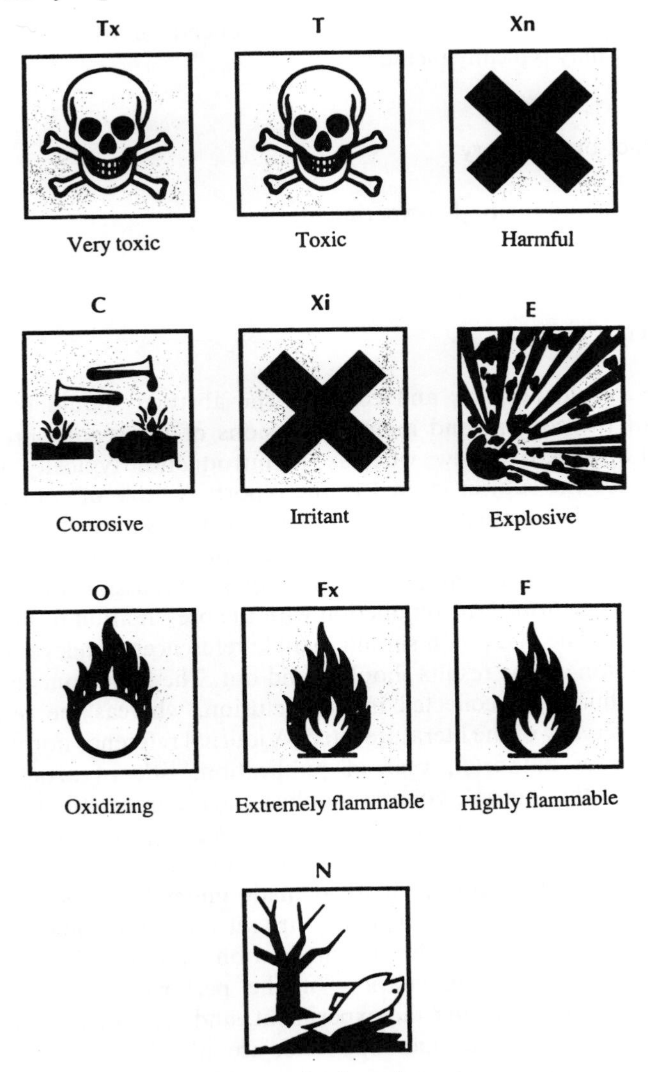

Figure 1.3 *Labelling of hazardous chemicals. This labelling is valid within EU*

In general, your notes should be recorded parallel to the work or at least before the end of work each day. The notes should be reviewed frequently, with comments on possible repetition of experiments, revision of procedures, missing experiments, future plans and ideas, and so on. The notes should be clear and easily understood, so that the work can be repeated by others with appropriate knowledge of the field in question.

The final report of the analytical results will be based on the daily laboratory notes. The frame of the final report of course depends on what or for whom it is intended, but may typically include:

- Title
- Abstract or summary
- Introduction
- Materials and methods (experimental)
- Results
- Discussion
- Conclusion
- References

The title should be clear and concise. The abstract should be a succinct statement of the content and main conclusions of the report, providing the reader with a quick overview, whereas the introduction typically outlines the background for the subject treated in the report. A table of contents may be included before the abstract. The material and methods part should contain all the relevant information, based on the report notes, and the results should be presented clearly in text, tables, and/or graphs. Discussion of the obtained results may be performed in conjunction with the presentation of the data or in a separate part. The discussion should include references to relevant literature, and your opinion on the results should stand out. The main conclusions of the work should finally be collected in a conclusion, whereas the reference list contains an overview of the literature cited. A journal reference normally should include: name of author(s), year of publication, title of article, accepted abbreviation of the journal, volume number, page number (first and last). It should be noted that most journals have their own guide to authors, stating the style of the reference list, which sections to include, *etc.*

In cases where the laboratory work is made under GLP (good laboratory practice) procedures, very strict demands are set for the documentation of all steps in the study, including the documentation of the performance of the analytical equipment, documentation of the performance of the applied methods, reports made during the experiment, and reporting. GLP regulates all non-clinical safety studies that support or are intended to support applications for research or marketing permits for products regulated by the FDA (Food and Drug Administration, USA) or similar national legislation. This includes medicinal and veterinary drugs, aroma and colour additives in food, nutrition supplements for livestocks, and biological products. The GLP procedures were set up by the authorities for drug approval because one or more of the following problems often were found in applications for drug approval: poor experimental design, poor study management and conduct, changes/corrections in data without explanation, incomplete documentation, missing data, incomplete reporting, and no verification of accuracy and completeness of scientific data.

Performance and presentation of most experimental work, including chromatographic analyses, demands accessibility to chemical literature: basic textbooks, handbooks, and monographs, which should be supplemented with original journal articles and reviews. A literature search may be performed by the use of reference lists in relevant reviews, articles, *etc.* or bibliographies, *e.g.* *Chemical Abstracts*, *Analytical Abstracts*, *Biological Abstracts*, *etc.*

Most bibliographies are now accessible in data bases, and an on-line literature search is a valuable tool. It should be noted that the amount and quality of the obtained material is very much dependent on the profile of words used in the search.

6 Selected General and Specific Literature

1. Quantities, Units, and Symbols in Physical Chemistry, International Union of Pure Applied Chemistry, Physical Chemistry Division, prepared for publication by I. Mills, T. Cvitaš, K. Homann, N. Kallay and K. Kuchitsu, 2nd Edition, Blackwell Scientific, Oxford, 1993.
2. IUPAC Information Bulletin, Appendix No. 46, September, reproduced in *Biochemistry*, **14**, 449, 1975.
3. Enzyme Nomenclature, Recommendations of the Nomenclature Committee of the International Union of Biochemistry, Academic Press, New York, 1979.
4. Nomenclature of Organic Chemistry, prepared for publication by J. Rigandy and S.P. Klesney. 4th Edition, Pergamon Press, Oxford, 1979, p. 193.
5. P.A. Carson and N.J. Dent, eds, 'Good Laboratory and Clinical Practices – Techniques for Quality Assurance for Professionals', Heinemann Newnes, Oxford, 1990.
6. D.S. Moore and G.P. McCabe, 'Introduction to the Practice of Statistics', 2nd Edition, Freeman, New York, 1993.
7. Principles of Chemical Nomenclature, A Guide to IUPAC Recommendations, prepared for publication by G.J. Leigh, H.A. Faure and W.V. Metanomski, Blackwell Scientific, Oxford, 1997.
8. Compendium of Chemical Technology, prepared for publication by A.D. McNaught and A. Wilkinson, 2nd Edition, Blackwell Scientific, Oxford, 1997.
9. A Guide to IUPAC Nomenclature of Organic Compounds, prepared for publication by R. Panico, J.-C. Richer and W.H. Powell, Blackwell Scientific, Oxford, 1994.

CHAPTER 2

Buffers and Micelles

1 Introduction

The native structure, stability, and activity of biomolecules in living cells depend to a great extent on the properties of the solutions in which they occur. This implies aqueous buffers; solutions containing mixtures of protolytically active compounds and their conjugated bases or acids, which must have pK_a-values close to the pH of the solutions. Results from investigations of biomolecules *in vitro* and from various types of liquid chromatographic methods of analyses, especially high-performance capillary electrophoresis (HPCE) and specialized techniques hereof, also depend on the type of buffers, ionic strength, modifiers, detergents, and micelles in the solutions.

2 Protolytic Activity, Acids, and Bases

Water has unique properties compared with other solvents owing to its polar nature and its hydrogen bonding potential. The properties of water and aqueous buffers can, however, be changed by molecules and ions dissolved therein. This also affects the properties of dissolved molecules and their activity, including their potential protolytic activities, *i.e.* their acid or base properties. According to the Brønsted theory, an acid is defined as a substance which donates a proton to another substance, whereas a base is a substance which accepts a proton from another substance. Acids react with bases, or *vice versa*, with the formation of salts, *e.g.*:

$$HCl + NaOH \rightarrow NaCl + H_2O$$

Strong acids (HCl), strong bases (NaOH), and salts (NaCl) are completely ionized in aqueous solutions, whereas less strong or weak acidic or basic compounds are only partly deprotonated or protonated in aqueous solution, to an extent defined by their pK_a-values. The proton does not exist freely in aqueous solution, but forms part of the water's structure (Figure 2.1). A proton in water can thus be described as H^+, H_3O^+, $H_5O_2{}^+$ ..., *etc.* but for simplicity, H^+ is most often used.

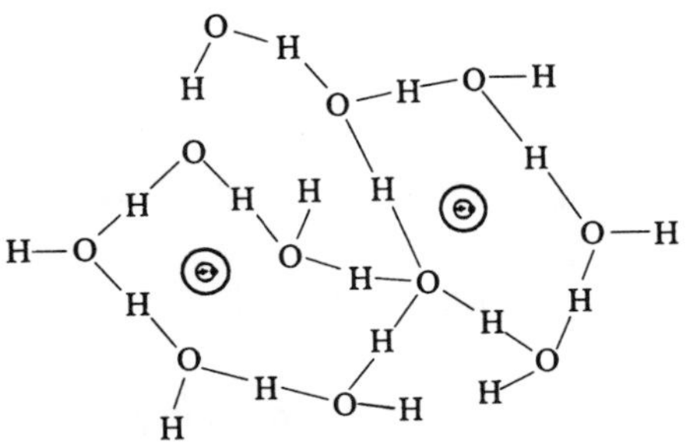

Figure 2.1 *The 'organized structure' of water, in which the flickering clusters of molecules are held together by hydrogen bonds that continually break and reform*

Water can act either as an acid or a base, depending on whether it reacts with a base or an acid. Water is thus an amphoteric molecule, as it is capable of either donating or accepting a proton. For the dissociation:

$$H_2O \underset{k_2}{\overset{k_1}{\rightleftharpoons}} H^+ + HO^-$$

the velocity from left to right is:

$$\underset{\rightarrow}{v} = k_1[H_2O] \qquad (2.1)$$

and in the opposite direction:

$$\underset{\leftarrow}{v} = k_2[H^+][HO^-] \qquad (2.2)$$

At equilibrium, $\underset{\rightarrow}{v} = \underset{\leftarrow}{v}$; the equilibrium constant (K_{eq}) at room temperature is:

$$K_{eq} = \frac{k_1}{k_2} = \frac{[H^+][HO^-]}{[H_2O]} = 1.8 \times 10^{-16} \qquad (2.3)$$

As with all other equilibrium constants (thermodynamic constants), K_{eq} varies with temperature. The concentration of water in water is 55.5 M and thereby the ion product (K_w) of water is:

$$K_w = [H^+][HO^-] = K_{eq}[H_2O] = 1.8 \times 10^{-16} \times 55.5 = 10^{-14} \qquad (2.4)$$

With $p = -\log$ we have pH + pOH = 14.

For the ionization of a weak acid (HA) or base (B:) in water, the concentration of water is not changed appreciably, and thus we have:

$$\underset{\text{acid}}{HA} + H_2O \rightleftharpoons H_3O^+ + \underset{\text{conjugate base}}{A^-} \quad \text{or} \quad A^- + H_2O \rightleftharpoons HA + HO^-$$

$$K'_a = K_{eq}[H_2O] = \frac{[H^+][A^-]}{[HA]} \quad \text{and} \quad K'_b = K_{eq}[H_2O] = \frac{[HO^-][HA]}{[A^-]} \tag{2.5}$$

$$K'_a K'_b = [H^+][OH^-] = K_w = 10^{-14} \tag{2.6}$$

$$pK'_a + pK'_b = 14 \tag{2.7}$$

Instead of considering the base constant K_b- and pK_b-values, we can consider the equilibrium involving dissociation of a proton from the protonated base, *i.e.* the conjugate acid.

All biomolecules, including the proton (Figure 2.1) and other ions and uncharged molecules are, however, surrounded by an 'ordered water structure' (solvation). This will change the activity of the molecules and the practical constants K'_a, K'_b, pK'_a, and pK'_b will be different from the thermodynamic constants defined by activities (a):

$$K_a = \frac{a_{H^+} a_{A^-}}{a_{HA}} \tag{2.8}$$

3 Activity Coefficients and Ionic Strength

In water as well as in other solvents, individual atoms, ions, and molecules dissolved in solutions will affect each other, owing to their solvation and attraction or repulsion due to electrical forces. These effects will increase with increased solute concentration, and the activity of a solute (a_s) will be different from its concentration (X_s). The relation between activity and concentration can be introduced by use of the activity coefficient for the solute (γ_s), which is defined by its chemical potential (μ_s):

$$\mu_s = \mu_s^* + RT \ln a_s = \mu_s^* + RT \ln \gamma_s X_s \tag{2.9}$$

where μ_s^* = the chemical potential of the solute in pure solvent under standard conditions at temperature (T) and pressure (P), and R = the gas constant. The activity in ideal solutions is more simply equal to the product of the activity coefficient and the concentration of the solute:

$$a_s = \gamma_s X_s \tag{2.10}$$

This equation corresponds to Henry's law, developed for vapour pressure, but is widely used for the relation between activities and concentrations of solutes. The activity coefficient for the solute is taken as approaching unity as its concentration approaches zero: $\gamma_s \rightarrow 1$ as $X_s \rightarrow 0$.

All ions in water will be strongly solvated, and the activity coefficients for an ion (γ_i) will change appreciably with the concentration of ions in the solution,

Box 2.1 *Calculation of the ionic strength for 0.1 M Na_2SO_4 in water*

Ions in solution

2 Na^+;	$z = 1$;	$[Na^+] = 0.2$ M
1 SO_4^{2-};	$z = 2$;	$[SO_4^{2-}] = 0.1$ M

$I = 0.5\,[(1^2 \times 0.2\text{ M}) + (2^2 \times 0.1\text{ M})] = 0.3$ M

i.e. with the ionic strength (I):

$$I = 0.5 \sum z_i^2[\text{i}] \tag{2.11}$$

where z_i = the charge of the ion (i) and [i] = the molar concentration of i. All ions (but not uncharged molecules) in the solution must be included in the calculations of I. Box 2.1 shows an example of the calculation of I for 0.1 M Na_2SO_4, in water.

In various liquid chromatographic techniques, especially electrophoresis and HPCE, the ionic strength will be an important factor. The proportion of current carrried by the buffer will thus increase whereas the current carried by the sample will decrease with increasing I, slowing the rate of sample migration. Effects like this will thus determine the electro-endoosmosis or electroosmotic flow (EOF) (Chapter 9).

In dilute aqueous solutions, the ionic strength can be used to calculate activity coefficients using the Debye–Hückel equation:

$$-\log \gamma_{\pm} = kz_+z_-\sqrt{I} \tag{2.12}$$

where k = a temperature-dependent constant that is 0.507 at 20 °C, 0.512 at 25 °C, and 0.524 at 38 °C. An acceptable approximation for the ion i in water is:

$$-\log \gamma_i = kz_i^2 \frac{\sqrt{I}}{1 + b\sqrt{I}} \tag{2.13}$$

where b is a constant defined by the type of ion. The size of b depends on the radius of the solvated ion. Most often it is an unknown size, which can also vary with the type of solution, but b = 1 can generally be used without problems. In buffer solutions, we consider dissociation/association of a proton, and as it is the activity of the proton that is determined by use of a pH-meter, it gives the following relations:

$$H_2A^- \rightarrow H^+ + HA^{2-}$$

$$z_{H_2A^-} - z_{HA^{2-}} = 1 \tag{2.14}$$

$$K_a = a_{H^+} \frac{a_{HA^{2-}}}{a_{H_2A^-}} \tag{2.15}$$

$$pK'_a = pK_a + \log \frac{\gamma_{HA^{2-}}}{\gamma_{H_2A^-}} = pK_a + 0.5(z^2_{H_2A^-} - z^2_{HA^{2-}}) \frac{\sqrt{I}}{1 + \sqrt{I}} \tag{2.16}$$

$$pK'_a = pK_a + 0.5(1 + 2z_{HA^{2-}}) \frac{\sqrt{I}}{1 + \sqrt{I}} \tag{2.17}$$

As revealed from this equation, pK'_a approaches pK_a when I approaches zero:

$$pK'_a \rightarrow pK_a \quad \text{as} \quad I \rightarrow 0$$

Several biochemistry textbooks do not mention the differences between the thermodynamic pK_a-values and the 'practical' pK'_a-values; however, there may be a considerable difference. For the phosphate buffer corresponding to:

$$H_2PO_4^- \rightleftharpoons H^+ + HPO_4^{2-}$$

pK'_a determined for 0.1–0.2 M total phosphate in aqueous solutions will be about 0.5 below the pK_a of 7.21.

The general equations for polyprotic anionic and cationic polymers are:

$$HA^{n-} \rightleftharpoons H^+ + A^{(n+1)-}$$

$$pH = pK_a + \log \frac{[A^{(n+1)-}]}{[HA^{n-}]} + \log \frac{\gamma_{A^{(n+1)-}}}{\gamma_{HA^{n-}}} \tag{2.18}$$

$$HA^{n+} \rightleftharpoons H^+ + A^{(n-1)+}$$

$$pH = pK_a + \log \frac{[A^{(n-1)+}]}{[HA^{n+}]} + \log \frac{\gamma_{A^{(n-1)+}}}{\gamma_{HA^{n+}}} \tag{2.19}$$

When the net charge, z, for polyprotic anionic and cationic polymers increases, the correction will increase. Anions give a negative value of z and the correction will be negative: $pK'_a < pK_a$. The positive values of z for cations give a positive correction and $pK'_a > pK_a$.

4 Electrolytes and Ampholytes

Protolytic activity is possible in solutions with a pH close to the pK_a (pK'_a)-value of the protolytically active group. Macromolecules often contain many protolytically active groups (polyprotic anionic and cationic polymers), thereby they can be polyelectrolytes or polyampholytes. A molecule that contains both acidic and basic groups is called an ampholyte (*e.g.* water, Section 2.2). Knowledge of

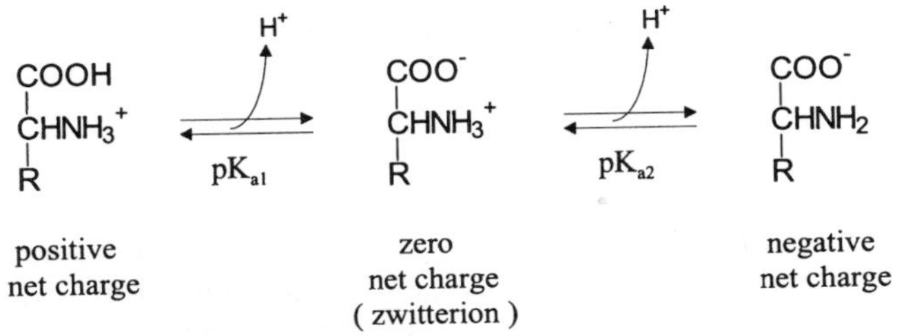

COOH | $CHNH_3^+$ | R ⇌ (H^+, pK_{a1}) COO⁻ | $CHNH_3^+$ | R ⇌ (H^+, pK_{a2}) COO⁻ | $CHNH_2$ | R

positive net charge — zero net charge (zwitterion) — negative net charge

Figure 2.2 *Possible net charges of a neutral α-amino acid*

the protolytically active groups of any electrolytes is a keypoint for success in analyses and studies of these compounds. Amino acids, as an example, possess both carboxylic and amino groups, and they will be ionized at all pH-values. An example of a neutral α-amino acid is shown in Figure 2.2.

pK_{a1}-Values for α-amino acids are close to 2, and for their α-amino group, pK_{a2}-values are about 9–10. In aqueous solutions with low pH-values, an amino acid exists as a cation, at high pH as an anion. At a particular intermediate pH, the amino acid carries no net charge and is then called a *zwitterion*. The corresponding pH is called the *isoelectric point* (pH_i or pI). The isoelectric point for an amino acid is identical with its *isoionic point*, as the number of negative charges is equal to the number of positive charges. Without a net charge, the molecule will not be bound to either anion or cation exchangers (Chapter 7) and it is electrophoretically immobile (Chapter 9).

Generally, proteins have an ability to bind both ions and uncharged molecules (Chapter 3), resulting in 'salting in' and 'salting out' effects (Chapter 4), which gives rise to the possibility of hydrophobic interaction chromatography (HIC; Chapter 6). Unlike the situation for amino acids, for proteins pI is not identical to the isoionic point, which is equal to the pH at which the protein possesses an equal number of positive and negative protolytically active groups. Proteins will thus have pI-values that vary with the type of buffer, its concentration, and the methods used for pI-determination. Studies of proteins, enzymes, and immunochemistry are always carried out in buffered solutions, and it is therefore the pI of the molecules in these buffers that is important in such methods of analyses.

The numerical value of pH = pI for neutral protein amino acids can be calculated from:

$$\mathrm{pI} = \frac{\mathrm{p}K_{a1} + \mathrm{p}K_{a2}}{2} \tag{2.20}$$

Acidic amino acids contain more acidic than basic groups, and for such protein amino acids pK_{a3} for the amino group is not included in the pI calculation (see above). Basic amino acids contain more basic than acidic groups, and for such protein amino acids pI is calculated from:

$$pI = \frac{pK_{a2} + pK_{a3}}{2} \quad (2.21)$$

Proteins contain many basic and acidic groups in their amino acid side chains, and they are therefore polyampholytes. Calculations of pI will be more complicated for such compounds, and, as mentioned above, their pI will also reflect the amount and types of associated ions.

Polyelectrolytic macromolecules such as nucleic acids, pectic acids in dietary fibres, and other phosphorylated or carboxymethylated polysaccharides carry only negative net charges. Nucleic acids are ionized at pH-values higher than 2, whereas pectic acids will have their carboxylates protonated at pHs below 3–4.

5 Buffer Solutions and Buffer Capacity

Buffers are solutions containing both an acid and its conjugate base:

$$\underset{\text{acid}}{HA} \rightleftharpoons H^+ + \underset{\text{conjugate base}}{A^-}$$

Such solutions will tend to resist pH-changes in the pH-range near the pK'_a-value of the acid. The capacity (β) of the buffer is defined by the buffer concentration in the equation:

$$\beta = \frac{d[HA] \text{ or } d[A^-]}{dpH} = \frac{dx}{dpH} \quad (2.22)$$

where x = no. equivalents of HO^- or H^+ per litre.

The buffer capacity depends both on the total buffer concentration (C_t) and the degree of dissociation (α) or protonization (θ):

$$C_t = [A^-] + [HA] \quad (2.23)$$

$$\alpha = \frac{[A^-]}{[A^-] + [HA]} = \frac{[A^-]}{C_t} \quad \text{and} \quad \theta = \frac{[HA]}{C_t} \quad (2.24)$$

$$\alpha + \theta = 1 \quad (2.25)$$

It follows that the equations above can be used to evaluate β, which determines the quantity of strong acid or base, that may be added to the solution without producing a significant change in pH:

$$pH - pK'_a = \log\frac{[A^-]}{[HA]} = \log\frac{\alpha}{1-\alpha} = \log\frac{1-\theta}{\theta} \quad (2.26)$$

θ and α can be written as:

$$\theta = \frac{1}{1 + 10^{(\text{pH}-\text{p}K'_a)}} \tag{2.27}$$

$$\alpha = \frac{10^{(\text{pH}-\text{p}K'_a)}}{1 + 10^{(\text{pH}-\text{p}K'_a)}} \tag{2.28}$$

The graphs obtained from these equations will be symmetrical around the pK'_a-value. When pK'_a is calculated from pK_a with a correction for the activity coefficients (Section 2.3), the figures for $|\text{pH}-\text{p}K'_a|$ can be used for a general determination of α and θ (Table 2.1), and thereby of buffer composition $[\text{A}^-]/[\text{HA}]$ at a defined pH and pK'_a. The corresponding figures for θ and α can be used to construct 'Bjerrum diagrams' (Figure 2.3) and titration curves, illustrating the buffer range, *i.e.* the pH-range where the buffer is effective.

From $[\text{A}^-]/[\text{HA}] = 10^{(\text{pH}-\text{p}K'_a)}$ it can be seen that a plot of log $([\text{A}^-]/[\text{HA}])$ against $|\text{pH} - \text{p}K'_a|$ will give a straight line, which can be used to calculate the buffer composition $[\text{A}^-]/[\text{HA}]$ required for obtaining a defined pH determined from the pK'_a of the buffer (Figure 2.4).

Table 2.1 *α and θ values for different values of $|pH-pK'_a|$*

	$\|pH-pK'_a\|$								
	0*	0.2	0.4	0.6	0.8	1.0	1.2	1.4	1.8
θ 100 (%)	50	61	72	80	86	91	94	96	98.5
α 100 (%)	50	39	28	20	14	9	6	4	1.5

*$|\text{pH}-\text{p}K'_a| = 0$ when $[\text{HA}] = [\text{A}^-]$

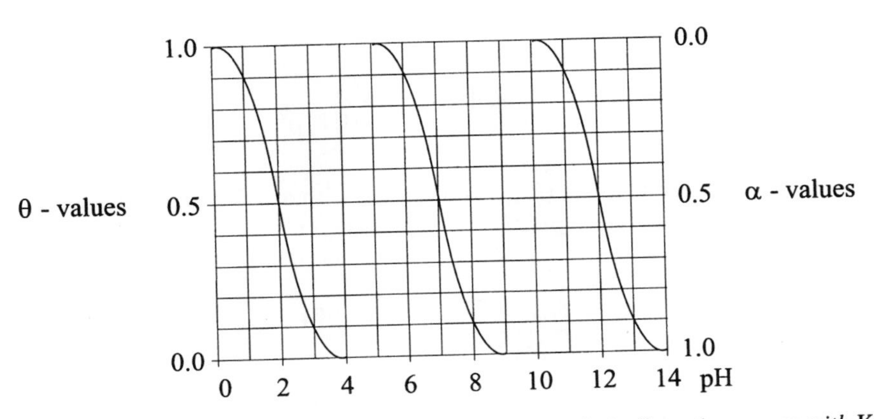

Figure 2.3 *Bjerrum diagram for an acid with three protolytically active groups with K'_a values 2, 7, and 12*

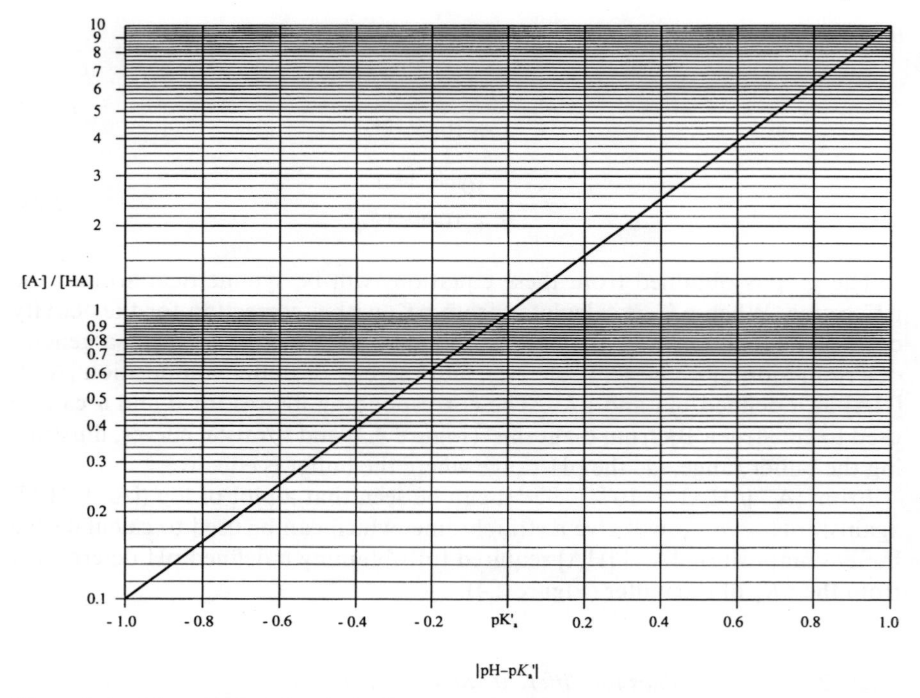

Figure 2.4 *Plot of values for the buffer composition against* $|pH - pK'_a|$ *on semilog paper. A pH-value of* $1 + pK'_a$ *will, for example, be obtained with* $[A^-] = 10\,[HA]$

The amount of protonated buffer acid (HA) transformed into the conjugate base (A^-) is equivalent to the amount of strong base added to (or released from) the buffer:

$$\beta = \frac{dx}{dpH} = \frac{d\alpha}{dpH} \tag{2.29}$$

$$\beta = 2.3C_t[H^+]\frac{K'_a}{(K'_a + [H^+])^2} \tag{2.30}$$

Maximum buffer capacity (β_{max}) is obtained with $\alpha = \theta = 0.5$, corresponding to $[HA] = [A^-]$ and $pH = pK'_a$:

$$\beta_{max} = \frac{2.3}{4}C_t = 0.576C_t \tag{2.31}$$

Total ionization or protonization of the buffer (α and $\theta = 0$ or 1) results in loss of buffer capacity.

Mixtures of buffers give additive values for the buffer capacity and with a 1:1

mixture, as seen for amino acids (Section 2.4), the result will be:

$$\mathrm{pH} = \tfrac{1}{2}(\mathrm{p}K'_{\mathrm{a}1} + \mathrm{p}K'_{\mathrm{a}2}) \tag{2.32}$$

The β-value for a mixture of two protolytically active groups with $\mathrm{p}K'_{\mathrm{a}1}$ and $\mathrm{p}K'_{\mathrm{a}2}$ is:

$$\beta = 2.3C_{\mathrm{t}}[\mathrm{H}^+]\left[\frac{K'_{\mathrm{a}1}}{(K'_{\mathrm{a}1} + [\mathrm{H}^+])^2} + \frac{K'_{\mathrm{a}2}}{(K'_{\mathrm{a}2} + [\mathrm{H}^+])^2}\right] \tag{2.33}$$

Water as solvent results in a pseudo-buffer capacity from H_2O, as it can act both as a weak base ($\mathrm{p}K'_{\mathrm{a}} \cong 0$) and a weak acid ($\mathrm{p}K'_{\mathrm{a}} \cong 14$). Water solutions of strong acids or strong bases give:

$$\beta = \frac{\mathrm{d}x}{\mathrm{dpH}} = C_{\mathrm{t}}2.3\left[[\mathrm{H}^+] + \frac{K_{\mathrm{w}}}{[\mathrm{H}^+]}\right]^2 \tag{2.34}$$

With the high concentration of water in water (55.5 M), a high buffer capacity exists at extreme pH-values:

$$\beta_{\mathrm{total}} = \beta_{\mathrm{buffer}} + \beta_{\mathrm{H_2O}} = 2.3\left[\frac{C_{\mathrm{t}}K'_{\mathrm{a}}[\mathrm{H}^+]}{(K'_{\mathrm{a}} + [\mathrm{H}^+])^2} + [\mathrm{H}^+] + [\mathrm{HO}^-]\right] \tag{2.35}$$

Maximal buffer capacity is obtained with pH $=$ $\mathrm{p}K'_{\mathrm{a}}$ or $[\mathrm{H}^+] = K'_{\mathrm{a}}$, resulting in:

$$\beta_{\mathrm{total}} = 2.3\left(\frac{C_{\mathrm{t}}K'^2_{\mathrm{a}}}{(2K'_{\mathrm{a}})^2} + K'_{\mathrm{a}} + \frac{K_{\mathrm{w}}}{K'_{\mathrm{a}}}\right) = 2.3\left(\frac{C_{\mathrm{t}}}{4} + K'_{\mathrm{a}} + \frac{K_{\mathrm{w}}}{K'_{\mathrm{a}}}\right) \tag{2.36}$$

At pH in the 3–11 range, $\beta_{\mathrm{H_2O}}$ will be without effect, whereas it will be a dominant part of β_{total} outside these pH-values (Table 2.2).

In the pH-interval 3–11, the buffer capacity will thus only be influenced by C_{t}:

$$\beta_{\mathrm{buffer}} = 2.3C_{\mathrm{t}}\frac{10^{(\mathrm{pH}-\mathrm{p}K'_{\mathrm{a}})}}{(1 + 10^{(\mathrm{pH}-\mathrm{p}K'_{\mathrm{a}})})^2} \tag{2.37}$$

Table 2.3. shows the influence of C_{t} on the buffer capacity for various values of $|\mathrm{pH} - \mathrm{p}K'_{\mathrm{a}}|$.

The effects from $\beta_{\mathrm{H_2O}}$ and from C_{t} are illustrated in Figure 2.5 using the above-mentioned figures for C_{t} = 0.1 and 0.01 M.

Table 2.2 *Buffer capacity of water at extreme pH*

pH	$[H^+]$	$[OH^-]$	$\beta_{H_2O} = 2.3\ ([H^+] + [OH^-])$
1	10^{-1}	10^{-13}	0.230
2	10^{-2}	10^{-12}	0.023
3	10^{-3}	10^{-11}	0.002
11	10^{-11}	10^{-3}	0.002
12	10^{-12}	10^{-2}	0.023
13	10^{-13}	10^{-1}	0.230

Table 2.3 *Buffer capacity at various* $|pH-pK'_a|$ *and* C_t *values*

	β_{buffer}		
$\lvert pH-pK'_a\rvert$	C_t = *1.0* M	C_t = *0.1* M	C_t = *0.01* M
2	0.02	0.002	0.0002
1.5	0.07	0.007	0.0007
1.0	0.19	0.019	0.0019
0.5	0.42	0.042	0.0042
0	0.57	0.057	0.0057

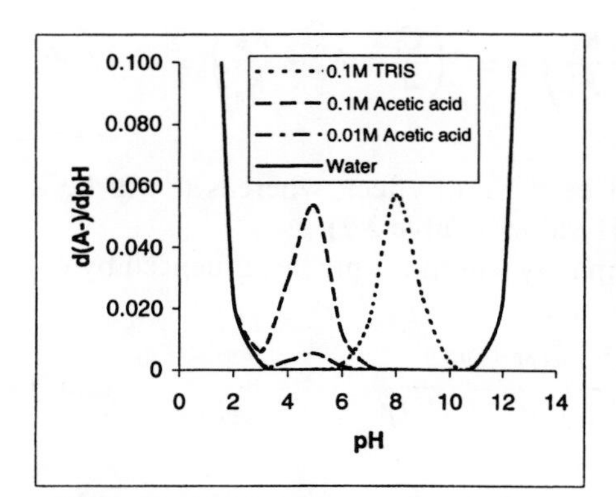

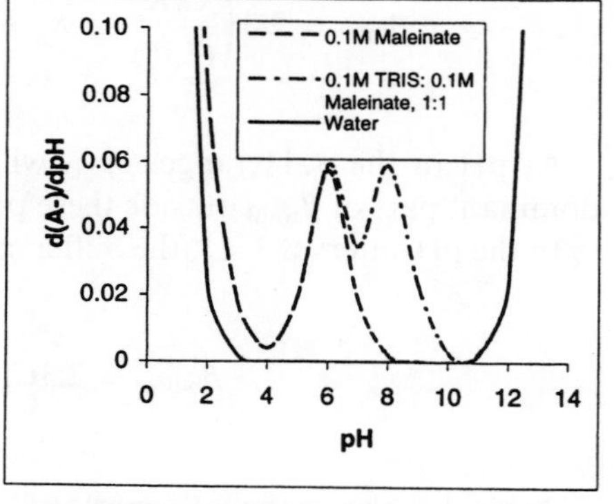

Figure 2.5 *The capacity of buffers depends on* C_t, *and for mixtures of buffers it gives additive* β_{buffer}. *All will be influenced by* β_{H_2O} *at pH below 3 and pH above 11. 0.1* M *TRIS,* $pK'_a = 8.1$; *0.1* M *acetic acid* ($pK'_a = 4.76$); *0.01* M *acetic acid* ($pK'_a = 4.76$) *and 0.1* M *maleinate* ($pK'_{a1} = 1.9$, $pK'_{a2} = 6.0$)

6 Preparation of Buffers

Buffers are generally defined by their pH and total concentration:

$$C_t = [HA] + [A^-] \tag{2.38}$$

For a buffer composed by an acid (HA) and its conjugate base (A^-), the composition can be calculated from:

$$HA \rightleftharpoons H^+ + A^-$$

$$pH - pK'_a = \log\frac{[A^-]}{[HA]} \tag{2.39}$$

We thus need to know K'_a or pK'_a, which can be calculated from K_a or pK_a if we know the ionic strength (Section 2.3). For this purpose it is necessary to know the counterions to the buffer ions, as monovalent ions will make a different contribution to the ionic strength than will di- or polyvalent ions (Box 2.1). Furthermore, these will also have different effects on solutions containing detergents, proteins, or other biomolecules, as discussed below in relation to 'salting out' and 'salting in', Hofmeister's series, chaotrope and antichaotrope compounds/ions (page 81). In addition, such effects are of importance in relation to various advanced types of liquid chromatography, including HPCE.

The principles used for buffer preparation can be divided into several methods (Box 2.2).

Box 2.2 *Methods for buffer preparation*

Method 1

Use of stock solutions defined and described in tables such as those in the Handbook of Biochemistry, ed. H.A. Sober, Chemical Rubber Co., Cleveland, Ohio, 1970. Mix as recommended and determine the pH

Method 2

Start with calculated C_t, of either HA or A^-, dissolve in water, titrate to wanted pH, and dilute to final volume and, finally, determine the pH

Method 3

Calculate the required amount of acid (salt) and conjugate base (salt) from the equations given in the text, dissolve in water, determine pH, dilute to final volume and, finally, determine the pH

Buffers prepared as more concentrated stocks permit smaller storage volume, and it is then more convenient to add antimicrobial agents such as 0.02% sodium azide; the antimicrobial agent is then diluted upon use. However, dilution will change the pH owing to changed ionic strength, *e.g.* a 100-fold dilution of 0.2 M phosphate buffer, pH 6.6, will change the pH to about 7.1

Table 2.4 *The dependence of* pK'_a*-values on temperature for various buffers*

Name (abbreviation)	*Structural formula*	pK'_a (20 °C)	$\Delta pK'_a$ (per °C)
2-Morpholinoethanesulfonic acid (MES)	O(morpholine ring)$N^+HCH_2CH_2SO_3^-$	6.15	−0.11
N-(Carbamoylmethyl)iminodiacetic acid (ADA)	$H_2NCOCH_2HN^+(CH_2COO^-)(CH_2COONa)$	6.60	−0.11
Piperazine-*N*,*N'*-bis(ethanesulfonic acid) (PIPES)	$NaO_3SCH_2CH_2N$(piperazine ring)$N^+HCH_2CH_2SO_3^-$	6.80	−0.0085
N-(Carbamoylmethyl)-2-aminoethanesulfonic acid (ACES)	$H_2NCOCH_2N^+H_2CH_2CH_2SO_3^-$	6.90	−0.020

N-Tris(hydroxymethyl)-2-aminoethanesulfonic acid (TES)	$(HOCH_2)_3N^+HCH_2CH_2SO_3^-$	7.50	−0.020
N-Tris(hydroxymethyl)methylglycine (Tricine)	$(HOCH_2)_3CN^+H_2CH_2COO^-$	8.15	−0.021
Tris(hydroxymethyl)methylamine (TRIS)	$(HOCH_2)_3CNH_2$	8.30	−0.031
N,*N*-Bis(hydroxyethyl)glycine (Bicine)	$(HOCH_2CH_2)_2N^+HCH_2COO^-$	8.35	−0.018
Glycylglycine (Gly-Gly)	$H_3N^+CH_2CONHCH_2COO^-$	8.40	−0.028

A correct description of the final buffer requires in all cases information on:

- buffer type
- C_t
- pH
- ionic strength
- types of counterions
- titration reagent

Depending on the buffer type, the temperature may have an appreciable effect on pH, as will any modifiers. A pH-meter measures the activity of H^+ and calculates therefrom the pH as the difference between the pH of an unknown solution (x) and that of standard solution(s):

$$\mathrm{pH}_{(x)} = \mathrm{pH}_{(s)} + \frac{(E_x - E_s)F}{RT \ln 10} \quad (2.40)$$

where E_x and E_s = electroosmotive forces, F = Faraday constant, R = gas constant and T = thermodynamic temperature. The pH will thus vary with temperature, which therefore should be the same for the standard and unknown solution. In practice, the buffer should be prepared and the pH adjusted at the temperature it will be used at. The temperature effects on buffer pH may be large and differ for various types of buffers as shown in Table 2.4. TRIS thus has pK'_a = 8.1 at 25 °C but at 0 °C, pK'_a = 8.9.

When pH is measured in solutions of proteins, there may be deposits on the electrodes. Careful cleaning is therefore important. Some general rules for handling electrodes are collected in Box 2.3.

Box 2.3 *General rules for handling pH-electrodes*

When pH is measured in solutions of proteins, there may be deposits on the electrodes. It is therefore important to clean them carefully with water, a detergent solution, or 1 M HCl containing pepsin. In the majority of pH measurements, glass electrodes are used, and they should, as with other electrodes, be treated according to the manufacturer's instructions. New glass electrodes, or those that have been allowed to dry out, should be conditioned by soaking in water, often for several hours. When used in different solutions, and/or before and after use, sample cups and the electrodes should be washed carefully with deionized water and gently dried with clean, soft paper.

7 Selection of Buffers

The first criterion for choice of a particular buffer is its pK'_a-value, which determines the pH-range for the buffer capacity (Section 2.5). In addition to the pK'_a, it is important to consider several other chemical properties of buffers. The UV–Vis absorption in the area of λ = 200–700 nm is important in connection with various types of liquid chromatographic procedures, including HPCE.

Furthermore, the temperature effects on pK'_a and thereby pH of the buffer solution should be considered, as this may vary considerably for different buffers (Table 2.4). There may thus be appreciable effects on liquid chromatographic methods such as HPLC and HPCE, especially in connection with the large temperature variations possible in these techniques.

In work with proteins, it may be important to avoid buffers that interfere with the protein determination. The buffers mentioned in Table 2.4 cannot be used when protein determinations are based on nitrogen determination, and several of these buffers will also interfere with the Lowry method. In enzymatic applications, one must be cautious so as to select a buffer that does not react with or inhibit the enzyme under consideration. Such interferences can be a result of binding between buffer ions and the protein/enzyme or its cofactors, *e.g.* metal ions. The interactions with inorganic cations can be evaluated from a knowledge of metal–buffer binding constants (K_m or log K_m; see Table 2.5 and Table 3.3) or determined in binding studies (Chapter 3).

In fact one of the most serious problems with some buffers is the possible formation of coordination complexes between buffers and di- or trivalent metal ions, resulting in proton release (lower pH), chelation of the metal (enzyme inhibition), and formation of insoluble complexes. Buffers composed of inorganic compounds (*e.g.* phosphate, borate, hydrogencarbonate) may also interact with enzymes, their substrates, and for borate–phosphate mixtures at alkaline pH, phosphate will precipitate cations, *e.g.* Ca^{2+}. Borate forms complexes with carbohydrates, glycoproteins, nucleic acids, and glycerol, especially at alkaline pH. Citrate also has the ability to bind to some proteins and to form metal complexes. Therefore, such properties also need consideration in the evaluation of buffers for enzyme/protein purification, analyses, and assays. The complexation properties of buffers can, however, in some cases, such as liquid chromatographic procedures, be utilized with advantages, and buffers with low metal binding constants such as PIPES, ACES, TES, and TRIS

Table 2.5 *Metal-buffer binding constants (K_m) for various buffers*

		Log K_m			
*Buffer**	*Molarity*†	Mg^{2+}	Ca^{2+}	Mn^{2+}	Cu^{2+}
MES	0.65	0.8	0.7	0.7	Negl.‡
ADA	–	2.5	4.0	4.9	9.7
PIPES	2.30	Negl.	Negl.	Negl.	Negl.
ACES	0.22	0.4	0.4	Negl.	4.6
TES	2.6	Negl.	Negl.	Negl.	3.2
Tricine	0.8	1.2	2.4	2.7	7.3
TRIS	2.4	Negl.	Negl.	Negl.	–
Bicine	1.1	1.5	2.8	3.1	8.1
Glycylglycine	1.1	0.8	0.8	1.7	5.8

*Abbreviation, see Table 2.4. Compare with section on page 44; † of saturated solution at 0 °C; ‡Negligible binding

are generally recommendable in studies of proteins/enzymes with metal requirements. Finally, the conductivity of buffer solutions is of very great importance for HPCE methods.

8 Organic Solvents as Modifiers of Aqueous Solutions

Organic solvents are required as modifiers of aqueous buffer solutions used in liquid chromatographic procedures, especially reversed-phase HPLC and in various HPCE methods (see section on pages 172 and 257). The organic solvents used need to be miscible with water and thus have an appropriate polarity. The choice may be made from the chromatographic series known as the eluotrope series, in which the polarity increases from left to right:

Chloroform – Toluene – Dioxane – Acetonitrile – Acetone – Butanol – Propanol – ⟶

Isopropanol – Ethanol – Methanol – Acetic acid – Formic acid – Water ⟶

The first members of this series, chloroform and toluene, are nearly insoluble in water, whereas dioxane and butanol are partly soluble in water at room temperature. Solvents of these types are often of value in mixtures used as solvents for extractions of membrane-bound proteins/enzymes and of biomolecules from lipid-rich materials (Section 2.9).

Limitations in the possible choice of organic solvents for liquid chromatographic purposes may be caused by insufficient purity of available solvents and/or inexpedient absorptions of UV-light (Table 5.2). Furthermore, too strongly acidic solvents may cause destruction (hydrolysis) of the stationary phases used in some reversed-phase HPLC columns. In liquid chromatographic procedures such as HPLC and HPCE, where elevated temperatures are required, it is advisable to select organic modifiers with boiling points close to that of water.

The polarity is an important property of organic solvents used in liquid chromatography and so is the *dipole moment* (μ) and *relative permittivity*, also known as the *dielectric constant*, (ε_r) in relation to discussion of polar–non-polar solvents. Dipole moments will generally be found in molecules where two different atoms of different electronegativities are connected. The polar character of a molecule with fractional charges $+\Delta e$ and $-\Delta e$ (esu) separated by a distance of l (cm) is the dipole moment:

$$\mu = l\Delta e \tag{2.41}$$

A dipole of an electron separated from a positive charge unit by 1 Å will give $\mu = 4.8 \times 10^{-10}$ esu $\times\ 10^{-8}$ cm $= 4.8 \times 10^{-18}$ esu cm or 4.8 D (debye units), as 10^{-18} esu cm is called a *debye unit* (D). In addition to the polar bond caused by the separated charges, the electrons must also have an asymmetrical distribution in the molecule to create a dipole moment. For such molecules, an increase in the difference in electronegativities of atoms will give an increased dipole moment and thereby increased attraction between the molecules. All non-covalent interactions are in fact electrostatic in nature and will thus increase with the dipole moment. If we have the charges q_1 and q_2 separated in vacuum by a distance l, the force F between them will be given by Coulomb's law:

$$F = k\frac{q_1 q_2}{l^2} \tag{2.42}$$

where k = constant. In aqueous solutions, the charges q_1 and q_2 will always be separated by the solvent or other molecules in the solution and the screening effect is denoted the dielectric constant (ε_r):

$$F = \frac{k}{\varepsilon_r}\frac{q_1 q_2}{l^2} \tag{2.43}$$

With l in metres (m), q_1 and q_2 in coulombs (C) (charge on one electron = 1.6 $\times$ 10^{-19} C), k = 8.99 $\times$ 10^9 J m C^{-2} (J (joule) = N m), F in newtons (N), the dielectric constant ε_r is a dimensionless figure. The force per mole is $F\,N_0$ (N_0 = Avogadro constant = 6.023 $\times$ 10^{23} mol^{-1}).

The dielectric constant of a solution will depend on the polarity of the molecules in the solution and on the temperature. Water has a high dielectric constant and is a polar solvent, whereas the non-polar organic solvents found at the other end of the eluotrope series have low dielectric constants, *e.g.* at 25 °C the ε_r-values for selected solvents from the eluotrope series are:

Chloroform –	Toluene –	Dioxane –	Acetonitrile –	Acetone –	Ethanol –	Methanol –	Water
4.8	2.4	2.2	20.7	21.4	24.3	33.1	78.5

With mixtures of water and organic modifiers, the dielectric constant of the solutions will be a function of the amount of modifier included. A 1:1 mixture of H_2O–dioxane will have ε_r = 30 whereas 20% dioxane in water will result in ε_r = 55. As electrostatic forces are so important for interactions between molecules in solutions and for the induction of changes in the conformations of charged molecules, the changes in ε_r by organic modifiers will be important for various liquid chromatographic procedures and methods of analyses.

An important aspect of electrostatic interactions in aqueous solutions is the hydration of ions, by which water molecules are oriented around the ions owing to the attraction of the water dipoles to the charged ions (Figure 2.1). The structure of polymers, proteins, and polysaccharides, *e.g.* pectic acids in dietary fibres, are thus influenced by the hydration of their hydrophilic groups (polar groups). Divalent ions such as Ca^{2+} are quite often important for such interactions, which, for example, also can be the case for micelles in milk, calcium caseinates and interactions with carboxylate groups in proteins and in carbohydrates. The Ca^{2+} can form a bridge between two carboxylate groups (negatively charged) or other polar groups (Figure 2.6), sometimes converting

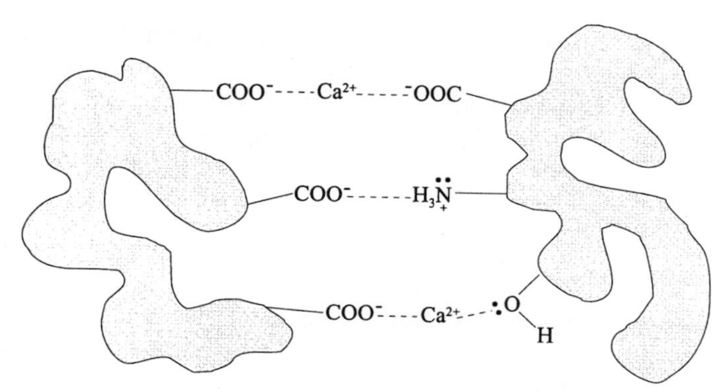

Figure 2.6 *Electrostatic interactions are important in binding between proteins, in protein structure stabilization, and for binding of proteins to other biomolecules. Divalent metal ions such as* Ca^{2+} *are often involved in these interactions*

solutions of such macromolecular solutions into rigid 'gels', *e.g.* agarose and pectinates.

The hydration of ions has a strong influence on all aspects of electrostatic interactions and plays a dominant role in determining the strength of acids and bases as well as the binding efficiency of metal ions to negatively charged groups. The pK'_a-values for acids will therefore also be influenced by the presence of organic modifiers in aqueous solutions. The $pK'_a = 4.5$–5.0 for carboxylic acids in water will increase to about 8 with changes of $\varepsilon_r = 78.5$ for water to about $\varepsilon_r = 10$ for water + organic modifiers, whereas the energetically favourable hydration of both the proton and the anion will favour dissociation in strong polar solvents, and thereby a low pK'_a-value. These effects are used in various methods of analyses, group separations, and ion-exchange procedures, HPLC and HPCE, as described in the following chapters.

9 Amphipathic Molecules and Micelles in Aqueous Solution

The amphipathic properties of phospholipids as well as of other detergents cause these compounds to arrange as ordered structures when suspended in aqueous solutions. These structures include monolayers (at the surface), bilayers, bilayer vesicles, or micelles when the monomer concentrations are higher than the c.m.c.-value (c.m.c. = critical micellar concentration) (Figure 2.7).

The type of micelle formed depends on the type, structure, and concentration of detergents, although the exact structure of the micelle and details of processes leading to micellization are usually unknown. The types of surfactant systems most commonly used in experimental biochemistry can be divided into ionic (anionic, cationic, zwitterionic), non-ionic (Triton X-100, octylglucoside), chiral, and steroidal (cholate, deoxycholate) types as illustrated by selected examples in Figure 2.8.

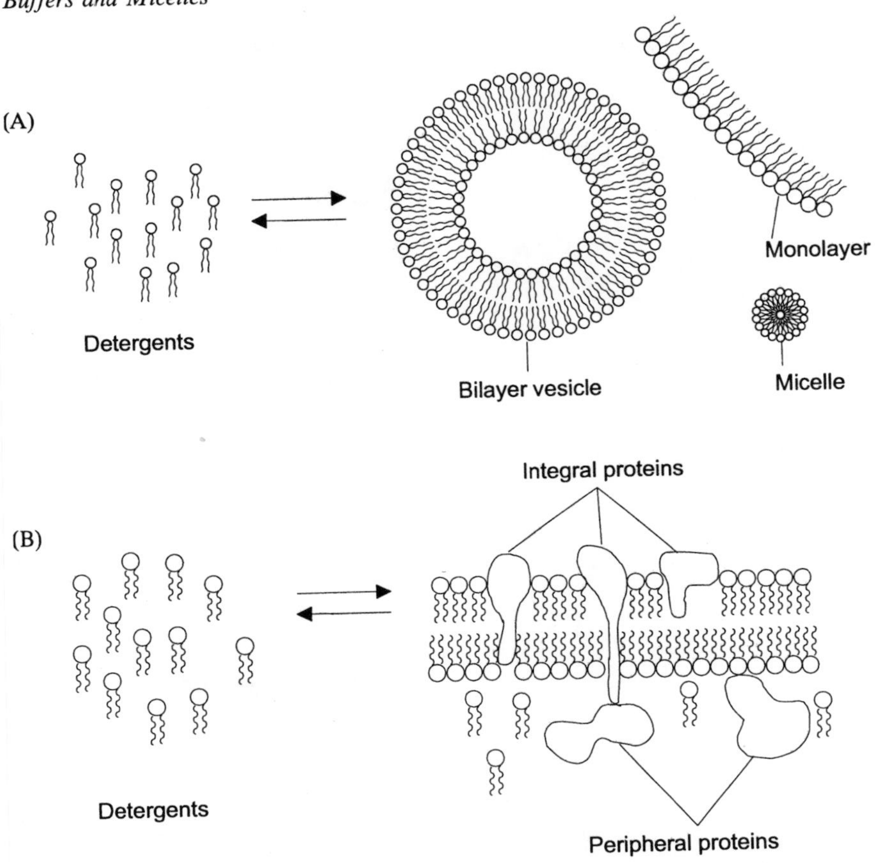

Figure 2.7 (A) *Detergents dispersed in water at concentrations > c.m.c. aggregate into micelles, resulting in a monomeric detergent concentration = c.m.c. at equilibrium.* (B) *Detergents will also disrupt the membrane and make complexes with the hydrophobic part of the integral membrane proteins and lipids*

c.m.c.-Values can be determined by measuring various physical properties of the solutions such as changes in the UV–Vis absorption of special chromophores at c.m.c. or the electrical conductivity of ionic surfactants (Figure 2.9).

The structures of micelles in aqueous solution most often comprise an interior region containing the hydrophobic parts of the amphipathic molecules, with the hydrophilic parts on the surface (Figure 2.7). The ionic groups on the surface of ionic micelles contain associated hydrated counterions, giving a hydrophilic extension into the aqueous phase. Most often, it is sufficient to consider the

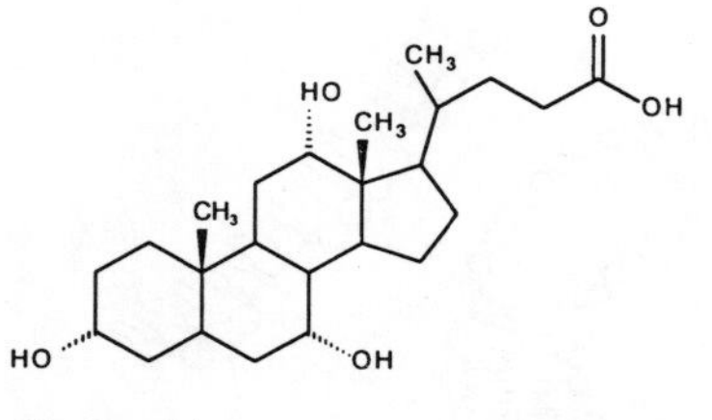

Cholic acid

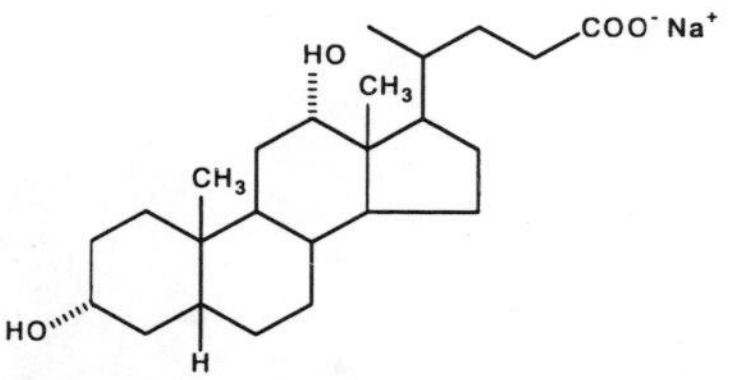

Sodium deoxycholate

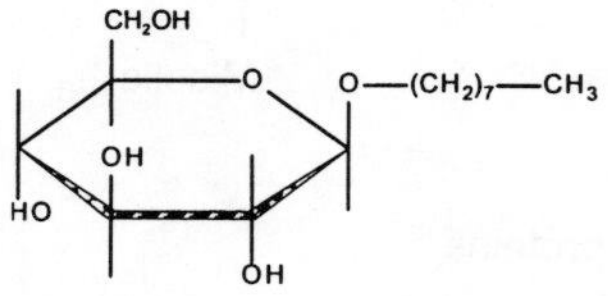

Octylglucoside
(octyl - β - D - glucopyranoside)

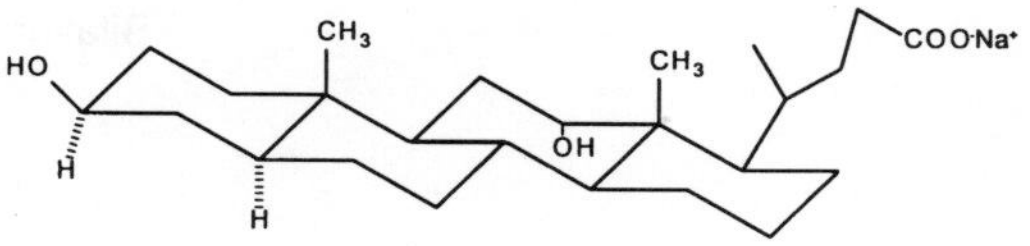

Sodium cholate

$CH_3(CH_2)_{11}OSO_3^-Na^+$

Sodium dodecyl sulfate (SDS)

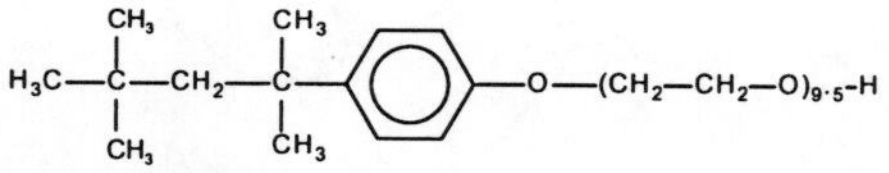

$H_3C—C(CH_3)_2—CH_2—C(CH_3)_2—C_6H_4—O—(CH_2—CH_2—O)_{9.5}—H$

Triton X - 100
(polyoxyethylene(9.5)p-t-octylphenol)

$CH_3(CH_2)_{15}N^+(CH_3)_3Br^-$

Cetyltrimethylammonium bromide (CTAB)

Figure 2.8 *Structures of selected surfactants often used in experimental biochemistry*

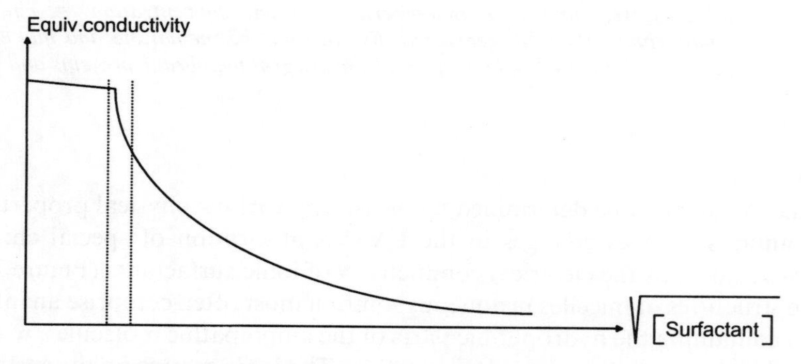

Figure 2.9 *Changes in equivalent conductivity (specific conductivity per mole of surfactant) as a function of ionic surfactant concentration in aqueous solution for the determination of c.m.c.; the area between the dotted lines*

Table 2.6 *Critical micellar concentrations (c.m.c.) and other data for selected surfactants in aqueous media*

Surfactant	*c.m.c.* (mM)	*Temp.* (°C)	*Monomer* (M_r)
Anionic			
$CH_3(CH_2)_9SO_4^-Na^+$, Sodium decyl sulfate	33.0	40	274.5
$CH_3(CH_2)_{11}SO_4^-Na^+$, Sodium dodecyl sulfate (SDS)*	9.7	40	288.5
*Dodecyl sulfate, Li^+	8.9	25	272.4
Dodecyl sulfate, Na^+	8.2	25	288.5
Dodecyl sulfate, K^+	7.8	40	304.6
Dodecyl sulfate, $\frac{1}{2}Ca^{2+}$	3.4	70	
Dodecyl sulfate, $N(CH_3)_4^+$	5.5	25	339.5
Dodecyl sulfate, $N(C_3H_7)_4^+$	2.2	25	451.5
Dodecyl sulfate, $N(CH_3)_4^+$, 3 M urea	6.9	25	–
Dodecyl sulfate, Na^+, 10 mM NaCl	5.6	21	–
Dodecyl sulfate, Na^+, 30 mM NaCl	3.2	21	–
Dodecyl sulfate, Na^+, 100 mM NaCl	1.5	21	–
$CH_3(CH_2)_{13}SO_4^-Na^+$, Sodium tetradecyl sulfate (STS)	2.2	40	302.5
$CH_3(CH_2)_{15}SO_4^-Na^+$, Sodium hexadecyl sulfate Sodium cetyl sulfate (SCS)	0.6	40	316.5
$CH_3(CH_2)_{17}SO_4^-Na^+$, Sodium octadecyl sulfate (SOS)	0.2	40	330.5
Cationic			
$CH_3(CH_2)_9N^+(CH_3)_3Br^-/Cl^-$, Decyltrimethylammonium bromide/chloride	68/61	25	280/235.5
$CH_3(CH_2)_{11}N^+(CH_3)_3Br^-/Cl^-$, Dodecyltrimethylammonium bromide/chloride (DTAB/DTAC)	16/20	25	294/249.5
DTAC + 500 mM NaCl	3.8	30	–
$CH_3(CH_2)_{13}N^+(CH_3)_3Br^-/Cl^-$, Tetradecyltrimethylammonium bromide/chloride (TTAB/TTAC)	3.5/4.5	30/25	308/263.5
$CH_3(CH_2)_{13}N^+(C_3H_7)_3Br^-$	2.1	30	392
$CH_3(CH_2)_{15}N^+(CH_3)_3Br^-/Cl^-$, Cetyltrimethylammonium bromide/chloride (CTAB/CTAC)	0.9/1.3	25/30	322/277.5
Bile salts			
Sodium cholate	14	25	430
Sodium deoxycholate	5	25	414
Non-ionic			
Octylglucoside	25	25	292.4
Triton X-100	0.3	25	628

micelles as being roughly spherical at surfactant concentrations not far from the c.m.c. (Table 2.6). In cases with organic modifier added to the aqueous solution containing micelles, and in solutions with surfactant concentrations at least ten times the c.m.c., micelles can have a non-spherical form and may therefore aggregate. This aggregation will increase with increased dissimilarity between the surfactant and the solvent, *e.g.* through an increase in the hydrophobicity of the surfactant.

c.m.c.-Values are affected by the hydrophobicity of surfactants, an increase resulting in a decrease in c.m.c., whereas introduction of a polar group into the surfactant gives a significant increase in c.m.c. Ionic surfactants thus have much higher c.m.c.-values than non-ionic surfactants with an otherwise equal hydrophobic part. An illustration of the trends in c.m.c.-values and other parameters for selected detergents is shown in Table 2.6.

The effect of a temperature increase is to disrupt the structured water surrounding the micelles and surfactant monomers, displacing the equilibrium toward the monomers. The concurrent decrease in hydration of hydrophilic groups also favours, however, micelle formation. It is the relative magnitude of these effects that determines the c.m.c.-value as function of the temperature, resulting in minimum values around 25 °C for ionic surfactants and around 50 °C for non-ionic surfactants.

The presence of ions and especially organic material may produce appreciable changes in the micellar structures and the c.m.c. in aqueous media. These changes are of great importance both for theoretical and practical purposes where micelles or surfactants are used in analytical procedures, *e.g.* in MECC. It is therefore most important to have a well-defined sample matrix and that appropriate purifications and group separations are carried out prior to such types of analyses. An increase in the sodium chloride concentration from 10 to 100 mM changes the c.m.c. for SDS by a factor of 4, and the type of counterion in ionic surfactants also changes the c.m.c. (Table 2.6). Increasing the number of carbons in tetraalkylammonium ions used as counterion to alkyl sulfates decreases the c.m.c., and compounds such as urea, formamide, and guanidinium salts increase the c.m.c., whereas organic solvents that can be incorporated into the micelles will reduce the c.m.c.

Ion-pair chromatography also comprises the use of ionic surfactants, a technique which is useful for separating natural products by reversed-phase HPLC and MECC.

10 Selected General and Specific Literature

1. C.R. Dillard and D.E. Goldberg, 'Chemistry. Reactions, Structures and Properties', MacMillan, New York, 1971.
2. D. Eisenberg and D. Crothers, 'Physical Chemistry with Application to Life Science', Benjamin-Cummings, Redwood City, CA, 1979.
3. S.F.Y. Li, 'Capillary Electrophoresis. Principles, Practice, and Applications', Journal of Chromatography Library, Elsevier, Amsterdam, 1992, vol. 52, Ch. 5, p. 201.
4. V.R. Williams, W.L. Mattice, and H.B. Williams, 'Basic Physical Chemistry for the Life Sciences', Freeman, San Francisco, 1978, Ch. 5.

CHAPTER 3

Binding, Association, Dissociation, and Kinetics

1 Introduction

Interaction between biomolecules or ions is the basis for chemistry, biochemistry, and for the analytical methods behind these disciplines, including natural product chemistry and studies of food composition and quality. The interactions can be studied as binding, association, or dissociation reactions that are given quantitative expression in thermodynamic terms as equilibrium constants (K_{eq}), changes in standard free energy (ΔG°), enthalpy (ΔH), entropy (ΔS), or by kinetics. This way, the strength of binding can be expressed in thermodynamic terms, which imply a dependency on temperature (T = absolute temperature; R = gas constant):

$$\Delta G = \Delta H - T\Delta S = \Delta G^{\circ} + RT \ln K_{eq} \tag{3.1}$$

Different types of forces may be involved in the binding: Van der Waals forces/hydrophobic interactions, hydrogen-, dipole-, or ion-binding, with binding between hydrated/solvated molecules or ions. The types of binding to be considered will comprise:

- Binding of protons (*e.g.* acid–base equilibria, buffers and microscopic ionization constants.
- Binding of anions and metal ions to buffer components (Chapter 2) and biomolecules (*e.g.* ADP–Mn^{2+} binding), to macromolecules [*e.g.* $(NH_4)_2SO_4$ precipitation], and in colorimetry/staining procedures (Coomassie Brilliant Blue, silver stain and various other types of staining) (*vide infra*).
- Binding of charged/uncharged ligands to macromolecules as serum protein (*e.g.* BSA–Trp binding), enzymes (inhibitor–enzyme binding) or dietary fibres [*e.g.* toxic/antinutritional compounds (xenobiotica)].

The theory behind binding is essential in connection with experimental biochemistry and thereby also for understanding the problems related to the extraction of biomolecules from their native bindings in membranes/micelles, in cell homogenates, and solutions/emulsions such as blood and milk.

2 Binding Forces – Position of Equilibrium

Below is given the equilibrium reaction of association of a ligand (L) to a molecule (P) resulting in the complex PL. The compounds L and P may be proteins, polysaccharides (dietary fibres), nucleic acid, metal ion, or other relative low molecular mass compounds (low-M_r compounds; vitamins, co-factors).

$$\mathrm{P} + \mathrm{L} \underset{k_2}{\overset{k_1}{\rightleftharpoons}} \mathrm{PL}$$

The PL complex is formed with the velocity

$$\underset{\rightarrow}{v} = k_1 \times p \times l \tag{3.2}$$

and dissociated with the velocity

$$\underset{\leftarrow}{v} = k_2 \times pl \tag{3.3}$$

At equilibrium: $\underset{\rightarrow}{v} = \underset{\leftarrow}{v}$, resulting in the equilibrium constant:

$$K'_{\mathrm{eq}} = \frac{k_1}{k_2} = \frac{pl}{p \times l} \tag{3.4}$$

For simplicity, small letters will be used as symbols for concentration, *e.g.* [P] = p, as also used in equations 3.2 and 3.3. The prime is used to indicate that it is an apparent constant based on concentrations instead of the thermodynamically correct values based on activities (Section 2.3), and this symbol may correspondingly be used for related thermodynamic denominations such as G, H, and S based on concentrations.

Under standard conditions, the position of the equilibrium could also be described by:

$$\Delta G^{\circ\prime} = -RT \ln K'_{\mathrm{eq}} = -2.303RT \log K'_{\mathrm{eq}} \tag{3.5}$$

This means that both $\Delta G^{\circ\prime}$ and K'_{eq} can be used to evaluate the position of the binding equilibrium, which is temperature dependent.

At 25 °C and R = 8.315 J K^{-1} mol^{-1} (1.987 cal K^{-1} mol^{-1}) this gives

$$\Delta G^{\circ\prime} = -5706 \text{ J mol}^{-1} \times \log K'_{\mathrm{eq}} = -1364 \text{ cal mol}^{-1} \times \log K'_{\mathrm{eq}}$$

which means that a tenfold change in K'_{eq} at 25 °C will give a change in $\Delta G^{\circ\prime}$ of 5.71 kJ mol^{-1} (1.36 kcal mol^{-1}). With consideration of the kinetic energy for motion of molecules, using Boltzmann's constant (k_B = 1.3807 × 10^{-23} J K^{-1}), Avogadro's constant (N_A = 6.022 × 10^{23}) and kinetic energy $E_{kin} = \frac{3}{2} k_B T$, we have at 25 °C:

$$E_{kin} = \tfrac{3}{2} 1.3807 \times 10^{-23} \text{J K}^{-1} \times 6.022 \times 10^{23} \text{mol}^{-1}\ 298 \text{ K}$$
$$= 3.717 \text{ kJ mol}^{-1} = 0.89 \text{ kcal mol}^{-1}$$

With a binding constant K'_f = 10, the above mentioned example shows that this K'_f gives only an unstable complex (PL) at room temperature with a $\Delta G^{\circ\prime}$ of 1.36 kcal mol^{-1} compared with the kinetic energy of 0.89 kcal mol^{-1}. A decrease in temperature will, however, give an increase in complex formation, and also an increase in K'_f or a more negative $\Delta G^{\circ\prime}$ corresponds to stronger binding. For many enzymes, the binding of substrates and products is weak, allowing the reaction to proceed easily from both left to right and right to left. In contrast, strong binding (high negative $\Delta G^{\circ\prime}$) is often found for enzyme cofactors, lac repressors and many antinutrients (Avidin), protein proteinase inhibitors (Chapter 12), and toxicants in food. Strong binding is also observed for various compounds, which can be used in several methods of analyses, *e.g.* colorimetric and staining procedures using the binding of the fluorescent molecule 1-anilinonaphthalene-8-sulfonate (ANS) (Table 3.1).

As seen from Table 3.2, an increase in concentration of P or L (for an

Table 3.1 *Selected examples of $\Delta G^{\circ\prime}$ for formation of protein–ligand complexes*

Protein	*Ligand*	$\Delta G^{\circ\prime}$ (kcal mol^{-1})*
Avidin	Biotin	−20
Lac repressor	DNA	−22
Alcohol dehydrogenase	NADH	−9
Lactate dehydrogenase	NADH	−8.8
Serum albumin (BSA)	ANS	−7.9
Alcohol dehydrogenase	CH_3CH_2OH	−2.5
Alcohol dehydrogenase	CH_3CO_2H	−4.7
Fumarase	Fumarate	−3
Fumarase	Malate	−2.5

* 1 cal = 4.184 J

Table 3.2 *Complex formation (%) at different equilibrium conditions (T = 25 °C)*

	Initial conc.		*Equilibrium conc.*	
K'_f (M^{-1})	*Ligand (mM)*	*Protein (mM)*	*Complex (mM)*	*% complex*
10	0.1	0.1	10^{-4}	0.1
10	10	10	0.84	8.4
10^4	0.1	0.1	26	38

equilibrium reaction with a given K'_f) results in an increase in the amount of complex. Changing K'_f from 10 to 10^4 (concentrations of P and L unchanged) leads to an increase of the molecules P and L present as the complex PL from 0.1 to 38%. If we assume $K'_f = 10^7$, an extremely strong binding is the result, with 97% of P and L in the complex PL.

3 Specificity of Proteins/Enzymes Towards Ligands and Antinutrients

The normal or undisturbed metabolism in all living cells implies the unique properties of native biomolecules to specifically bind to other biomolecules, substrates, inhibitors, and allosteric effectors. Interference with this binding by antinutritional compounds or toxicants is most often specific and may range from reversible to irreversible binding, depending on K'_{eq} (Section 3.2), and may thereby disturb the normal metabolism, resulting in antinutritional or toxic effects. The effects of ligand binding and the subsequent processes depend on the type of binding:

- Catalytic proteins/enzymes; chemical change of ligand (substrate → product).
- Immunoproteins; antibody–antigen.
- Contractile proteins; contraction–relaxation.
- Membrane proteins; transport processes or signals through membranes.
- Transport proteins; transport–dissociation.

Binding of ligands to proteins, including enzymes, may occur at one or more binding sites, active sites, and effector sites. This binding of ligands at different sites can be either independent of (non-cooperative) or dependent on each other (cooperative) (page 49).

The effector A is a homotrope effector when it is identical with the substrate (S) in the active site. If A ≠ S, it is a heterotrope effector. With respect to specificity of active sites in proteins, these can be:

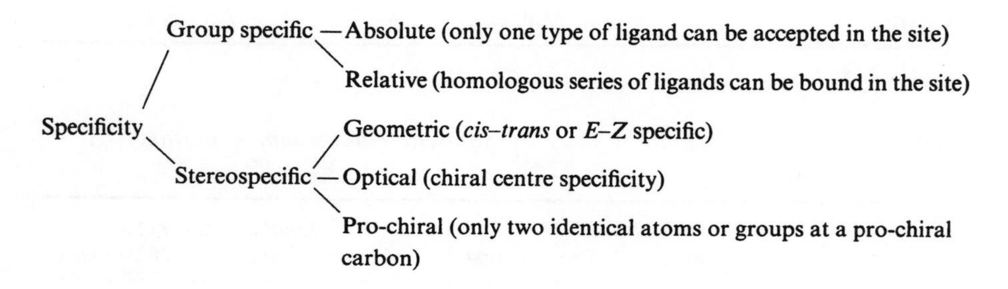

4 Analysis of Binding Experiments

Quantitative measurements of binding for equilibria with not too high numerical values of $\Delta G^{\circ\prime}$ ($= -RT\ln K'_{eq}$) (Section 3.2) are based on nearly the same type of theory as used for enzyme kinetics. The experimental studies are based on:

- use of defined symbols and abbreviations;
- put forward an assumption (hypothesis);
- draw up the mathematical relationships;
- convert the relations into linear functions;
- perform appropriate and required experiments;
- introduce the results into the mathematical relationships/linear functions;
- evaluation and conclusions based on the obtained data;
- re-evaluate the conclusions by use of alternative experiments.

Binding of Ligands to Molecules with One Binding Site

Consider the binding of a ligand (L) to a molecule (P) with one binding site.

$$\mathrm{P} + \mathrm{L} \rightleftharpoons \mathrm{PL}$$

$$K'_{f} = K'_{eq} = \frac{pl}{p \times l} \tag{3.6}$$

As defined in connection with acid–base discussions (Section 2.5), we use the association degree (v) corresponding to:

$$v = \frac{\text{Concentration of bound L}}{\text{Concentration of P}_{\text{total}}} = \frac{pl}{p + pl} \tag{3.7}$$

$$= \frac{K'_{f} \times p \times l}{p + K'_{f} \times p \times l} = \frac{l}{K'_{d} + l}\left(K'_{d} = \text{dissociation; } K'_{d} = \frac{1}{K'_{f}}\right) \tag{3.8}$$

Linear functions of these equations are:

$$\frac{1}{v} = 1 + \frac{K'_{d}}{l} \quad \text{(Hughes–Klotz = Lineweaver–Burk plot)} \tag{3.9}$$

$$\frac{v}{l} = \frac{1}{K'_{d}} - \frac{v}{K'_{d}} \quad \text{(Scatchard plot)} \tag{3.10}$$

$$\frac{l}{v} = K'_{d} + l \quad \text{(Hanes plot)} \tag{3.11}$$

$$v = 1 - K'_{d}\frac{v}{l} \quad \text{(Eadie–Hofstee plot)} \tag{3.12}$$

Binding of Ligand to Molecules with Two Equivalent and Independent Sites

L+P(L)
k_1 k_3
P + 2L $\overset{K_1}{\rightleftarrows}$ (PL) + L $\overset{K_2}{\rightleftarrows}$ P(L)(L)
k_2 k_4
L+P(L)

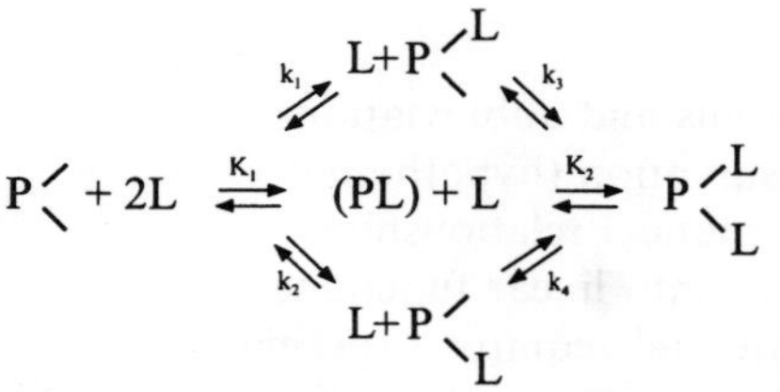

Figure 3.1 *Illustration of ligand binding to two different sites on a molecule*

$$k_1 = \frac{p^{/l}_{\backslash}}{l \times p^{/}_{\backslash}};\quad k_2 = \frac{p^{/}_{\backslash l}}{l \times p^{/}_{\backslash}};\quad k_3 = \frac{p^{/l}_{\backslash l}}{l \times p^{/l}_{\backslash}};\quad k_4 = \frac{p^{/l}_{\backslash l}}{l \times p^{/}_{\backslash l}}$$

$$K'_{\text{total}} = K'_1 K'_2 = k_1 k_3 = k_2 k_4 = \frac{p^{/l}_{\backslash l}}{l^2 \times p^{/}_{\backslash}} \tag{3.13}$$

$$K'_1 = k_1 + k_2 \quad \text{and} \quad K'_2 = \frac{k_3 k_4}{k_3 k_4} \tag{3.14}$$

With the microscopic constants $k_1 = k_2 = k_3 = k_4 = k$ (equivalent and independent sites, see also on page 42), we have for the macroscopic constants $K'_1 = 2k$; $K'_2 = \frac{1}{2}k$; $K'_1/K'_2 = 4$. In such cases, $K'_1 = 4K'_2$.

$$v = \frac{pl + 2pl_2}{p + pl + pl_2} = \frac{2kl}{(1 + kl)} = \frac{2l}{K'_d + l} \tag{3.15}$$

Linear functions:

$$\frac{1}{v} = \frac{1}{2} + \frac{K'_d}{2l} \quad \text{(Hughes–Klotz = Lineweaver–Burk plot)} \tag{3.16}$$

$$\frac{v}{l} = \frac{2}{K'_d} - \frac{v}{K'_d} \quad \text{(Scatchard plot)} \tag{3.17}$$

$$\frac{l}{v} = \frac{K'_d}{2} + \frac{l}{2} \quad \text{(Hanes plot)} \tag{3.18}$$

$$v = 2 - K'_d \frac{v}{l} \quad \text{(Eadie–Hofstee plot)} \tag{3.19}$$

Binding of Ligands to Macromolecules with *n* Equivalent, Independent Sites

$$\mathrm{P} + n\mathrm{L} \overset{K'_1}{\rightleftharpoons} \mathrm{PL} + (n-1)\mathrm{L} \overset{K'_2}{\rightleftharpoons} \mathrm{PL_2} + (n-2)\mathrm{L} \overset{K'_3}{\rightleftharpoons} \ldots \overset{K'_n}{\rightleftharpoons} \mathrm{PL}_n$$

Using

$$K'_1 = \frac{pl}{p \times l}; \quad K'_2 = \frac{pl_2}{pl \times l}; \ldots K'_n = \frac{pl_n}{pl_{(n-1)} \times l} \tag{3.20}$$

we have

$$v = \frac{pl + 2pl_2 + 3pl_3 + \ldots npl_n}{p + pl + pl_2 + pl_3 + \ldots pl_n} \tag{3.21}$$

Introducing K' for each step and recognizing a binominal distribution this seemingly complex equation reduces very neatly to:

$$v = \frac{nl}{K'_\mathrm{d} + l} \tag{3.22}$$

Linear functions correspond to the case with two ligands ($n = 2$):

$$\frac{1}{v} = \frac{1}{n} + \frac{K'_\mathrm{d}}{nl} \quad \text{(Hughes–Klotz = Lineweaver–Burk plot)} \tag{3.23}$$

$$\frac{v}{l} = \frac{n}{K'_\mathrm{d}} - \frac{v}{K'_\mathrm{d}} \quad \text{(Scatchard plot)} \tag{3.24}$$

$$\frac{l}{v} = \frac{K'_\mathrm{d}}{n} + \frac{l}{n} \quad \text{(Hanes plot)} \tag{3.25}$$

$$v = n - K'_\mathrm{d}\frac{v}{l} \quad \text{(Eadie–Hofstee plot)} \tag{3.26}$$

Binding of Analytes to Adsorbent – Langmuir's Theory of Adsorption

Langmuir considered the adsorption of gas molecules to a solid surface,[1,2] work which resulted in an equation as shown for the association degree (*vide supra*). This theory can also be used in connection with some simple chromatographic systems and thus for analytes in solutions (mobile phase) and with the solid phase = stationary phase it results in:

$$A_\mathrm{s} = \frac{kA_\mathrm{m}}{1 + k'A_\mathrm{m}} \tag{3.27}$$

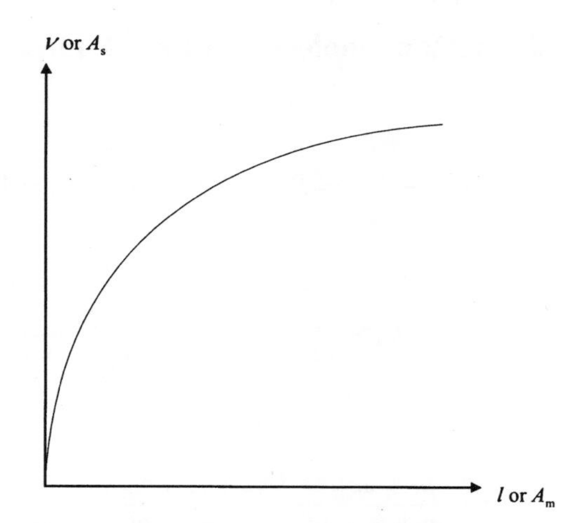

Figure 3.2 *Binding of ligand to macromolecules* $[v = nl/(K'_d + l)]$ *or analytes distributed in the mobile and stationary phases* $[A_s = kA_m/(1 + k'A_m)]$

where A_s = amount of analyte on the stationary phase, A_m = amount of analyte in the mobile phase and k and k' = constants.

This equation corresponds to a hyperbolic curve for A_m plotted against A_s as the case is for a ligand (l) plotted against the degree of association (v) (Figure 3.2). In connection with more complex chromatographic systems, the results will, however, often correspond to a sigmoidal curve and thus follow the Freundlich equation for low concentrations of analyte.[3]

The theory behind binding studies seems thus to follow the physico-chemical equations and theories developed some time ago. These theories are again in agreement with data obtainable by chromatography, in terms of a distribution of analytes between the mobile and stationary phase.

Binding at Two Different but Independent Sites – Microscopic Constants

Microscopic association/dissociation constants are defined as the equilibrium constant, which applies to a single binding site, as introduced in connection with Equations 3.13 and 3.14. A neutral amino acid has at least two protolytically active groups and thus two macroscopic equilibrium constants; pK'_{a1} *ca.* 2–2.4 and pK'_{a2} *ca.* 9.5–10. Such pK'_a-values can easily be determined from titration curves (Chapter 2). However, for pK_a-values which differ by less than two units, other methods are required, which is the case for amino acids such as cysteine and tyrosine as well as with several other natural products, *e.g.* pyridoxine (vitamin B_6) which has two pK'_a-values in the 8.2–8.8 range. The dissociation/association corresponds to the binding shown in Figure 3.2 and is illustrated as dissociation of protons from cysteine when the equations are considered from left to right, as shown in Figure 3.3.

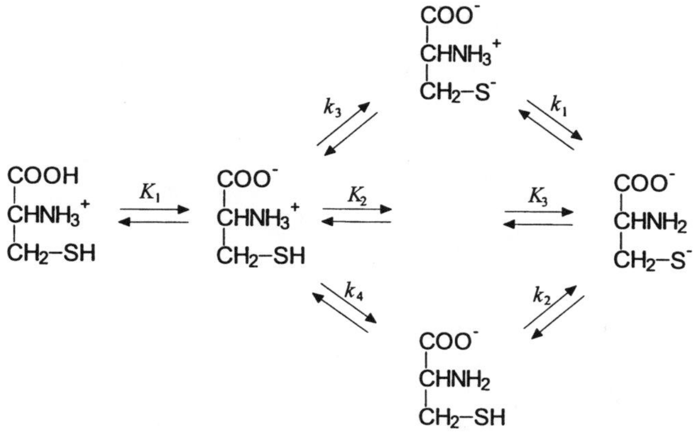

COOH–CHNH$_3^+$–CH$_2$–SH $\underset{}{\overset{K_1}{\rightleftharpoons}}$ COO$^-$–CHNH$_3^+$–CH$_2$–SH $\overset{K_2}{\rightleftharpoons}$ ... $\overset{K_3}{\rightleftharpoons}$ COO$^-$–CHNH$_2$–CH$_2$–S$^-$

k_3: COO$^-$–CHNH$_3^+$–CH$_2$–S$^-$; k_1

k_4: COO$^-$–CHNH$_2$–CH$_2$–SH; k_2

Figure 3.3 *Dissociation of protons from protonated cysteine or association of protons to the cysteine di-anion*

As revealed from above (page 40) using the constants in Figure 3.3, we have $K'_2 = k_3 + k_4$ and $K'_3 = \dfrac{k_1 k_2}{k_1 + k_2}$ and $K'_2 \times K'_3 = k_3 k_1 = k_4 k_2$. Values for the dissociation constants K'_1, K'_2, and K'_3 can be obtained from titration curves, but an alternative method of analysis is required for one of the microscopic constants k_1, k_2, k_3, or k_4. The degree of ionization for the thiol group (k_3 or k_2) can be determined by quantitative UV spectroscopy (Table 5.4) as the ionized thiol absorbs at $\lambda = 232$ nm, which is not the case for the protonated thiol group. This allows the determination of the degree of dissociation (α) from the absorption ($A_{232\,\mathrm{nm}}$) at different pH values:

$$\alpha = \frac{[\mathrm{CH_2-S^-}]}{[\mathrm{CH_2SH}] + [\mathrm{CH_2-S^-}]} = \frac{A_{232(\mathrm{pH\,8\text{–}11})} - A_{232(\mathrm{pH\,4.6})}}{A_{232(\mathrm{pH\,12})} - A_{232(\mathrm{pH\,4.6})}} \tag{3.28}$$

$$\alpha \cdot ([\mathrm{H^+}] + K'_2) = K'_2 K'_3 \frac{(1-\alpha)}{[\mathrm{H^+}]} + k_3 \tag{3.29}$$

With pairs of α and pH ([H$^+$]) belonging together, k_3 can be determined from a plot of $(1-\alpha)/[\mathrm{H^+}]$ against $\alpha([\mathrm{H^+}] + K'_2)$, and all of the microscopic constants can then be calculated from the equations shown above. Box 3.1 outlines the experimental procedure for determination of microscopic dissociation constants for cysteine.

A knowledge of the microscopic constants both for thiols and phenolics are important as the properties of such functional groups depend on their degree of ionization or protonation. In the ionized form, both of these groups have increased sensitivity towards oxidation, which needs to be considered in relation

Box 3.1 *Determination of microscopic dissociation constants for cysteine*

Prepare the following stock solutions and record UV spectra (200–300 nm) immediately after preparation.

Cysteine, HCl	*Buffer solution**	*Resulting pH†*
1 ml, 1.6 mM	9 ml 50 mM acetate, pH 4.6	*ca.* 4.6
1 ml, 1.6 mM	9 ml 25 mM borax, pH 8.0	*ca.* 7.9
1 ml, 1.6 mM	9 ml 25 mM borax, pH 9.0	*ca.* 8.9
1 ml, 1.6 mM	9 ml 25 mM borax, pH 10.0	*ca.* 9.9
1 ml, 1.6 mM	9 ml 25 mM borax, pH 10.8	*ca.* 10.5
1 ml, 1.6 mM	9 ml 10 mM NaOH‡	*ca.* 11.7

*All buffer solutions with 0.15 M NaCl
†Record the exact pH-value
‡25 ml 10 mM NaOH by dilution of 100 mM NaOH with 0.14 M NaCl

Prepare a solution of 1 mmol Cys, HCl, H_2O (176 mg) dissolved in 25 ml of 0.11 M NaCl and produce a titration curve from this solution using 0.1 M NaOH. Determine K'_1, K'_2, and K'_3 from the titration curve. Calculate k_3 from the UV data and corresponding pH of the solutions, and finally all of the microscopic constants using the equations given in this section

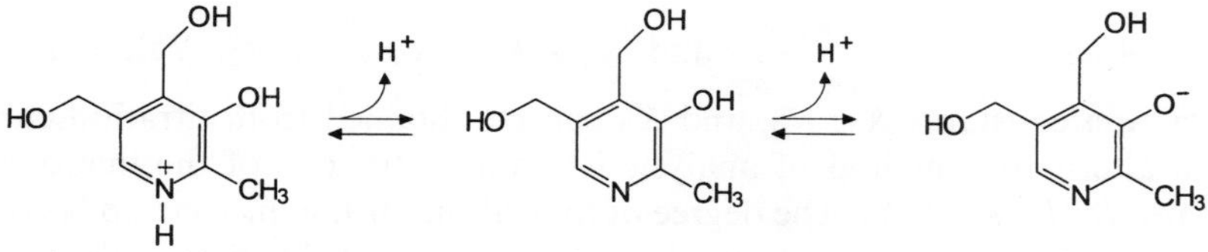

Figure 3.4 *Pyridine N as an electron sink*

to the methods of extraction, group separation, and various chromatographic techniques used for the determination of natural products, including proteins/enzymes. Information on microscopic constants are also valuable in relation to understanding the reactivity of functional groups in the active sites of enzymes. This will apply to cofactors, *e.g.* pyridoxine or pyridoxalphosphate cofactors, as their reactivity depends on the protonization/ionization of the pyridine N (resulting in an efficient electron sink) and phenol group (Figure 3.4).

The pK'_a-values for these two protolytically active groups are in the 8.2–8.8 range and it is thus necessary to use the techniques described above for the determination of the microscopic constants.

Binding of Ligands to Metal Ions

Metal ions combine with ligands as anions or neutral molecules with available electron pairs to form complexes in accordance with the coordination number of the metal. The more nucleophilic, and therefore the more basic, ligands tend to bind metal ions more tightly, just as they do protons. To form complexes, the metal ion must usually lose most of its hydration sphere (Chapter 2). For this reason, the larger, less hydrated metal ions often bind ligands more strongly than do the smaller more hydrated ones. Generally, Ca^{2+} binds more tightly to ligands than does Mg^{2+}, as also shown in Table 2.5, but with nucleotide phosphates (ATP, ADP), Mg^{2+} binds more strongly than does Ca^{2+}, and with protolytically active ligands the binding activity strongly depends on their ionization or pK'_a-values (Chapter 2). For the transition metals, the stability of their complexes in general follows the sequence: $Mn^{2+} < Fe^{2+} < Co^{2+} < Ni^{2+} < Cu^{2+}$.

The heavier metal ions of this type, including Zn^{2+}, Cu^{2+}, Fe^{2+}, Fe^{3+}, Ca^{2+}, and Ni^{2+} are often central atoms in cofactors, *e.g.* porphyrins, where their d orbitals may participate in covalent bond formation. The alkali metal ions Na^+ and K^+ are mostly free in solution or only in part bound to specific sites in proteins. Ca^{2+} and Mg^{2+} are partially free (cf. Section 2.7) but form complexes with various phosphates and carboxylates (Figure 2.6).

The binding of ligands (L) to metal ions (M) can be treated with the general theory for binding equilibria (section on page 38):

$$n\mathrm{L} + \mathrm{M} \overset{K'_1}{\rightleftharpoons} \mathrm{ML} + (n-1)\mathrm{L} \overset{K'_2}{\rightleftharpoons} \mathrm{ML_2} + (n-2)\mathrm{L} \overset{K'_3}{\rightleftharpoons} \ldots \overset{K'_n}{\rightleftharpoons} \mathrm{ML}_n$$

$$K'_{\mathrm{total}} = K'_1 K'_2 K'_3 \ldots K'_n \tag{3.30}$$

$$v = \frac{\mathrm{ml} + 2\mathrm{ml_2} + 3\mathrm{ml_3} + \ldots n\mathrm{ml}_n}{\mathrm{m} + \mathrm{ml} + \mathrm{ml_2} + \ldots \mathrm{ml}_n} \tag{3.31}$$

Table 3.3 shows some selected binding constants $K'_m = K'_f$ (shown as $\log K'_f$) which are supplemental to those shown in Table 2.5.

Various compounds in food, including dietary fibres (Chapter 14) and especially phytate and other inositol phosphates, bind strongly to metal ions. Phosphate esters, phosphate anhydrides, and nucleotide phosphates will bind

Table 3.3 *Metal–ligand binding constants shown as* $\log K'_f$

*Ligand**	H^+	Mg^{2+}	Ca^{2+}	Mn^{2+}	Fe^{2+}	Co^{2+}	Ni^{2+}	Cu^{2+}	Zn^{2+}
CH_3COO^-	4.7	0.5	0.5				0.7	2.0	1.0
$(COO^-)_2$	4.2	3.4	3.0	3.9	4.7	4.7	5.3	6.2	4.9
NH_3 K'_1	9.3		−0.2			2.1	2.8	4.2	2.4
K'_2						1.6	2.2	3.5	2.4
Ethylendiamine	10.1			2.7	4.3	5.9	7.7	10.6	5.7
Imidazole	7.0		0.1	1.7		2.4	3.3	4.3	2.6
His	9.2			1.8		6.9	8.7	10.4	6.6
Gly	9.7	3.4	1.4	3.4		5.2	6.2	8.6	5.5
GSH	8.9			1.9		3.7	4.0		5.0
Cys	10.5			4.1	6.2	9.3	10.5		9.9

* Compare with Section 2.7 and Table 2.5

Table 3.4 *Selected data for Mn^{2+}-nucleotide complexes (log K'_m), at two pH values close to pK'_a for the phosphate*

Nucleotide	*log K'_m (pH = 8.05)*	*log K'_m (pH = 5.01)*
ADP	4.2	2.6
GDP	4.0	2.7
UDP	3.8	2.7

metals, strongly dependent on pH, and thereby their reactivity with nucleophilic reagents increases appreciably. The pH-dependence for Mn^{2+} binding is illustrated by the data shown in Table 3.4.

Binding constants can be determined by various techniques, which allow separation of ligand and complex:

$$\text{ADP} + \text{Mn}^{2} \rightleftharpoons \text{ADP–Mn}^{2+}$$

$$K'_{\mathrm{f}} = K'_{\mathrm{m}} = \frac{[\text{ADP–Mn}^{2+}]}{[\text{ADP}][\text{Mn}^{2+}]} \quad (3.32)$$

$$v = \frac{[\text{ADP–Mn}^{2+}]}{[\text{Mn}^{2+}] + [\text{ADP–Mn}^{2+}]} \quad (3.33)$$

$$\frac{v}{l} = nK'_{\mathrm{f}} - vK'_{\mathrm{f}} \quad (3.34)$$

Gel filtration with use of a Sephadex G-10 column (Chapter 6) allows separation of Mn^{2+} from the complex ADP–Mn^{2+}, and with the column equilibrated with an appropriate Mn^{2+} concentration in TRIS buffer (Table 2.5), such techniques will allow the determination of K'_m if it is 1 : 1 binding. For determination of the number of Mn^{2+} (n) bound to ADP, experiments with different [Mn^{2+}] are required, and the data can then be used in linear functions (pages 39–41). The results will then appear as shown in Figure 3.5 when both UV-absorption and [Mn^{2+}] (from atomic spectroscopy; Section 5.2) are recorded.

The amount of Mn^{2+} bound in ADP–Mn^{2+} corresponds to the amount represented by the 'top' in Figure 3.5, which is equal to the amount lacking in the 'trough'. The volume representing the 'top' and the 'trough' may be different, but as the complex is in the 'top'-fractions these represent the volume, which shall be used to calculate the concentration (Box 3.2).

Binding of Ligands to Macromolecules

The binding of ligands to macromolecules is simple to determine if we have equilibria (weak binding) and fulfillment of the assumptions given in pages 39–40. Experimental binding results may thus be plotted according to the equations

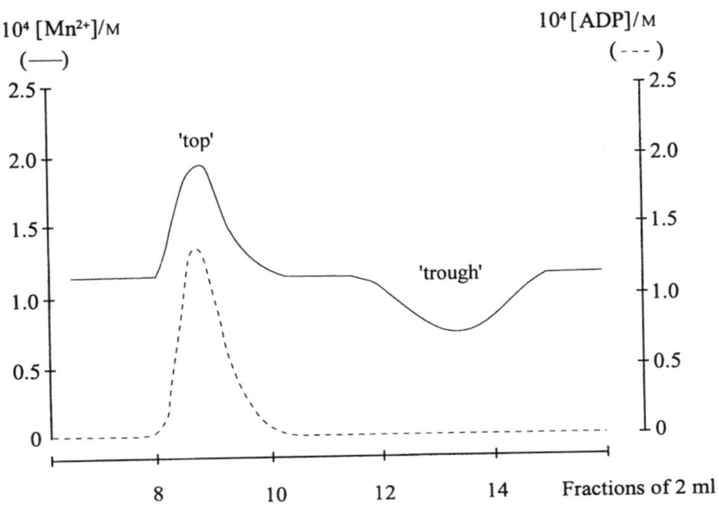

Figure 3.5 *Plot of results from ADP–Mn^{2+} binding experiments based on Sephadex G-10 column separations; UV-absorption peak (lower curve), and record of Mn^{2+} concentration in effluent (upper curve) with 'top' and 'trough'*

Box 3.2 *Determination of Mn^{2+} binding to ADP*

1. Prepare 1 l TRIS buffer, pH 8.0, 108 mM, containing $MnSO_4$ (0.114 M)
2. A Sephadex G-10 column (25 × 1 cm) is equilibrated with the buffer (1.)
3. Dissolve 10 mg ADP in 25 ml buffer (1.)
4. Transfer 500 μl ADP-solution (3.) to the column and elute with buffer (1.)
5. Collect the eluate in fractions of 2 ml using a fraction collector (elution rate: 1 ml min^{-1})
6. Record the UV-absorption spectrum (280 nm)
7. Prepare a standard graph from five 50 ml stock solutions containing 2, 4, 6, 8, and 10 ppm Mn^{2+} in the fractions, respectively, using atomic absorption spectroscopy
8. Determine the concentration of Mn^{2+} in the fractions (5.) by atomic absorption spectroscopy (7.)
9. Prepare a figure such as Figure 3.5 and use the baseline for calculation of $[Mn^{2+}]_{free}$. [ADP–Mn^{2+}] is calculated from the Mn^{2+} amount in the 'top'-fractions minus the amount corresponding to the base line. Alternatively, also from the deficient amount in the 'trough'. $[ADP]_{free}$ = $[ADP]_{total}$ − [ADP–Mn^{2+}] where $[ADP]_{total}$ is calculated from the ADP amount applied and the volume in the 'top'-fractions
10. Calculate K'_f using the equations corresponding to this binding process

in the section on pages 39–41, revealing the size of K'_f and K'_d as well as the number of binding sites involved. Methods of analyses for determination of free and bound ligand include dialysis and gel filtration as shown in the previous section and other alternatives, *e.g.* HPCE (pages 389–390).

An example of the experimental conditions employed for determination of

Box 3.3 *Determination of tryptophan binding to serum albumin (BSA)*

1. Prepare four ethanolamine buffer solutions (1 L), 125 mM, pH 9, containing 32, 60, 82, and 100 μM Trp, respectively
2. Prepare four Sephadex G-25 columns (0.9 × 60 cm), each of them equilibrated with one of the buffer solutions (1.)
3. Connect each of the columns (2.) to a peristaltic pump which gives 1.5 ml min^{-1}, a UVicord (280 nm), and a fraction collector (2 ml $fraction^{-1}$; *ca.* 30 fractions will be needed for each experiment)
4. Dissolve 4.0 mg BSA in 1 ml of the Trp-containing buffer (1.) required for the experiment and transfer this solution to the column
5. Elute with the buffer selected for the experiment (1.), record and collect fractions according to 3.
6. Determine the volume of the effluent corresponding both to 'top' and 'trough' (Figure 3.6)
7. Repeat the experiments with the other three buffer solutions (1.)
8. Perform calculations and plot data as shown for the example in Table 3.5 and Figure 3.7

the number of binding sites (n) and K'_f for binding of a chromophoric low-M_r compound (L) to a protein (P) based on column chromatographic gel filtration is outlined in Box 3.3.

The general appearance of the UV-absorption spectrum obtained in this kind of binding experiment is shown in Figure 3.6.

The 'top' fractions contain P and PL_n, whereas the 'trough' fractions represent the missing L bound in PL_n. Determination of the volume and UV-absorption of the 'top' and 'trough', respectively, allow calculation of the concentrations of both L_{free} and PL_n as the concentration of P + PL_n is known from the start concentration of P. A plot of binding curves according to the linear functions given in Equations 3.23 and 3.24 will now be possible to perform and gives the basis for determination of binding constants and number

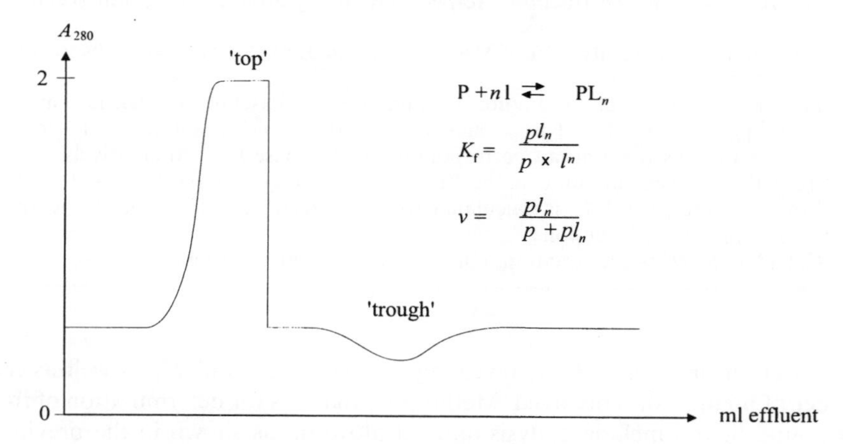

Figure 3.6 *Separation of the macromolecules P and PL_n from L by gel filtration and record of UV-absorption at 280 nm for effluent from the column*

Table 3.5 *Results from binding experiments between a protein (P) and a low-M_r compound (L)*

$p_{total} = p + pl_n$ (μM)	11	11	11	11	11
l_{free} (μM)	2.3	4.8	11.3	27.4	45.8
pl_n (μM)	2.9	5.6	9.5	14.2	16.6
l_{total} (μM)	5.2	10.4	20.8	41.6	62.4
$v = \frac{pl_n}{p_{total}}$	0.264	0.510	0.864	1.291	1.510
$1/v$	3.793	1.964	1.158	0.775	0.663
$1/l_{free}$ (μM^{-1})	0.435	0.208	0.088	0.036	0.022
v/l_{free} (μM^{-1})	0.115	0.106	0.076	0.047	0.033

of binding sites. An example of such calculations and the corresponding plots and curves are shown in Table 3.5. and Figure 3.7.

Tight Binding of Ligands to Macromolecules: Protein-type Hydrolase Inhibitors and Antibody–Antigen Binding

In the cases where binding or dissociation constants (K'_f or K'_d) or $\Delta G^{\circ\prime}$ (Section 3.2) have high numerical values, *e.g.* for avidin–biotin (Table 3.1), a tight binding will be the result. This means that the equations and techniques for studies of equilibria corresponding to weak binding at independent binding sites are not useful. In such cases, every ligand molecule is bound if the number of ligand molecules is less than the total number of binding sites on the macromolecule. If the number of ligands exceeds the number (n) of binding sites, each of the n sites are occupied and the additional ligand molecules remain unbound. Binding of this type can efficiently be studied by HPCE[4] (page 389) as shown for binding of the soybean inhibitors Bowman–Birk inhibitor (BBI) and Kunitz soybean trypsin inhibitor (KSTI), or the proteinase inhibitors in pea, to trypsin or chymotrypsin:

$$\text{KSTI} + \text{Trypsin} \rightarrow \text{KSTI–Trypsin}$$

$$\text{BBI} + \text{Chymotrypsin} \rightarrow \text{BBI–Chymotrypsin}$$

This technique has also been successful for determination of binding between an antibody (mAb) and an antigen (L): mAb + L → mAb–L (page 390).

Knowing the amount (μmol) of added ligand and the amount (μmol) of macromolecule, the number of binding sites can be determined (Figure 3.8).

Binding of Ligands to Macromolecules with Cooperative and Non-identical Sites

In cases where ligand molecules bind to different sites on a macromolecule, *e.g.* a protein (P), the binding of ligand (L) to one site can increase or decrease the

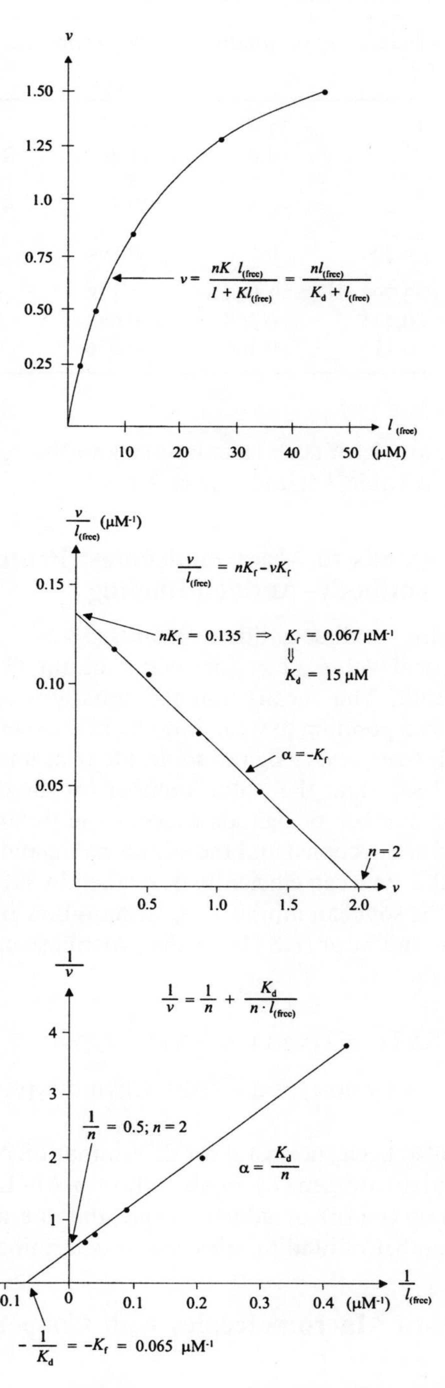

Figure 3.7 *Binding curve and plot of data from Table 3.5 experiments performed as shown in Figure 3.6*

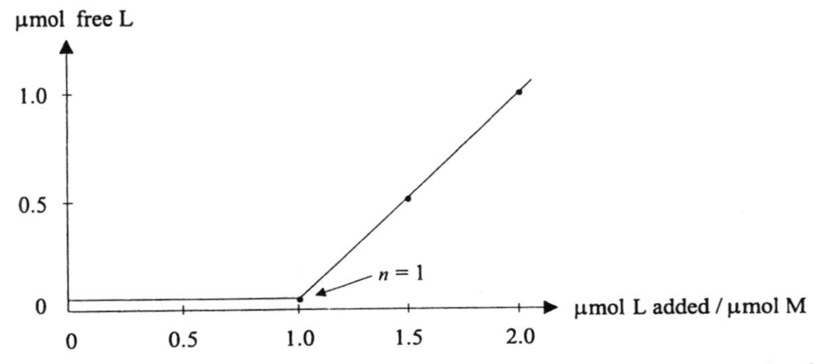

Figure 3.8 *Binding curve illustrating tight binding of ligand (L) to a macromolecule (M) with one binding site* (n = 1) *as found for binding of proteintype inhibitors to trypsin, chymotrypsin, and mAbs by use of HPCE*

affinity of L to other sites, meaning that cooperation between sites (Section 3.3) can be either positive or negative. Cooperativity (that is when the proportion between successive $K'_f \neq 4$; page 40) is, for example, negative for the binding of H^+ to PO_4^{3-} with successive K'_fs of about 10^{12}, 10^7, and 10^2. An example of positive cooperativity may be the binding of O_2 to haemoglobin, which has successive K'_fs in the relative proportion 1:4:24:9.

A plot of data for cooperative binding according to the equations (3.35–3.37) will not follow a hyperbolic curve (Figure 3.2) or a line corresponding to the linear functions (Equations 3.23 – 3.26) (page 41).

In a relatively simple situation where a protein with four cooperative sites has two sites which bind L tightly (K_{d1}) and two other sites that bind less tightly (K_{d2}), the Scatchard plot will be as shown in Figure 3.9.

If the sites have different cooperativity, a sigmoidal curve will be the result of a plot of ligand concentration against the association degree (v) (page 41) and the Scatchard plot will be convex (Figure 3.10).

Highly cooperative binding of ligands (L) to a protein (P) gives binding data (Figure 3.10) which needs an alternative treatment:

$$P + nL \rightleftharpoons PL_n$$

$$K'_f = \frac{pl_n}{p \times l^n}; \quad K'_d = \frac{p \times l^n}{pl_n}; \quad v = \frac{npl_n}{p + pl_n} = \frac{nl^n}{K'_d + l^n} \tag{3.35}$$

The Hill equation is often useful in this connection:

$$\frac{v}{n - v} = \frac{l^{\tilde{n}}}{K'_d} = \frac{Y}{1 - Y} = \frac{\text{fraction of occupied sites}}{\text{fraction of free sites}} \tag{3.36}$$

$$\log \frac{v}{n - v} = \log K'_f + \tilde{n} \log l \tag{3.37}$$

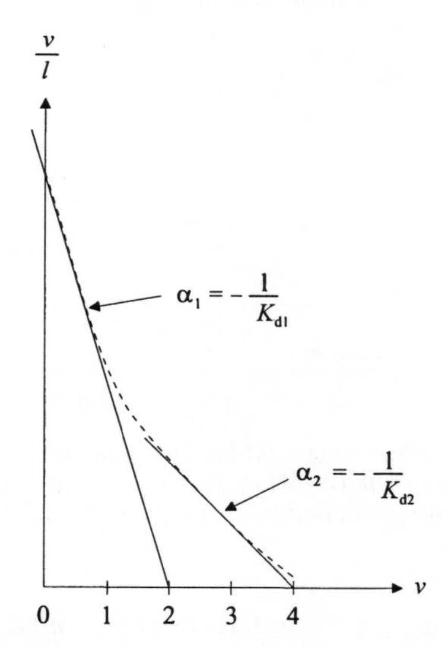

Figure 3.9 *Scatchard plot (Equation 3.24; page 41) for tight binding of ligands to four different sites with two different* K'_d

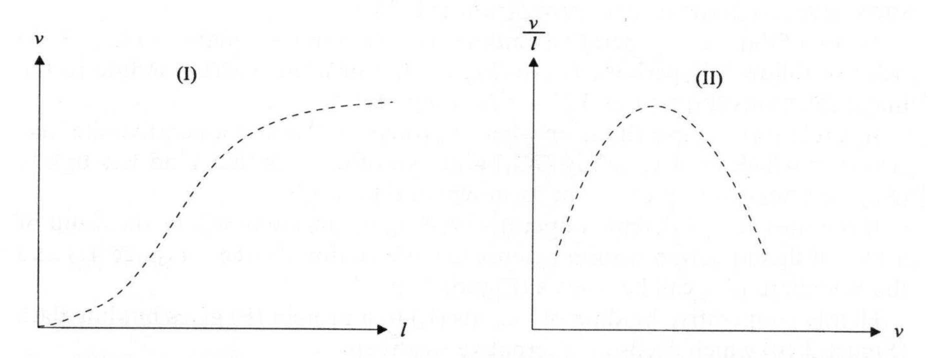

Figure 3.10 *Illustration of data plotted according to equations (page 41) for the binding of ligands (L) to a protein with positive cooperativity between sites, resulting in a sigmoid curve* (I) *and a Scatchard plot* (II) *that is a convex curve*

where $\tilde{n}$ = the Hill coefficient. A value of $\tilde{n} = 1$ indicates that there is no cooperativity or interaction between sites, $0 < \tilde{n} < 1$ corresponds to negative cooperativity (negative effects), and $1 < \tilde{n} < n$ corresponds to positive cooperativity (positive effects).

A protein with three cooperative binding sites will give a curve corresponding to the Hill plot shown in Figure 3.11.

More complicated systems will result if we have many different sites on the protein, *e.g.* formed from protolytic active groups from amino acid side chains,

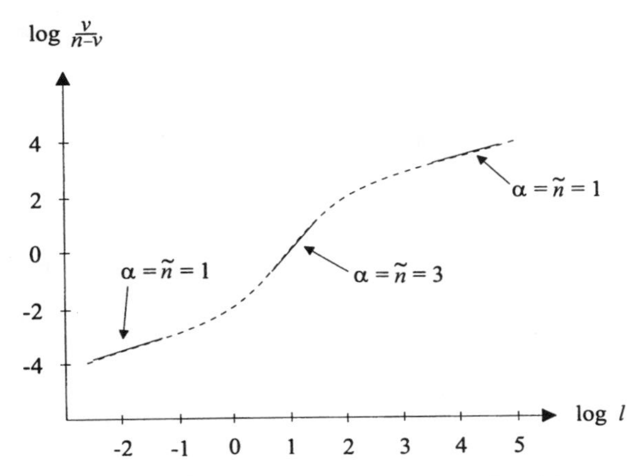

Figure 3.11 *Binding of ligand to a protein with three sites of positive cooperativity*

and when we have different types of ligands, *e.g.* in connection with ammonium salt precipitation and hydrophobic interaction chromatography (HIC). However, such techniques also have several advantages in connection with various types of food analyses.

Enzyme Kinetics, Velocity, Order of Reaction, and Pseudoreactions

Experimental biochemistry with its various methods of detection such as chromatographic methods, and the theory behind methods used in food analyses, involves in several cases work with enzymes. This includes testing enzyme activity or appropriate enzyme inactivation in the initial step of homogenization and extraction procedures to avoid unwanted formation of artefacts, which may create serious problems in otherwise acceptable methods of food analyses. Therefore, it is necessary to know the most important aspect of the basis for enzyme structure, activity, and function – enzyme kinetics, including theories on binding, which in every respect is, to some degree, associated with enzyme kinetics. Enzymes are specific proteins (Section 3.3) with catalytic activities toward the substrates bound in the active sites. Catalysts, such as enzymes, do not alter the position of the equilibrium (K'_{eq} or $\Delta G^{\circ\prime}$), they only increase the velocity of the reaction by reducing the energy of activation in the rate-limiting step (Figure 3.12). However, products in the reaction are substrates for the reverse reaction (k_2, k_4) and K'_{eq} are related to the kinetic parameters, as revealed from Haldane's relationships (*vide infra*).

To avoid interference from products through the reverse reactions, it is important to work with the initial kinetics [v_0: initial reaction velocity (time = 0); s_0 initial substrate concentration (time = 0)], especially for reactions where $\Delta G^{\circ\prime}$ is not too far from zero (K'_{eq} close to 1). The rate of a reaction is defined as the change in concentration as function of time and is

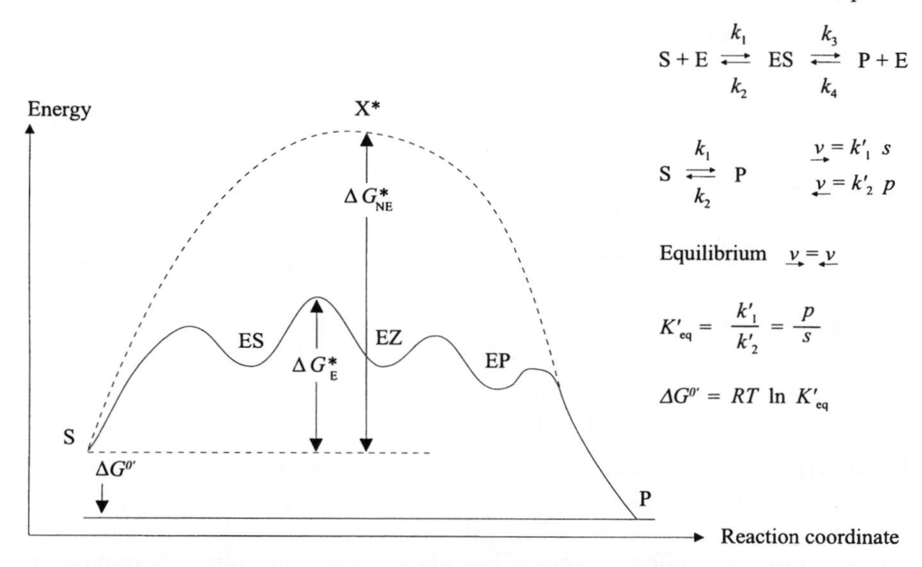

Figure 3.12 *An enzyme increases the reaction velocity v in both directions of the equilibrium by reducing the energy of activation ($\Delta G^{\circ}{}_{E}$) after binding and twisting of the substrate into an optimal position for reaction. $\Delta G^{\circ\prime}$ and K'_{eq} are unchanged. E = enzyme, S = substrate, ES = enzyme–substrate complex, EZ = intermediate enzyme–substrate complex, EP = enzyme–product complex, P = product*

proportional to the substrate concentration:

$$v = \frac{\text{concentration change}}{\text{time}} = \frac{\mathrm{d}c}{\mathrm{d}t} = \frac{-\mathrm{d}s}{\mathrm{d}t} = \frac{\mathrm{d}p}{\mathrm{d}t} = ks \quad (3.38)$$

The velocities should therefore always be expressed as M s^{-1} (concentration unit per time unit) and not as μmol s^{-1}, which does not specify the concentration. However, an enzyme unit (U) is the amount of enzyme that catalyses the transformation of 1 μmol of substrate into product per min under optimal conditions. The International Union of Biochemistry recommends U expressed in katal (kat) based on mol s^{-1}, where 1 kat = 6×10^7 U or 1 U = 16.67 nkat (nanokatals) (Chapter 1).

The reaction order is defined by the exponent of the concentration in: $v = ks^n$. This means that $n = 1$ corresponds to a first order reaction; $n = 2$ gives a second order reaction, and for reactions with more than 1 substrate it could be:

$$\mathrm{S} + \mathrm{R} \rightarrow \mathrm{P}; \quad v = ksr \quad (3.39)$$

which is a second order reaction in total but first order with respect to both

S and R. The order of reactions can be obtained from a plot corresponding to:

$$\ln v = k + n \ln s \tag{3.40}$$

A pseudoreaction order will be the result if the concentration of a substrate is too high for the concentration changes to be detected. Correspondingly, if one of two substrates in a bisubstrate reaction has a high concentration relative to the other substrate, such that changes in its concentration can not be detected, the reaction will appear to be a monosubstrate reaction (pseudomonosubstrate reaction).

As the velocity of the reactions is proportional to the substrate concentration, the velocity will be highest when determined as v_0 (Figure 3.13).

The time of reaction (τ) taken for a defined part (x) of the substrate (s) to be transformed into products is found as the time where s = s_0 $(1-x)$. Consequently, τ is found by use of this relation in the equations for the corresponding order of reaction (n) and is generally proportional to $(s_0)^{1-n}$.

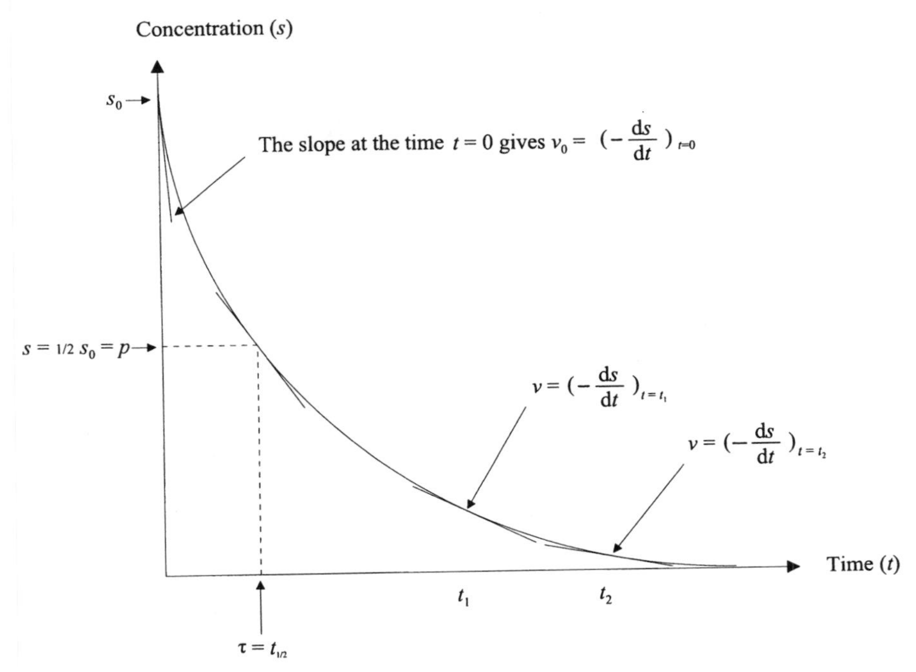

Figure 3.13 *Determination of velocities at different times for the reaction. The reaction time $\tau = t_{1/2}$ corresponds to the time where the concentration of substrate is reduced to $\frac{1}{2}s_0$*

Kinetics of Enzyme-catalysed Monosubstrate Reactions

The dependence of reaction velocity (v_0) on the substrate concentration (s_0) is analogous to that of ligand binding (*vide supra*), which implies that catalysis occurs only after binding of the substrate, with one of the twisted conformations giving the rate-limiting step (Figure 3.12). For monosubstrate reactions, considering v_0–s_0 (Figure 3.14), we have the simple situation:

$$E + S \underset{k_2}{\overset{k_1}{\rightleftharpoons}} ES \overset{k_3}{\rightarrow} E + P$$

$$v = -\frac{ds}{dt} = k_1 e \times s - k_2 es \tag{3.41}$$

$$v = \frac{des}{dt} = k_1 e \times s - (k_2 + k_3)es \tag{3.42}$$

$$v = \frac{dp}{dt} = k_3 es \tag{3.43}$$

$$v = -\frac{de}{dt} = k_1 e \times s - (k_2 + k_3)es \tag{3.44}$$

and with the conservation of matter;

$$e_0 = e + es \tag{3.45}$$

There is, however, no exact solution to the equations, but it can be shown that

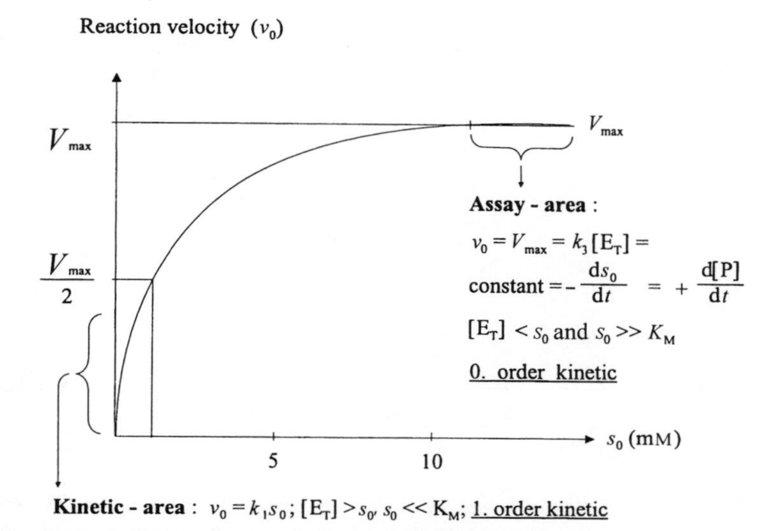

Figure 3.14 *Reaction velocity* (v_0) *as function of initial substrate concentration* (s_0)

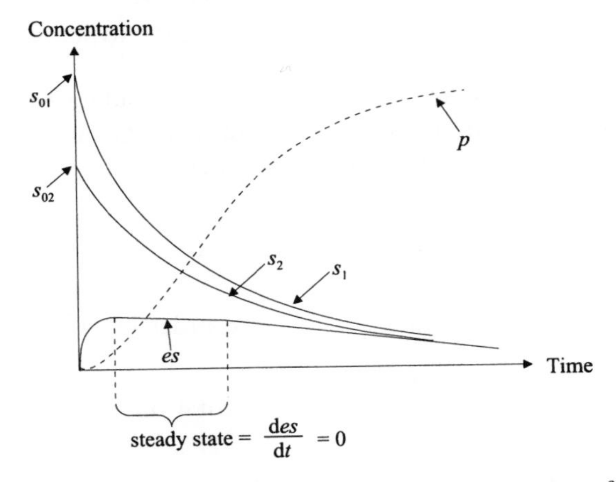

Figure 3.15 *Illustration of the steady-state area and progress curves for the concentrations of substrates* ($s_{01} > s_{02}$)*, product* (p) *and enzyme-substrate complex* (es)

we can utilize the 'steady state validity area' ($des/dt = 0$) (Figure 3.15), which therefore is used in the experimental design of the experiments. This means that we want a short pre-steady-state, where $(des/dt) \gg 0$ and $(de/dt) \ll 0$ and this is obtained by the use of a relatively high s_0/e_0, resulting in a fast saturation of the enzyme (E) by substrate (S), corresponding to $e_0 \ll s_0$ in all of the experiments.

Use of steady-state assumptions and the Equations 3.42 and 3.43 give:

$$es = \frac{k_1 e \times s}{k_2 + k_3} = \frac{k_1 s(e_0 - es)}{k_2 + k_3} = \frac{k_1 s \times e_0}{k_1 s + k_2 + k_3} \tag{3.46}$$

$$v_0 = \frac{k_3 e_0 \times s_0}{s_0 + \dfrac{k_2 + k_3}{k_1}} \tag{3.47}$$

with $K_m = \dfrac{k_2 + k_3}{k_1}$ and $V_{max} = k_3 e_0$.

The Michaelis–Menten equation $v_0 = V_{max} s_0/(K_m + s_0)$ defines the kinetic parameters K_m (Michaelis–Menten constant) and V_{max} (maximal reaction velocity) for:

(a) only monosubstrate reactions
(b) $v_0 - s_0$ determined for different s_0 and with $v_0 = \left(\dfrac{-ds}{dt}\right)_{t=0} = \left(\dfrac{dp}{dt}\right)_{t=0}$
(c) $s_0 \gg e_0$ for all s_0 and the same e_0 in all experiments.
(d) all other variables affecting the reaction (temperature, pH, ionic strength, *etc.*) kept constant.

The equation developed for v_0 thus corresponds to the equation for binding (v)

(page 41) and plot of corresponding data will give the hyperbolic curve (Figure 3.14 and Figure 3.2).

It can be seen from Figure 3.14 and the corresponding equation for v_0–s_0 that special conditions should be used to determine the enzyme concentration ($e_0 = V_{max}/k_3$) and the kinetic parameters K_m and V_{max}:

$$v_0 = \frac{V_{max}}{2} \Rightarrow s_0 = K_m \tag{3.48}$$

$$s_0 \gg K_m \Rightarrow v_0 = V_{max} \quad \text{(zero order kinetics)} \tag{3.49}$$

$$s_0 \ll K_m \Rightarrow v_0 = \frac{V_{max}}{K_m} s_0 = k s_0 \quad \text{(first order kinetics)} \tag{3.50}$$

In the case of first order kinetics, we thus have $v_0 = s_0 V_{max}/K_m = k_3 e_0 s_0/K_m$ and the formation of ES becomes rate limiting and k_3 is then often discussed as k_{cat}.

Determination of K_m and V_{max} from s_0–v_0, covering ideally at least 80% of the hyperbolic curves (Figure 3.14), can be performed by use of computer programs and/or, as is most often the case, by use of the linear functions corresponding to those described on pages 39–41.

$$v_0 = \frac{V_{max} s_0}{K_m + s_0} \tag{3.51}$$

$$\text{(Double reciprocal)} \quad \frac{1}{v_0} = \frac{1}{V_{max}} + \frac{K_m}{V_{max}} \frac{1}{s_0}; \quad \text{Lineweaver–Burk} \tag{3.52}$$

$$\text{(Single reciprocal)} \quad \frac{v_0}{s_0} = \frac{V_{max}}{K_m} - \frac{1}{K_m} v_0; \quad \text{Scatchard} \tag{3.53}$$

$$\text{(Single reciprocal)} \quad \frac{s_0}{v_0} = \frac{K_m}{V_{max}} + \frac{1}{V_{max}} s_0; \quad \text{Hanes} \tag{3.54}$$

$$\text{(Single reciprocal)} \quad v_0 = V_{max} - K_m \frac{v_0}{s_0}; \quad \text{Eadie–Hofstee} \tag{3.55}$$

The experimental results obtained should be evaluated with these linear functions (Figure 3.16), and the data should preferably be tested in all equations to determine how they appear in the graphs. Uncertainty of experimental data concentrated at one or the other end of a line may thus have a great effect on the calculated parameters.

As the enzymes are catalysts (page 53), they will also be able to catalyse the transformation of products into substrates:

$$E + S \underset{k_2}{\overset{k_1}{\rightleftharpoons}} ES \rightleftharpoons EP \underset{k_4}{\overset{k_3}{\rightleftharpoons}} E + P$$

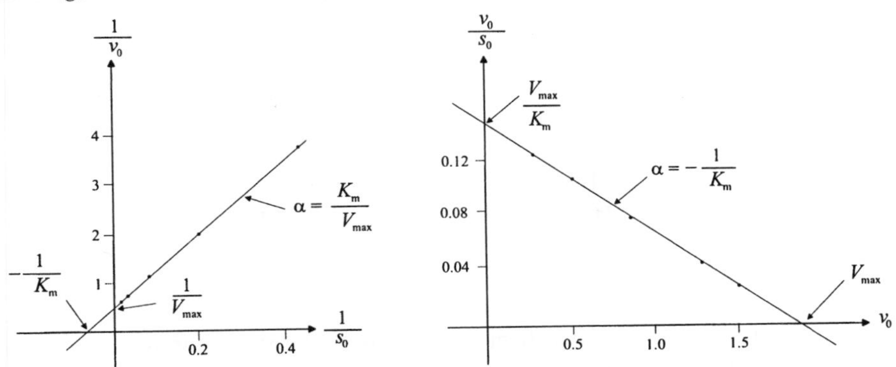

Figure 3.16 *Illustration of double reciprocal (left) and Scatchard plot (right) used to determine* K_m *and* V_{max}

Therefore, we will also have an enzyme–product complex (EP) and $K^{\mathrm{p}}_{\mathrm{m}}$ and $V^{\mathrm{r}}_{\mathrm{max}}$ for the reverse reaction where k_2 corresponds to $k^{\mathrm{r}}_{\mathrm{cat}}$. The substrate and product will thus compete for binding to the free enzyme and the equation for expressing this is as illustrated by the Haldane relationship:

$$K'_{\mathrm{eq}} = \frac{p}{s} = \frac{K^{\mathrm{s}}_{\mathrm{m}} k_2}{K^{\mathrm{p}}_{\mathrm{m}} k_3} \tag{3.56}$$

This type of enzyme kinetics can also be used in studies of the transport of metabolites through membranes where ES and EP then represent binding to proteins at each of the two sites of the membranes.

Multi-substrate Enzyme-catalysed Reactions

In several cases enzymes catalyse reactions with more than one substrate. Three or more substrates are, however, rare and therefore we will only consider bisubstrate kinetics. The kinetic theory for bisubstrate reactions is more complicated than that shown for monosubstrates (page 57), but for the majority of actual cases it can be limited to the following five types.

(1) Sequential ordered mechanism:

$$\mathrm{A + E \rightleftarrows AE \overset{B}{\underset{B}{\rightleftarrows}} EAB \rightleftarrows\rightleftarrows\rightleftarrows EQP \overset{P}{\underset{P}{\rightleftarrows}} EQ \rightleftarrows E + Q}$$

The substrates A or Q will always be bound to E before binding to B or P, and will affect these bindings, resulting in $K_{\mathrm{A}} \neq K^{\mathrm{A}}_{\mathrm{m}}$ and $K_{\mathrm{B}} \neq K^{\mathrm{B}}_{\mathrm{m}}$. This type of mechanism is often found for oxidoreductases where the redox cofactors correspond to A and Q.

(2) Random mechanism:

$$
\begin{array}{ccccc}
A + E \leftrightarrows AE & B & & P & EQ \leftrightarrows E + Q \\
& B \nwarrow\!\!\searrow & & \swarrow\!\!\nearrow P & \\
& & EAB \leftrightarrows\leftrightarrows\leftrightarrows EQP & & \\
& A \swarrow\!\!\nearrow & & \searrow\!\!\nwarrow Q & \\
B + E \leftrightarrows BE & A & & Q & EP \leftrightarrows E + P
\end{array}
$$

Binding of A will not always affect the binding of B, resulting in $K_A = K_m^A$ and binding of B will not always affect the binding of A, resulting in $K_B = K_m^B$.

$$K_m^A K_B = K_A K_m^B \tag{3.57}$$

The rate-limiting steps in mechanisms (1) and (2) are considered to be k_3 at $EAB \xrightarrow{k_3} EQP$, for which

$$K_A = \frac{e \times a}{ea}; \quad K_B = \frac{e \times b}{eb}; \quad K_m^A = \frac{eb \cdot a}{eab}; \quad K_m^B = \frac{ea \cdot b}{eab}; \quad e = \frac{K_A \cdot K_m^B \cdot eab}{a \times b} \tag{3.58}$$

$$e_0 = e + ea + eb + eab = eab\left(\frac{K_A K_m^B}{a \times b} + \frac{K_m^B}{b} + \frac{K_m^A}{a} + 1\right) \tag{3.59}$$

$$v_0 = k_3 eab = \frac{k_3 e_0}{1 + \frac{K_m^A}{a} + \frac{K_m^B}{b} + \frac{K_A K_m^B}{a \times b}} = \frac{V_{max} ab}{a \times b + K_m^A b + K_m^B a + K_A K_m^B} \tag{3.60}$$

(3) Ping-Pong mechanism:

$$A + E \rightleftarrows EA \rightleftarrows\rightleftarrows E'P \underset{P}{\overset{P}{\rightleftarrows}} E' \underset{B}{\overset{B}{\rightleftarrows}} E'B \rightleftarrows\rightleftarrows EQ \rightleftarrows E + Q$$

In this mechanism, one substrate is transformed into a product and the enzyme is modified before it can bind and catalyse the transformation of the second substrate into another product, thus regenerating E. The enzyme thus has two forms and the reaction does not involve an enzyme–bisubstrate complex (EAB)/(EQP). The rate equation reduces to:

$$v_0 = \frac{k_3 e_0}{1 + \frac{K_m^A}{a} + \frac{K_m^B}{b}} = \frac{V_{max} a \times b}{a \times b + K_m^A b + K_m^B a} \tag{3.61}$$

(4) Theorell–Chance mechanism:

$$A + E \rightleftarrows EA \underset{B}{\overset{B}{\rightleftarrows}} (EAB \rightleftarrows EQP) \underset{P}{\overset{P}{\rightleftarrows}} EQ \rightleftarrows E + P$$

This type of reaction is often seen, for example, with peroxidases and lactic acid dehydrogenases. It has very low steady-state *eab* at some substrate concentrations and will consequently sometimes give data corresponding to the Ping-Pong mechanism (3). Otherwise it will be as mechanism (1).

(5) Reactions with covalent ES-intermediates:

AX EAX A

E AX YX A EX

Y

YX EYX Y

e.g. $$E + S \underset{k_2}{\overset{k_1}{\rightleftarrows}} \xrightarrow{k_3} ES \rightleftarrows EP \xrightarrow[k'_3]{H_2O} E + P_2$$ (with P_1 released at the k_3 step)

These reactions are often found with transferases and hydrolases. For the latter case, with H_2O as the second substrate, the reaction occurs as a pseudomonosubstrate reaction, often with a fast step corresponding to k_3 and a slow step corresponding to k'_3.

The experimental investigations of bisubstrate reactions are performed as for pseudomonosubstrate reactions, with one of the substrates present in a relatively high (constant) concentration so that first order kinetics apply to variations of the other substrate, enabling the use of Michaelis–Menten kinetics. The kinetic parameters/constants are then determined by use of the following equations, including secondary plots of *y*-axes intercepts (Y) and slopes (α) from the double reciprocal plots (Table 3.6 and Figure 3.17).

Effect of Inhibitors

Inhibitors reduce the rate of enzyme-catalysed reactions. Covalently bound inhibitors give effects corresponding to a reduced amount of available enzyme and this is also the case for inhibitors exhibiting strong binding, large $\Delta G^{\circ\prime}$ and K'_f (Section 3.2, Table 3.1, and pages 36–37). In this section, we will only

Table 3.6 *Equations for intercept and slope for determination of bisubstrate kinetic parameters*

	Sequential ordered mechanism	*Random mechanism*	*Ping-Pong mechanism*
$v_0 =$	$\dfrac{V}{1 + K_m^A + K_m^B/b + \dfrac{K_A K_m^B}{a \times b}}$	$\dfrac{V}{(1 + K_m^A/a)(a + K_m^B/b)}$ for $K_m^A = K_A$ and $K_m^B = K_B$	$\dfrac{V}{(1 + K_m^A/a + K_m^B/b)}$
$\dfrac{1}{v_0}$ against $\dfrac{1}{a_0}$	$\dfrac{1}{v_0} = \dfrac{K_m^A}{Va}\left(1 + \dfrac{K_A K_m^B}{K_m^A b}\right) + \dfrac{(1 + K_m^B/b)}{v}$	$\dfrac{1}{v_0} = \dfrac{K_m^A}{Va}\left(1 + \dfrac{K_m^B}{b}\right) + \dfrac{(1 + K_m^B/b)}{v}$	$\dfrac{1}{v_0} = \dfrac{K_m^A}{Va} + \dfrac{(1 + K_m^B/b)}{V}$
$\dfrac{1}{v_0}$ against $\dfrac{1}{b_0}$	$\dfrac{1}{v_0} = \dfrac{K_m^B}{Vb}\left(1 + \dfrac{K_A}{a}\right) + \dfrac{(1 + K_m^A/a)}{V}$	$\dfrac{1}{v_0} = \dfrac{K_m^B}{Vb}\left(1 + \dfrac{K_m^A}{a}\right) + \dfrac{(1 + K_m^A/a)}{V}$	$\dfrac{1}{v_0} = \dfrac{K_m^B}{Vb} + \dfrac{(1 + K_m^A/a)}{V}$
$\dfrac{1}{a_0} = 0$	$Y = \dfrac{(1 + K_m^B/b)}{V} = \dfrac{1}{V} + \dfrac{K_m^B}{Vb}$	$Y = \dfrac{1}{V} + \dfrac{K_m^B}{Vb}$	$Y = \dfrac{1}{V} + \dfrac{K_m^B}{Vb}$
$\dfrac{1}{b_0} = 0$	$Y = \dfrac{1}{V} + \dfrac{K_m^A}{Va}$	$Y = \dfrac{1}{V} + \dfrac{K_m^A}{Va}$	$Y = \dfrac{1}{V} + \dfrac{K_m^A}{Va}$
$\dfrac{1}{v_0}$ against $\dfrac{1}{a_0}$	$\alpha = \dfrac{K_m^A}{V} + \dfrac{K_A K_m^B}{Vb}$	$\alpha = \dfrac{K_m^A}{V} + \dfrac{K_m^A K_m^B}{Vb}$	$\alpha = \dfrac{K_m^A}{V}$
$\dfrac{1}{v_0}$ against $\dfrac{1}{b_0}$	$\alpha = \dfrac{K_m^B}{V} + \dfrac{K_A K_m^B}{Va}$	$\alpha = \dfrac{K_m^B}{V} + \dfrac{K_m^A K_m^B}{Va}$	$\alpha = \dfrac{K_m^B}{V}$

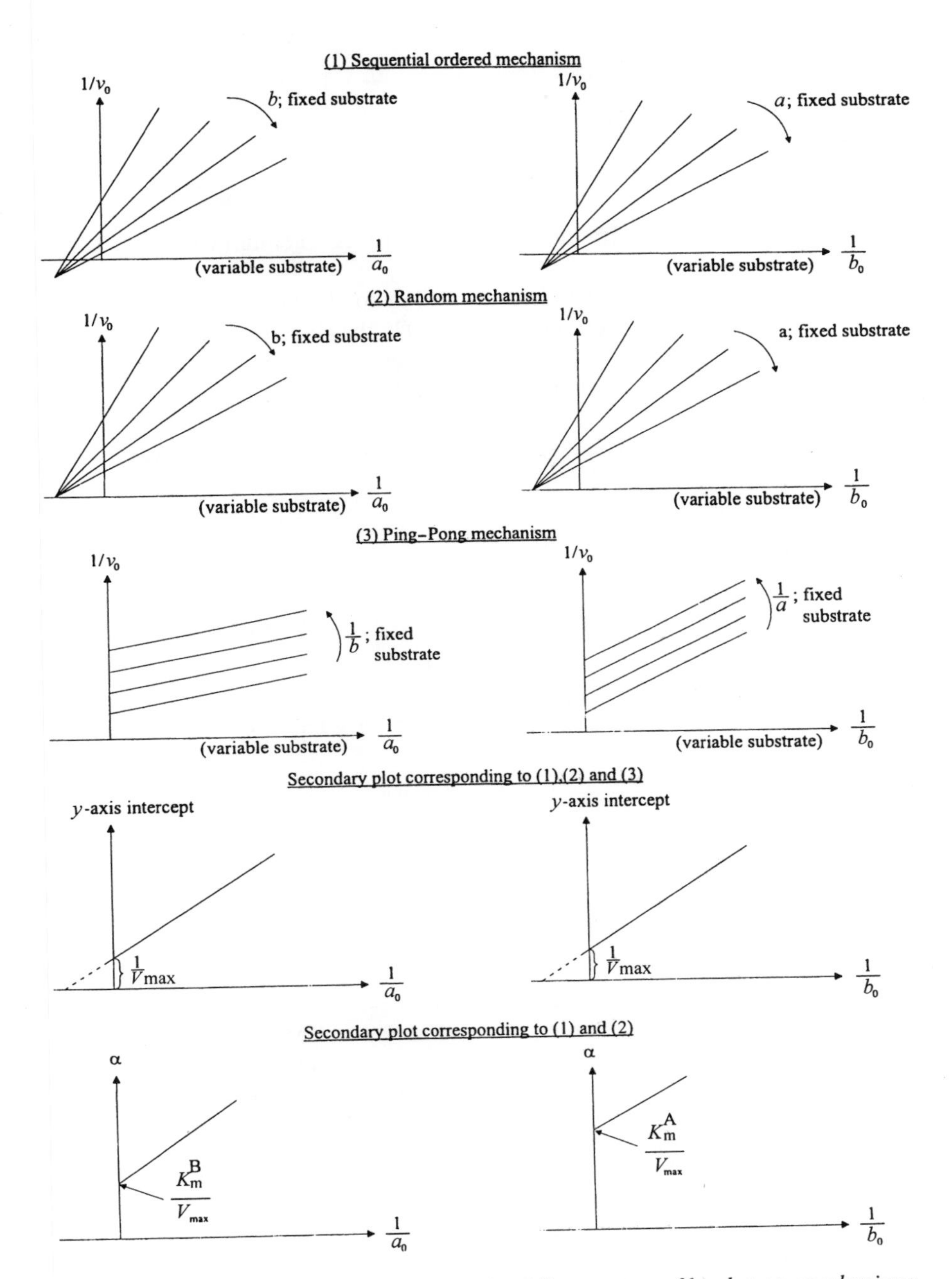

Figure 3.17 *Illustration of curves obtained by different types of bisubstrate mechanisms*

consider the cases which give reversible binding of the inhibitors:

$$\begin{array}{ccccc} E+S & \underset{k_2}{\overset{k_1}{\rightleftarrows}} & ES & \xrightarrow{k_3} & E+P \\ \swarrow\!\nearrow \quad \nwarrow\!\searrow & & \nwarrow\!\searrow & & \\ e'E \qquad s'S & & es'ES & & \end{array}$$

where e', s', and es' are the fractions of inactive forms in equilibrium with active forms.

The inhibition types are classified according to the kinetics which are seen from the effects in double reciprocal plots on (1) the slope (α), (2) the intercepts with x- and y-axes or (3) both α and intercepts resulting in (i) competitive inhibition, (ii) uncompetitive inhibition, and (iii) 'general' or simple non-competitive inhibition.

Employing the basis of Michaelis–Menten kinetics (page 57) we have:

$$v_0 = \frac{dp}{dt} = k_3 es; \quad V_{max} = k_3 e_0; \quad K_m = \frac{k_2 + k_3}{k_1} = \frac{e \times s}{es} \tag{3.62}$$

$$e_0 = e + e'e + es'es + es = e(1 + e') + es(1 + es') \tag{3.63}$$

$$s_0 = s + s's = s(1 + s'); \quad es = \frac{e_0 s_0}{K_m(1 + e')(1 + s') + s_0(1 + es')} \tag{3.64}$$

$$v_0 = k_3 es = \frac{V_{max} s_0}{K_m(1 + e')(1 + s') + s_0(1 + es')} \tag{3.65}$$

$$\frac{1}{v_0} = \frac{K_m(1 + e')(1 + s')}{s_0' V_{max}} + \frac{(1 + es')}{V_{max}} \tag{3.66}$$

It is seen that without inhibition $e' = s' = es' = 0$, the equations will correspond to the Michaelis–Menten equation and the corresponding linear functions (page 58). If we exclude binding of the inhibitor to the substrate S ($s' = 0$) and consider binding of the inhibitor (I) to either E or ES, this gives the dissociation constants $K_i = e \times i/ei$ and $K_i' = esi/es \times i$.

The inhibition types and their kinetics are:

Competitive inhibition: $es' = s' = 0$; $e' \neq 0$

$$v_0 = \frac{V_{max} s_0}{K_m\left(1 + \frac{i}{K_i}\right) + s_0}; \quad \frac{1}{v_0} = \frac{K_m\left(1 + \frac{i}{K_i}\right)}{s_0 V_{max}} + \frac{1}{V_{max}} \tag{3.67}$$

Table 3.7 *Equations for intercept and slope for determination of inhibition type and constants*

	Inhibition types ($s' = 0$)				
				Non-competitive	
	None	*Competitive*	*Uncompetitive*	*General*	*Simple*
$e' =$	0	$e' = \frac{i}{K_i}$	0	$e' = \frac{i}{K_i}$	$e' = \frac{i}{K_i}$
$es' =$	0	0	$es' = \frac{i}{K_i}$	$es' = \frac{i}{K'_i}$	$es' = \frac{i}{K_i}$
Intercept with y-axis $\left(\frac{1}{s_0} = 0\right)$	$\frac{1}{V_{max}}$	$\frac{1}{V_{max}}$	$\frac{1}{V_{max}}\left(1 + \frac{i}{K_i}\right)$	$\frac{1}{V_{max}}\left(1 + \frac{i}{K'_i}\right)$	$\frac{1}{V_{max}}\left(1 + \frac{i}{K_i}\right)$
Intercept with x-axis $\left(\frac{1}{v_0} = 0\right)$	$-\frac{1}{K_m}$	$\frac{-1}{K_m\left(1 + \frac{i}{K_i}\right)}$	$\frac{-\left(1 - \frac{i}{K_i}\right)}{K_m}$	$\frac{-\left(1 + \frac{i}{K_i}\right)}{K_m\left(1 + \frac{i}{K'_i}\right)}$	$-\frac{1}{K_m}$
Slope (α)	$\frac{K_m}{V_{max}}$	$\frac{K_m\left(1 + \frac{i}{K_i}\right)}{V_{max}}$	$\frac{K_m}{V_{max}}$	$\frac{K_m\left(1 + \frac{i}{K_i}\right)}{V_{max}}$	$\frac{K_m\left(1 + \frac{i}{K_i}\right)}{V_{max}}$

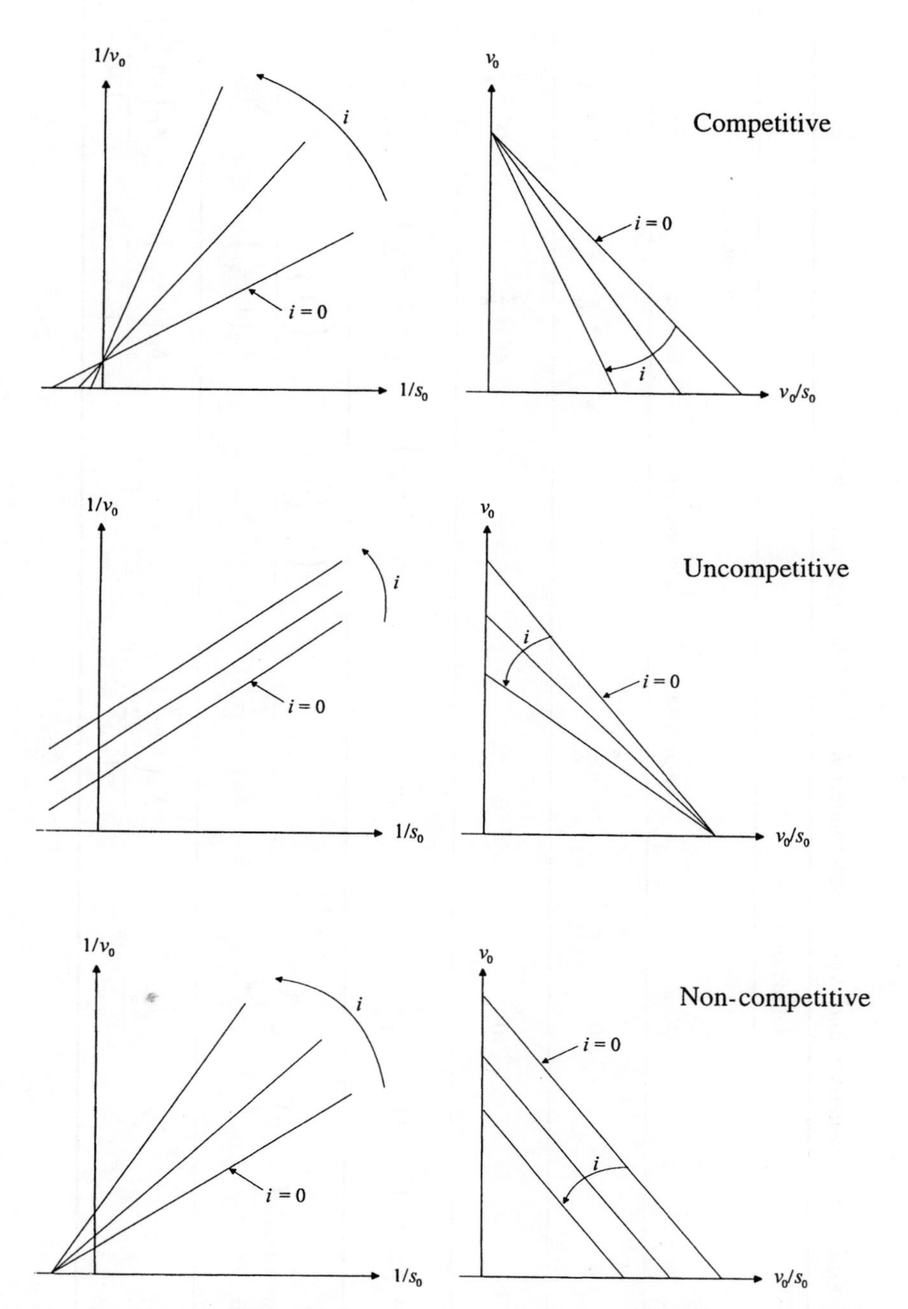

Figure 3.18 *Identification of inhibition type based on kinetics: Lineweaver–Burk plots (left) or Eadie–Hofstee plots (right)*

Uncompetitive inhibition: $e' = s' = 0$; $es' \neq 0$

$$v_0 = \frac{V_{\max} s_0}{K_m + \left(1 + \frac{i}{K_i'}\right) s_0}; \quad \frac{1}{v_0} = \frac{K_m}{s_0 V_{\max}} + \frac{\left(1 + \frac{i}{K_i'}\right)}{V_{\max}} \tag{3.68}$$

General non-competitive inhibition: $s' = 0$; $e' \neq 0$ and $es' \neq 0$

$$v_0 = \frac{V_{\max} s_0}{K_m\left(1 + \frac{i}{K_i}\right) + \left(1 + \frac{i}{K_i'}\right) s_0}; \quad \frac{1}{v_0} = \frac{K_m\left(1 + \frac{i}{K_i'}\right)}{s_0 V_{\max}} + \frac{\left(1 + \frac{i}{K_i'}\right)}{V_{\max}} \tag{3.69}$$

Simple non-competitive inhibition: as general non-competitive inhibition but $K_i = K_i'$

$$v_0 = \frac{V_{\max} s_0}{\left(1 + \frac{i}{K_i}\right)(K_m + s_0)} \tag{3.70}$$

Secondary plots of intercepts or α from double reciprocal plots give the basis for determination of K_i/K_i'. The type of inhibition is defined by the curves resulting from either double reciprocal or single reciprocal plots (page 58 and Table 3.7) as shown in Figure 3.18.

Prediction of inhibition type or site of inhibitor binding based on steady-state kinetics should be supplemented with direct studies, *e.g.* HPCE, as mentioned in the sub-section on page 49, as some inhibitors bind to different sites and measurements based on kinetics alone may give erroneous conclusions. Binding studies as discussed on pages 46–53 could be a recommendable supplement to kinetics, and could also be useful in the determination of metal ion complexation (pages 44–46) and possible activation of enzyme-catalysed reactions. Knowledge of the various methods of efficient types of liquid chromatography, especially HPCE, raises the possibility of progress in the area of food analyses.

5 Selected General and Specific Literature

[1] I. Langmuir, *J. Am. Chem. Soc.*, 1916, **38**, 2267.
[2] I. Langmuir, *J. Am. Chem. Soc.*, 1918, **40**, 1361.
[3] G.D. Halsey and H.S. Taylor, *J. Chem. Phys.*, 1947, **15**, 624.
[4] H. Frøkiær, K. Mortensen, H. Sørensen, and S. Sørensen, *J. Liq. Chromatogr. & Relat. Technol.*, 1996, **19**, 57.
[5] N. Agerbirk, C. Bjergegaard, C.E. Olsen, and H. Sørensen, *J. Chromatogr.*, 1996, **U745**, 239.

[6] S.P. Colowick and N.O. Kaplan, eds, 'Methods in Enzymology', Academic Press, New York.
[7] T.E. Creighton, 'Proteins. Structure and Molecular properties', 2nd Edition, Freeman, New York, 1993.
[8] D. Freifelder, 'Physical Biochemistry. Application to Biochemistry and Molecular Biology', Freeman, San Francisco, 1982.
[9] D.M. Hirst, 'A Computational Approach to Chemistry', Blackwell Scientific, Oxford, 1990.
[10] K.E. Van Holde, 'Physical Biochemistry', Prentice-Hall, Englewood Cliffs, NJ, 1985.
[11] Tze-Fei. Wong, 'Kinetics of Enzyme Mechanisms', Academic Press, New York, 1975.
[12] E.J. Wood, 'Microcomputer in Biochemical Education', Taylor and Francis, London, 1984.

CHAPTER 4

Extraction of Native low-M_r and high-M_r Biomolecules

1 Introduction

All living cells contain a mixture of low relative molecular mass (low-M_r) and high-M_r compounds that are well organized in organelles or membranes (*compartmentalization*), which protect the compounds from unwanted reactions and artefact formation. Based on weight, the dry matter is dominated by high-M_r compounds, whereas low-M_r compounds generally are dominant in terms of the number of molecules. Appreciable differences are, however, found between different cells, even within the same organism, and standard – or typical – cells do not exist.

The types of compounds occurring in different organisms are *chemotaxonomically* defined. Consequently, in addition to the ubiquitous natural products that take part in the general metabolic pathways, considerable variations are found with respect to other types of compounds, as biochemical systems or specific biochemical pathways vary in accordance with the classical evolutionary tree. These compounds are *allelochemicals*, which in food are called *xenobiotics*. The more highly developed organisms are the most simple ones with respect to the type of compounds they are able to synthesize and, therefore, they have a need for essential nutrients. In general, the variation in types of compounds present in animal tissue is more limited than that in plants. Some of these plant compounds are quite different from those present in fungi and other micro-organisms. Furthermore, appreciable differences in types and amounts of compounds will generally be found among different types of tissues and as a function of developmental stage and growth conditions. A marked difference between constituents occurring in animals and plants is illustrated by the occurrence in plants of large amounts of different types of phenolics and dietary fibres (Chapter 14).

From a chemotaxonomic point of view, appreciable differences are also found in the type of low-M_r compounds occurring in different plant orders. Specific compounds are found within plant families, with variations in concentration levels and types of individual compounds found in extracts from different genera, whereas only minor differences are generally found between varieties of the same genera. The order Centrospermae (Chenopodiales) thus has betalains as plant pigments instead of anthocyanins/flavonoids, which are the common pigments in other higher plants. Glucosinolates are another well-defined group of plant products, occurring in all plants of the order Capparales but only in a few other plants. The comprehensive group of non-protein amino acids (*ca.* 900 naturally occurring compounds) and heteroaromatics also contributes to many distinct chemotaxonomic details, as is the case for various other low-M_r biomolecules.

With food dominated by animal and plant products, it is important to know what type of compounds occur in different organisms (chemotaxonomi) as the methods of analyses required are defined by the type of compounds subjected to analysis. Furthermore, information on potentially interfering compounds is essential. The range and number of discrete molecular structures produced as natural products are huge, and so is our knowledge of them (*ca.* 5000 alkaloids; *ca.* 5000 flavonoids). Progress in natural product chemistry and thereby also in food analyses has been aided enormously by the development of rapid and accurate methods for the determination of individual compounds in each group of natural products. In recent decades, the emphasis has mainly been on liquid chromatography techniques, especially HPLC, FPLC (fast polymer liquid chromatography), and HPCE. Successful use of such high efficiency techniques increases demands on extraction and group separation procedures.

The structure and thereby properties of biomolecules determine the procedures needed for extraction and preparation of crude extracts. These extracts should be appropriate for the group separations necessary prior to use of the more advanced liquid chromatography procedures. Usually, it is not possible to solve all problems using a single standard procedure. For liquid chromatography analyses of native high-M_r biomolecules, it is necessary to use non-denaturing extraction procedures with protection against unwanted binding of the analytes to other biomolecules, membranes or dietary fibres. In addition, the procedure should minimize chemical or enzymatic destruction of the native biomolecules. For analyses of low-M_r biomolecules, it is often necessary to carry out an efficient denaturation of the high-M_r compounds present by use of boiling alcohol during the homogenization.

For all biomolecules, the first step in extraction procedures includes disruption of the tissue, cells, cellular tissues, and membranes either by homogenization or by use of disrupting solvents. This step is thus 'destructive' with respect to the compartmentalization. The following step will then be a reorganization of the biomolecules according to their properties. Consequently, it is crucial to have an appropriate strategy to develop fast, simple, and cheap methods of analyses for individual compounds, including chiral separation.

2 Extraction of Low-M_r Biomolecules

The choice of extraction procedure will depend on the type of starting material: dry matter content, emulsions, liquids (milk, blood, urine, soft drinks, *etc.*).

Membrane or dietary fibre associated compounds may sometimes require special solvents with detergent or enzymes included for efficient extraction. Essentially, the extraction procedure will depend on the type of compounds that are to be isolated/extracted.

Sampling and Pretreatment of Samples

An important point in all analytical procedures is the sampling and scaling down technique, which allows use of a small sample that is representative of the total batch for extraction/analysis. A knowledge of the dry matter (DM) content in the starting material is most often needed to allow the expression of determined individual compounds as μmol per g DM. However, the sample materials left after standard methods of DM determination (Box 4.1) may not be acceptable as starting material for extraction of native biomolecules as a number of natural products are too reactive/unstable at drying temperatures of 100 °C. In the case of fresh vegetative plant materials or other biological tissues with a high water content, freezing of the material is also detrimental owing to disruption of cells and membrane structures and mixing of otherwise separated biomolecules. During destruction of the compartmentalization various enzymes with high k_{cat} (page 58) can transform native biomolecules into artefacts; for example, C-S-lyases transform β-thioalkyl amino acids into 'garlic oil' and β-thioglucosidases (myrosinases) transform glucosinolates into various products, including isothiocyanates and thiocyanates. Consequently, lyophilized material will generally be a recommendable starting point for extraction.

Box 4.1 *Determination of dry matter in flour (Air oven method)*

Determination of dry matter (or moisture; indirect method) is basically dependent on the material to be analysed. The method described is applicable for determination of dry matter in flour. Consult AOAC, Association of Official Analytical Chemists, Official Methods of Analyses (32.1.03), Washington D.C. 1995, vol. 16 for specific procedures.

1. Dry metal dish (*ca.* 55 mm in diameter, *ca.* 15 mm in height, with inverted slip-in cover fitting tightly inside) at 130 ± 3 °C, cool in a desiccator and weigh soon after reaching room temperature
2. Accurately weigh *ca.* 2 g well-mixed sample in a covered dish
3. Uncover the sample and dry the dish, cover, and contents 1 h in an oven provided with opening for ventilation and maintained at 130 ± 3 °C
4. Cover dish while still in oven, transfer to desiccator and weigh soon after reaching room temperature
5. Dry matter content (%) is calculated as:

$$\frac{\text{Wt of dish plus sample after heating}}{\text{Wt of dish plus sample before heating}} \times 100$$

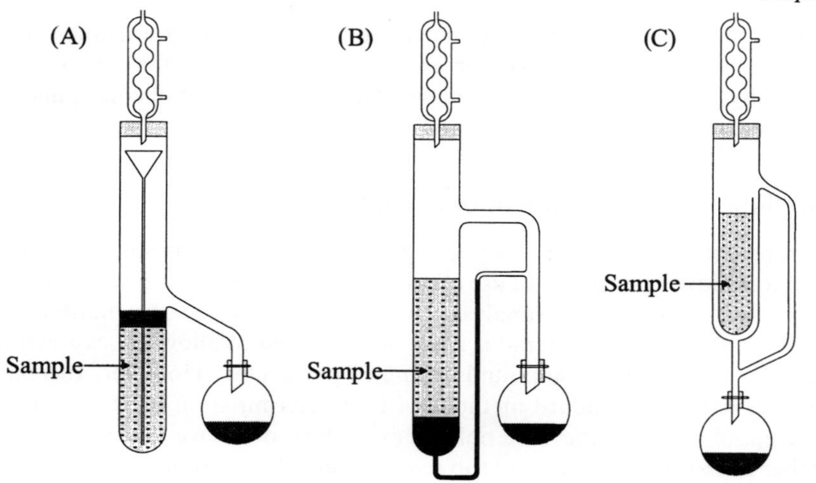

Figure 4.1 *Extraction of lipids from liquid by perforation;* (A) *with an organic solvent which has a specific gravity below that of water (light petroleum; bp range* $<50\,^{\circ}C$*);* (B) *organic solvent with a specific gravity higher than that of water.* (C) *A Soxhlet extractor for extraction of lipids from solid material*

Lipid Extraction

Lipid-containing materials will generally require an initial lipid extraction depending on the amount of lipid in the material and the type of analysis. Fat is often extracted by use of diethyl ether or for Stoldt fat determination by diethyl ether extraction from acidic media resulting in extraction of compounds as fatty acids to the organic phase. Alternatively, light petroleum (pentanes, hexanes or heptanes) is now often used instead of diethyl ether, one of the reasons being the explosion hazard connected with the use of diethyl ether. Depending on the type and amount of material and type of organic solvent, the extraction can be performed as a perforation or as a Soxhlet extraction (Figure 4.1) or by supercritical fluid extraction (SFE) (page 74).

Box 4.2 describes the extraction procedure for lipids using the Soxhlet technique.

Extraction of amphiphilic compounds as phospholipids, glycolipids, and dolicholphosphates (Section 4.3) can be performed after light petroleum extraction of fat/fatty acids using a more polar organic solvent/solvent mixture, *e.g.* chloroform–methanol (2:1), or with another solvent from the eluotrope series (Section 2.8) as an alternative to chloroform.

Extraction of Hydrophilic low-M_r Compounds

Optimal conditions for extraction of hydrophilic low-M_r compounds depend on the type of material, its texture and water content and on the type of compounds that are being isolated.

In general, it is important to carry out an efficient enzyme inactivation, *e.g.* by

Box 4.2 *Example of Soxhlet extraction of lipids*

1. Weigh fat extraction tube (FET)
2. Accurately weigh *ca.* 5 g of well-mixed sample to FET and cover with cotton at the top to avoid waste of material
3. Place FET including sample in a Soxhlet apparatus and fill with light petroleum; bp range $<50\,^{\circ}\mathrm{C}$ (*ca.* 100 ml)
4. Extract for minimum 4 h or until the light petroleum is no longer coloured after passing through the sample material
5. After extraction, remove FET including the extracted sample and leave to dry in a fume cupboard. Discard the cotton
6. After drying, weigh FET including the extracted sample
7. Lipid content (%) is calculated as:

$$\frac{\text{Wt of FET incl. sample before extraction} - \text{Wt of FET incl. sample after extraction}}{\text{Wt of sample}} \times 100$$

denaturation, as soon as the compartmentalization is destroyed. This will protect against unwanted enzyme-based reactions caused by C-S-lyase, myrosinase (*vide supra*), oxidoreductases and hydrolases as lipases, proteinases, and glycosidases. Homogenization in boiling methanol–water (7:3) is a good all-purpose type of extraction but other alcohol mixtures such as ethanol–water may also be useful. In addition to inactivation of enzyme activities in the initial extraction step, boiling methanol–water also efficiently extracts various types of low-M_r compounds that may be bound in membranes. These extraction conditions will also create the desired possibility for denaturation of most high-M_r biomolecules, thus facilitating their separation from low-M_r compounds by precipitation/centrifugation or filtration. The temperature and time required for optimal low-M_r extraction with boiling methanol–water will most often be sufficient for denaturation of high-M_r biomolecules without destruction of native low-M_r biomolecules (Box 4.4). Depending on the structure and properties of the low-M_r compounds examined and their binding to other cellular constituents, proteins, dietary fibres, or membranes it will in some cases be necessary to carry out repeated extractions (often five times) and/or use surfactants or enzymatic treatments.

Internal–External Standards

Quantitative methods of analyses require standard compounds to generate a reference value or a standard graph. Bovine serum albumin (BSA) or $E^{1\%}_{1\,\mathrm{cm}}$ (page 99) are often used for quantification of proteins. In connection with quantitative determination of individual low-M_r compounds, a standard of the same structural type as the compounds under consideration is preferable and if possible as internal standard. The internal standard has to be selected so that it has a retention, or migration, time comparable to the analytes it is to represent. This should be confirmed by a blank trial which will also give information on the appropriate concentration. Internal standards for analyses of low-M_r compounds could be trigonellinamide (1-methylnicotinamide), agmatine and/or 3,4-

dimethoxyphenylethylamine for biogenic amines, aromatic choline ester, alkaloids and basic amino acids; norvaline, 3-iodotyrosine, 3-carboxytyrosine, or α-aminopimelic acid for neutral and acidic amino acids; 3,4-dimethoxycinnamic acid or malonic acid for carboxylic acids; glucotropaeolin, sinigrin, or glucobarbarin for glucosinolates; and appropriate monosaccharides or glycosides for carbohydrates. The internal standards should be added at the start of the extraction and in well-defined concentrations expressed as μmol per gram of dry matter material to be extracted. *Spiking* and the recovery of internal standards added in different amounts can be used for evaluation of the extraction efficiency.

Boxes 4.3 and 4.4 give some examples of procedures for the extraction of low-M_r compounds from liquid samples and solid samples/cellular material.

Solid Phase Extraction and Supercritical Fluid Extraction

In recent years, solid phase extraction (SPE) and supercritical fluid extraction (SFE) procedures have been developed as alternatives to traditional solvent extraction procedures. SPE in fact builds on chromatographic procedures mainly based on column materials developed in connection with HPLC, with small prepacked columns available from various commercial firms. SPE has advantages compared with traditional solvent extraction owing to the great variations in selectivity it exhibits because different types of column materials can be used. SPE is also quicker, and it is possible to use several SPE columns simultaneously when commercially available instruments with vacuum suctions are used. A detailed description of the use of these instruments and the available columns will follow the instrument. With SPE the compounds of interest must be available in solution, *e.g.* through one of the described extraction procedures, and the successful use of SPE requires also knowledge of the properties of the column material and the principles behind liquid chromatographic procedures (Chapters 6–8).

SFE is another promising technique which allows the gentle extraction of various types of compounds. It is based on the use of gases which can be transformed into supercritical fluids at an appropriate pressure and temperature. Carbon dioxide is often used in SFE, and exists as a fluid at 31.1 °C and 7.38 MPa $= 7.38 \times 10^6$ N m^{-2} ($= 7.38 \times 10^6$ kg m^{-1} s^{-2} $= 73.8 \times 10^6$ dyn cm^{-2} $= 72.85$ atm). Under these conditions, carbon dioxide is a solvent with a polarity comparable to that of hexane, but the properties of the extraction solvent can be changed by use of another appropriate solvent selected from the eluotrope series (Section 2.8). The procedure requires special instrumentation and a detailed description of SFE and supercritical fluid chromatography (SFC) is presented in the Appendix with references to more detailed descriptions of supercritical fluid techniques (SFT = SFE + SFC).

3 Amphipathic Membrane Constituents

For all living cells, a major part of their dry matter is found in biological membranes, and for plants appreciable amounts of their dry matter are present

Box 4.3 *Extraction of low-M$_r$ compounds from liquid samples*

Various extraction procedures are available depending on the type of analytes, their concentration, the analytical methods to be used, *e.g.* HPLC, FPLC, HPCE, and the type of liquid sample: milk, soft drink, wine, beer, blood/serum, urine or other more or less complex liquid samples. The extraction procedure will generally be adapted to the separation of hydrophilic and lipophilic compounds.

A. Extraction with acid-induced protein denaturation

A1. Add internal standard and trichloroacetic acid (50 μl; 50% trichloroacetic acid in water) to the liquid sample (200 μl). Mix for 10 min
A2. Centrifuge at 10000 *g* for 5 min
A3. The supernatant represents the crude extract with low-M_r compounds and may require neutralization with NaOH and/or SPE (solid phase extraction) purification and group separation prior to analyses (HPLC or HPCE)

B. Extraction with organic-solvent-induced protein denaturation

B1. Add internal standard and boiling MeOH (1.2 ml) to the liquid sample (500 μl) in an appropriate centrifuge tube and homogenize with Ultra-Turrax, keeping the extraction solution/centrifuge tube in a boiling water bath for few minutes, depending on the starting material and the compounds to be extracted
B2. Centrifuge at 10000 *g* for 5 min and evaporate the supernatant to an appropriate volume to give the crude extract
B3. If required, purify the crude extract by SPE and group separation prior to liquid chromatographic analyses (HPLC or HPCE)

C. Extraction with organic solvent under gentle conditions

Different types of organic solvents/mixtures could be:

(i) n-butanol–H_2O (1:2)
(ii) $CHCl_3$–H_2O (1:3)
(iii) $CHCl_3$–EtOH–H_2O (1:1:4)

C1. Add internal standard to the sample (500 mg or 500 μl) and then the organic solvent/mixture (4.5 ml)
C2. Homogenize carefully with Ultra-Turrax, 5 min
C3. Centrifuge at 12000 *g* (4 °C; 30 min)
C4. This separation will generally result in:

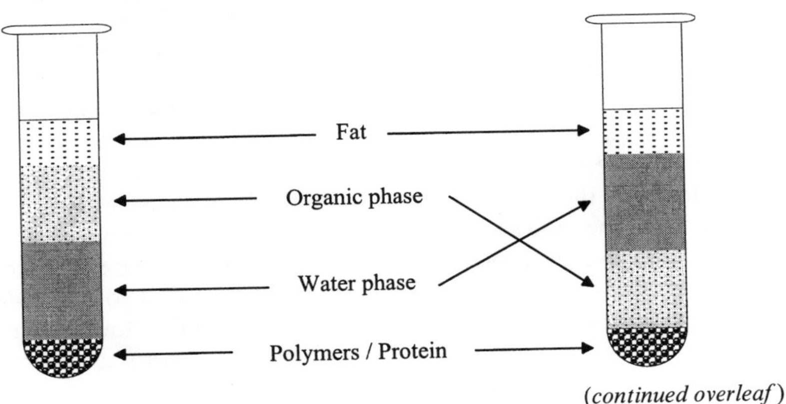

(*continued overleaf*)

C5. Analyses for hydrophilic low-M_r compounds (water phase) may require concentration and purification as mentioned in B2 and B3.

These methods of extraction are in several cases also well suited for studies of special proteins/enzymes.

Box 4.4 *Extraction of low-*M_r *compounds from cellular materials/solid samples*

Extraction from solid samples, cellular materials, and food may present special problems for compounds firmly bound in membranes, dietary fibres, or to other parts of the cellular material. In these cases, it is necessary to destroy the cellular structure and liberate the compounds of interest without having these compounds transformed into artefacts. For some foodstuffs, which may be complex mixtures of many different plant and animal products, this may lead to special requirements of the method of extraction. With simpler foodstuffs that are dominated by a limited number of compounds, more simple methods of extraction may be sufficient.

1. Weigh out 0.5 g sample (freeze dried) in a centrifuge tube with an accuracy of ± 0.1 mg
2. Add 3 ml boiling extraction solution [MeOH–H_2O (7:3)] and internal standard (*vide supra*) and homogenize for a few minutes, keeping the centrifuge tube in a boiling water bath
3. Centrifuge (3000 *g*) then transfer the supernatant to an appropriate flask, repeat twice the homogenization and extraction of the sediment, and combine the supernatants
4. If required for subsequent analyses, proceed with steps as mentioned in Box 4.3, B2 and B3

If other amounts of sample need to be extracted, steps 1.–4. can be performed by appropriate changes of the extraction volumes. Some types of plant materials/food may also require use of more extraction solution.

With extraction of fresh vegetative plant materials, a higher % of MeOH should be used in the first extraction step.

Various types of solid materials may need defatting or alternatively one of the procedures in Box 4.3, part C, could be advantageous.

in cell walls, which in analytical terms are treated as dietary fibres (Chapter 14). Studies of the membrane and cell wall constituents of living organisms with their many and varied functions require knowledge of the molecules included in these structures. Thereby appropriate methods of analyses for the individual molecules need to be chosen, both in connection with biochemical studies and for food analyses.

Membranes are made up largely of lipids and proteins with varying amounts of carbohydrates on the outside of the membranes. In plant cell walls or the dietary fibre fraction, the polysaccharides are quantitatively dominant, and these structures also contain proteins, lipids, and phenolics (lignins), whereas membranes of animals generally contain less than 5% carbohydrates. Membrane proteins possess special characteristics that distinguish them from storage proteins and the peripheral globular proteins. The peripheral proteins are weakly bound to membrane structures and as such are often considered as 'soluble proteins'. The membrane proteins or integral membrane proteins

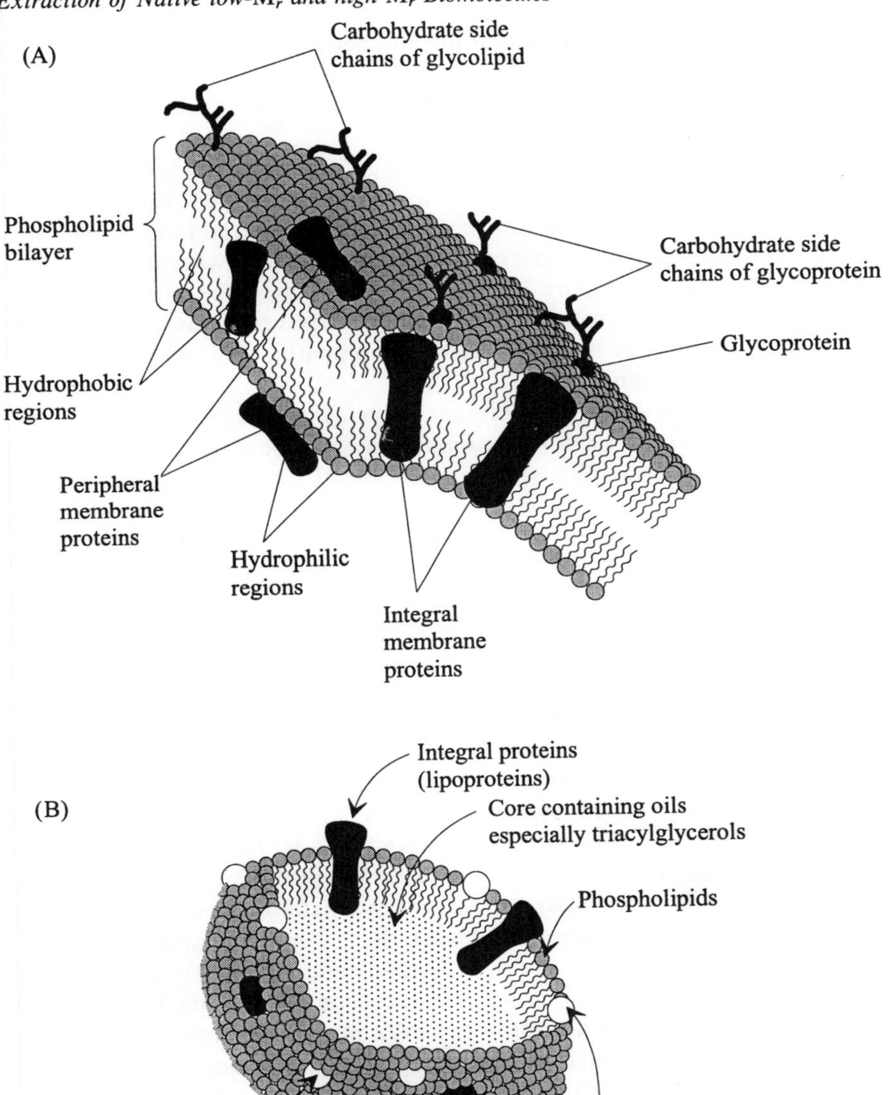

Figure 4.2 (A) *Illustration of a membrane containing both integral and peripheral membrane proteins in addition to lipids, which are especially present in the phospholipid layer.* (B) *Illustration of an oil body with lipoproteins and lipids with their hydrophobic parts on the inside, surrounding the lipids, and with their hydrophilic parts outside the micelle*

(Figure 4.2) are deeply embedded in membranes and contain therefore a high proportion of hydrophobic amino acids. These parts of the proteins in the interior of the membranes are thus in a lipophilic area, where there is a high content of lipids.

The protein content varies greatly among different kinds of membranes, including dietary fibres, and it appears to be directly related to the functions carried out by the specific type of membrane. Mitochondrial inner membranes and bacterial cell wall membranes contain 60–80% protein, whereas membranes surrounding oil bodies (Figure 4.2) and plasma lipoproteins contain about 10–50% protein. The content of lipids in membranes varies inversely with the protein content, and the lipids are mainly amphipathic molecules with both a lipophilic and a hydrophilic part.

The isolation and study of membrane proteins can be problematic for the integral proteins (next section) as they must first be dissociated from the membrane matrix. Generally, this requires use of detergents and/or organic solvents, as such proteins are associated with phospholipids or steroids, *e.g.* saponins in some plants. The protein–lipid complexes (micelles) must be dissociated before further analyses are possible. Detergents (Section 2.9) are also amphipathic molecules, as are the membrane constituents, and this gives rise to a need to consider the properties both of synthetic and natural detergents and their ability to form micelles (*vide infra*). The formation of micelles are of the upmost importance for various advanced analytical methods such as MECC, but should be avoided during extraction and solubilization of membrane lipids and proteins.

When solubilized, the individual membrane proteins and other amphiphilic compounds can be isolated by a variety of separation techniques, provided appropriate detection systems for the compounds have been developed (Chapter 5 and 12). Various types of ion-exchange chromatography techniques are generally useful as an initial step for group separation and isolation of the individual integral membrane proteins, if the solvent contains a non-ionic detergent, *e.g.* octylglucoside or Triton X-100 (Figure 2.8). Octylglucoside has a considerably higher critical micellar concentration than Triton X-100. HIC is another technique which is advantageous in connection with the separation of proteins on the basis of their relative hydrophobicities, especially when fast techniques such as FPLC (Chapter 8) are used. Procedures such as gel filtration, where separation is based on size and shape of integral membrane proteins, are often problematic. This may be caused by the ability of the proteins together with amphipathic molecules/detergents to form micelles that can mask individual size differences, or because the detergents are separated from the macromolecules, resulting in solubility problems. However, the M_r of membrane proteins can be estimated by traditional SDS-PAGE (sodium dodecyl sulfate – polyacrylamide gel electrophoresis) or capillary SDS-PAGE (Chapters 9 and 10) after solubilization of the membrane in a solvent with SDS as detergent. A drawback of methods based on surfactants/detergents with sulfate or trimethylammonium as the hydrophilic surfactant part seems to involve problems with the maintenance of biologically active structures of isolated membrane proteins.

These proteins usually require a hydrophobic environment, and techniques based on cholate as surfactant (Figure 2.8) often have advantages compared with other anion surfactants. Removal of detergent and residual phospholipids from these proteins, however, often leads to inactivation of enzymatic or other biological activities, and results in ready aggregation or precipitation of the proteins.

Extraction of Proteins

Living cells include a great number of structurally different proteins (native proteins) that interact with each other and with other biomolecules in various binding reactions (Chapter 3). These interactions/bindings have numerous effects on the physical, physiological, and biological properties of proteins, and the great variations in the strength of protein binding in membranes (Chapter 3) and dietary fibres (Chapter 14) imply a need for different methods of extractions.

Extraction of proteins from biological sources will generally require mechanical destruction or homogenization of the cells, organelles, and cell walls as mentioned for the procedures in Boxes 4.3 and 4.4. However, to avoid denaturation of proteins/enzymes several precautions should be considered, *e.g.* selection of an appropriate buffer solution (Chapter 2), avoiding too high a temperature during homogenization, and in some cases thiols and proteinase inhibitors should be added to the extraction buffer to inactivate specific enzymes.

The solubilities and stabilities of proteins in aqueous solutions vary enormously and so do the requirements of the extraction procedure. The interactions/binding of proteins with/to other molecules, ions, and solvent are determined by the protein structure/surface and the solvent. Proteins may have several natural conformations or conformers, one of which will be more stable and soluble than the others. The most favourable interactions with aqueous solvents are provided by the hydrophilic groups on the protein surface. Proteins have, as do amino acids, the lowest solubility in water at their pI (Section 2.4), and low ionic strengths will generally destabilize the proteins. The solubilities of proteins are often used in analytical laboratories to divide them into groups even if it is a division without well-defined borderlines. The subclasses in such a division are often:

- *Albumins*: proteins soluble in water and dilute aqueous salt solutions. High salt concentrations are required for 'salting out' (page 81).
- *Globulins*: like albumins, soluble in dilute aqueous salt solutions but only sparingly soluble in deionized water. Globulins will often be salted out of aqueous solutions with 50% $(NH_4)_2SO_4$ saturation.
- *Prolamins*: plant proteins which are insoluble in water, but soluble in 50–90% aqueous ethanol.
- *Glutelins*: plant proteins insoluble in the above-mentioned solvents but soluble in dilute solutions of acids or bases.
- *Protamins* and *histones*: proteins with high pI and often low M_r (4–8×10^3), resulting in solubility in dilute acid solutions.

Membrane proteins or integral membrane proteins (Figure 4.2) usually require the presence of some detergents in order to be solubilized. A number of detergents may be used and, as none of them are best suited for all purposes, examination of optimal detergent concentrations is required. In the following, some general guidelines for solubilization and stabilization of membrane proteins are presented.

As discussed elsewhere (Section 2.9), an important property of detergents is their ability to form micelles, and solubilized membrane proteins may form micelles with detergents. The hydrophobic part of the released membrane protein is expected to be protected against contact with the aqueous buffer by binding to detergent molecules. This will then increase the protein solubility and, at appropriate detergent concentrations, it may stabilize the protein, although high detergent concentrations generally will result in protein denaturation. Ionic detergents such as SDS and cationic detergents (Table 2.6) generally produce more denaturation than non-ionic detergents, and in some cases extractions with organic solvents are preferable (Box 4.3, part C).

Specific examples of extraction procedures are presented in Chapter 13. The solubility of proteins/protein subclasses (see above) as well as other types of molecules in the starting material determines the need for the addition of *cosolvents* as salt, detergent, and organic solvents. The pH and temperature are also of importance. An example strategy for protein extraction/solubilization is given in Box 4.5.

Box 4.5 *Strategy for protein extraction/solubilization*

The solvent giving the highest protein concentration and preventing protein denaturation is often selected. However, the choice of solvent should be balanced against the general advantage of low cosolvent addition, which often facilitates the subsequent purification strategy and methods of analyses. Extractions of albumins and globulins are thus possible in water with low salt concentrations, with a buffer with pH ≠ pI (10–20 mM). Otherwise, initial tests of protein solubility are needed, including addition of different cosolvents:

- NaCl (0.1–2.0 M)
- Glycerol or sucrose (10–30%, v/v)
- Thiol compounds (dithiothreitol; 1–2 mM)
- Protease inhibitors (75 mg PMSF/l; PMSF = phenylmethylsulfonyl fluoride)
- Detergents in concentration close to c.m.c. (Table 2.6)

1. Use the minimum amount of solvent (+ cosolvents); 3–20 times the volume of the starting material (v/v)
2. Homogenize and stir gently for 0.5 h at room temperature (or 4 °C) avoiding foaming
3. Centrifuge or filter
4. Determine protein concentrations and enzyme activity in the supernatant
5. Repeat steps 1.–4. with the pellet from 3

Salt Effects on Protein Solubility

In aqueous solution proteins, like other molecules, will be surrounded by more or less tightly bound water in an ordered structure, thus affecting the organized structure of pure water (Figure 2.1). Water is a unique solvent for proteins and other biomolecules owing to its high dielectric constant (Figure 2.7). The dielectric constant changes with the addition of ions, *e.g.* low concentrations of NaCl give an increase in the dielectric constant, following the effects on the activity coefficients (Section 2.3). The structure and properties of water vary, however, both with concentration and type of ions, thus affecting the solubility of proteins in water. The effect of salts on protein solubility is often discussed with reference to the Hofmeister series (= lyotrop series) which for selected cations and anions is:

Anions: PO_4^{3-} > SO_4^{2-} > HPO_4^{2-} > acetate > citrate > Cl^- > NO_3^- > ClO_4^- > I^- > ^-SCN
Cations: NH_4^+ > K^+ > Na^+ > Li^+ > Mg^{2+} > Ca^{2+} > Ba^{2+} > guanidium
'salting out' ←
→ 'salting in' and chaotropic effects

The effect of anions and cations is usually independent and additive, with anions having the largest effect on protein solubility. The first ions of each series disrupt the structure of water, and decrease the solubility of proteins ('salting out' effect). The last ions of each series generally increases the solubility of non-polar molecules ('salting in' effects). This means that different salts exhibit effects on protein solubility that will be a combined effect of those of the anions and cations, as shown in Figure 4.3.

For some ions, such as Ca^{2+}, Mg^{2+}, and Ba^{2+}, their effect on protein solubility is often determined by their binding to the proteins (Chapter 3). To keep proteins in solution, NaCl is a recommendable choice, and is often used in various types of protein extraction procedures and in liquid chromatography procedures. The ions/compounds in the series that have a low 'salting out' effect will also change the structure and properties of water. The result will be an increase in the solubility of non-polar molecules such as proteins ('salting in') owing to reduced hydrophobic interactions (reduced denaturation of many proteins), which also can be caused by some neutral molecules such as urea. These compounds are known as *chaotropic compounds* and include perchlorate, trifluoroacetate, thiocyanate, urea and guanidine. With urea and guanidine in high concentrations (4–8 M), hydrogen bonding to the peptide groups in addition to effects on the water structure will occur, and the result will be destabilized native protein structures and reduced solubility. Conversely, $(NH_4)_2SO_4$ is an *antichaotropic* salt, often preferred for salt fractionation and used in HIC, where it effectively promotes ligand–protein binding as it gives reversible denaturation precipitation with many proteins.

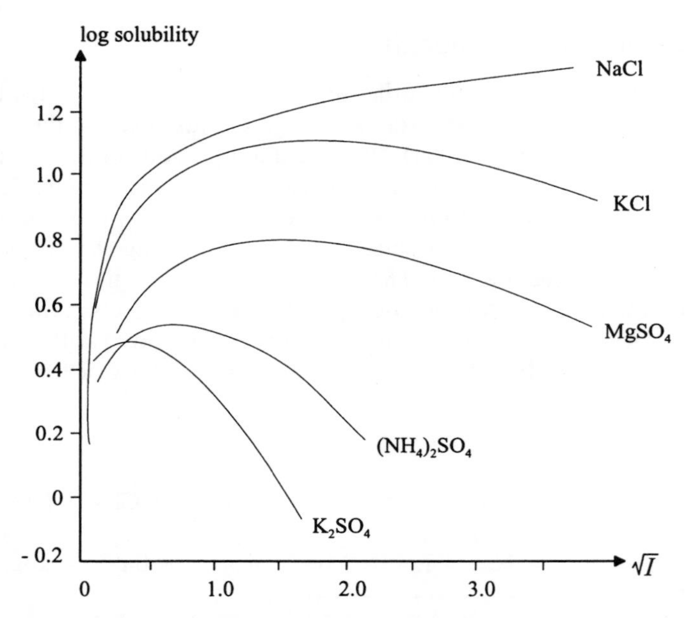

Figure 4.3 *Solubility of isoelectric carboxyhaemoglobin as a function of the ionic strength (I) produced by different salts*
(Reproduced by permission from A.A. Green, *J. Biol. Chem.*, 1932, **95**, 47)

Ammonium Salt Fractionation of Proteins

Ammonium sulfate is the most commonly used salt for fractional precipitation of proteins owing to its properties as an antichaotropic salt, its solubility, and its effectiveness in 'salting out' of proteins (Figure 4.3). The salt concentration required for protein precipitation is described in terms of '% saturation' (see Box 4.6 and Table 4.2), and this is again temperature dependent (Table 4.1). The density of saturated ammonium sulfate is thus close to the density of many protein aggregates (1.29–1.37 g mL^{-1}), which means that more efficient centrifugations are required for protein sedimentation from such solutions than from low ionic strength water solutions.

The amount of ammonium sulfate necessary to bring 1 litre of solution from one '% saturation' to another can be calculated for each temperature as shown in Table 4.2.

The experimental procedure for protein precipitation with $(NH_4)_2SO_4$ is described in Box 4.6.

Isoelectric Precipitation of Proteins

The isoelectric point (pI; Section 2.4) is an important piece of information in relation to proteins. At pH = pI, proteins are most easily precipitated from an aqueous solution. Consequently, such information is important in relation to

Table 4.1 *Data for saturated ammonium sulfate solutions at different temperatures (°C)*

	Temperature (°C)			
	0	*20*	*25*	*30*
g $(NH_4)_2SO_4$ to 1000 cm^3 H_2O	707	761	767	777
Volume of solution (cm^3)	1375	1425	1433	1444
g $(NH_4)_2SO_4$ per 1000 cm^3 H_2O	515	533	538	542
Molarity (M)	3.90	4.03	4.07	4.10
Density (g cm^{-3})	1.241	1.235	1.233	1.231

Table 4.2 *Amount of $(NH_4)_2SO_4$ required to bring a 1 litre solution from one defined '% saturation' to another at 25 °C. The initial concentration of ammonium sulfate is given in the left hand column*

	Final concentration of ammonium sulfate (% saturation)																
	10	*20*	*25*	*30*	*33*	*35*	*40*	*45*	*50*	*55*	*60*	*65*	*70*	*75*	*80*	*90*	*100*
	Grams solid ammonium sulfate to be added to 1 L of solution																
0	56	114	144	176	196	209	243	277	313	351	390	430	472	516	561	662	767
10		57	80	119	137	150	183	216	251	288	320	365	406	449	494	592	694
20			29	59	78	91	123	155	189	225	262	300	340	382	424	520	619
25				30	49	61	93	125	158	193	230	267	307	348	390	485	583
30					19	30	63	94	127	162	198	235	273	314	356	449	546
33						12	43	74	107	142	177	214	252	292	333	426	522
35							31	63	94	129	164	200	238	278	319	411	506
40								31	63	97	132	168	205	245	285	375	469
45									32	65	99	134	171	210	250	339	431
50										33	66	101	137	176	214	302	392
55											33	67	103	141	179	264	353
60												34	69	105	143	227	314
65													34	70	107	190	275
70														35	72	153	237
75															36	115	198
80																77	157
90																	79

Box 4.6 *Protein precipitation with $(NH_4)_2SO_4$*

1. Place the protein solution in a beaker in ice/water and stir gently (avoid foaming)
2. Produce a fine powder of the required amount of $(NH_4)_2SO_4$, which then is added slowly to the solution during *ca.* 30 min, always with all the added salt solubilized
3. Continue stirring for an additional *ca.* 30 min
4. Centrifuge (10000 *g*, 10 min, 4 °C). The supernatant is used in 5. and the precipitate resuspended in buffer (*ca.* 2 times pellet volume)
5. Repeat 2–4 with the supernatant to the next '% saturation'

Table 4.3 *Isoelectric points for selected proteins*

Protein class	*Protein*	*pI*	*Ionic strength* (I)
Albumins	Ovalbumin	4.59	0.10
		4.71	0.01
	Serum albumin	4.7–4.9	0.02
Globulins	β-Lactoglobulin	5.1–5.3	0.10
		4.7–5.1	0.01
	γ_2-Globulin (human)	7.3	0.10
		8.2	0.01
Mucoproteins	α-Ovomucoid	3.8–4.4	
	Lysozyme	11.0–11.2	0.01
Chromoproteins	Myoglobin (horse)	6.85	
	Haemoglobin (horse)	7.35	
	Cytochrome *c*	9.8–10.1	
Histones	Thymohistone	10.8	
Other proteins			
Pepsin		*ca.* 1	
Amyloglucosidase		3.50	
Carbonic anhydrase B (bovine)		5.85	
Carbonic anhydrase B (human)		6.55	
Trypsin inhibitor (KSTI)		4.55	
Trypsinogen		9.30	
Lectins (lentil)		8.15; 8.45; 8.65	

protein isolation based on ion-exchange chromatography (Chapter 7). The pIs for selected examples of proteins are shown in Table 4.3.

The pI values are generally obtained from isoelectric focusing (Chapter 9) and/or chromatofocusing (Chapter 8). As seen from Table 4.3, the values will depend on the ionic strength (I) as the isoelectric point for most proteins differs from their isoionic point (Section 2.4) owing to the binding of ions to the proteins (Chapter 3). An illustration of protein solubility as a function of pH is shown in Figure 4.4.

Preparation of Acetone or Ethanol Powder of Proteins

Proteins associated with membranes may in some cases be extracted to organic solvents without denaturation (Box 4.3, part C), and with the use of acetone or ethanol, native or biologically active precipitates can in some cases be produced. To avoid denaturation of proteins, it is important to keep the temperature in the solutions close to 0 °C. Ionic strengths in the 0.05–0.2 M range are recommended, as is immediate washing of the precipitate with diethyl ether to remove residual acetone or ethanol. The majority of proteins with $M_r > 10–15 \times 10^3$ will precipitate with 50% (v/v) organic solvent.

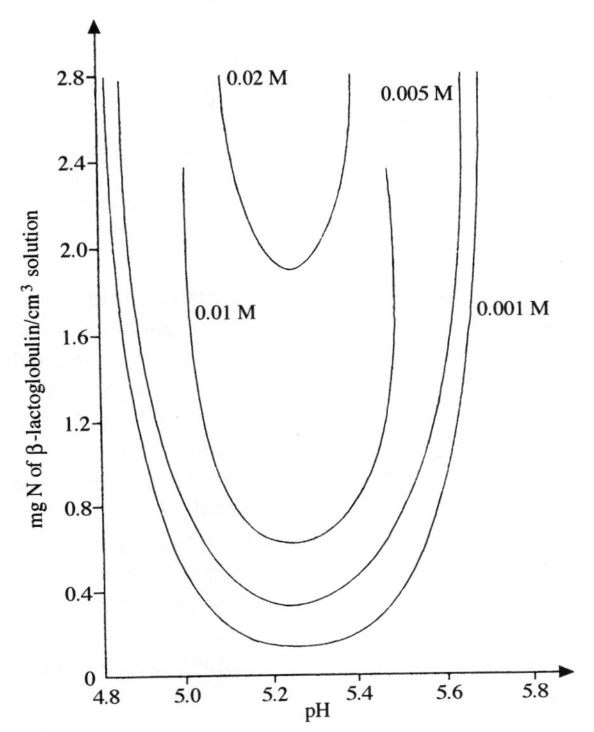

Figure 4.4 *Solubility of β-lactoglobulin as a function of pH and at different molar concentrations of NaCl*
(Adapted from S. Fox and J.F. Forster, 'Introduction to Protein Chemistry', Wiley, New York, 1957)

Box 4.7. *Organic solvent (acetone or ethanol) precipitation of proteins*

1. Place the organic solvent in a deep freeze to obtain a solvent temperature of −10 to −20 °C
2. 1 ml protein solution at 0–4 °C is mixed with 10 ml acetone or ethanol with continuous stirring
3. Continue stirring at 0–4 °C for 10–15 min
4. Centrifuge (10 000 *g*, 10 min, 0–4 °C)
5. Decant and wash the precipitate with diethyl ether before it is resuspended in cold buffer. Undissolved material is removed by filtration and/or centrifugation

4 Dialysis

Dialysis is a frequently used technique for removing low-M_r compounds from high-M_r compounds (desalting), but may also be used for various biochemical experiments, *e.g.* polymer–ligand binding studies (Chapter 3).

Commercially available dialysis tubings are typically made of cellulose. They are available in various diameters (*ca.* 10–120 mm) and pore sizes (cut-off values

Box 4.8 *Example of soaking procedure for dialysis tubings prior to dialysis of compounds sensitive to tubing contaminants*

1. Boil the dialysis tubing in 10 mM sodium hydrogencarbonate ($NaHCO_3$)/1 mM EDTA for minimum of 30 min
2. Boil (10 min) or wash the boiled dialysis tubing thoroughly in deionized water
3. Store the boiled and washed dialysis tubing at 4 °C in 1 mM EDTA. Approximate storage time: 3–6 months. Check for microbial contamination prior to use

of M_r up to 5×10^5), the most common being impermeable to molecules with M_r above $12–14 \times 10^3$. Dialysis tubings may contain chemical contaminants from the manufacturing process, *e.g.* glycerol, heavy metals, and sulfur-containing compounds. Working with easily denatured proteins or other sensitive compounds thus requires some pre-treatment of the dialysis tubings (Box 4.8); in most cases, however, soaking for a few minutes in deionized water will be sufficient. Once soaked, the dialysis tubings should not be allowed to dry.

Dialysis is usually performed as illustrated in Figure 4.5. The soaked dialysis tubing is turned into a bag by making two tight knots at one end of the tubing or alternatively by use of commercially available dialysis clips. The sample to be dialysed is transferred to the dialysis bag by use of a funnel and the bag is closed (knots, adhesive tape or clips). Make sure that there is enough space (* in Figure 4.5) to allow volume increases due to influx. Place the dialysis bag in a beaker, an Erlenmeyer flask or the like with stirred deionized water/buffer.

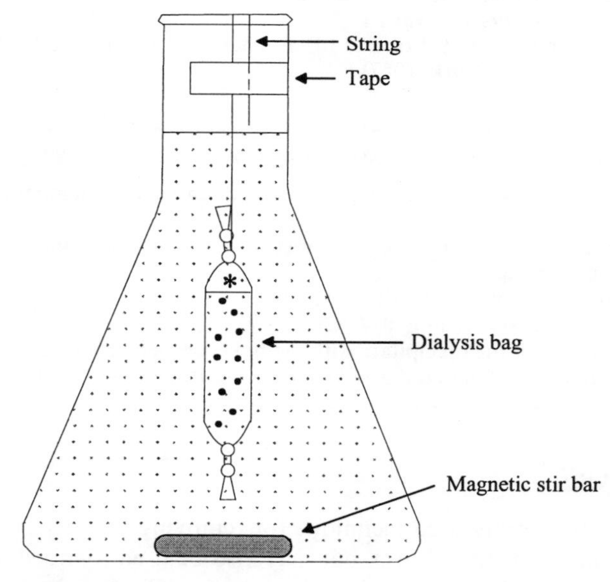

Figure 4.5 *Experimental arrangement for dialysis. A glass bead can be dropped into the bag together with the sample to ensure a vertical position during dialysis. The magnetic stir bar must not be in contact with the dialysis bag*

Box 4.9 *Donnan equilibrium, osmotic pressure, and virial coefficient*

Proteins at pH $\neq$ pI and several other macromolecules are polyelectrolytes (Section 2.4), and this gives a need for counterions to the polyelectrolytes (P^{n-} and nNa^+). If the polyelectrolyte occurs in solutions together with low-M_r salts, *e.g.* NaCl, only the low-M_r ions will be able to pass through the semipermeable membrane and the equilibrium will result in:

$[Na^+]_i[Cl^-]_i = [Na^+]_o[Cl^-]_o$ (i and o indicating inside and outside the dialysis bag, respectively)

Requirement for electrical neutrality gives:

$$[Na^+]_i = n[P^{n-}]_i + [Cl^-]_i \text{ and } [Na^+]_o = [Cl^-]_o$$

These equations are then combined into:

$$[Na^+]_o^2 = [Na^+]_i([Na^+]_i - n[P^{n-}]_i) = [Na^+]_i^2 - n[P^{n-}]_i[Na^+]_i$$
$$[Cl^-]_o^2 = [Cl^-]_i(n[P^{n-}]_i + [Cl^-]_i) = [Cl^-]_i^2 + n[P^{n-}]_i[Cl^-]_i$$

At equilibrium, the counterion to the polyelectrolyte will be concentrated on the polymer side, while the opposite will be the case for Cl^- (the low-M_r ion with the same type of charge as the polyelectrolyte):

$$n[P^{n-}]_i[Na^+]_i = [Na^+]_i^2 - [Na^+]_o^2 = ([Na^+]_i + [Na^+]_o)([Na^+]_i - [Na^+]_o)$$
$$n[P^{n-}]_i[Cl^-]_i = [Cl^-]_o^2 - [Cl^-]_i^2 = ([Cl^-]_o + [Cl^-]_i)([Cl^-]_o - [Cl^-]_i)$$

If the equations are used in combination with the expression for osmotic pressure:

$$\Pi = RT\sum c = RT([P^{n-}]_i + ([Na^+]_i - [Na^+]_o) + ([Cl^-]_i - [Cl^-]_o))$$
$$\Pi = RT\left[[P^{n-}]_i + \frac{n^2[P^{n-}]_i^2}{[Na^+]_i + [Na^+]_o + [Cl^-]_i + [Cl^-]_o}\right] \cong RT\left[[P^{n-}]_i + \frac{n^2[P^{n-}]_i^2}{4[NaCl]}\right]$$

This corresponds to a non-ideal solution with the last part of the equation corresponding to the virial coefficient proportional to $(n[P^{n-}]_i)^2$ and, as seen, high salt concentrations will reduce the effects of the virial coefficient. The Donnan effect should be considered if dialysis is used in binding experiments with ions bound to the polymers (Chapter 3). Furthermore, relatively high concentrations (0.1–1 M) of salts like NaCl (compare section on pages 81–82) will therefore be useful in work with polyelectrolytes (polymers/proteins) as this makes the solution more ideal, and it will generally be required for the correct determination of M_r for proteins in procedures like gel filtration.

During dialysis, the small sample molecules (M_r < cut-off value for the chosen dialysis tubing) will leave the dialysis bag by diffusion until their concentration is the same at both sides of the cellulose membrane, if the polymers are uncharged molecules. If the high-M_r compounds are polyelectrolytes, it may be necessary to make corrections for the Donnan equilibrium effect by, for example, performing ion-binding experiments (Box 4.9).

Prior to a dialysis, one has to consider the following:

- dialysis tubing pore size (cut-off value);
- dialysis tubing diameter;
- type of solution to dialyse against;
- temperature;
- how often the solution is to be changed;
- time of dialysis.

The dialysis tubing pore size is determined by the actual sample to be dialysed, the sample molecules of interest having a molecular weight above the cut-off value.

The dialysis tubing diameter should be considered in relation to the sample volume. In general, a large diameter results in a small surface-to-volume ratio and thus a slow dialysis, as no stirring occurs inside the dialysis bag. A small diameter results in a correspondingly faster dialysis (quicker attainment of equilibrium); however, the risk of adsorption problems to the relatively large surface has to be considered.

Dialysis is usually performed against deionized water or possibly a buffer solution. Alternatively, solutions of high-M_r compounds may be concentrated by dialysis against polymeric solutions, *e.g.* polyethylene glycol or Sephadex materials (Chapter 6). The polymer will pull water out of the dialysis bag, which will often be seen as a folding of the otherwise distended bag. In such cases, one has to be aware of the risk for denaturation of protein samples, and a shorter dialysis time/less concentrated polymer solution should be considered. Generally, 6–10% (w/v) solutions of polymer are suitable. A different situation arises if the osmotic pressure of the sample is higher than that of the solution outside the dialysis bag. In this case there will be an increase in sample volume and thus a dilution.

The temperature used for the dialysis is totally dependent on the tolerance of the sample solution. If possible, dialyses are performed at room temperature, otherwise in a refrigerator at 4 °C.

The efficiency (and time) of dialysis are dependent on the volume of the solution outside the dialysis bag (preferably > 10 times the sample volume) and on when and how often the solution is changed and whether it is stirred. Optimally, the solution should be changed continuously, making the volume and stirring of less importance. This is, however, seldom possible. The second best conditions comprise a large volume of solution outside the dialysis bag compared with the sample, effective stirring, and frequent changes of the solution. If there is only limited possibilities for changing the outside solution, it should preferably be done most frequently at the beginning of the period of dialysis.

5 Ultrafiltration

Concentration of protein solutions is often performed by ultrafiltration, which is also applicable for desalting purposes. In contrast to dialysis, ultrafiltration implies the use of a pressure differential to force the liquid through a semipermeable membrane, retaining the compounds of interest. Inlet gas pressure at the solution side of the membrane by use of, for example, N_2 is a possibility as well as reduced pressure (vacuum) applied on the filtrate side. Alternatively, a pressure differential may be obtained by centrifugation (*vide infra*). An example of an Ultrafree MC filter unit is shown in Figure 4.6.

The semipermeable membranes applicable for ultrafiltration have typical cut-off values in the $M_r = 1–3 \times 10^4$ range. The maximum possible pressure depends

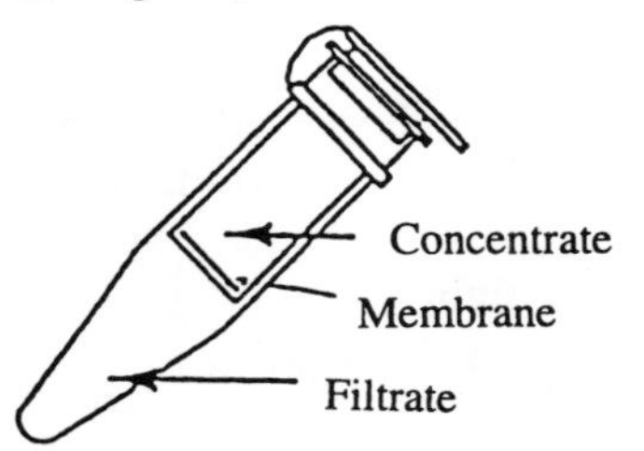

Figure 4.6 *Disposable Ultrafree MC filter unit placed in an Eppendorf tube, which should be placed in a centrifuge. The filter membrane of non-disposable ultrafiltration units can be reused until the membrane is so clogged up that the filtration rate is too slow*

Box 4.10 *Example of cleaning procedure for an ultrafiltration semipermeable membrane*

Check whether there is any specific procedures defined for your particular ultrafiltration membrane
If not:

1. Wash with 0.9% NaCl for 10 min (by use of pressure, vacuum, *etc.*)
2. Wash with 0.1 M NaOH for 30 min as above
3. Wash with deionized water for 5 min as above
4. Alternatively, enzyme cleaning can be performed with very good results
   The flux through the filter should always be registered after cleaning

on the membrane used. Magnetic stirring is performed at the inlet side during ultrafiltration to prevent clogging; however, stirring is not possible when using centrifugal forces. It is very important to clean the membrane thoroughly after each experiment (Box 4.10).

The membrane must not be allowed to dry. In general, the membrane can be kept in water for up to 48 h. If a longer time is needed, use a 0.5% formalin solution for storage.

Concentration of proteins and other macromolecules by ultrafiltration is a consequence of the retainment of compounds with M_r higher than the membrane cut-off value combined with a reduced volume of the solution. The concentration of small solute molecules will, however, remain unchanged but may be reduced by repeated dilution of the solution during ultrafiltration. The recovery of proteins may be less than 100% owing to losses to the membrane. The maximum volume sample depends on the actual ultrafiltration unit.

6 Centrifugation

Centrifugation is widely used for the separation of various particles in chemistry and biochemistry, and one should be familiar with the following symbols and equations related to this technique:

Sedimentation rate = v_s (cm s^{-1})
Sedimentation coefficient = s (s); S = Svedberg unit = 10^{-13} s
Velocity of the centrifugation rotor = ω (radian s^{-1})
Rotor radius = r (cm)
Revolutions per min = rpm (min^{-1})

$$\text{Centrifugal field} = \text{G} = \omega^2 r = \frac{4\pi^2 \times \text{rpm}^2 \times r}{3600} \ (\text{cm s}^{-2})$$

Gravitational constant = g = 980.7 cm s^{-2}

$$\text{Relative centrifugal field} = \text{RCF} = G/g = \frac{4\pi^2 \times \text{rpm}^2 \times r}{3600 \times 980.7}$$

The relationship between G, RCF, rpm, and rotor radius are often shown in tables. The sedimentation rate is dependent on factors other than G, *e.g.*:

- particle size
- particle shape
- particle charge
- density of the particle compared with the centrifugation medium
- viscosity of the medium
- temperature

Some of these relationships are expressed by Stokes' law, which describes the sedimentation of a rigid spherical particle:

$$v_s = \frac{2}{9} \frac{r_p^2(\rho_p - \rho_m)}{\eta} \times g$$

where v_s = sedimentation rate, ρ_p = density of the particle, ρ_m = density of the centrifugation medium, r_p = radius of the particle, η = viscosity of the centrifugation medium, g = gravitational constant, and $\frac{2}{9}$ = shape constant factor for a sphere.

Theory and practice in centrifugation may be somewhat contradictory as several factors not included in Stokes' law may affect sedimentation. Moreover, non-spherical particles have properties different from spherical particles.

Centrifugation should normally be performed according to the instrument's instruction manual. There are, however, some general rules and factors which should be considered. In some centrifuges it is possible to change between rotors of different sizes, allowing centrifugation of various volumes of sample. The volume in the centrifugation tube should in general not exceed $\frac{2}{3}$ of the total volume. The balancing of centrifugation tubes is very important, the allowable weight difference between two opposite tubes depending on the actual equipment. One has to be aware of the material used for the centrifugation tubes as this in practice will determine the centrifugation rate. The chemical resistance of the centrifugation tubes should also be considered. The optimal centrifugation time, temperature, and rate of course depend on the particles to be separated.

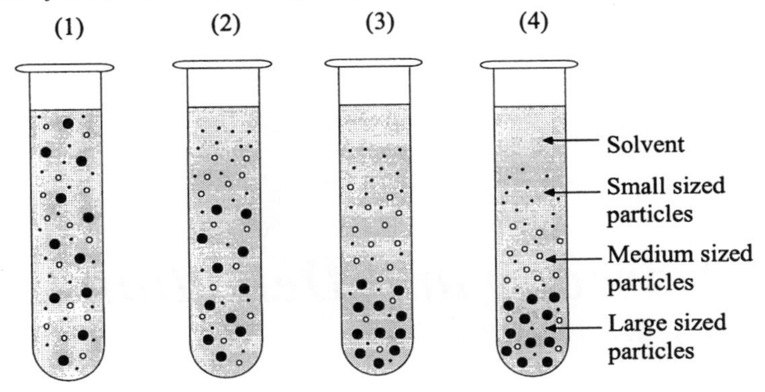

Figure 4.7 *Centrifugation.* (1) *At start of centrifugation particles will be uniformly distributed.* (2)–(4) *Sedimentation of particles during centrifugation*

Centrifugation rate and time are of particular interest in specialized centrifugation techniques where complex mixtures of particles, *e.g.* cell homogenates, should be separated according to differences in their densities and/or sizes (Figure 4.7).

7 Selected General and Specific Literature

1. H.U. Bergmeyer, 'Methods of Enzymatic Analysis', Verlag Chemie, Weinheim, Germany, 1983, vols 1–10.
2. B. Bjerg and H. Sørensen, in 'World Crops: Production, Utilization, Description. Glucosinolates in Rapeseeds: Analytical Aspects', ed. J.-P. Wathelet, Martinus Nijhoff Publishers, Boston, 1987, ch. 13, p. 59.
3. L. Buchwaldt, L.M. Larsen, A. Plöger, and H. Sørensen, *J. Chromatogr.*, 1986, **363**, 71.
4. T. Børresen, N.K. Klausen, L.M. Larsen, and H. Sørensen, *Biochim. Biophys. Acta*, 1989, **993**, 108.
5. S. Clausen, O. Olsen, and H. Sørensen, *Phytochemistry*, 1982, **21**, 917.
6. T.E. Creighton, 'Proteins. Structure and Molecular Properties', 2nd Edition, Freeman, New York, 1993.
7. B.O. Eggum and H. Sørensen, in 'Absorption and Utilization of Amino Acids', ed. M. Friedman, CRC Press, Boca Raton, FL, 1989, vol. 3, ch. 17, p. 265.
8. B.O. Eggum, N.E. Hansen, and H. Sørensen, in 'Absorption and Utilization of Amino Acids', ed. M. Friedman, CRC Press, Boca Raton, FL, 1989, vol. 3, ch. 5, p. 67.
9. J. Frias, K.R. Price, G.R. Fenwick, C.L. Hedley, H. Sørensen, and C. Videl-Valverde, *J. Chromatogr. A*, 1996, **719**, 213.
10. L.M. Larsen, T.H. Nielsen, A. Plöger, and H. Sørensen, *J. Chromatogr.*, 1988, **450**, 121.
11. M.D. Luque de Castro, M. Valcáecel, and M.T. Tena, 'Analytical Supercritical Fluid Extraction', Springer Verlag, Berlin, 1994.
12. J. Mann, R.S. Davidson, J.B. Hobbs, D.V. Banthorpe, and J.B. Harborne, 'Natural Products. Their Chemistry and Biological Significance', Longman Scientific and Technical, Harlow, UK, 1994.
13. H. Sørensen, in 'Canola and Rapeseed: Production, Chemistry, Nutrition, and Processing Technology', ed. F. Shahidi, Van Nostrand Reinhold, New York, 1990, ch. 9, p. 149.
14. H. Sørensen and P.M.B. Wonsbek, in 'Advances in the Production and Utilization of Cruciferous Crops', ed. H. Sørensen, Martinus Nijhoff/Dr. W. Junk Publishers, Boston, 1985, p. 127.

CHAPTER 5

Spectroscopy and Detection Methods

1 Introduction

Methods of analysis used in experimental biochemistry, natural product chemistry, and for feed and food analyses are to a great extent based on spectroscopy. The basis for spectroscopy is the interaction between electromagnetic radiation and atoms or chromophoric groups in the molecules. These interactions imply absorption of energy (ΔE) from the electromagnetic radiation, which is only possible when ΔE corresponds to allowed energy states of the molecules (Figure 5.1).

Excitation is a fast process, occurring in about 10^{-15} s, and the exited state corresponds to unstable conditions with lifetimes of about 10^{-9}–10^{-7} s, which are utilized in relaxation methods. Spectroscopic methods based on determination of energy required for excitation are known as *absorption spectroscopy*. It is also possible to measure the energy released on the return to the ground state, which is denoted *emission spectroscopy*. If some energy is lost before the final return to the ground state, this emission corresponds to less energy (Figure 5.1), giving rise to the spectroscopic phenomena of *fluorescence*, a property of a limited number of molecules, *e.g.* calcium fluorite from which the phenomenon got its name. *Phosphorescence* is closely related to fluorescence as both processes give emission of light at higher wavelength (λ_2) than λ_1, which corresponds to excitation. However, phosphorescence is distinguished from fluorescence on the basis of the mechanism of the emission of the light; a phosphorescing molecule after loss of energy falls from the excited state to an intermediate energy level from which it can not undergo direct transition to the ground state, unlike the case for fluorescence (Figure 5.1). Therefore, it will then go to another vibrational or rotational energy level from where it finally falls to the ground state.

Various types of spectroscopy are useful for both quantitative and qualitative methods of analyses, including identification of the biomolecules. The light is in all cases electromagnetic radiation, which is considered to have both wave and

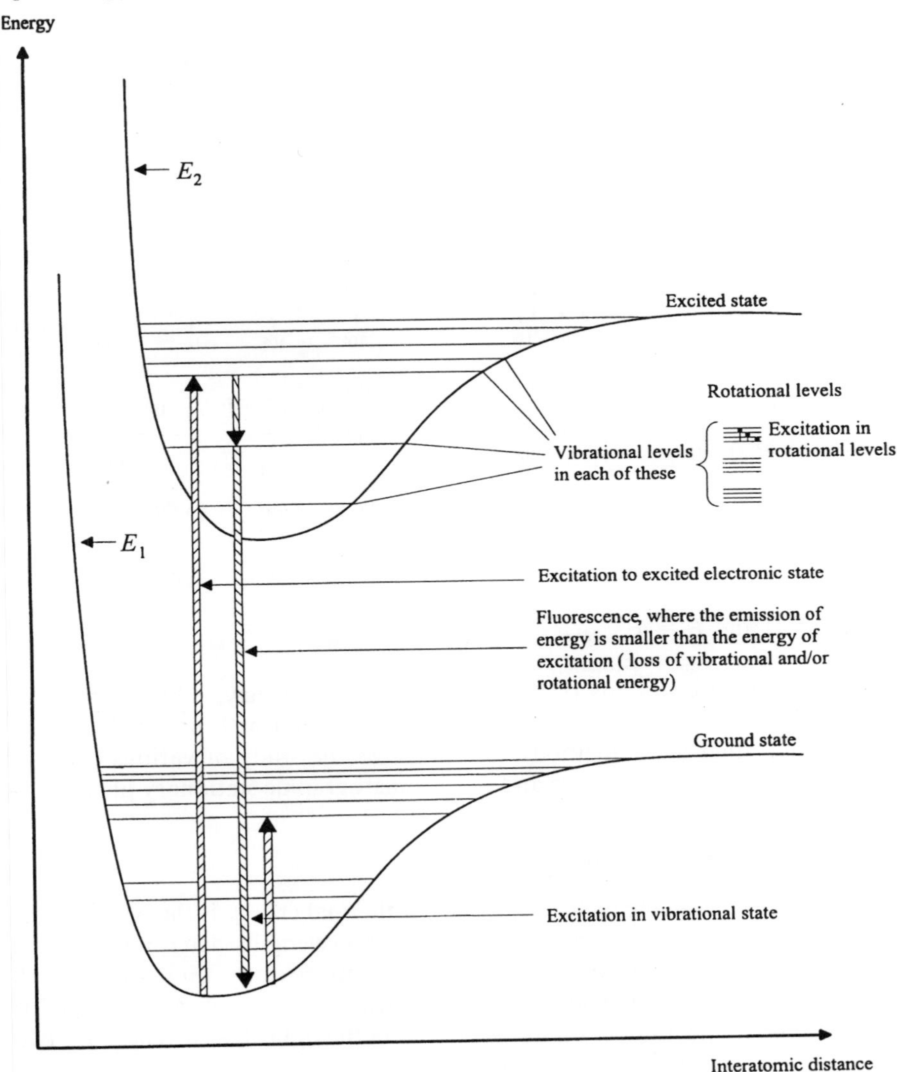

Figure 5.1 *Principles of absorption spectroscopy, illustrating quantified energy levels;* $\Delta E = E_2 - E_1 = h\nu = hc/\lambda$ *for excitations between electronic, vibrational, and rotational levels;* h = *Planck constant,* ν = *frequency,* λ = *wavelength, and c = velocity of light (electromagnetic radiation, which for light in a vacuum is* $2.9979 \times 10^{10}\ cm\ s^{-1} \cong 300\,000\ km\ s^{-1}$*)*

particle properties and is divided into arbitrary areas of wavelengths, energy, and the type of spectroscopy useful to the analyst: X-ray, UV, Vis, IR, ESR, NMR, MS (Figure 5.2). The light absorbed or emitted from a molecule can only have permitted frequencies (Box 5.1), determined by $\Delta E = E_2 - E_1 = h\nu$

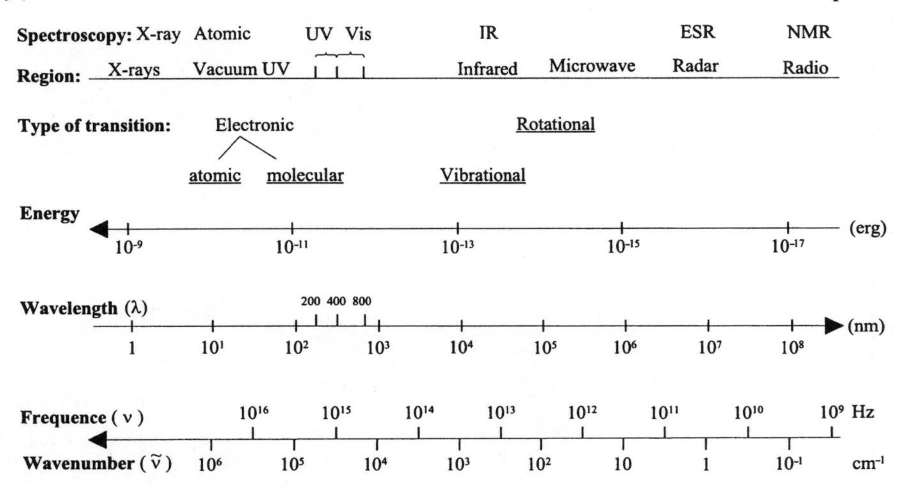

Figure 5.2 *The electromagnetic spectrum of continuous radiation, spectral characteristics, and associated spectroscopy*

(Figure 5.1) and $E = E_{\text{nuclear}} + E_{\text{electron}} + E_{\text{vibration}} + E_{\text{rotation}}$ and generally $E_{\text{electron}} \gg E_{\text{vibration}} \gg E_{\text{rotation}}$.

Absorption spectroscopy (UV-Vis) including colorimetry, is by far the most commonly used spectroscopic technique in connection with liquid chromatography and other assay methods, and it allows the study of various types of molecules in the dissolved state and on chromatographic stationary phases.

2 Atomic Spectroscopy

Placing atoms in a flame or the use of electrothermal energy is the basis for the absorption of light corresponding to specific wavelengths determined by the type of atom being analysed. Atomic/flame spectroscopy thus allows both identification of the atoms and their quantification as the energy absorbed is proportional to the number of atoms present in the optical path. There are two techniques: emission flame spectroscopy and atomic absorption spectroscopy, which measures the absorption of a beam of monochromatic light by atoms in the flame (Figure 5.3).

Special lamps, which produce the appropriate wavelengths, are available in commercial instruments, and different lamps are used for different elements (Table 5.1).

3 UV–Vis Spectroscopy

Solvents and Cuvettes

UV-Vis spectroscopy is performed with the analytes in solution, which may be limited by the availability of suitable solvents, especially for analytes where the

Box 5.1 *Characteristics of electromagnetic radiation*

The wave and particle properties of electromagnetic radiation are described by various symbols and functions. Unpolarized light consists of waves vibrating in all planes perpendicular to the direction in which the light is travelling, whereas plane polarized light, used in determination of the stereochemistry of biomolecules, only has waves vibrating in a single plane.

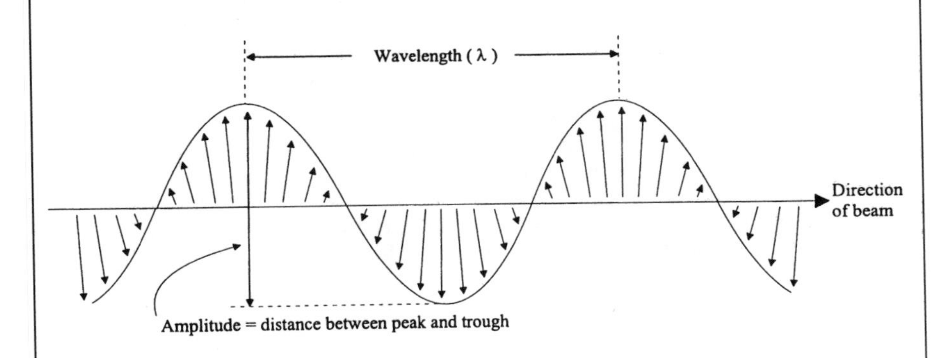

λ = wavelength of radiation (nm)
c = speed of light; in a vacuum = $2.9979 \times 10^{10}\,\text{cm s}^{-1} \simeq 300000\,\text{km s}^{-1}$
ν = frequency of radiation, $\nu = c/\lambda\ (\text{s}^{-1})$
$\tilde{\nu}$ = wave number of the radiation in waves $\times$ cm^{-1}; $\tilde{\nu} = 1/\lambda = \nu/c$
h = Planck constant = 6.624×10^{-27} erg s = 6.624×10^{-34} J s = 1.58×10^{-37} kcal s
N_0 = Avogadro constant = $6.022 \times 10^{23}\,\text{mol}^{-1}$
$h\nu$ = photon
The energy in a photon is, according to Planck:
$E = h\nu = hc/\lambda$
in 1 mole of photons:
$E = N_A h\nu = 6.022 \times 10^{23}\,\text{mol}^{-1}\ 1.58 \times 10^{-37}\,\text{kcal s}\ 3 \times 10^{10}\,\text{cm s}^{-1}\ 1/\lambda = 1/\lambda\ 2.86 \times 10^{-3}\,\text{mol}^{-1}\,\text{kcal cm}$
An increase in wavelength gives thus a decrease in energy (Figure 5.2).
Energy of UV-Vis light: with λ = 200 nm = 200×10^{-7} cm:
$E = 0.5 \times 10^5\,\text{cm}^{-1}\ 2.86 \times 10^{-3}\,\text{mol}^{-1}\,\text{kcal cm} = 143\,\text{mol}^{-1}\,\text{kcal}$
with λ = 400 nm = 4×10^{-5} cm:
$E = 0.25 \times 10^5\,\text{cm}^{-1}\ 2.86 \times 10^{-3}\,\text{mol}^{-1}\,\text{kcal cm} = 71.5\,\text{mol}^{-1}\,\text{kcal}$
with λ = 800 nm = 8×10^{-5} cm:
$E = 0.125 \times 10^5\,\text{cm}^{-1}\ 2.86 \times 10^{-3}\,\text{mol}^{-1}\,\text{kcal cm} = 35.8\,\text{mol}^{-1}\,\text{kcal}$
Energy of IR light with $\lambda = 10^4$ nm = 10^{-3} cm:
$E = 10^3\,\text{cm}^{-1}\ 2.86 \times 10^{-3}\,\text{mol}^{-1}\,\text{kcal cm} = 2.86\,\text{mol}^{-1}\,\text{kcal}$
Energy of microwave with $\lambda = 10^6$ nm = 0.1 cm:
$E = 10\,\text{cm}^{-1}\ 2.86 \times 10^{-3}\,\text{mol}^{-1}\,\text{kcal cm} = 0.029\,\text{mol}^{-1}\,\text{kcal}$

The unit used in the microwave (ESR)–radio–(NMR) area is often the frequency in hertz (Hz) and for IR spectroscopy the wavenumber ($\tilde{\nu}$) in cm^{-1} is often used (Figure 5.2). In UV-Vis spectroscopy the wavelength (λ) is most often given in nm. Finally, if some wavelengths are removed from visible light it will result in colour. The colour will be complementary to the light removed.

(*continued overleaf*)

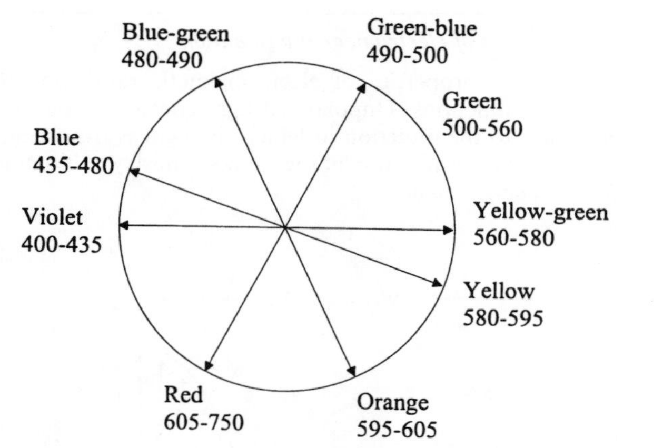

With more than one absorption peak in the Vis area, the colour will be complementary to the vectorial sum of the vectors for the absorption peak.

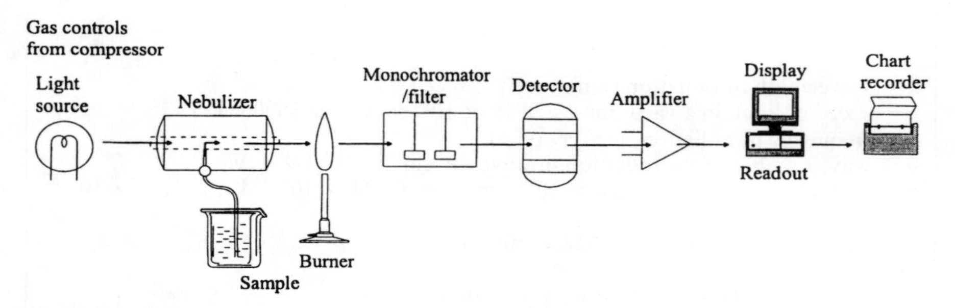

Figure 5.3 *Schematic illustration of an atomic/flame spectrophotometer*

Table 5.1 *Detection limits and wavelengths for absorption and emission of light for different atoms in flame spectroscopy*

	Absorption		*Emission*	
Element	*λ (nm)*	*Detection limit (ppm)*	*λ (nm)*	*Detection limit (ppm)*
Ca	442.7	0.1	442.7	0.005
Mg	285.2	0.01	285.2	0.1
Na	589.0	0.03	589.0	0.0001
K	766.5	0.03	766.5	0.001
Li	670.7	0.03	670.7	0.001
Mn	279.5	0.05	403.3	0.02
Fe	248.3	0.2	372.0	0.5
Cu	324.8	0.1	324.8	0.1
Pb	283.3	0.5	–	–
Hg	253.8	10.0	–	–

Table 5.2 *Solvent cut-off values for selected UV-solvents*

Solvent	*Cut-off (nm)*
Water	195
Methanol	210
Ethanol	210
Cyclohexane	210
Acetonitrile	190
Acetone	230
Chloroform	245
Carbon tetrachloride	265

actual absorption peak is in the first part of the UV-spectrum. The organic solvents required to dissolve various low-M_r biomolecules in solution most often have simple structures and they tend only to absorb light toward the short wavelength end of the UV-spectrum. This type of absorption for the solvent is often said to have a *cut-off* value, which is the wavelength below which the solvent absorption is greater than 0.3 in a cuvette with a 1 cm path-length (Table 5.2).

The material used for the cuvettes must be of special, optical transparent material. Plastic cuvettes are acceptable in the Vis area where high precision is not required, otherwise they should be made of glass, and for the UV-spectrum, quartz cuvettes are required. Typical cuvettes for routine work are *ca.* 4 cm in height with a 1 × 1 cm base area (1 cm path-length), but various types of special cuvettes with other path-lengths are available.

Instruments

Various types of instruments are available, with the recording double beam UV-spectrophotometers capable of both scanning a predetermined spectrum and measuring the change in extinction at predetermined wavelengths as function of time, as used in kinetic experiments (Chapter 3). They contain a beam splitting arrangement which enables a simultaneous comparison of the absorption by both the sample and reference cuvettes, as illustrated in Figure 5.4. The data are

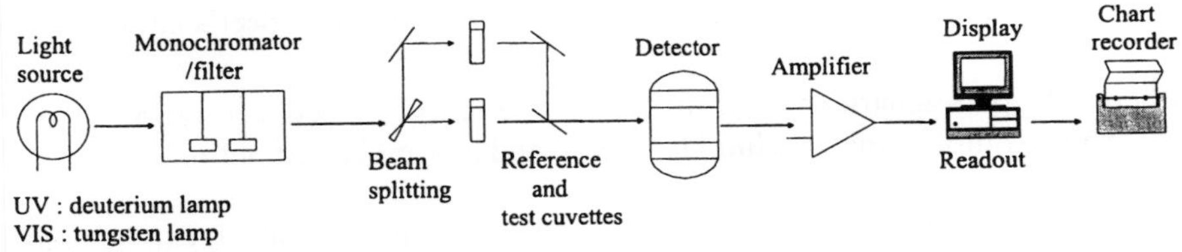

Figure 5.4 *Block diagram of a double-beam spectrophotometer*

commonly recorded on a paper chart, shown on a display unit, and/or transferred to a computer for additional data treatment.

Relatively simple single-beam spectrophotometers/colorimeters are cheaper and of more limited performance than the more advanced double beam spectrophotometers, but they give adequate precision for many routine applications. In addition highly advanced multibeam recording spectrophotometers are also commercially available. For a more detailed description of the various types of instruments, the reader is referred to the relevant technical literature.

Absorbance and Transmittance

The basic laws governing the transmittance of light through cuvettes (Figures 5.4 and 5.5) are inverse relations, where doubling of one of the variables produces a halving of the other. For a uniform absorbing medium, the proportion of the radiation passing through the cuvette is called the transmittance (T) where $T = I/I_0$ (I_0 = intensity of the incident radiation and I = intensity of transmitted radiation).

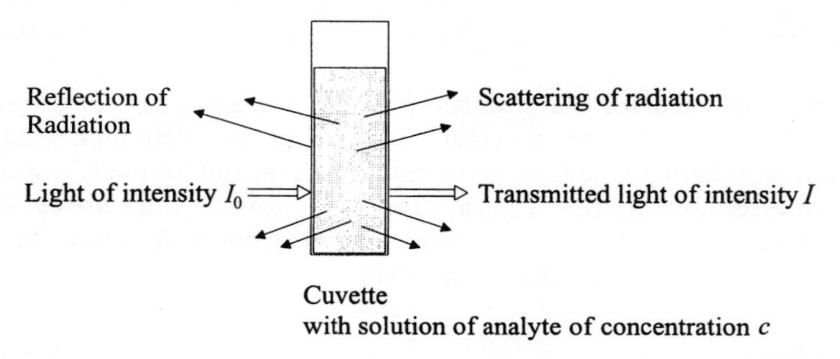

Figure 5.5 *Illustration of light absorption and scattering or reflection of radiation not absorbed but scattered or reflected by particles in suspension*

A graph of $T (= I/I_0)$ plotted against the concentration (c) in the cuvette will give a hyperbolic curve. Therefore, the extent of radiation absorption is more commonly treated as absorbance (A) or extinction (E), defined by:

$$A = E = \log 1/T = \log(I_0/I) = \varepsilon c l \quad \text{(Bouguer–Lambert–Beer's law)} \qquad (5.1)$$

where l = path-length of light in the cuvette (cm), c = concentration of the absorbing solution/analytes [mol L^{-1} (M)], and ε = molar extinction coefficient (M^{-1} cm^{-1})

For compounds with unknown M_r the concentration (g/100 ml) can be used, and the corresponding extinction of a 1% solution in a cuvette with a light path of 1 cm will then give the reference value $E^{1\%}_{1\,cm}$, which is often quoted for proteins. This will then give the relation:

$$A = E = E^{1\%}_{1\,cm}\, c'l \quad (5.2)$$

where c' = concentration in g/100 ml and l = 1 cm and this gives:

$$c' = E/E^{1\%}_{1\,cm}\ (\text{g}/100\,\text{ml}) \quad \text{or} \quad c = E\,10/E^{1\%}_{1\,cm}\ (\text{g L}^{-1}\ \text{or mg ml}^{-1}) \quad (5.3)$$

The size of $E^{1\%}_{1\,cm}$ is determined from experiments where A is measured at different c', followed by plot of A against c'.

Determination of Protein Concentration by UV Spectroscopy

Proteins absorb light in the UV-region at 200–220 nm and at 270–290 nm (Figure 5.6 and Table 5.3). The absorption in the first region is a function of all protein amino acids, especially the peptide groups and the aromatic side chains. The absorption at 270–290 nm is mainly determined by the proteins content of Tyr and Trp ($\varepsilon_{280\,nm} \cong 1400\,\text{M}^{-1}\,\text{cm}^{-1}$ and $5600\,\text{M}^{-1}\,\text{cm}^{-1}$, respectively). The other aromatic amino acid, Phe, has λ_{max} at 257.6 nm ($\varepsilon_{257.6\,nm} = 195\,\text{M}^{-1}\,\text{cm}^{-1}$), and only gives a limited contribution to the protein absorbance at 280 nm.

Several proteins have $E^{1\%}_{1\,cm}$-values which give the basis for a mean value of 10 as reference, resulting in a simple determination of protein concentration from the UV absorption at 280 nm:

$$A = E = cE^{1\%}_{1\,cm}10^{-1} = c \text{ in mg protein ml}^{-1} \text{ for } E^{1\%}_{1\,cm} = 10 \quad (5.4)$$

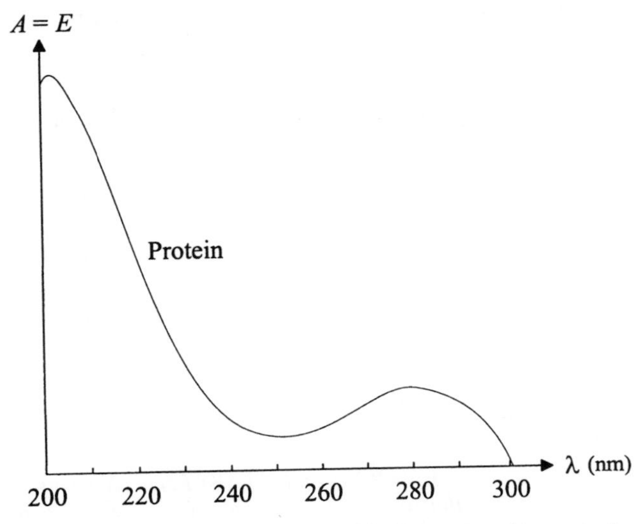

Figure 5.6 *Illustration of the two regions in the UV-spectrum where proteins absorb light*

Table 5.3 *Values of $E^{1\%}_{1\,cm}$ for selected proteins*

Proteins	$E^{1\%}_{1\,cm}$ *(280 nm)*
Bovine serum albumin (BSA)	6.8
Ribonuclease A	7.7
Ovalbumin	7.9
γ-Globulin	13.8
Trypsin	16.0
Chymotrypsin	20.2
α-Amylase	24.2

In various methods of liquid chromatography, including HPLC, FPLC, and HPCE, it is advantageous to determine the concentration of proteins by utilization of the more sensitive detection obtainable at 200–220 nm (Figure 5.6), which is possible when UV-impurities are separated from the proteins in such procedures. However, various types of UV-absorbing compounds may occur in crude extracts and thereby interfere with protein determinations performed directly on such solutions. This is especially a serious problem with many plant extracts containing appreciable amounts of phenolics and/or heteroaromatics, *e.g.* pyridine, pyrimidine, purine, and isoxazoline derivatives. Interference may also be caused by nucleic acids present in the samples, as the absorption of heteroaromatics at about 260 nm gives overlapping peaks with the protein absorptions at about 280 nm (Figure 5.7).

The UV-absorption of nucleic acids at 280 nm is about 25 times stronger than found for BSA, and the maximum for both DNA and RNA are at 260 nm, but still with appreciable absorption at 280 nm, which is the maximum for BSA. In the Warburg–Christian method[1], tables are produced for corrections of the nucleic acid contribution to the protein values determined by UV spectroscopy corresponding to use of the equation:

$$\text{Protein (mg ml}^{-1}) = 1.55A_{280\,\text{nm}} - 0.76A_{260\,\text{nm}} \tag{5.5}$$

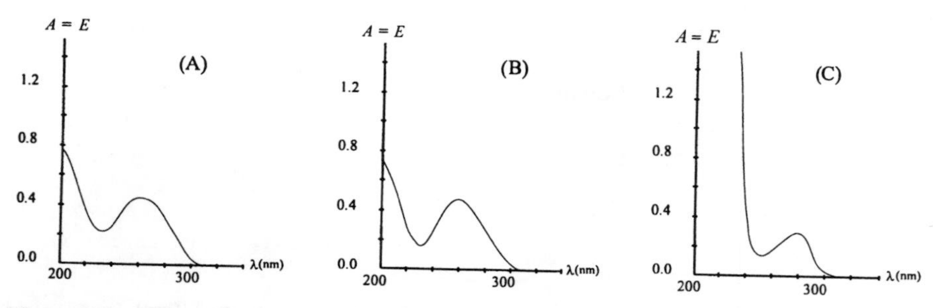

Figure 5.7 *UV-spectra of* (A) *DNA (20 μg ml^{-1}),* (B) *RNA (20 μg ml^{-1}) and* (C) *BSA (0.48 mg ml^{-1}), illustrating the strong absorption of nucleic acids (DNA and RNA) compared with the absorption of the protein (BSA)*

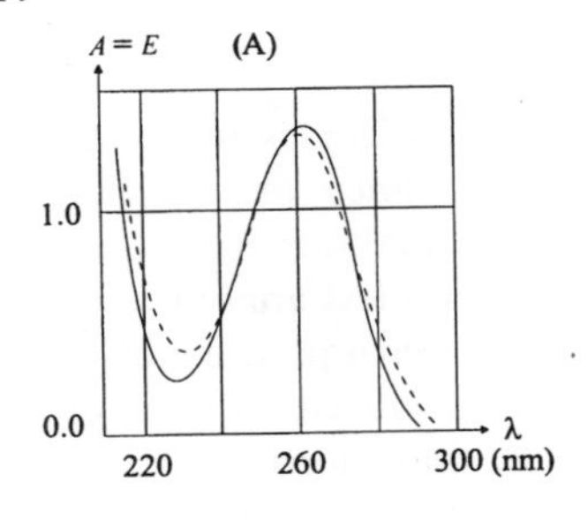

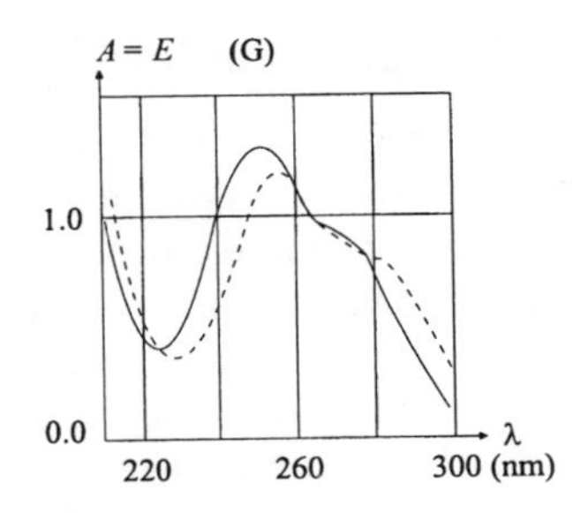

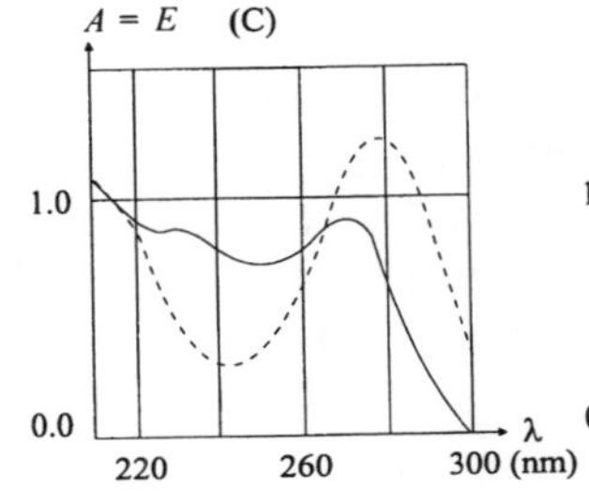

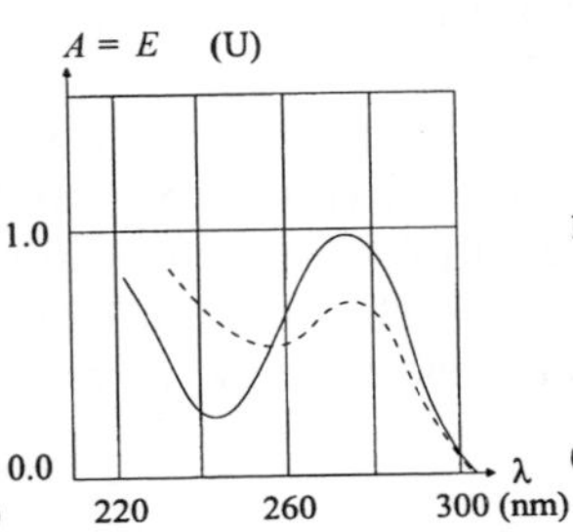

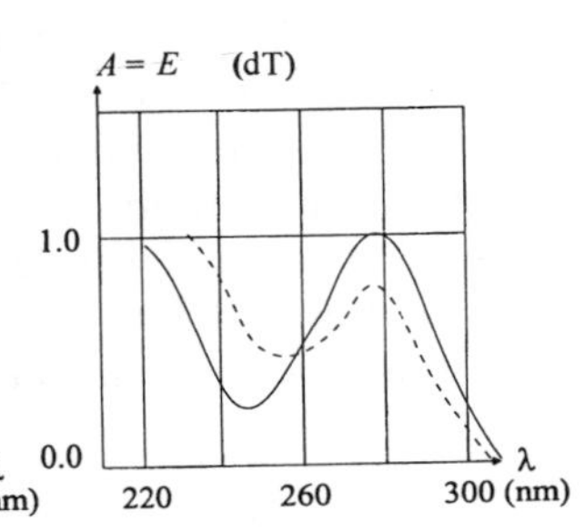

Figure 5.8 *UV-spectra of 10^{-4} M aqueous solutions of the nucleosides:* (A) *adenosine,* (G) *guanosine,* (C) *cytidine,* (U) *uridine, and* (dT) *thymidine.* (——) *pH 7,* (– – – –) *pH 1. The ratio of absorption at 280/260 nm for pH 7 solutions* (A) *0.15,* (G) *0.55,* (C) *0.94,* (U) *0.39, and* (dT) *0.74*

The validity of such corrections is questionable as the side chains of aromatic amino acids buried in protein interiors will give changes of λ_{max}, and changes in the relative amount of purine and pyrimidine bases in the nucleic acids will give changed ratios of 280/260 nm absorptions (Figure 5.8). Box 5.2 gives guidance for an experimental determination of $E_{1\,cm}^{1\%}$ and ε.

Box 5.2 *UV-determination of protein,* $E_{1\,cm}^{1\%}$ *and ε-values*

A double beam UV spectrophotometer and quartz cuvettes should be used

1. Prepare stock solution of:
   Phe: 10.0 mg in 10 ml H_2O
   Tyr: 1.4 mg in 10 ml H_2O
   Trp: 1.6 mg in 50 ml H_2O
   DNA: 1.0 mg in 10 ml 15 mM NaCl, pH 8
   RNA: 1.0 mg in 10 ml 15 mM NaCl, pH 8
   BSA: 4.8 mg in 10 ml H_2O
   0.1 mM solutions in H_2O of the nucleosides in Figure 5.8
2. Set instrument to zero and record the spectra from 190–400 nm for each of the compounds from 1. with the solvent, H_2O or 15 mM NaCl in the reference cuvette
3. Calculate ε at λ_{max} for the low-M_r compounds and $E_{1\,cm}^{1\%}$ for the high-M_r compounds
4. Calculate the ratio of absorption at 280 nm/260 nm for each of the compounds. This ratio is often used for characterization of heteroaromatic compounds

UV–Vis Absorptions of low-M_r Biomolecules

Knowledge of the absorption spectra of biomolecules in the UV-Vis region (Figure 5.2 and Box 5.1) is of the upmost importance for analytical methods, as a great number of liquid chromatographic methods are based on UV-Vis detections or the evaluation of colours. The simplest analytical procedures use chromophoric groups in a molecule. If such groups are not present in the molecules to be analysed, they are often introduced by pre- or postcolumn derivatization (before or after the analytical separations).

Simple functional groups are generally weak chromophores. An illustration of this is given in Table 5.4, which shows some selected examples of chromophore groups or systems.

The type of chromophore determines the energy required for transitions from the ground state to the exited state (Figure 5.1), and thereby the position of the absorption peak or wavelength for light absorption (Box 5.1). If the chromophore consists of lone pairs of electrons on oxygen and nitrogen, electronic transitions require relatively much energy, corresponding to absorption peaks

Table 5.4 *Structure–chromophore relations with λ_{max} and the corresponding ε*

Chromophore	*Structure*	λ_{max} *(nm)*	ε *(M^{-1} cm^{-1})*
Nitrile	$—C{\equiv}N$	<180	–
Alcohol	$—CH_2—OH$	180	100
Ether	$—CH_2—O—CH_2—$	185	1000
Amine	$—CH_2—NH_2$	194	2800
Ketone	$—CH_2—CO—CH_2—$	205	1000
Aldehyde	$—CH{=}O$	210	5000
Carboxylic acid	$—COOH$	200–210	70–100
Ester/amide	$—COOR$/$—CONH_2$	200–215	100–200
Thiol	$—SH$	190	100
Disulfide	$—CH_2—S—S—CH_2—$	204	2000
Thiolate	$—S^-$	230–235	5000
Imidazole	[imidazole ring structure: H, N, N]	211	5800
Alkene	$—(CH{=}CH)—$	190	10 000
	$—(CH{=}CH)_2—$	220	20 000
	$—(CH{=}CH)_3—$	260	35 000
Benzene (Ar)	band 1	<190	–
	band 2	205	9000
	band 3	255	190
Phenol (4-hydroxyalkylbenzene)	—Ar—OH band 2	223	8000
	band 3	275	1400
Phenolate	$—Ar—O^-$ (band 3)	293	2400
4-Hydroxy-3-iodo-alkylbenzene/ionized		283/305	2700/4100
4-Hydroxy-3-nitroalkylbenzene/ionized		360/428	2800/4200

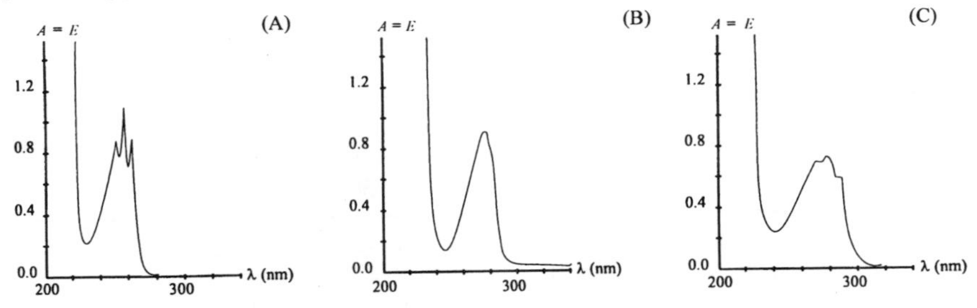

Figure 5.9 *Absorption spectra (200–390 nm) of the aromatic amino acids* (A) *Phe,* (B) *Tyr, and* (C) *Trp dissolved in water at concentrations of 5.5, 0.64, and 0.123 mM, respectively. The absorptions at 280 nm are: Phe, $A = 0$; Tyr, $A = 0.798$; Trp, $A = 0.716$*

at λ below 200 nm. Isolated double bonds (C=C, C=N, C=O) are also weak chromophores with absorption peaks in the lower wavelength end of the UV-spectrum. However, conjugations of double bonds and lone pair electrons lower the energy required for electronic transitions and hence causes an increase in the λ_{max} at which the chromophore absorbs light (Figure 5.2). The chromophores in aromatic compounds have three absorption peaks (bands 1, 2 and 3; Table 5.4), where band 1 in an unsubstituted benzene ring is at λ below 200 nm. Therefore, only bands 2 and 3 are recorded in UV-Vis spectra. Conjugation of an aromatic ring with hydroxy, amine, nitro (—NO_2), halide (—I) or azo- (—N=N—) groups gives more efficient chromophores (Table 5.4; Figure 5.9).

Absorption of light in chromophore systems with an increasing number of chromophore groups in conjugation requires less energy and occurs at higher wavelengths, as illustrated in Figure 5.9. This phenomenon is known as *bathochromic shift*, whereas a shift of the absorption peak to lower wavelengths, *e.g.* by protonization of aromatic amines or heteroaromatic compounds, is known as a *hypsochromic shift*. Functional groups with lone pairs of electrons (—OH, —OR, —NH_2, —I) in conjugation with chromophoric systems are known as *auxochrome* (colour increasing) groups. Values of ε can increase, as seen for the ionization of phenolics (Table 5.4) or denaturation of DNA, which is known as a *hyperchromic shift*, whereas a decrease, *e.g.* by renaturation of DNA, is known as a *hypochromic shift*.

Amino Acids

It is seen from Figure 5.9 that the λ (nm) values of band 3 absorptions for the aromatic amino acids, with the corresponding ε ($M^{-1}\,cm^{-1}$) shown in parentheses, are: Phe 257.6 (195), Tyr 275.4 (1400); Trp 279.8 (5600). It is also revealed that Phe does not contribute significantly to the 280 nm absorption peak often used for determining protein concentrations. The corresponding band 2 absorptions [λ_{max} (nm), ε_{max} ($M^{-1}\,cm^{-1}$)] are Phe (206.0, 9340), Tyr (223.2, 8260), and Trp (217.6, 46700). The UV-absorptions of the aliphatic protein amino acids are, however, less pronounced, but they give weak absorptions at

Table 5.5 *Molar extinction coefficients (ε/M^{-1} cm^{-1}) for amino acids at λ = 205 nm**

Phe	*Tyr*	*Trp*	*His†*	*Arg*	*Met*	*Cys‡*	*CysSSCys*	*Gln*	*Asn*	*Asp*	*Glu*
9360	5600	19600	5170	1300	1860	730	1050	400	400	230	230

* Lys, Pro, Thr, Ser, Gly, Ala, Val, Leu, and Ile all have ε-values ≅ 100 at 205 nm
† The imidazole group in His has λ_{max} = 211.3 nm with ε = 5860 M^{-1} cm^{-1}
‡ The ionization of the thiol group in Cys gives absorptions as shown in Table 5.4

λ = 205 nm (Table 5.5). Thereby, they also contribute to the absorption peak at λ = 205 nm, which quite often is used in liquid chromatographic procedures as a more sensitive detection wavelength for proteins than 280 nm (Figure 5.6).

Determination of aliphatic amino acids through the direct UV-detections used in liquid chromatographic procedures necessitates special requirements for the solvents and buffer systems (Table 5.2). To increase the sensitivity to detection of such molecules with weak chromophores, derivatizations (post- or precolumn) with various types of chromophores are often used in colorimetric methods (*vide infra*) or reversed UV-detection may be used in some HPCE procedures (Chapter 10).

Phenolics and Conjugation in Chromophore Systems

Phenolics embrace a wide range of structurally different plant substances, which may have a great influence on the quality of food and feed, *e.g.* as antinutritional compounds, as antioxidants, or as colours. Native phenolics occur most frequently as glycosides or esters, *e.g.* combined with carbohydrates (Chapter 14), and their chromophore systems give advantages in connection with analytical methods used for their detection or quantification. Examples of simple but widely distributed phenolics are benzoic and cinnamic acid derivatives (Figure 5.10).

Bathochromic and hyperchromic shifts seen for phenolics are a result of chromophore groups in conjugation. Within the groups of flavonoids (Figure 5.11), these effects are even more pronounced. In this big group of plant constituents, which contains *ca.* 5000 known compounds, anthocyanins have conjugated systems which give absorption in the visible area of the electromagnetic spectrum, resulting in various scarlet, red, mauve, and blue colours or pigments. In plants of the order Centrosperma, these colours are produced by betalains (betanins and betacyanins) instead of anthocyanins. Selected examples of these types of natural products are shown in Figure 5.11.

The flavonols have λ_{max} both at 250–270 nm and at 350–380 nm. Anthocyanins have λ_{max} at 260–280 nm and at 510–560 nm. The compounds in both of these flavonoid groups also have absorptions in the first part of the UV spectrum. Isoflavones have the B-ring attached to C-3 instead of C-2 (Figure 5.11) and flavones (3-deoxyflavonols, *e.g.* apigenin and luteolin glycosides) have

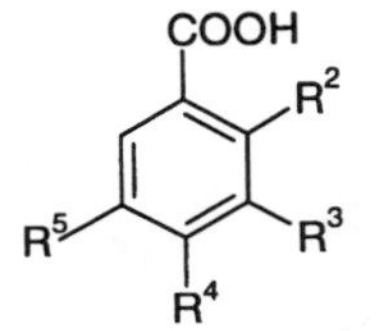

COOH, R^2, R^3, R^4, R^5

Benzoic acid derivatives

Name	R^2	R^3	R^4	R^5	*ca.* λ_{max} (nm)
Salicylic acid	-OH	-H	-H	-H	235, 305
4-Hydroxybenzoic acid	-H	-H	-OH	-H	265
Isovanillic acid	-H	-OH	$-OCH_3$	-H	258, 288
Vanillic acid	-H	$-OCH_3$	-OH	-H	260, 290
Gallic acid	-H	-OH	-OH	-OH	272

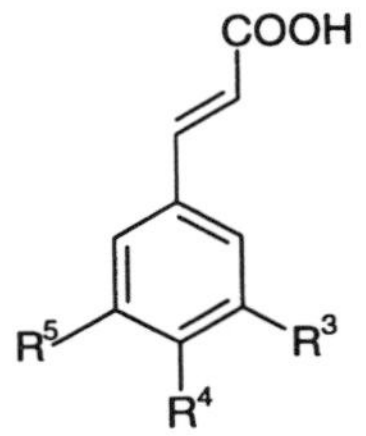

COOH, R^3, R^4, R^5

Cinnamic acid derivatives

Name	R^3	R^4	R^5	*ca.* λ_{max} (nm)
Coumaric acid	-H	-OH	-H	227, 312
Ferulic acid	$-OCH_3$	-OH	-H	235, 330
Isoferulic acid	-OH	$-OCH_3$	-H	240, 325
Sinapic acid	$-OCH_3$	-OH	$-OCH_3$	240, 333
Caffeic acid	-OH	-OH	-H	240, 328

Figure 5.10 *Structure of benzoic and cinnamic acid derivatives and approximate values for λ_{max} of bands 2 and 3. The ε-values for band 3 are ca. 11 000–16 000 $M^{-1}\,cm^{-1}$ for the benzoic acid derivatives and ca. 18 000–23 000 $M^{-1}\,cm^{-1}$ for the cinnamic acid derivatives*

chromophore systems closely related to, but slightly different from, that of flavonols. Naringin (flavanone) and catechine (flavanol) are examples of 2,3-dehydroflavones and 2,3-dehydroflavonols, respectively.

Another important group of natural colours and antioxidants are carotenoids, a group of compounds for which about 400 different natural products have been described. These compounds vary in colour depending on the chromophoric system or, principally, on the number of conjugated double bonds. An example is shown in Figure 5.12 together with examples of chlorophylls. *β*-Carotene ($C_{40}H_{56}$; M_r 536.9) has λ_{max} at about 450 and 478 nm with $E^{1\%}_{450} = 2590$ and $E^{1\%}_{478} = 2280$, respectively. Chlorophyll a ($C_{55}H_{72}N_4O_5Mg$; M_r 893.5) and chlorophyll b have λ_{max} at about 420–460 nm and 640–670 nm with molar extinction coefficients (E^{M}_{nm}) of $E^{M}_{428} = 11.17 \times 10^4\,M^{-1}\,cm^{-1}$ and $E^{M}_{662} = 8.63 \times 10^4\,M^{-1}\,cm^{-1}$ for Chl a and $E^{M}_{452} = 15.9 \times 10^4\,M^{-1}\,cm^{-1}$ and $E^{M}_{664} = 5.61 \times 10^4\,M^{-1}\,cm^{-1}$ for Chl b.

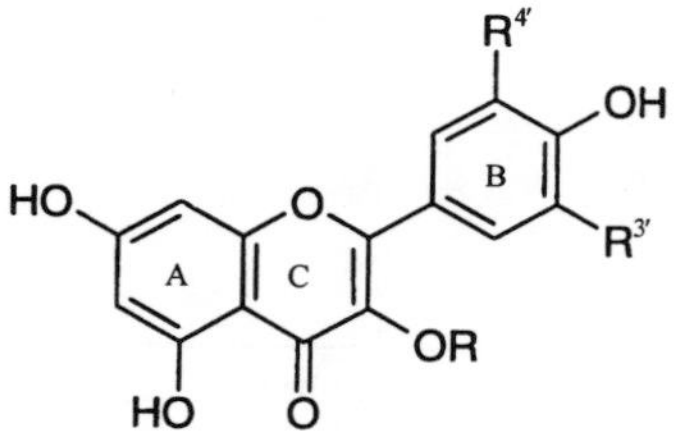

Flavonol-glycosides with the aglycones:
Kaempferol ($R^{3'} = R^{4'} = H$)
Quercetin ($R^{3'} = OH, R^{4'} = H$)
Isorhamnetin ($R^{3'} = OCH_3, R^{4'} = H$)
Myrecetin ($R^{3'} = R^{4'} = OH$)

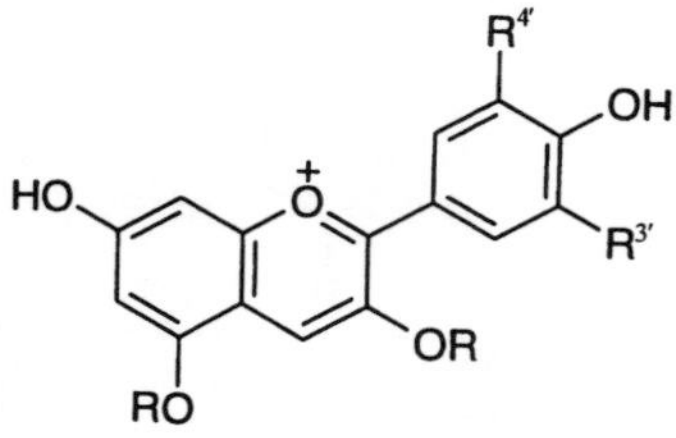

Anthocyanins with the aglycones:
Pelargonidin ($R^{3'} = R^{4'} = H$)
Cyanidin ($R^{3'} = OH, R^{4'} = H$)
Peonidin ($R^{3'} = OCH_3, R^{4'} = H$)
Delphinidin ($R^{3'} = R^{4'} = OH$)

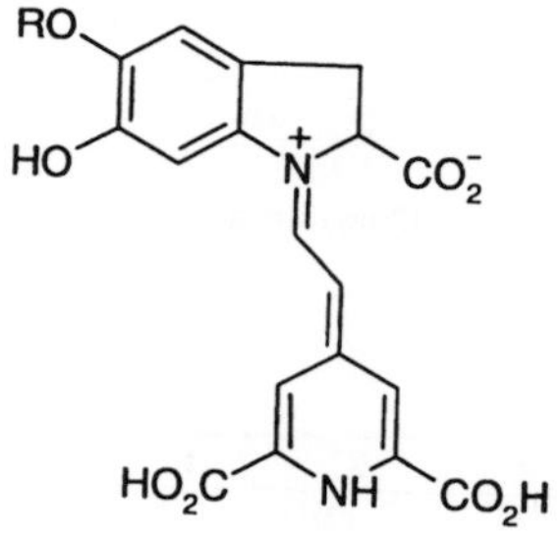

Betalains:
R = glycoside

Figure 5.11 *Structures and names of flavonoid, anthocyanin, and betalain aglycones. Various types of mono- and oligosaccharides, often substituted with ester-bound benzoic and cinnamic acid derivatives, are generally attached as glycosides at the R groups in the aglycones. For some compounds in these groups, other phenolic groups may also be glycosylated or occur as methoxy groups*

The tetrapyrrole haeme structures in chlorophylls are also found in several other coloured natural products. In cobalamins, the central atom is Co instead of the Mg in chlorophylls, and iron porphyrins with Fe as the central atom are ubiquitous in biological materials. This group of compounds comprises the cytochromes, haemoproteins, peroxidases, and cytochrome *P*-450. Blood and red meat owe their colour to the haem groups carried by haemoglobin, myoglobin, and other haem proteins.

The oxidoreductases cytochrome *P*-450 are a family of haemoproteins, involved in hydroxylation of a large variety of ingested foreign compounds (xenobiotics). Hydroxylations of xenobiotics, often followed by glycosylation, usually increase their solubility, whereby they may be detoxified and/or easily excreted. Although the cytochromes *P*-450 resemble cytochrome oxidase in many respects, they are also different from these proteins, especially in their binding properties, *e.g.* in binding of carbon monoxide to the reduced form of

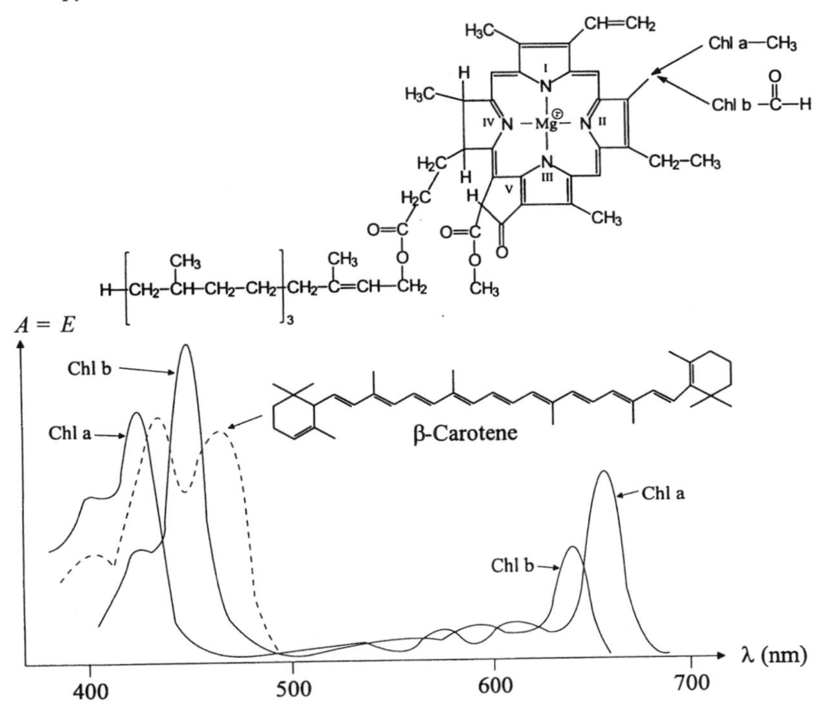

Figure 5.12 *Structure and absorption spectra for β-carotene, chlorophyll a (Chl a), and chlorophyll b (Chl b)*

the haeme group, resulting in the characteristic strong light absorption at $\lambda = 450$ nm for cytochrome *P*-450. The inducible cytochromes *P*-450 participate in a wide variety of important reactions as detoxifications occurring in the liver, activation/deactivations of carcinogenic substances, epoxidation, peroxygenation, desulfuration, *etc.* Utilization of their spectroscopic properties are, therefore, important in many analytical studies of these compounds both in direct assays and in combination with HPLC/FPLC and HPCE based studies of the individual proteins/isoenzymes.

The haem group in haemoglobin, myoglobin, catalase, and peroxidase is ferroprotoporphyrin IX [Figure 5.13(A)] and for cytochromes of class *c* this prosthetic group is covalently bound to the protein part [Figure 5.13 (B)]. The hexacoordinate octahedral complexes of Fe^{2+} contain four coplanar strong-field ligands bound to Fe^{2+} through the pyrrole-N and perpendicular to this are the two additional ligands (X or Y; Figure 5.13). The fifth ligand is another strong-field ligand, most often the imidazole group from the His side chain of the protein, and if the sixth position is occupied by a relatively weak-field ligand (H_2O, Cl^-, CO_2) it is possible to replace it with a ligand capable of binding more strongly (O_2, SCN^-, CO, CN^-). Thereby high-spin complexes may be transformed into low-spin complexes, depending on the protein part and other

(A)

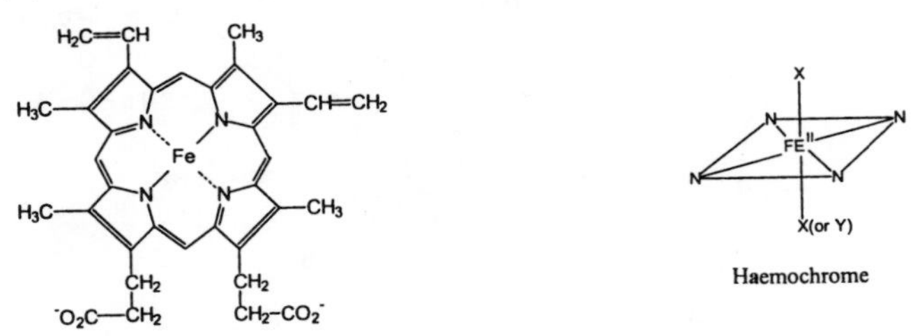

(B)

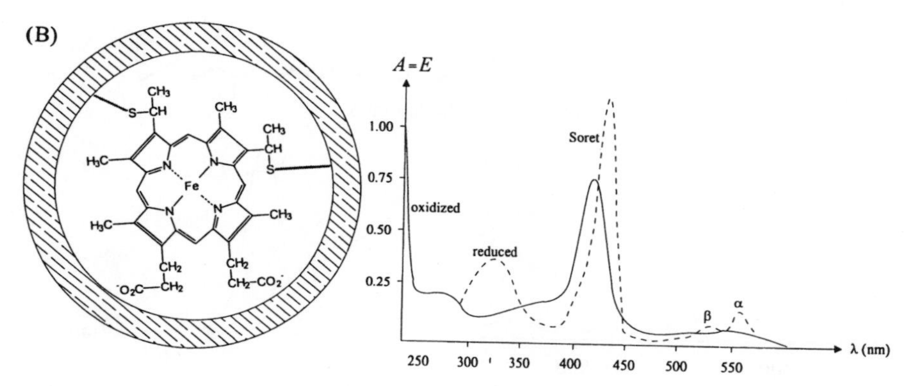

Figure 5.13 *Structure of the haeme group in haemoglobin and peroxidases* (A) *with an illustration of the octahedral complex, the covalently bound protein in cytochrome* c *and* (B) *the absorption spectra of the haem group*

external conditions, which affects the binding constants and rate of association (Chapter 3). As one of the most characteristic features of haemoproteins is their UV-Vis spectra (Figure 5.13) this is, as mentioned above, widely used in various analytical connections. The low-spin Fe^{2+} complexes generally give three absorption peaks in the Vis area, α, β and γ (Soret), with the Soret peak as the most intense at λ around 400–420 nm. This peak occurs in all haemoproteins, whereas the other two (α and β) are more variable. The light absorptions of the Soret peak may also be used in studies of the purity of haemoproteins (Reinheits Zahl; $RZ = A_\gamma/A_{280}$).

Heteroaromatics and Redox Cofactors

In addition to nucleic acids (Figure 5.8), heteroaromatic compounds also occur in various cofactors, the structural parts of vitamins, and other biomolecules (low-M_r compounds). Different plants, especially legumes, often accumulate appreciable amounts of heteroaromatics, which as xenobiotics may behave as antinutritional-toxic compounds when they occur in too high a concentration in food. Isoxazoline derivatives (Figure 5.14) (lathyrism), the pyrimidine

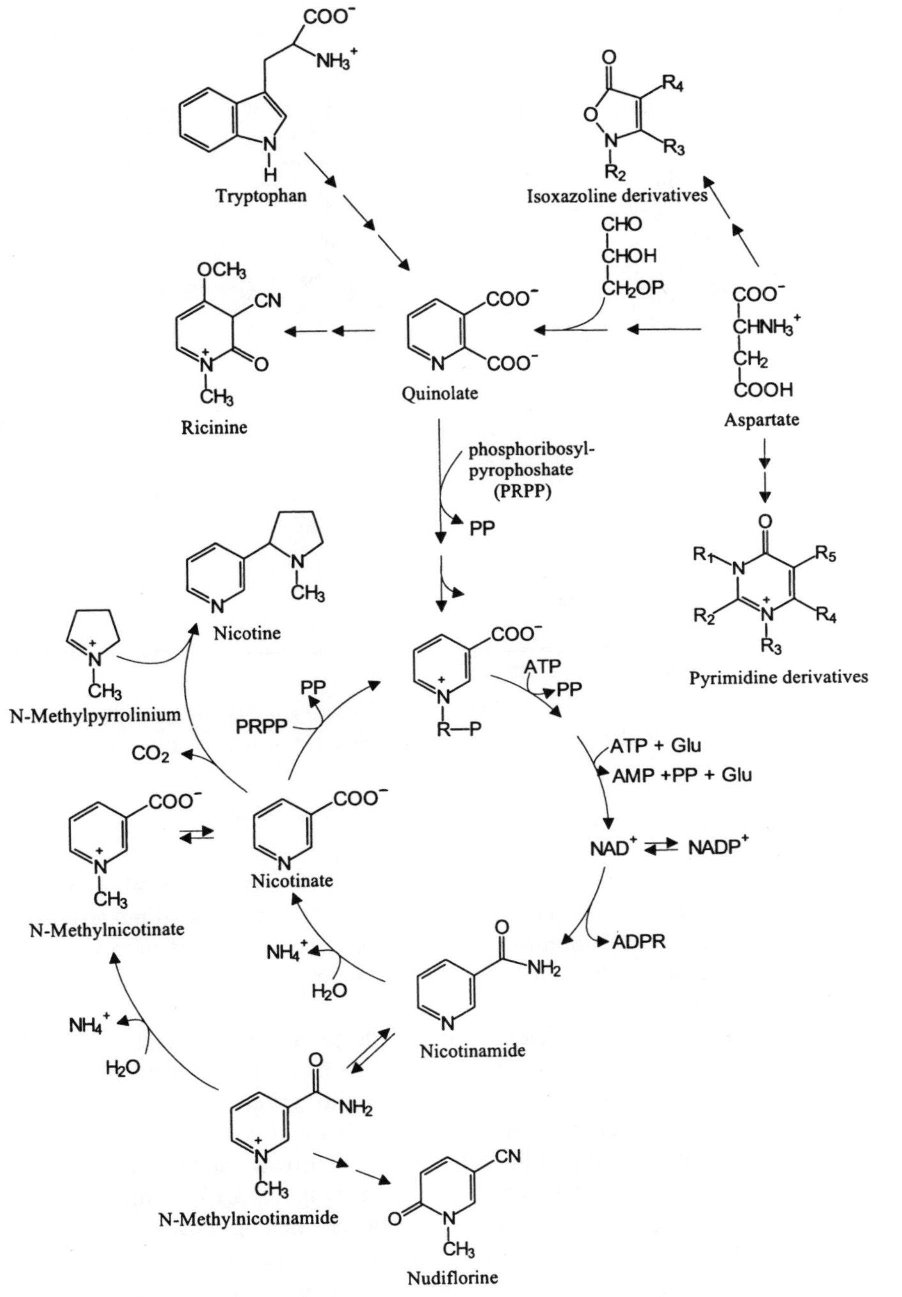

Tryptophan

Isoxazoline derivatives

Ricinine

Quinolate

Aspartate

phosphoribosyl-pyrophoshate (PRPP)

PP

Pyrimidine derivatives

Nicotine

N-Methylpyrrolinium

ATP

PP

PRPP

CO_2

ATP + Glu

AMP +PP + Glu

NAD^+ ⇄ $NADP^+$

N-Methylnicotinate

Nicotinate

ADPR

NH_4^+

H_2O

Nicotinamide

N-Methylnicotinamide

Nudiflorine

Figure 5.14 *Structures and metabolic relations between some selected heteroaromatics which may occur as xenobiotics in food*

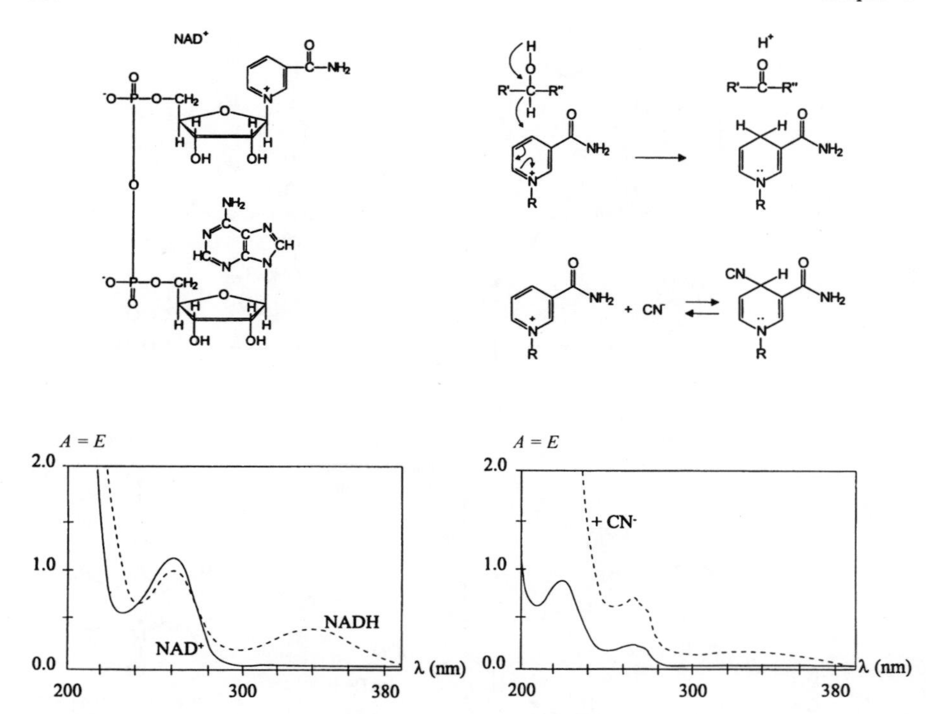

Figure 5.15 *UV-spectra of NAD^+, NADH, and trigonellinamide (right) before and after reaction with CN^-, which gives the product with an absorption at 338 nm*

derivatives as vicine ($R_1 = R_3 = H$; $R_2 = R_4 = NH_2$; $R_5 = \beta$-D-glucopyranoside) and convicine ($R_2 = OH$ otherwise as vicine) (favism), pyridine derivatives (trigonelline and trigonellinamide, *i.e.* N-methyl derivatives of nicotinate and nicotinamide, respectively) and other heteroaromatics are examples of such compounds (Figure 5.14).

Heteroaromatics have efficient chromophore systems, as illustrated in Figure 5.8, which are useful in connection with analytical studies of these compounds by HPLC or HPCE. The functional group in the nicotinamide coenzymes I and II (NAD^+/$NADP^+$) is identical with the structure in trigonellinamide, which gives a basis for nearly identical reactions and changes in the spectra following changes in structure (Figure 5.15). This illustrates the sensitivity of these heteroaromatics towards nucleophiles, the possibility of spectroscopy in studies of such redox-reactions, the possibilities of nucleophiles as inhibitors and the risk of interference in assays based on spectroscopy. For a redox cofactor that can be used by the corresponding enzymes, CH_3 as R-group (Figure 5.15) is not sufficient, whereas the ADP-part of coenzyme I and II is crucial, as revealed from simple spectroscopic experiments (Box 5.3).

Box 5.3 *Spectroscopic, chemical, and biological properties of pyridines*

Recording double beam UV spectrophotometer and quartz cuvettes. M_r (D) and approximate λ_{max} (nm) are shown in parentheses after the names. For structures see Figures 5.14 and 5.15. Nicotinic acid (123.11; 261), nicotinamide (122.12; 261), trigonelline, HCl (173.61; 265), trigonellinamide, I^- (264.05, 265); NAD^+ (663.4; 260); $NADP^+Na_2$ (787.4; 260). The ε-values ($M^{-1}\,cm^{-1}$) at about 260 nm are: NAD^+, 17.6×10^3; NADH, 14.1×10^3; trigonellin 9.3×10^3; trigonellinamide, 4.1×10^3

1. Prepare 5 ml stock solutions of the compounds with $A_{260} \simeq 0.8–1.0$ after 20 × dilution, using equation 5.1 (page 98)
2. Set instrument to zero and record the spectra from 200–400 nm, with H_2O in the reference cuvette, and dilutions giving spectra as shown in Figure 5.15
3. Add a few grains of $NaBH_4$ to each of the solutions and record the spectra again when air bubbles have disappeared
4. Repeat as in 2. and then mix 1:1 with 2 M KCN and record the spectra again
5. Repeat as in 2. but with a 1:1 mix of the compounds in water and 0.5 M EtOH in TRIS buffer (pH 10; 0.1 M)
6. Add a few grains of alcohol dehydrogenase and follow the possible reactions at $\lambda_{max} \simeq 340$ nm. Only the solution with the oxidized coenzymes will give an absorption at 340 nm, illustrating that the redox process (Figure 5.15) has proceeded

4 Colorimetry

Colorimetry based on utilization of the Vis area of the electromagnetic spectrum (Box 5.1) is of great importance for experimental biochemistry both in qualitative and quantitative methods of analysis. The colour seen in the Vis area of the spectrum is complementary to the vectorial sum of the absorbing peaks in the Vis area (Box 5.1). Yellow samples such as carotenes (Figure 5.12) have thus absorbing peaks in the blue region, the blue–violet–red anthocyanins have absorbing peaks in the green–yellow area, and the haemoproteins (Figure 5.13) absorb in the blue–green area.

Detection of many biomolecules with only weak chromophore groups (Table 5.4) often requires derivatization of functional groups with chromophore reagents, which results in colours. Such more or less specific staining can be utilized in qualitative detections, spot tests, or in quantitative methods of analyses based on Equation 5.1 (page 98) as for direct UV spectroscopy. A comprehensive collection of derivatization or staining procedures, often developed in connection with chromatographic methods (TLC, PC; HPLC, HPCE), is described in Ref. 2. Selected examples are shown in Table 5.6, and more specific descriptions for some of these procedures are presented in connection with various analytical methods, assays, and staining procedures (Chapters 9–14).

5 Fluorimetry

Fluorescence (Figure 5.1) occurs when chromophore groups lose only part of their excitation energy by transfer to an intermediate vibrational or rotational

Table 5.6 *Selected colorimetric methods used in staining procedures*

Biomolecules	*Derivatization reagents*	*ca. Vis absorbance of product (nm)*
Amino acids and amines (primary and, partly, secondary aliphatic amino groups)	Ninhydrin; often blue-violet colours, but various other colours for many particular amines and amino acids.	570 420 (Pro)
	2,4-Dinitrofluorobenzene (Sanger's method)	Yellow
Alkaloids	Dragendorff reagent	Orange
Guanido compounds	Sakaguchi reaction	Orange
Proteins	Folin (Lowry method)	660 and 750
	Coomassie Blue	595
	BCA (bicinchoninic acid)	562
	Biuret	540
	Silver	Black
Carbohydrates	Phenol–H_2SO_4	Varies *ca.* 480–500
	Thymol–H_2SO_4	500
	Orcinol	500
	Periodate	White/blue
Reducing low-M_r carbohydrates	Aniline–phthalate	Brown
	Dinitrosalicylate	540
Ketones	Diphenylamine	635
	2,4-Dinitrophenylhydrazine	Yellow
Hexosamines	Dimethylaminobenzaldehyde (Ehrlich)	530
Phenolics (tannin)	Folin	660 and 750
	Diazotized sulfanilic acid (Pauly reaction	Red/brown
Carboxylic acids	Aniline–xylose	Brown
Steroids	Liebermann–Burchard reagent	425
Lipids	Ammonium sulfate (heat)	Dark/black
Inorganic phosphate	Ammonium molybdate	600

energy level. The final transfer to the ground state therefore releases less energy than required for excitation, resulting in a higher wavelength for the fluorescence spectrum compared with the absorption spectrum. Fluorescence spectroscopy is thus limited to molecules with the required properties, but compounds that do not fluoresce can be labelled with a fluorescent molecular group before analysis/detection (Figure 5.16).

Fluorescamine can be bound to amine groups and thereby give strong fluorescence to the product. This label may be used for detection of proteins after electrophoresis and in amine and amino acid analyses[3,4]. Ethidium bromide is often used for sensitive detection of nucleic acids after electrophoretic separation, and the other molecules in Figure 5.16 are correspondingly often used for the detection of proteins.

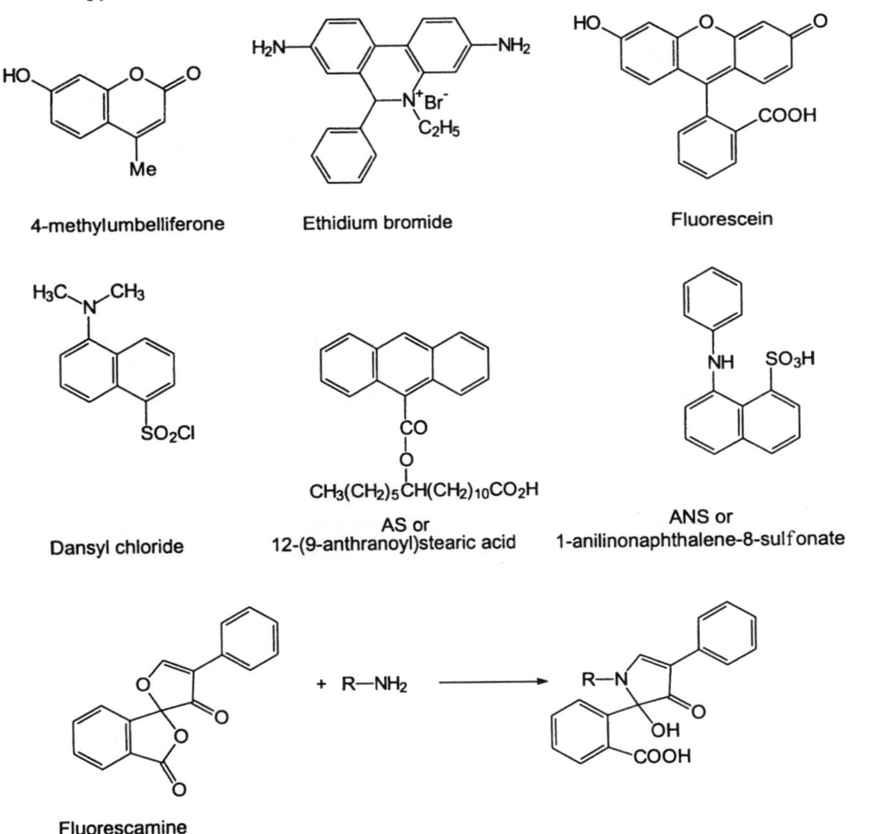

Figure 5.16 *Selected examples of chromophore groups with fluorescence properties that are often used in experimental biochemistry*

In proteins, tyrosine and tryptophan are native major fluorescent groups, and the fluorescence spectrum for tyrosine has its maximum around 300 nm which is about 20 nm higher than the corresponding absorption spectrum (Figure 5.9). The environment of the phenol and indolyl groups in proteins can, however, greatly modify their fluorescence properties, and the fluorescence process is in general susceptible to quenching, which makes the emitted intensity less than predicted, sometimes to such an extent that the fluorescence is virtually extinguished.

NADH and NADPH (Figure 5.15) are fluorescent compounds that allow many redox-enzymes that utilize these coenzymes to be followed by spectrofluorimetry. Flavoproteins also give fluorescence, which can be measured at 525 nm after excitation at 345 nm. Riboflavin and other chromophore groups may in some cases be photochemically decomposed, and exposure of the samples to the high energy of UV radiation can thus lead to sample decomposition while the measurements are being made. However, fluorimeters often have a device blocking the light path to the sample except at the time it is being measured. Compared with UV–Vis spectrophotometers, the main difference in a fluorimeter is the need for two monochromators, one for excitation and one for emission, and it is recommendable with a scanning instrument to facilitate the determination of both the optimum excitation and emission wavelengths.

6 Diode-array Detection

Diode-array detection is a technique that allows the simultaneous determination of absorbance at many or all wavelengths within few milliseconds. Therefore, such detectors have been found to be usable as HPLC and HPCE detection systems. In these instruments, a lens system focuses the light into the measuring area and the beam is then dispersed by a diffraction grating where after the whole spectrum of wavelengths falls on an array of light-sensitive diode detectors, allowing effective and simultaneous monitoring of all wavelengths. The array is read into a computer with appropriate software and capacity for treatment of the data in such a way that the presentation can be in a three-dimensional form with the absorption *versus* both wavelength and analysis time. This technique is thus able to give very useful information on otherwise complex chromatograms of unknown compounds.

7 Infrared Spectroscopy (IR)

IR has only limited interest in connection with liquid chromatographic procedures and will, therefore, not be considered here.

8 Nuclear Magnetic Resonance (NMR)

NMR spectrometry is a special form of absorption spectrometry in which nuclei of certain isotopes in a magnetic field absorb electromagnetic radiation in the radio frequency region (Figure 5.2). A plot of the frequencies of the absorption peaks *versus* peak intensities constitutes an NMR spectrum. The essential properties of the nuclei, which make them suitable to NMR spectrometry studies, where they behave like minute magnets, are their spin number $I = \frac{1}{2}$ and a spherical charge distribution. The nuclei of special biochemical interest having these properties are ^{1}H, ^{3}H, ^{13}C, ^{15}N, and ^{31}P, and of these by far the most widely used in NMR spectrometry are ^{1}H and ^{13}C.

If a sample of this type is placed in a strong external magnetic field, the different orientations of the nuclear spin will have different energies (Figure 5.17). The energy between these two stages corresponds to microwave radiation which will be sufficient to flip the nuclei from one energy state to another.

The original ^{1}H NMR spectrophotometers, commercially available about four decades ago, had electromagnets corresponding to 60 MHz, whereas the far more efficient instruments now available are 100–600 MHz instruments. Such instruments are generally based on helium-cooled superconducting magnets and operate in the pulsed Fourier transform (FT) mode. High-field

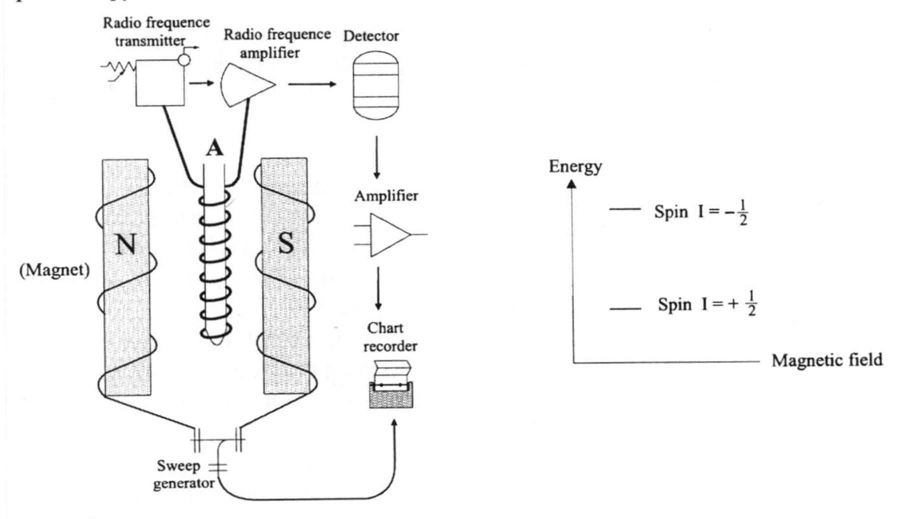

Figure 5.17 *Schematic illustration of an NMR spectrometer with the sample tube* (A) *in a strong magnet (N–S). To the right is illustrated the energy differences (ΔE) between the two spin states $I = +\frac{1}{2}$ and $I = -\frac{1}{2}$*

magnets have dramatically improved the sensitivity and chemical shift separations and allowed new applications of NMR to biological problems, with possibilities for studies of larger molecules and molecules which previously were too complicated for NMR studies.

The energy levels of the nuclei (^{1}H or ^{13}C) in a magnetic field are very sensitive to the environment surrounding the atom in question. The atoms are to a small extent shielded by their electron clouds, the densities of which vary with chemical structure. These variations give rise to different absorption positions, usually within a range of about 750 Hz for protons in 60 MHz instruments or 3750 Hz in 300 MHz instruments. The absorption position in relation to a reference is called the chemical shift (δ) which is expressed in dimensionless units of parts per million (ppm) defined with respect to a reference compound, and is independent of the operating frequency (MHz). Values of δ are calculated by dividing the difference in resonance frequencies of the sample (ν_{s}) and the reference (ν_{ref}) in Hz by the operating frequency (MHz):

$$\delta = \frac{\nu_{\mathrm{s}} - \nu_{\mathrm{ref}}}{\text{operating frequency}} \tag{5.6}$$

With $\nu_{\mathrm{s}} - \nu_{\mathrm{ref}} = 60$ Hz in a 60 MHz instrument 100 Hz in a 100 MHz instrument this gives $\delta = 1$ ppm. An often used reference compound is tetramethylsilane (TMS), with $\delta = 0$ ppm, or for ^{13}C NMR in aqueous solution, dioxane may be used, adjusting the instrument to a $\delta = 66.5$ ppm.

With the units for coupling constants (*J*) in Hertz, chemical shifts (δ) in ppm, and if the shift difference between two coupled protons is $\Delta\delta$ ppm, then at higher fields $\Delta\delta/J$ is larger, and the spectrum is closer to first order and readily analysable by inspection. This means that two groups of peaks with a separation of 1 ppm are 250 Hz apart when analysed with a 250 MHz instrument, whereas the same two groups when analysed with a 500 MHz instrument are 500 Hz apart.

NMR spectroscopy is thus a very powerful technique for structural identification of biomolecules, and one-dimensional ^{13}C NMR especially is useful in the determination of the chemical structure (constitution); ^{1}H NMR can give additional information on fine structures, including configuration and conformations. Examples are shown in Figures 5.18 and 5.19.

Various advanced NMR techniques are also useful in studies of more complicated biomolecules, where multidimensional NMR as correlated spectroscopy (COSY), totally correlated spectroscopy (TOCSY/HOHAHA), and nuclear Overhauser effect (NOE) or NOE spectroscopy (NOESY) are of great value. Development within the area of NMR techniques during the last decade has thus opened up new ranges of biological applications even in tissue and

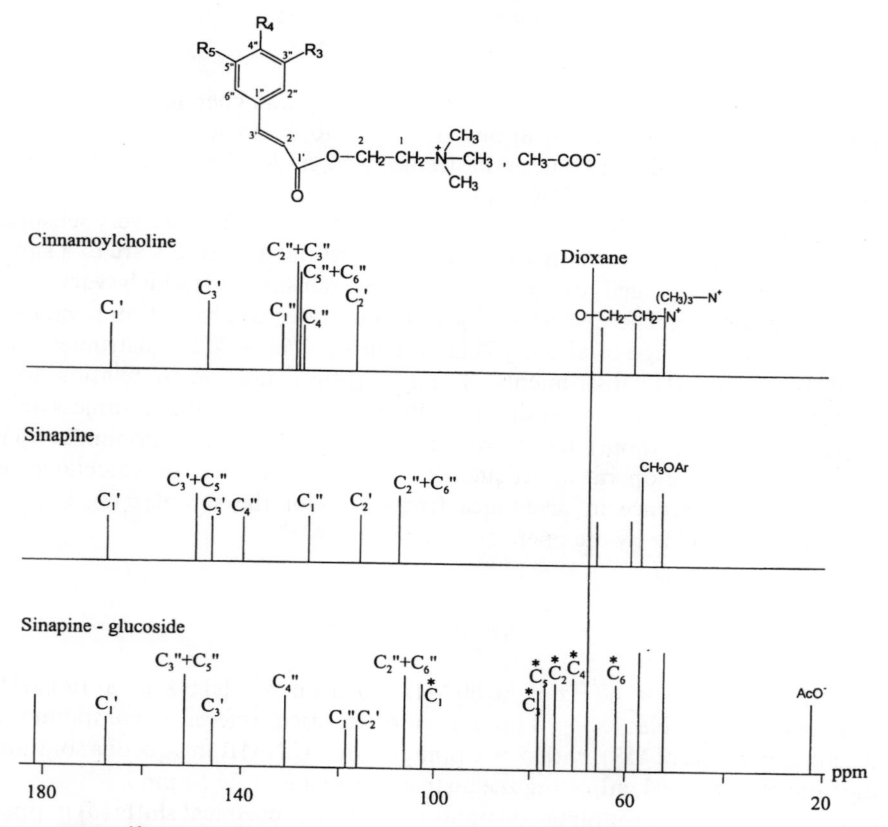

Figure 5.18 *^{13}C NMR data for selected aromatic choline esters dissolved in D_2O*

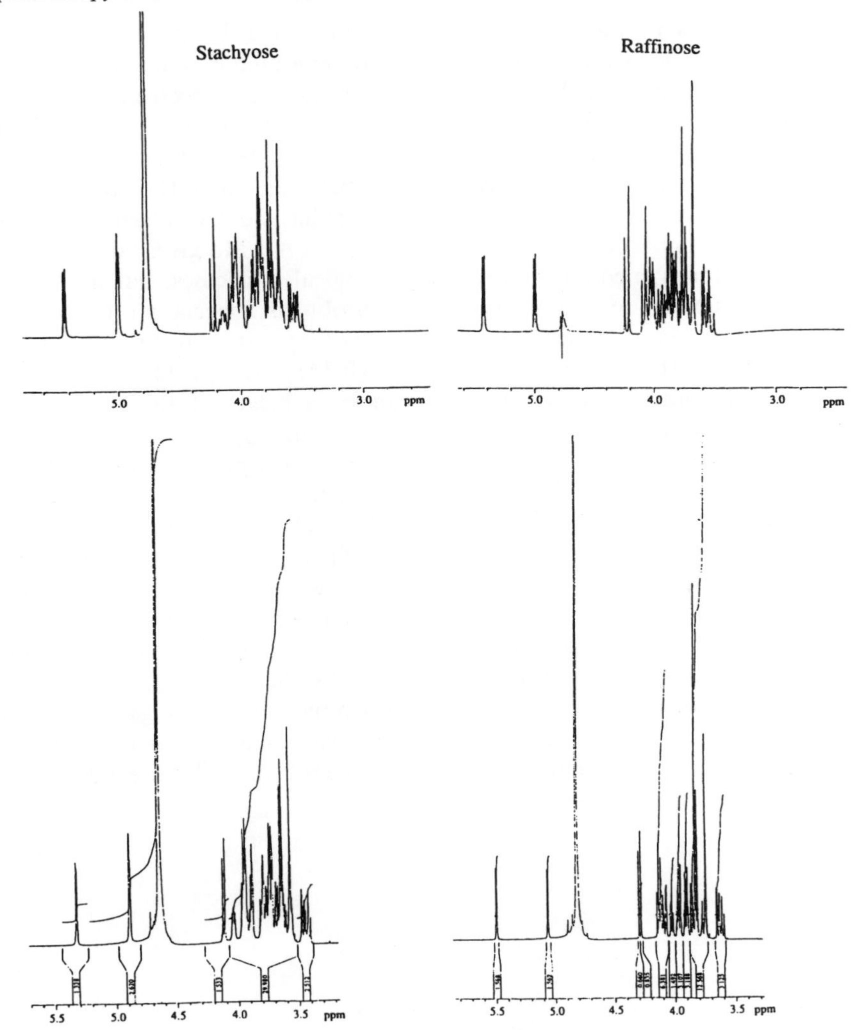

Figure 5.19 *^1H NMR for the carbohydrates raffinose and stachyose dissolved in D_2O and analysed at 250 MHz (top) and 500 MHz (bottom) instruments*

whole-animal studies, and for more detailed information the reader is referred to Section 5.10.

9 Mass Spectrometry

Mass spectrometry (MS) is not a spectroscopic technique like those given in Figure 5.2 as it does not involve electromagnetic radiation and quantum principles. It is essentially dependent upon the thermodynamic stability of the compounds or ions produced in the commonly used electron-impact (EI) mode, where molecules in the vapour phase are bombarded with a high-energy electron beam. The positive ions produced are separated in a magnetic field on the basis of their mass-to-charge (m/z) ratio and the mass spectrum recorded is presented as a computer-plotted bar graph of abundance (vertical peak intensity) *versus* m/z. Mass spectroscopy may also be performed with ions produced in different ways, *e.g.* by use of adducts such as NH_4^+ in a process known as chemical ionization (CI). More interest has, however, been devoted to the fast atom bombardment (FAB) ionization, as FAB-MS gives advantages in the study of many medium-to-large M_r biological molecules, but it also has disadvantages owing to various suppression problems, especially if FAB-MS is applied to mixtures or impure samples. By use of separation techniques such as electrophoresis, HPCE, or HPLC prior to FAB-MS, these problems are largely removed, and much greater efficiency can be achieved if these separation methods are interfaced on-line to the MS instruments.

In recent years, MS has also aroused interest in connection with studies of proteins, especially when it is based on the new technique known as MALDI-TOF (Matrix-Assisted-Laser Desorption of Ions, electrically accelerated into a Time Of Flight mass analyser). The proteins are placed on a suitable matrix and heated briefly by a laser flash. The source of the matrix may be radioactive californium (^{252}Cf) and sometimes additives such as nitrocellulose are included before the sample is placed in front of the laser source. The emitted plasma particles pass through the support foil and impart sufficient energy to the sample to cause ionization and project or desorb the sample ion into the gas phase. The ions are then electrically forced to fly down the evacuated mass analyser tube to a suitable detector where they are recorded as the m/z ratio (Figure 5.20).

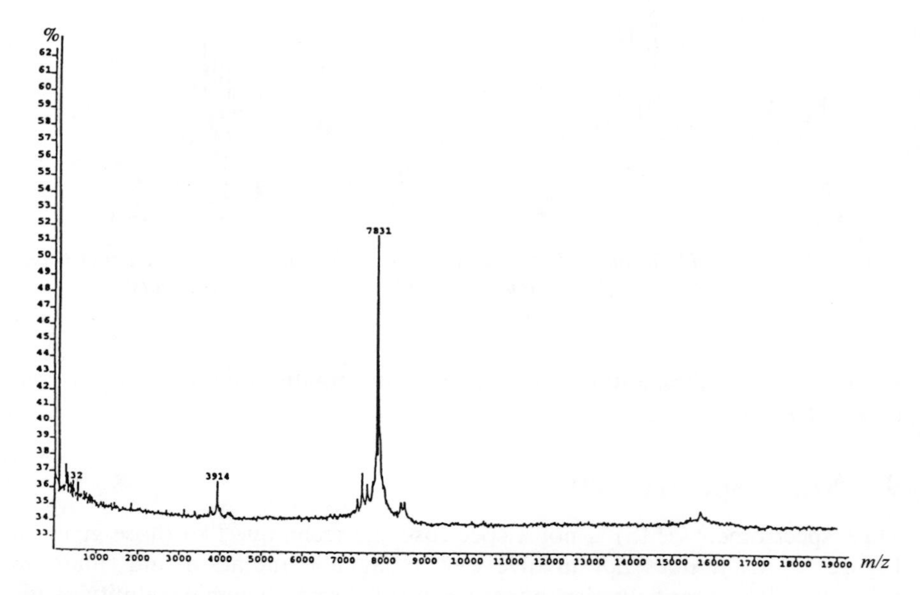

Figure 5.20 *MALDI-TOF used for determination of* M_r *of isoinhibitors in the BBI preparation of Sigma*

10 Selected General and Specific Literature

[1] O. Warburg and W. Christian, *Biochem. Z.*, 1941, **310**, 384.
[2] H. Jork, W. Funk, W. Ficher, and H. Wimmer, 'Thin-Layer Chromatography, Reagents and Detection Methods', WCH Verlagsgesellshaft, Weinheim, 1990.
[3] S. Udenfriend, S. Stein, P. Böhlen, W. Dairman, W. Leimgruber, and M. Weigele, *Science*, 1972, **178**, 871.
[4] J.R. Benson and P.E. Hare, *Proc. Natl. Acad. Sci. U.S.A.*, 1975, **72**, 619.
[5] B.W. Christensen, A. Kjær, J.-Ø. Madsen, C.E. Olsen, O. Olsen, and H. Sørensen, *Tetrahedron*, 1982, **38**, 353.
[6] B.O. Eggum and H. Sørensen, in 'Absorption and Utilization of Amino Acids', ed. M. Friedman, CRC Press, Boca Raton, FL, 1989, vol. 3, ch. 17, p. 265.
[7] D.D. McIntyre and H.J.J. Vogel, *J. Nat. Prod.*, 1989, **52**, 1008.
[8] O. Olsen and H. Sørensen, *J. Am. Oil Chem. Soc.*, 1981, **58**, 857.
[9] J.K.M. Sanders and B.K. Hunter, 'Modern NMR Spectroscopy. A Guide for Chemists', 2nd Edition, Oxford University Press, Oxford, 1994.
[10] R.M. Silverstein, G.C. Bassel, and T.C. Morrill, 'Spectrometric Identification of Organic Compounds' 5th Edition, Wiley, New York, 1991.
[11] H. Sørensen, in 'Canola and Rapeseed: Production, Chemistry, Nutrition, and Processing Technology', ed. F. Shahidi, Van Nostrand Reinhold, New York, 1990, ch. 9, p. 149.

CHAPTER 6

Liquid Chromatography

1 Introduction

Liquid chromatography (LC) comprises several methods or techniques for the separation of structurally different analytes by use of two phases. These two phases consist, in LC, of a liquid mobile phase and an immobile stationary phase which in micellar electrokinetic capillary chromatography (MECC; page 233) is a pseudostationary phase or 'mobile stationary phase'. In gas–liquid chromatography (GC or GLC), the mobile phase is gaseous, and GLC techniques are considered to be beyond the scope of LC methods considered in this book. Interested readers are referred to texts on GLC. LC techniques based on utilization of the charges on the analytes are considered in the following sections: ion-exchange chromatography (Chapter 7); HPLC and FPLC, including ion-pair chromatography (Chapter 8); electrophoresis (Chapter 9); and HPCE, including MECC (Chapter 10). In this chapter, focus will be placed on partition, adsorption, permeation or size-exclusion, and bonded or modified phase chromatography, comprising normal- and reversed-phase chromatography, affinity chromatography, HIC and solid phase extractions (SPE).

Basic concepts used in LC will initially be considered, with the purpose of making LC more than a semi-empirical exercise. The empirical approach has enabled many researchers to make some progress in their research, with use of LC relying on 'green fingers' and/or by use of information from colleagues that has allowed them to isolate some natural products or purify enzymes of interest. In more advanced food analyses, natural product chemistry, experimental biochemistry, efficient biotechnology or processing technologies, however, empirical approaches to LC are far from sufficient. An understanding of the basic principles and theoretical fundaments of LC enables obstacles to be overcome, and more efficient and fast LC procedures to be adapted to the analytical problems considered. This is also the case if LC procedures have to be used in large-scale separations and to ensure that analytical procedures are designed rationally and efficiently in accordance with good laboratory practice (GLP) (Section 1.4).

2 Separation of Analytes between the Two Phases in LC

Efficient separation of analytes in LC requires an appropriate solubility of the analytes in the system and neither too strong nor too weak binding/solubility of the analytes in the solvated stationary phase. The distribution of analytes in the LC phases can be described by the equilibrium constant (K_d) for the concentration of the analyte in the mobile phase (C_m) and in the solvated stationary phase (C_s):

$$K_d = \frac{C_s}{C_m} \tag{6.1}$$

For non-ideal solutions, it is necessary to introduce activity coefficients (Section 2.3), and in some cases the distribution or equilibrium constant depends markedly on concentration because the analytes may exist in dissociated or associated forms (Chapter 3) in one of the two phases. The binding–dissociation of protons in protolytically active groups will, for example, depend very much on the dielectric constant, and thereby on the amount of organic solvents in the aqueous solutions (Section 2.8). The choice of such solvents, addition of 'salting-out' solutes (page 81), and use of 'complexing or binding agents' (Chapter 3) are some of the parameters that can be used to increase the efficiency of separations.

The stationary phase may consist of solvated, often porous solid materials, polymers, or pseudostationary micelles/surfactants. If the stationary phase is hydrophilic, it corresponds to normal-phase LC (page 172) whereas a lipophilic stationary phase gives reversed-phase LC (page 172). The membrane and porous structures of the stationary phase require consideration of swelling, and capillary properties, as well as the size and shape of analytes in relation to the pore size.

The porous structure of chromatographic materials based on particles is shown in Figure 6.1, together with large high-M_r and small low-M_r molecules.

The interstitial volume (Figure 6.1; V_0) will be the compartment where large, but not too large, molecules are allowed to move whereas small molecules will also penetrate the inner volume (V_i), determined by the size of the channels or pores, of the chromatographic particles. The volume of the particles ($V_i + V_g$) will depend on the degree of swelling (Sections 6.2 and 7.3) and so will the pore size as well as the total volume (V_t) of the system:

$$V_t = V_0 + V_i + V_g \tag{6.2}$$

In LC, the stationary phase can in addition to porous particles also be made of membrane or gel structures. The result of this will be that low-M_r and high-M_r analytes which only separate according to their size and shape, without binding to the chromatographic materials, will show reversed retention times (t_R; page 133) in the two systems (Figure 6.2).

With chromatographic materials consisting of membrane or gel structures

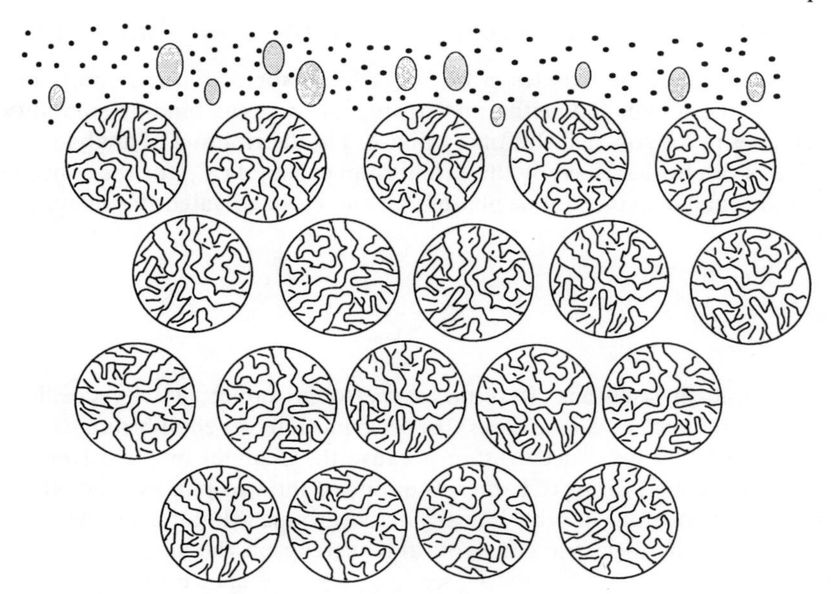

Figure 6.1 *Illustration of chromatographic material of porous particles with a layer of low-*M_r *and high-*M_r *analytes on top. The volume between the particles of the column material is the interstitial volume (*V_0*); the volume of mobile phase inside the particles is the inner volume (*V_i*); the volume of gel particles is the matrix volume (*V_g*)*

(paper, polysaccharides, polyacrylamide), high-M_r compounds will thus be retarded more than low-M_r compounds, and such systems are often inadequate for the separation of many high-M_r compounds. Porous particles of small sizes, such as those generally used in HPLC and FPLC columns (Chapter 8), may especially create problems when used for separating crude extracts containing very large high-M_r compounds, cell-membrane structures, and other impurities. It is therefore recommended that purification steps or group separations (Section 7.6) be included prior to use of such analytical procedures as the increased separation efficiency obtainable with chromatographic materials composed of high surface area, small particles, porous materials, and fine membrane structures only give optimal separations when pre-purified samples are used.

3 Surface Tension, Osmotic Pressure, Viscosity, Frictional Coefficient, and Diffusion

The properties of the surface layer of chromatographic materials are different from those of the mobile phase, although they are not independent, as also discussed for charged surfaces (Section 9.2). It is also evident that the effects of surfaces become greater when the surface area increases, *e.g.* with porous materials and when the particle sizes and capillary diameters reduces, and one

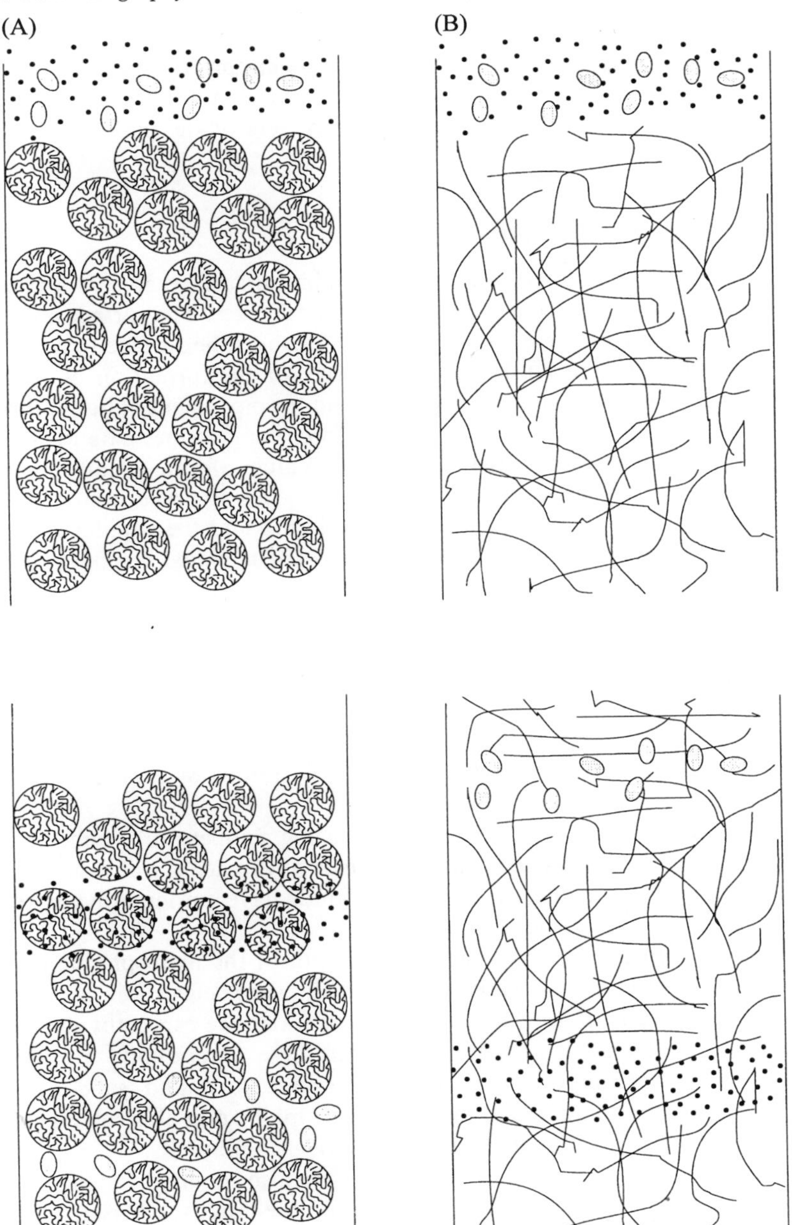

Figure 6.2 *Separations of large and small molecules in chromatographic materials consisting of porous particles* (A) *and membrane structures* (B), *when the separations only involve separations according to size of the molecules. The top part of each section illustrates the start of the separations, with small and large molecules in a mixture. The bottom sections illustrate separations obtained after the molecules have been allowed to move through part of the chromatographic systems*

of the most important results of this increased surface area is adsorption to the surface layer. The viscosity of the solution is as important in determining the behaviour of chromatographic systems, and macromolecules give increased viscosity which is a function of several parameters of the molecules and the frictional coefficient.

Surface Tension and Emulsions

A liquid surface tends to contract to the minimum surface area as a result of unbalanced forces of molecular attraction at the surface. The surface molecules are attracted by the centre of the liquid, which causes the surface to contract if it can. Thereby, surface tension is responsible for the formation of spherical droplets, for the rise of water in a capillary, and for movement of liquid through a porous solid. The surface tension (γ) is the force per centimetre on the surface, and opposes the expansion of the surface area. Values of γ are usually expressed in dyn cm^{-1}. In capillaries, the surface tension gives a curved miniscus which can be concave or convex depending on the liquid and the capillaries surface material, and is defined by the contact angle (θ). The actual height (h) (Figure 6.3) is defined by the upward forces ($2\pi r\gamma \cos \theta$) which at equilibrium is equal to the downward process ($\pi r^2 h\rho g$), resulting in:

$$\gamma = \frac{h\rho gr}{2\cos\theta} \tag{6.3}$$

where ρ = density of the liquid (g cm^{-3}), r = radius of capillary and g = acceleration due to gravity (9.8 m s^{-2}).

The surface tension decreases with increasing temperature, and capillary active compounds such as some aliphatic carboxylic acids will also reduce γ (Figure 6.3), and especially when the compounds have amphipathic or surfactant structures and properties (Section 2.9).

As seen in Figure 6.3(B), increasing the length of the hydrophobic parts of amphipathic molecules increases considerably their effect on the surface tension. Such effects on the surface tension of water are especially exerted by surfactants or detergents (Section 2.9) and these compounds are thereby '*capillary active*'. Compounds such as inorganic ions and salts of small hydrophilic organic compounds, *e.g.* some hydrophilic non-electrolytes such as glycerol and carbohydrates, are '*capillary inactive*' compounds.

Capillary active compounds or surfactants are also required for the production of emulsions, which are systems of immiscible liquids, where droplets of one of these liquids are dispersed in the other liquid. Stable emulsions will only be obtainable with use of surfactants which then are known as emulsifying agents. Such agents include, in addition to the capillary active compounds or surfactants mentioned above, proteins such as albumins, gelatins, and other lyophilic colloids. The liquid forming the drops is the dispersed liquid, and the other liquid is the dispersion medium. For water–oil emulsions, water is the dispersion medium when water mixes with the emulsion, and oil is the

(A)

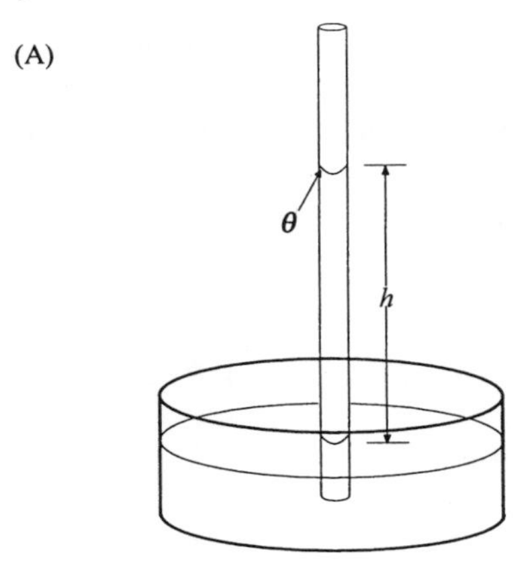

(B)

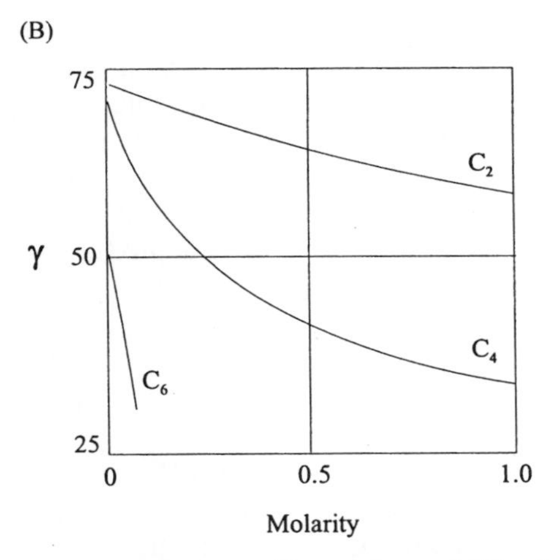

Figure 6.3 (A) *Rise of solvent in a capillary tube and* (B) *effects of surface active compounds on the surface tension (γ); C_2 = acetic acid, C_4 = butyric acid, C_6 = valeric acid*

dispersion medium when oil mixes with the emulsion. The ratio of the amounts of water and oil determines whether the result will be a water-in-oil or an oil-in-water emulsion. The latter is stabilized by sodium salts of anionic detergents whereas calcium and magnesium salts of these detergents stabilize water-in-oil emulsions, owing to the fact that sodium soaps are soluble both in water and in oil whereas calcium soaps only are soluble in oil. Emulsions will be destabilized or broken by the addition of capillary inactive compounds or by changes in the oil:water ratio, *e.g.* by evaporation.

Osmotic Pressure

When a solution is separated from the solvent by a semipermeable membrane in living cells, in dialysis (Section 4.4) and by ultrafiltration (Section 4.5), the solvent and low-M_r, but not high-M_r, compounds will flow through the membrane to the site with the lowest chemical potential (Section 2.3). This process is known as osmosis. The osmotic pressure (Π) is the pressure difference across the membrane required to prevent spontaneous flow in either direction across the membrane. For dilute solutions, an equation for Π analogous to the ideal gas law (Section 1.1) is obtained when using Raoult's law; $p_1 = x_1 p_1^0 = (1 - x_2)p_1^0$:

$$\Pi V = -n_1 RT \ln x_1 = -n_1 RT \ln(p_1^0/p_1) \tag{6.4}$$

and at sufficiently high dilution:

$$\Pi V = n_1 RT x_2 = n_2 RT \tag{6.5}$$

where V = volume of the solution, R = gas constant, T = temperature, p_1^0 = vapour pressure of the pure solvent, p_1 = vapour pressure of the solvent in solution, n_1 and n_2 = number of solvent and solute molecules, respectively, x_1 = the mol fraction of the solvent, and x_2 = the mol fraction of the solute. For non-ideal solutions, where macromolecules (*e.g.* proteins or nucleic acids) act as polyelectrolytes, a correction of Equation 6.5 is necessary. For a strong polyelectrolyte ($M_2X_z \rightarrow M_2^{z+} + zX^-$) in a mixture with a salt ($MeX \rightarrow Me^+ + X^-$), a virial coefficient is included in Equation 6.5, giving:

$$\Pi = RT\left[m_2 + \left(\frac{zm_2}{2}\right)^2 (mex)^{-1}\right] \tag{6.6}$$

after rearrangement of Equation 6.5 using $(n_2/V) = m_2$. Concentrations of M_2 (m_2) and MeX (mex) are in mol L^{-1}. An example with consideration of the osmotic pressure and Donnan equilibrium is given in Box 4.9, and all ion binding experiments may be corrected for these effects.

Viscosity, Frictional Coefficients, and Shape of Molecules

The viscosity (η) is a measure of the resistance a fluid offers to an applied shearing force. If liquids are thought of as a large number of layers, the frictional resistance is due to a transfer of momentum from one layer of liquid to the next layer. Each layer is thought of as sliding along the adjacent one, and the frictional resistance between the adjacent layers generates a velocity gradient, resulting in successively lower velocities and with the layer immediately adjacent to the stationary phase remaining stationary. With liquids flowing in capillaries, the result will be a laminar profile as shown in Figures 6.4 and 10.3 (right) whereas HPCE can give a flat flow profile.

The unit of viscosity is the poise (g cm^{-1} s^{-1}; the SI unit is Pa s), and it can be measured as the rate of settling of a sphere with radius (r) in a liquid following Stokes' law. Consequently, a sphere with frictional coefficient f, density ρ settling at the rate v, owing to the acceleration of gravity (g), in a liquid of

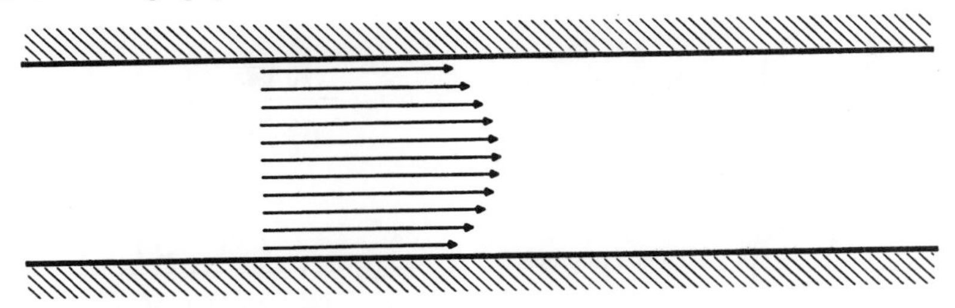

Figure 6.4 *Illustration of the shearing of a liquid in a capillary giving a velocity profile with a nearly stationary layer at the capillary surface*

density ρ_0, experiences at equilibrium a force due to gravity $[\frac{4}{3}\pi r^3(\rho - \rho_0)g]$ equal to the retarding force $[fv = 6\pi\eta rv]$:

$$v = \frac{2r^2(\rho - \rho_0)g}{9\eta} \tag{6.7}$$

$$f = 6\pi\eta r \tag{6.8}$$

The viscosity is an important factor in LC methods, especially when using chromatographic media with large surface area, porous particles (HPLC and FPLC procedures; Chapter 8), or capillaries (HPCE; Chapter 10). The viscosity increment of the solution is also a function of several parameters of the macromolecules. Each of these may be important, *e.g.* the volume of solution that is occupied, the solvation and the shape of molecules (Figure 6.5), and their adsorption properties.

When the viscosity changes with the flow velocity, the solution is called non-Newtonian. This is often encountered with colloidal systems. With increasing temperature, the viscosities of most liquids will decrease, whereas increased pressure will increase the viscosity. An increased viscosity leads to a decrease in both the mobility of sample molecules and, thereby, the separation efficiency in, for example, HPCE (pages 212 and 266).

Diffusion and Partial Specific Volume

Molecules move constantly, resulting in net flow of molecules from regions of high chemical potential (concentrations) to regions of low concentrations. This means that compounds separated by, for example, LC in sharp boundaries/bands or spots will move away from the high concentration area, resulting in a larger and more diffuse area. This diffusion will also occur during an LC procedure, if it is a technique other than those based on focusing (Sections 7.10 and 9.7). The driving force for diffusion is the chemical potential gradient, and the diffusion is defined in terms of the concentration gradient (dc/dx), where c is the concentration and x is the distance. Relating this to the flux J, which is the

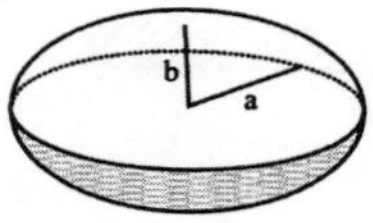

Oblate ellipsoids : $f_f = 6\pi r_o \eta \dfrac{\left(\frac{a^2}{b^2}-1\right)^{\frac{1}{2}}}{\left(\frac{a}{b}\right)^{\frac{2}{3}} \tan^{-1}\left(\frac{a^2}{b^2}-1\right)^{\frac{1}{2}}}$

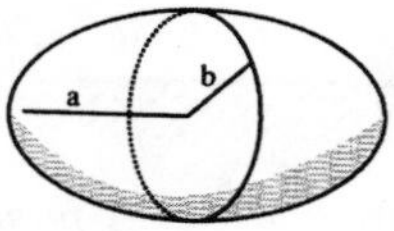

Prolate ellipsoids : $f_f = 6\pi r_p \eta \dfrac{\left(1-\frac{b^2}{a^2}\right)^{\frac{1}{2}}}{\left(\frac{b}{a}\right)^{\frac{2}{3}} \ln\left(\left(1+\left(1-\frac{b^2}{a^2}\right)^{\frac{1}{2}}\right)/\frac{b}{a}\right)}$

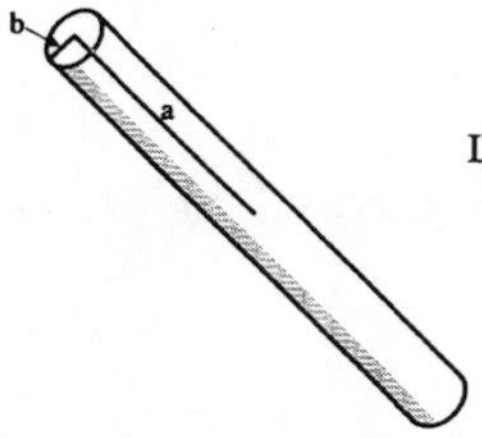

Long rods : $f_f = 6\pi r_r \eta \dfrac{\left(\frac{a}{b}\right)^{\frac{2}{3}}}{\left(\frac{3}{2}\right)^{\frac{1}{3}}\left(2\ln\left(2\left(\frac{a}{b}\right)\right)-0.11\right)}$

Where a is the half-length, b is the radius as shown, and r_r, r_o and r_p are the radii of spheres of equal volume.

Figure 6.5 *The variation in shape for high-*M_r *molecules can be divided into arbitrary groups: random coil, spherical molecules, and those shown in this figure, which give different effects on LC separations as seen for the modified equations for the frictional coefficient compared with* f = $6\pi\eta$r *for spherical molecules*

quantity of substance diffusing per second perpendicular to the direction of diffusion through an area of 1 cm^2, gives Fick's first law:

$$J = vc = -RT\left(\frac{\mathrm{d}c}{\mathrm{d}x}\right)(fN_0)^{-1} = -D\left(\frac{\mathrm{d}c}{\mathrm{d}x}\right) \tag{6.9}$$

$$D = \frac{RT}{N_0 f} \tag{6.10}$$

where D = diffusion coefficient, v = diffusion rate, c = concentration, R = gas constant, T = temperature, N_0 = Avogadro constant, and f = frictional coefficient.

Initially sharp peaks will, as a function of time, broaden as illustrated in Figure 6.6.

In theory, the measurement of D is simple, but in practice it is more difficult to obtain correct values of D for macromolecules as they are often small (10^{-8} to 10^{-5} cm^2 s^{-1}). It is also seen from Equation 6.10 that D is a function of the frictional coefficient and thereby of the shape of molecules (Figure 6.5). Taking shape and solvation of the high-M_r molecules into account, it is possible to obtain reasonable agreement between measured values and the M_r as seen in Table 6.1.

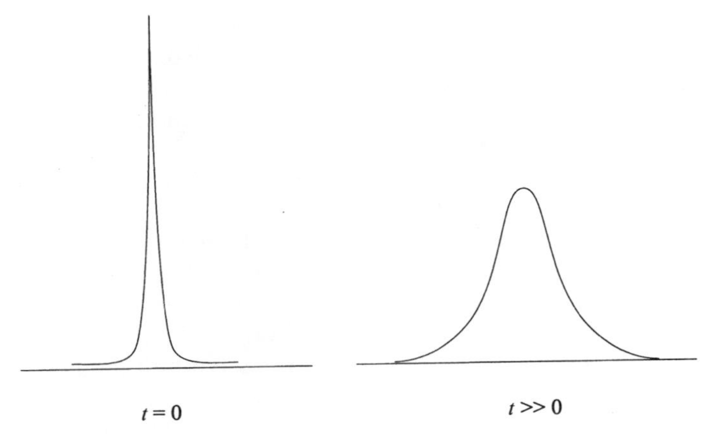

Figure 6.6 *Illustration of broadening of an initially sharp peak as a result of free diffusion as a function of time*

Table 6.1 *Diffusion coefficients (D), relative molecular mass (M_r), and partial specific volume ($\bar{v}$) for selected proteins* (1 atm, 20 °C)

Protein	10^{-3} M_r	$\bar{v}$ (cm^3 g^{-1})	10^7 $D_{20\,°C,\ water}$ (cm^2 s^{-1})
Trypsin inhibitor (BBI)	7.8	0.72	12.9
Cytochrome *c* (equine)	12.3	0.72	13.0
Ribonuclease	13.7	0.73	11.9
Lysozyme (hen egg-white)	14.1	0.69	10.4
Myoglobin (sperm whale)	17.8	0.75	11.3
Chymotrypsinogen	25.2	0.72	9.5
Trypsin (bovine)	23.8	0.73	9.3
β-Lactoglobulin	35.0	0.75	7.8
Carboxypeptidase A	35.0	0.73	9.2
Ovalbumin	45.0	0.75	7.8
Concanavalin A	51.3	0.73	6.3
Serum albumin	65.0	0.73	5.9
Haemoglobin	68.0	0.75	6.9
Lactate dehydrogenase	146.2	0.74	5.0
Catalase	250.0	0.73	4.1
Urease	480.0	0.73	3.5
Rod-shaped proteins			
Tropomyosin	93.0	0.71	2.2
Fibrinogen	330.0	0.71	2.0
Collagen	345.0	0.69	0.7
Myosin	493.0	0.73	1.2

The proteins that are more or less spherical give decreasing D for increasing M_r. Rod-shaped protein (Table 6.1) D values are quite different from those for the globular proteins. For low-M_r compounds, diffusion may be a serious problem for optimal separations, as discussed in the section on page 136.

The partial specific volume ($\bar{v}$) given for the proteins in Table 6.1 is the volume (V) increment produced in a solution when a unit mass (m) of solute is added to the solution; $\bar{v} = \mathrm{d}V/\mathrm{d}m$ in $\mathrm{cm^3\ g^{-1}}$.

The partial specific volume ($\bar{v}$) is sometimes, although not correctly, approximated to be the reciprocal of the density of the solute. The volume increment in solution differs from the volume of solid, and $\bar{v}$ is not an invariant parameter of a particular high-M_r compound. In addition, $\bar{v}$ varies with the solvent composition, *e.g.* the salt concentration, pH, and various types of modifiers (Section 2.8), as is the case for the stationary phases in LC (Section 6.2). The value of $\bar{v}$ varies with the proteins, and is in the 0.65–0.75 range (Table 6.1). For RNA it is about 0.51, and for DNA it is about 0.55.

4 Basic Concepts in Liquid Chromatography

Separation of analytes based on LC may be achieved using column chromatography as illustrated in connection with ion-exchange chromatography (Chapter 7; Figures 7.4 and 7.7) or by use of solid support media [paper- (PC) and thin-layer chromatography (TLC); next section] and as described in connection with zone electrophoresis (Section 9.3). A description of the performance obtainable with these LC techniques, including HPLC/FPLC (Chapter 8) and HPCE (Chapter 10), can be based on some basic concepts considered in this section.

The separation of two analytes (A and B) between the two LC phases, the solvated stationary phase and the mobile phase (Section 6.2), can thus be performed by use of columns or PC/TLC as illustrated in Figure 6.7. The separations based on PC are often obtained by the descending technique and those for TLC by the ascending technique.

When analytes separate in LC, the distribution in partition chromatography will in general be with different equilibrium concentrations in the two phases (Section 6.2), as shown for the analytes A and B in Figure 6.7. If the analytes distribute equally between the mobile (concentration C_m) and stationary phases (concentration C_s), the result will be a normal distribution (Figures 1.2 and 6.8, left). This distribution is different from situations with $K_d \neq 1$ (Section 6.2 and Figure 6.8), where the result then will be bands/peaks or spots (Figure 6.7) with fronting or tailing, reducing the separation efficiency, as also discussed in connection with HPCE (Figure 10.4; page 212).

The separation in Figure 6.8 (right) could be caused by an analyte which in addition to partition chromatography between the two phases is also partially adsorbed on to the stationary phase, or it could be an analyte that is only partly soluble in the mobile phase.

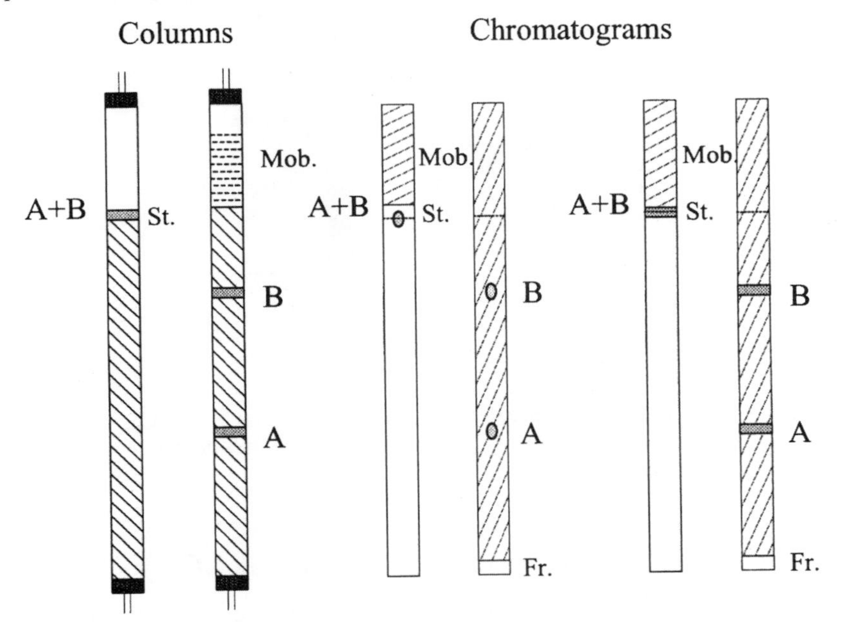

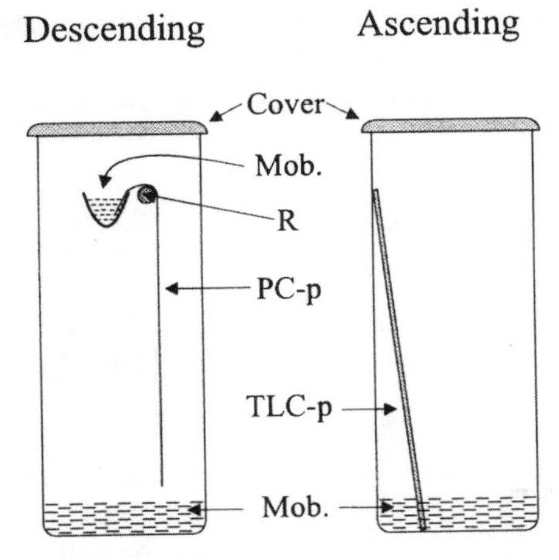

Figure 6.7 *Illustration of LC separation based on columns or chromatograms from PC or TLC, with separation of the analytes (A and B) in bands or spots. St. = starting line; Fr. = solvent front; Mob. = mobile phase, which in descending PC is also to saturate the local atmosphere; R = support rod: PC-p = paper; TLC-p = TLC plate*

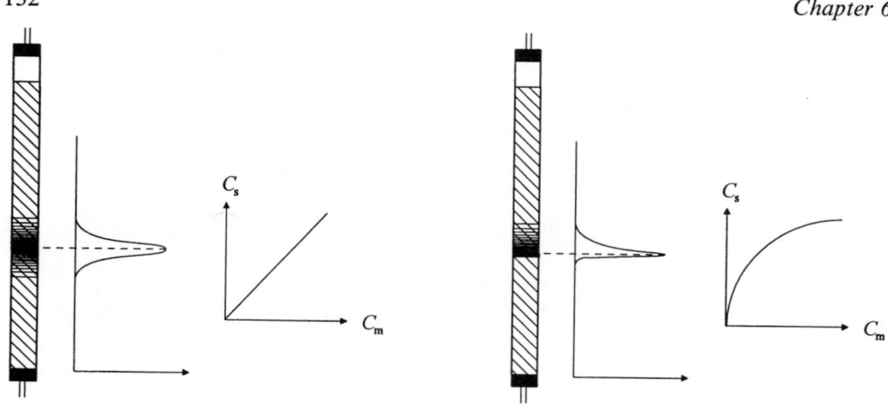

Figure 6.8 *LC-separation of analytes on column systems with* $K_d = C_s/C_m = 1$ *(left) and* $K_d > 1$ *(right),* C_s *and* C_m *represent concentrations in stationary and mobile phases, respectively*

Equations and Theory Describing Separations in LC

Descriptions of the separation in PC and TLC techniques are most often based on R_F-values as originally introduced in Ref. 1:

$$R_F = \frac{\text{distance travelled by the analyte}}{\text{distance travelled by the mobile phase}} \tag{6.11}$$

The R_F-values obtainable in PC and TLC techniques will vary more or less with temperature and composition of both the solvated stationary phase and the mobile phase. It is therefore advisable to use relative R_F-values, preferably based on a reference compound of the same type as the analytes and with R_F close to 0.5

$$\text{relative } R_F = \frac{R_F(\text{analyte})}{R_F(\text{reference compound})} \tag{6.12}$$

This is in agreement with recommendations for HPCE, where it is advantageous to use relative migration times (RMT) instead of migration times (MT) (page 218; Equation 10.8).

In column chromatographic techniques (HPLC/FPLC; Chapter 8), the time taken for each analyte peak to emerge from the column is referred to as the retention time (t_R). Under well-defined conditions, t_R will be a fairly characteristic figure of an analyte. However, it can be difficult to maintain constant conditions over long periods, resulting in variations in the time the analyte is in the mobile phase (t_m), in the stationary phase (t_s), and the total time, $t_R = t_m + t_s$ may also vary. An improved characterization could therefore be obtained by use of a correction obtained from analytes having $t_s = 0$ (Figure 6.9), and through the use of relative (t'_R) values corresponding to R_F and

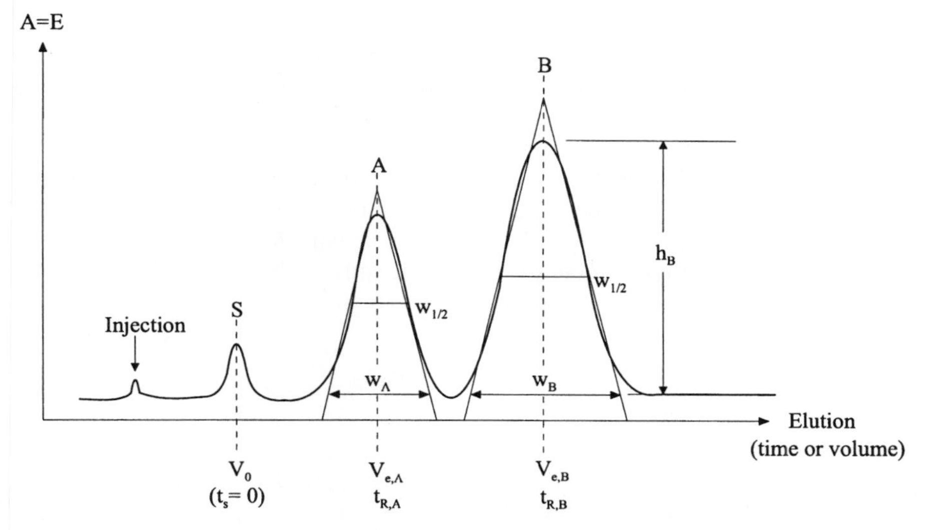

Figure 6.9 *Parameters for peaks in column chromatographic LC with elution measured as elution volumes* (V_0, $V_{e,A}$, $V_{e,B}$) *or retention times* ($t_{R,A}$, $t_{R,B}$) *based on detection (*E*) by, for example, UV.* $w_{1/2}$ *is the peak width at half peak height (peak height shown for analyte B =* h_B*)*

RMT (*vide supra*):

$$t'_R = t_R - t_{t_{s=0}} \tag{6.13}$$

The elution volume (V_e/ml) is often used in column chromatography, and with use of a linear flow rate (t_R/min) and known elution velocity (t_v/ml min^{-1}), the relation with t_R is:

$$V_e = t_R t_v \tag{6.14}$$

Analytes that are unable to penetrate into the stationary phase (Figure 6.2) and are found exclusively in the mobile phase ($t_s = 0$) will elute in the volume often denoted the 'solvent front' (S; Figure 6.9) at V_0.

Key parameters often used in connection with permeation or size-exclusion chromatography are:

$$\text{Retention coefficient; } R = \frac{V_0}{V_e} \tag{6.15}$$

$$K_d = \frac{V_e - V_0}{V_t - V_0 - V_g} = \frac{V_e - V_0}{V_i} \quad \text{(Section 6.2)} \tag{6.16}$$

$$K_{av} = \frac{V_e - V_0}{V_t - V_0} \tag{6.17}$$

where V_0 = interstitial volume (ml), V_e = elution volume (ml), V_t = total volume (ml), V_i = inner volume (ml), and V_g = matrix volume (ml). The terms K_d (K_d = diffusion coefficient) and K_{av} (K_{av} = approximate value of K_d) are differently defined equilibrium constants often used in connection with permeation or size-exclusion chromatography.

In this type of LC where molecules are excluded from the gel particles (Figure 6.2) $K_d = 0$. For small molecules able to utilize the total V_i, $K_d = 1$, and only if adsorptions are involved will analysis be possible with $K_d > 1$.

The capacity or retention factor (k') is a measure of the retention of analytes. For A and B in Figure 6.9, it is given by:

$$k'_A = \frac{V_{e,A} - V_0}{V_0} \quad \text{and} \quad k'_B = \frac{V_{e,B} - V_0}{V_0} \tag{6.18}$$

where k'_A and k'_B = the capacity/retention factor and $V_{e,A}$ and $V_{e,B}$ = elution volume (ml) for compounds A and B, respectively, and V_0 = interstitial volume (ml). For separation of A and B, the sample volume V_s should not exceed $V_s = V_{e,A} - V_{e,B}$. The values of k' are generally low in LC techniques such as size-exclusion chromatography, whereas ion-exchange chromatography can give high values of k'.

The ratio of capacity factors is known as the separation factor (α), shown here for compound A and B (Figure 6.9):

$$\alpha = \frac{k'_B}{k'_A} \tag{6.19}$$

The value of α specifies the relative separations between the maxima of the analytes. While α only describes the positions of A and B relative to one another, it may also be of value with descriptions of separations where the peak width is taken into account. Peak broadening is thus one of the most important concepts in LC.

The number of theoretical plates is a measure of peak and zone broadening and, therefore, also of the efficiency of separations, as described for HPCE (page 216; Equation 10.6). N can be calculated for each peak in LC (Figure 6.9) by:

$$N = \left(\frac{t_R}{\sigma}\right)^2 = 16\left(\frac{t_R}{w}\right)^2 = 5.54\left(\frac{t_R}{w_{1/2}}\right)^2 = 5.54\left(\frac{V_e}{w_{1/2}}\right)^2 \tag{6.20}$$

where t_R = retention time (time units), σ = standard deviation, w = peak width (time units), $w_{1/2}$ = peak width at half peak height (time units), and V_e = elution volume.

The position of the peak and its width or the variance σ^2 in the normal distribution function (Figure 1.2) thus determine the size of N, which is frequently stated as the number of theoretical plates per metre of chromatographic separation media. Large N/m indicate, that the performance of the system is good. Alternatively, the efficiency of the LC system can be expressed as

HETP (height equivalent to a theoretical plate), which is the length (L) of the separation media divided by the plate number (N):

$$\text{HETP} = \frac{L}{N} \tag{6.21}$$

Last, but not least, LC results can be evaluated by the resolution (R_s), which can be simply determined from the chromatogram (Figure 6.9), using the peak width $w = 1.699\ w_{1/2}$.

$$R_s = \frac{2(t_{R,B} - t_{R,A})}{w_A + w_B} = 1.177\frac{(t_{R,B} - t_{R,A})}{w_{A_{1/2}} + w_{B_{1/2}}} \tag{6.22}$$

where $t_{R,A}$ and $t_{R,B}$ = retention time (time units), w_A and w_B = peak width (time units), and $w_{A_{1/2}}$ and $w_{B_{1/2}}$ = peak width at half peak height (time units) for compound A and B, respectively.

The R_s is expressed in terms of three factors:

$$R_s = \text{(selectivity) (efficiency) (capacity)} = \left(\frac{\alpha - 1}{4\alpha}\right)(\sqrt{N})\left(\frac{k'}{1 + k'}\right) \tag{6.23}$$

where α = separation factor, N = number of theoretical plates, and k' = capacity/retention factor. These three factors, which constitute R_s, are the most important factors to control in LC. However, this form of R_s is an approximation that depends on a normal or Gaussian distribution function for the peaks A and B, as is the case for N (*vide supra*), and the two peaks (Figure 6.9) should be close together. Many other equations in LC theory also assume Gaussian peak shape, even if it is not the actual case. Furthermore, it is assumed that the forces underlying the separation of analytes follow a linear isotherm (Figure 6.8) and that elution is isocratic. Thereby it is important to realize that non-linear, non-ideal, and gradient elution systems do not follow simple mathematical models. For a practical approach, R_s = 1.0 corresponds to 98% and $R_s \geqslant$ 1.5 correspond to 99.7% purity of the peaks, provided they are Gaussian distributed.

Surface Area of the Stationary Phase and Flow Rate in Relation to LC Efficiency of Separations

The resolution and efficiency of LC separations depend on, among other parameters (*vide supra*), N, which again is related to the surface area of the stationary phase. Traditional column materials for LC (Section 6.6, Tables 7.1 and 7.2) often have particles with diameters $> 100\ \mu$m, which may be expressed as Mesh values where large figures correspond to small particles. In such systems it is common for N to lie in the 10–1000 m^{-1} range. In column materials for HPLC/FPLC, the particle diameters are often 3–10 μm, which may result in values of N in the 1–3 $\times$ 10^4 m^{-1} range, compared with values 10–35 times higher for HPCE systems. It may therefore seem that reducing particle size or increasing surface area would be the best way to improve LC performance, and the introduction of HPLC/FPLC clearly improved resolution compared with what was achievable with other traditional LC column media. Limitations are, however, met when the particle diameter is reduced as the backpressure of packed columns increases, which may lead to a mechanical problem with the system. The pressure drop [P (atm 10^6)] in columns is a function of viscosity (η; page 126), linear velocity [v (cm s^{-1})], column length [l (cm)], and permeability [k^o (cm^2)] which includes particle diameter [d (cm)]; $k^o \cong d^2 \times 10^3$:

$$P = \eta v l (k^o)^{-1} \tag{6.24}$$

With respect to the flow rate through the columns, this can be expressed as:

Volumetric flow velocity (V_v) = volume time^{-1}, *e.g.* ml min^{-1}

Linear flow rate = V_v (cross-section area of column)$^{-1}$, *e.g.* ml (min cm^2)$^{-1}$

Soft column materials, *e.g.* Sephadex G-150 and G-200, often create serious limitations with flow through columns, whereas newer porous column materials, *e.g.* Poros media and Mono Beads (Tables 7.1 and 7.2), give several advantages, including higher efficiency. HPLC/FPLC used with such materials gives the basis for higher speed and resolution compared with conventional LC, but the capacity is limited, making it more suited to analytical purpose than to preparative LC. HPLC/FPLC methods are normally not recommended for use with crude extracts without previous group separations (Figure 7.4) and/or other purifications.

5 Paper and Thin-layer Chromatography

Paper (PC) and thin-layer chromatography (TLC) are experimentally very simple techniques, as revealed in Figure 6.7, and these techniques only require limited resources. Both PC and TLC can be performed as one- and two-dimensional LC (Figure 6.10), with evaluation of results as described above. A comprehensive treatment of the techniques and a tremendous number of detection methods for the different types of functional groups in biomolecules are described in Refs 2–4.

TLC is most often performed with a cellulose or silica matrix on glass, plastic, or metal foil plates, which are commercially available. The matrix on the TLC plates consists of fine particles, which leads to the possibility of relatively high resolution compared with PC. Both PC and HVE (page 183) are, however, LC techniques that are well suited to qualitative analyses of hydrophilic compounds. These techniques also have other advantages as they only show limited sensitivity to impurities, which often cause severe problems when TLC and especially HPLC/FPLC are the methods of choice. This means that PC and HVE can be used with crude extracts, and give information on the native

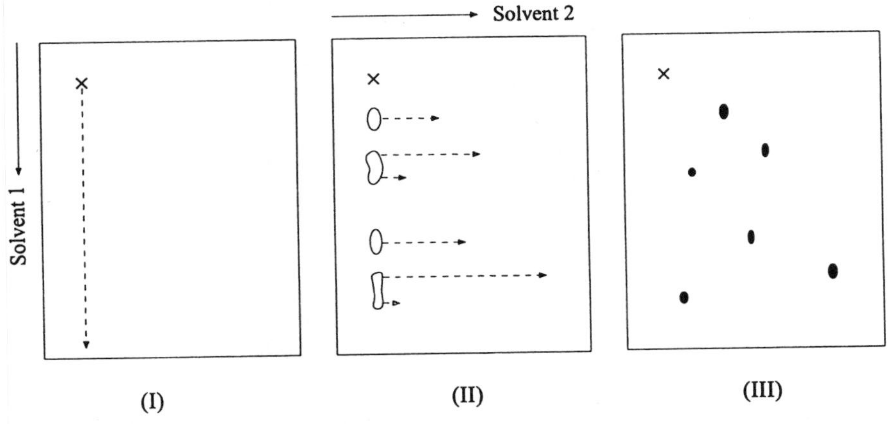

Figure 6.10 *Illustration of one- and two-dimensional PC and TLC. The sample mixture is applied (x) and after separation in one dimension with solvent 1* (I) *it can be used for separation in the second direction with solvent 2* (II), *giving results as shown in* (III)

natural products or constituents in crude extracts, helping to reveal possible loss of material and the risk of artefact formation during purification prodedures. For the separation of lipophilic or amphiphilic natural products and food constituents, *e.g.* phospholipids, lipid soluble vitamins, and steroid glycosides, TLC generally has advantages over PC. Otherwise, the mobile phases and relative separations of hydrophilic low-M_r compounds are often more or less identical for PC and TLC. Both techniques may give advantages compared with column chromatographic LC (*e.g.* HPLC) as a large number of samples can be studied simultaneously. Furthermore, both PC and TLC are easy to use for preparative isolation of reference compounds which may then be identified by, for example, MS, ^{1}H NMR and UV-vis spectroscopy (Chapter 5).

Detection of compounds separated by TLC and PC/HVE can be performed by inspection in the UV-vis region if they have appropriate chromophore systems (page 104). Alternatively, pre-derivatization can be performed, resulting in specific colours (Table 5.6). A more comprehensive collection of well-described and tested methods is presented in Refs 2–4. Some selected procedures for PC and TLC that are applicable to various low-M_r biomolecules are described in Box 6.1.

Results obtained by two-dimensional PC of amino acids using solvent (i) and solvent (ii) as described in Box 6.1 are shown in Figure 6.11.

The amino acids separate in PC according to their hydrophilic–lipophilic properties as seen for the homologous series of amino acids. This is the case both for protein amino acids and non-protein amino acids. Results from more comprehensive studies based on PC, HVE, and TLC are described elsewhere for amino acids, peptides,[5] biogenic amines,[6] glucosinolates,[7] aromatic choline esters,[8] carboxylic acids,[9] and flavonoids.[10] Results obtained with selected low-M_r carbohydrates are shown in Table 6.2.

Box 6.1 *Selected solvents and detection procedures for PC and TLC of some low-*M_r *biomolecules*

Use Whatman No. 1 paper 46 × 57 cm for analytical PC and No. 3 for preparative PC.

1. Fold the paper and indicate the starting line or starting spot in accordance with the chromatographic tank available (Figure 6.7), *e.g.* folding as indicated in this illustration:

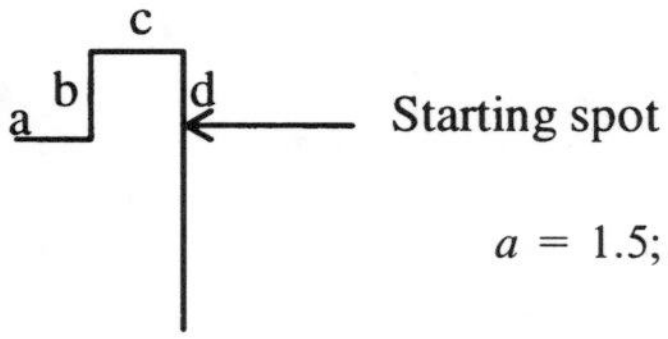

c
a b d ← Starting spot

$a = 1.5; b = 2.5; c = 2; d = 2$ cm

2. For PC, use descending technique as shown in Figure 6.7, and for one-dimensional PC, for example, one of the solvents:
   (i) n-butanol: acetic acid: deionized water (12:3:5)
   (ii) phenol: deionized water: concentrated NH_3–H_2O (120:30:1)
   (iii) isopropanol: deionized water: concentrated NH_3–H_2O (8:1:1)
   TLC of carbohydrates (Silica gel 60 F254, Merck), *e.g.* with solvent:
   (iv) ethyl acetate: EtOH (96%): acetic acid: boric acid solution (cold saturated) (50:20:10:10)
   TLC on cellulose DC-Alufolien (Merck) with the solvent:
   (v) pyridine: ethyl acetate: acetic acid: deionized water (9:9:2:5)
   TLC (silica gel 60) of phospholipids and steroid glycosides (saponins), *e.g.* with solvents:
   (vi) chloroform: MeOH: acetic acid: deionized water (59:29:8.4:3.6)
   (vii) chloroform: MeOH: conc. NH_3–H_2O (68:2:5)
   (viii) n-butanol: EtOH: conc. NH_3–H_2O (7:2:5)
3. For two-dimensional PC, use solvent (i) in the first dimension (Figure 6.10) and either (ii) or (iii) in the second dimension
4. After drying through total removal of solvent, the chromatograms are first examined in UV (shortwave and longwave) followed by detection of specific compounds with selected reagents and procedures
5. Detection of amino acids, peptides, and amines with ninhydrin as described in Box 12.7
6. Detection of carbohydrates. Aniline–phthalate for reducing low-M_r carbohydrates. The chromatogram is sprayed with a solution of aniline (0.93 g) + phthalic acid (1.66 g) in water-saturated n-butanol (100 ml). Heat the chromatogram at 105 °C for 5–10 min, resulting in brown spots for the carbohydrates
7. Detection of reducing carbohydrates by aniline–diphenylamine–H_3PO_4. The chromatogram is sprayed with a solution of aniline (1 ml) + diphenylamine (1 g) + H_3PO_4 (10 ml) in 100 ml acetone. Heat the chromatogram at max 80 °C for *ca.* 5 min
8. Detection of carbonyl compounds by *o*-phenylenediamine. The chromatogram is sprayed with a freshly prepared solution of 0.05% *o*-phenylenediamine in 10% CCl_3–CO_2H in deionized water. Heat the chromatogram for 2 min at 100 °C, and examine the chromatogram under UV light
9. Detection of both reducing and non-reducing carbohydrates with thymol–sulfuric acid. The chromatogram (only TLC on silica) is sprayed with or immersed in a freshly prepared solution of thymol (0.5 g) in 96% EtOH (95 ml) with continuous addition of concentrated sulfuric acid (5 ml). The chromatogram is then heated to 110–115 °C for 5–20 min *(continued opposite)*

10. Detection of carbohydrates/glycosides by periodate oxidation. The chromatogram is sprayed with 10 mM KIO_4, dried for *ca.* 10 min at room temperature, followed by spraying with solution A. Solution A is a saturated borax ($Na_2B_4O_7$) solution diluted to 35% and then 0.8% KI, 3% starch, and boric acid are added to pH 7 (*ca.* 0.9%). Oxidized compounds give white spots on a blue background.
11. Detection of carboxylic acids by aniline–xylose. The chromatogram is immersed (sprayed) in a solution of xylose (1 g in 3 ml deionized water) + aniline (1 ml) diluted to 100 ml with MeOH. After drying, heat the chromatogram at 105–110 °C in 5–10 min
12. Detection of glucosinolates. The chromatogram is immersed in a solution of $AgNO_3$ (1 g) + conc. ammonia (25 ml) diluted to 1 L with MeOH. Heating the chromatogram at 120 °C results in grey/black spots for glucosinolates. If needed repeat the procedure and finally immerse the chromatogram in 0.4 M HNO_3 and wash carefully with deionized water
13. Detection of alkaloids and choline esters with Dragendorff's reagent. The chromatogram is sprayed with solution A (2 ml) + acetic acid (4 ml) + deionized water (40 ml). Solution A contains alkaline bismuth nitrate [$Bi(OH)_2NO_3$; 0.85 g] dissolved in acetic acid (10 ml) and deionized water (40 ml) which then is mixed with an equal volume of KI (8 g) dissolved in deionized water (50 ml)
14. Detection of guanido groups with Sakaguchi reagent. The chromatogram is immersed in a solution of 8-hydroxychinoline (0.1%) in acetone, air-dried in 5 min, and then sprayed with 0.3 ml Br_2 in 25 ml 2 M NaOH + 75 ml deionized water
15. Detection of lipids, phospholipids, and fatty acids. The chromatogram (only TLC on silica) is sprayed with a solution of ammonium sulfate (5 g) dissolved in EtOH–H_2O (1:1; 100 ml). After air-drying, heat at 175 °C for 45–60 min
16. Detection of steroid glycosides and saponins with Libermann–Burchard reagent. The chromatogram is sprayed with a freshly prepared solution of sulfuric acid:acetic acid anhydride:chloroform (1:20:50). Heat at 110 °C for 10–15 min and then view the chromatogram in longwave UV light

Table 6.2 R_f *values obtained by PC as described in Box 6.1*

	Solvent	
Carbohydrates	*(i)*	*(ii)*
Glucose	0.18	0.39
Galactose	0.16	0.44
Mannose	0.20	0.45
Fructose	0.23	0.51
Xylose	0.28	0.44
Arabinose	0.21	0.44
Ribose	0.31	0.59
Rhamnose	0.37	0.59
Lactose	0.09	0.38
Maltose	0.11	0.36
Ascorbic acid	0.38	0.24

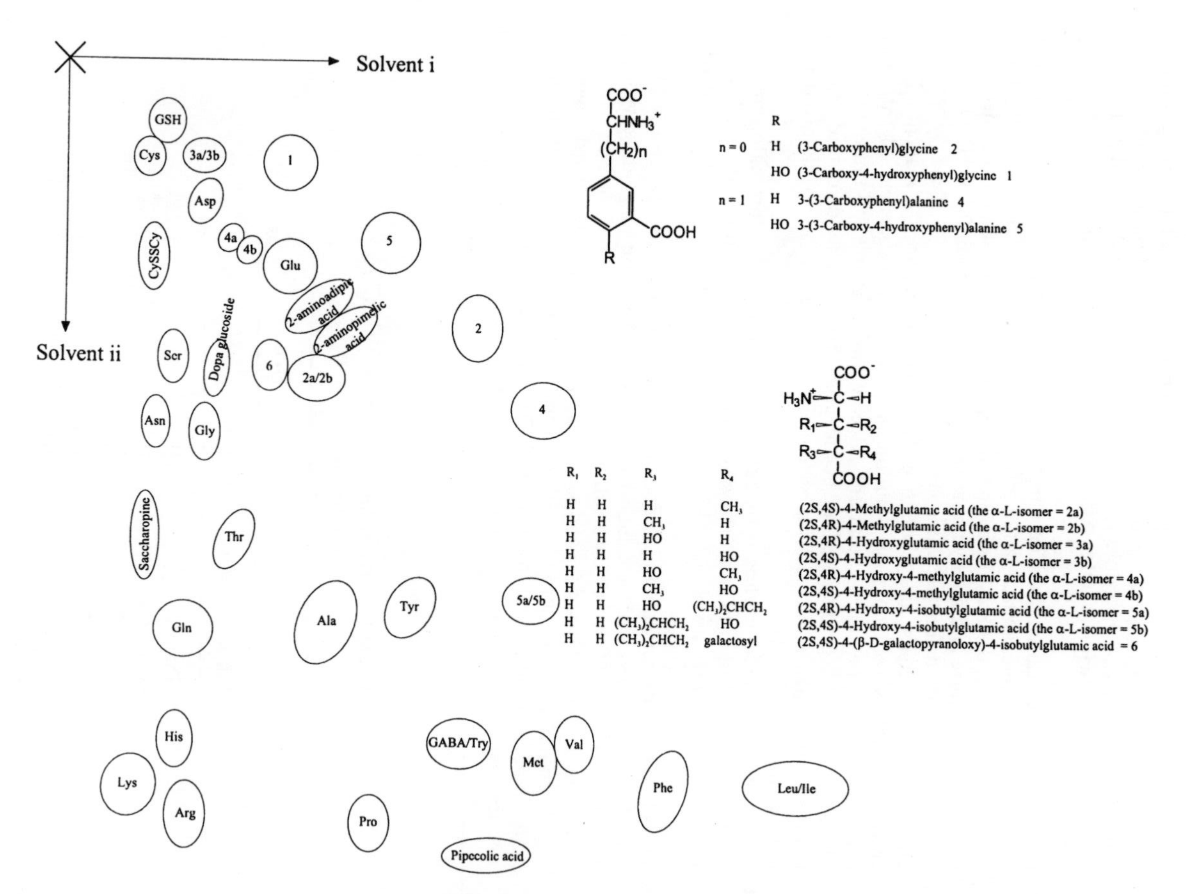

Figure 6.11 *Two-dimensional separation of amino acids using the procedure described in Box 6.1 using solvents (i) and (ii)*

Figure 6.12 shows the results obtained with PC as described in Box 6.1 for the crude extract used for HVE shown in Figure 9.4.

The hydrophilic glucosinolates[11] 20, 18, epi-18 and 17 (Figure 6.12) have much smaller R_f-values than aromatic choline esters [*t*-I, *t*-C, and (1)] and aromatic carboxylates (2). As expected, the glucosinolates with an isoferuloyl

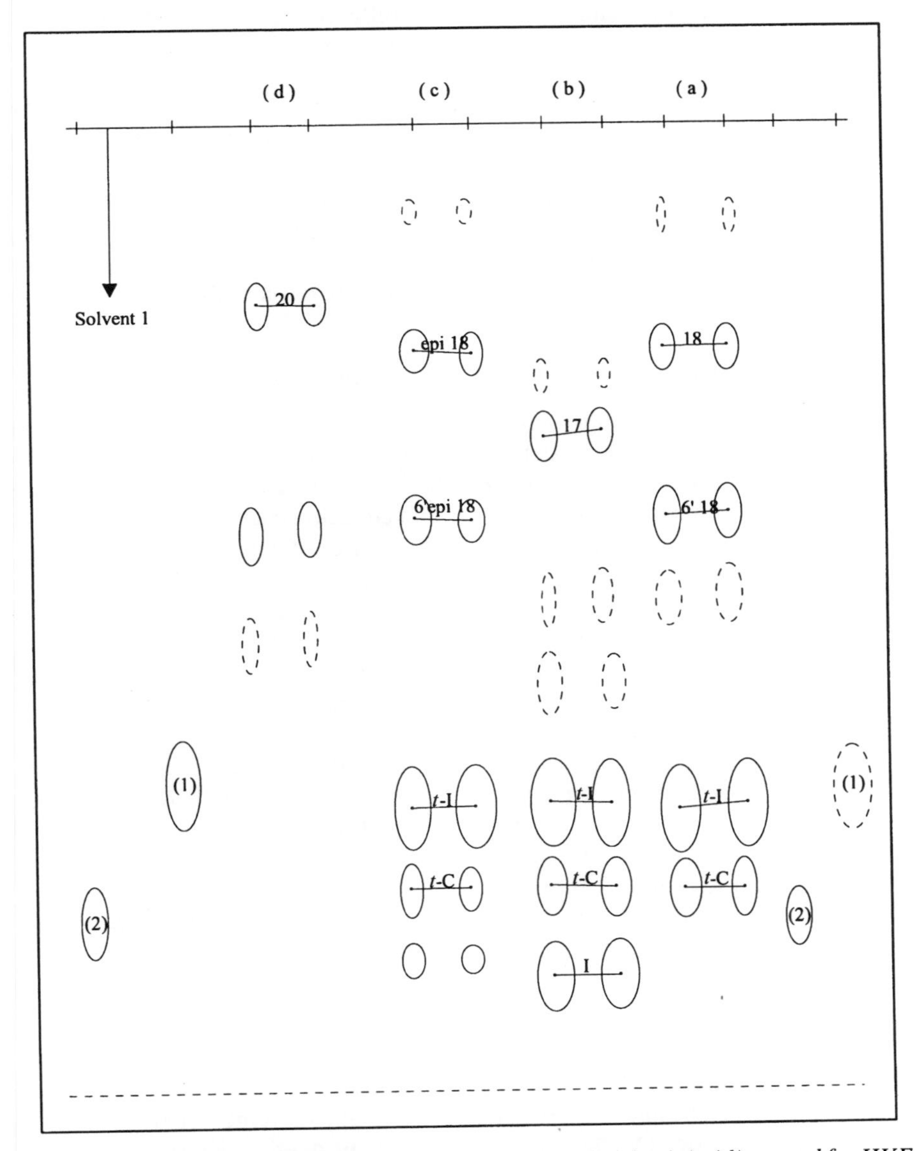

Figure 6.12 *PC in solvent 1 (Box 6.1) of crude extract samples (a)–(d) as used for HVE (Figure 9.4). The compounds are ones with isoferulic chromophores (I), which give quite different UV colours than seen for compounds with coumaric (C) and sinapic chromophores [(1) and (2)]*

substituted thioglucose part (6′-epi-18 and 6′-18) are well separated from the unsubstituted glucosinolates.

6 Permeation- or Size-exclusion Chromatography

LC separations of molecules according to differences in their size and shape (Figure 6.5) occur when they pass through a chromatographic material of porous particles (Figures 6.1 and 6.2). These techniques are known as permeation or size-exclusion chromatography or gel filtration. These techniques can be performed as TLC but most often column chromatography is the method of choice, *e.g.* as FPLC (Chapter 8 and 12). Column chromatography represents simple and gentle methods of analysis often used for determination of M_r or size of, in particular, proteins, for desalting or separation of low-M_r and high-M_r compounds (Figure 6.2). Furthermore, it is a well-suited method for initial purification prior to further isolation of enzymes (Chapter 12).

Equations and Key Parameters used in Gel Filtration Techniques

The ability of gel filtration to separate molecules on the basis of their size and shape is described by use of defined expressions. Illustrations of the experimental basis are shown in Figures 6.2 and 6.9, with an indication of elution volume and definition of key parameters and equations described in Section 6.2 and on page 133. In practice it is found that a series of compounds with similar molecular shape, f, $\bar{v}$, and D (pages 126–130) give a sigmoidal relationship between K_{av} and ln M_r. Calibration curves constructed in this way for a particular gel type (next subsection) are known as selectivity curves and a conveniently linear relationship between K_{av} and ln M_r is found for a certain M_r range (Figure 8.4). In gel filtration, where the separations only depend on the size and shape of the analytes, no molecules will be eluted with $K_{av} > 1$ or $K_{av} < 0$ (*vide supra*). For molecules with properties resulting in binding or adsorption to the stationary phase, separation is possible with $K_{av} > 1$. This can thus give more or less specific purifications, *e.g.* with glycoproteins (myrosinases and lectins) as well as flavonoids and pyrimidine glycosides. Deviation from K_{av}–ln M_r calibration curves may thus have important applications in the separations of various low-M_r biomolecules, *e.g.* by use of Sephadex G-10, G-15, and G-25 (coarse, medium, fine, or super fine).

Gel Filtration Media

Sephadex is bead-formed gel material prepared by cross-linking dextran with epichlorohydrin. The linear chains of α-(1→6)-D-glucose units are thereby more or less interlinked by $-O-CH_2-CHOH-CH_2-O-$ bridges, which thus determine the properties of the gel materials. A high degree of cross-linking, small G numbers, gives limited swelling as shown in Table 6.3, but the degree of swelling will also depend on the ionic strength, pH, and amount of organic modifiers.

Table 6.3 *Properties of Sephadex gel filtration media*

Product	*Bead size (dry)* (μm)	*Water regain* (ml g^{-1} $\pm$ 10%)	*Packed bed volume* (ml g^{-1} $\pm$ 10%)	*Fractionation range (approximate* M_r*)*		*Hydration time* (h)	
				Globular molecules	*Linear molecules*	*20 °C*	*100 °C*
Sephadex G10	4–120	1.0	2–3	<700	<700	3	1
Sephadex G15	40–120	1.5	2.5–3.5	<1500	<1500	3	1
Sephadex G25				100–5000			
Superfine	10–40	2.5	4–6		100–5000	6	2
Fine	20–80						
Medium	50–150						
Coarse	100–300						
Sephadex G50							
Superfine	10–40	5.0	9–11	1500–30000	500–10000	6	2
Fine	20–80						
Medium	50–150						
Coarse	100–300						
Sephadex G75							
Superfine	10–40						
Regular	40–120	7.5	12–15	3000–70000	1000–50000	24	3
Sephadex G100							
Superfine	10–40						
Regular	40–120	10.0	15–20	4000–150000	1000–100000	48	5
Sephadex G150							
Superfine	10–40		18–20				
Regular	40–120	15.0	20–30	5000–400000	1000–150000	72	5
Sephadex G200							
Superfine	10–40		20–25				
Regular	40–120	20.0	30–40	5000–800000	1000–200000	72	5

Sephadex LH-20, Sephacryl HR, and Sepharose CL are recommended for gel filtration in organic solvents.

Sephacryl HR is a composite gel consisting of dextran allyl and *N,N'*-methylene-bis-acrylamide (Figure 9.5). These high-resolution Sephacryl HR products (Table 6.4) are supplied pre-swollen and they can be used with mobile phases containing organic solvents, surfactants, *e.g.* 1% SDS, and chaotropic salts, *e.g.* 8 M urea or 6 M guanidine hydrochloride. Strong NaOH solutions (0.5 M) can be used to clean the gels if they are immediately washed after use with water and buffer. Sephacryl columns can also be used in HPLC or FPLC systems for LC.

Superdex of different product numbers are composite gels based on highly cross-linked agarose + dextran beads, suitable for fractionation of proteins with different M_r by use of relatively high flow rates. Superose consists of porous cross-linked agarose beads which in pre-packed HR 10/30 columns for FPLC have relatively high efficiency, with N/m of *ca.* 30000. Sepharose is another bead-formed agarose gel where most of the charged agarose present in naturally occurring agar has been removed (Table 6.5).

Bio-Gel A media is cross-linked agarose whereas Bio-Gel P media is based on cross-linked polyacrylamide and has the properties shown in Table 6.6.

Preparation of Gel Filtration Columns

Sephacryl, Superdex, Superose, and Sepharose are pre-swollen gel suspensions that are ready to use but contain 20% ethanol as preservative. The gels need, however, to be diluted with eluent before they are suitable for pouring into the selected column. Sephadex materials (Table 6.3) and Bio-Gel-P (Table 6.6) are powders which need to be swollen in the required amount of eluent for swelling times as indicated in the tables. When the gels are swollen and diluted they should be degassed and packed in the columns as described for ion-exchange materials (Section 7.8).

7 Affinity Chromatrography

Affinity chromatography is a special LC method of analysis based on selective binding of analytes to ligands (Section 3.3) immobilized on an insoluble matrix, often one of the gel filtration media mentioned above. The binding forces involved need to be of a size that allow the process to be an equilibrium, *e.g.* K_f is to be in the region 10^4–10^7 M^{-1} (Section 3.2), allowing eluent molecules to release the analytes from the ligand binding. Affinity chromatography thus exploits specific and reversible binding of analytes to the column material, where the specificities derive from the interacting molecules and correspond to the specificities of enzymes toward their substrates (Section 3.3). For this reason such methods of analyses can be used for purification of molecules from complex biological mixtures. The principle of these methods is illustrated in Figure 6.13.

Table 6.4 *Properties of Sephacryl gel filtration media*

Product	*Bead size (wet) (μm)*	*Fractionation range (approximate 10^{-3} M_r)*			*Bed volume in % of volume in water*		
		Globular proteins	*Dextran*	*DMSO*	*Methanol*	*Ethanol*	*Acetone*
Sephacryl S100 HR	25–75	1–100	–	100	100	100	85
Sephacryl S200 HR	25–75	5–250	1–80	110	100	100	85
Sephacryl S300 HR	25–75	10–1500	2–400	90	100	95	85
Sephacryl S400 HR	25–75	20–8000	10–2000	90	100	95	85
Sephacryl S500 HR	25–75	–	40–20000	95	100	100	90

Table 6.5 *Properties of Sepharose, Superose, and Superdex media*

Product	*Bead size (wet)* (μm)	*Fractionation range, approximate* $10^{-3}M_r$	
		Globular proteins	*Dextrans*
Sepharose 6B	45–165	10–4000	10–1000
Sepharose 4B	45–165	60–20000	30–5000
Sepharose 2B	60–200	70–40000	100–20000
Superose 12	8–12	1–300	–
Superose 6	11–15	5–5000	–
Superdex 30	22–44	1–10	–
Superdex 75	11–15	3–70	5–30
Superdex 200	11–15	10–600	1–100

Table 6.6 *Properties of Bio-Gel P gel filtration media*

Product	*Bead size (wet)* (μm)	*Packed bed volume* (ml g^{-1})	*Fractionation range (approximate* M_r*)*	*Hydration time* (h) *20* °C
Bio-Gel P-2 gel, fine	45–90	3	100–1800	4
Bio-Gel P-2 gel, extra fine	<45			
Bio-Gel P-4 gel, medium	90–180	4	800–4000	4
Bio-Gel P-4 gel, fine	45–90			
Bio-Gel P-4 gel, extra fine	<45			
Bio-Gel P-6 gel, medium	90–180	6.5	1000–6000	4
Bio-Gel P-6 gel, fine	45–90			
Bio-Gel P-6 gel, extra fine	<45			
Bio-Gel P-6DG gel	90–180			
Bio-Gel P-10 gel, medium	90–180	7.5	1500–20 000	4
Bio-Gel P-10 gel, fine	45–90			
Bio-Gel P-30 gel, medium	90–180	9	2500–40 000	4
Bio-Gel P-30 gel, fine	45–90			
Bio-Gel P-30 gel, medium	90–180	11	3000–60 000	4
Bio-Gel P-60 gel, fine	45–90			
Bio-Gel P-100 gel, medium	90–180	12	5000–10 000	4
Bio-Gel P-100 gel, fine	45–90			

Matrix, Spacer Arms, and Ligands

The type of matrix recommendable for affinity chromatography should be without appreciable interaction with the analytes considered. In addition, it should exhibit acceptable flow properties, contain suitable functional groups for ligand binding, and be sufficiently stable during the processes required for covalent coupling of spacer arms and ligands as well as during the binding–release processes required. The media mentioned above, *e.g.* Sepharose, are often used as acceptable matrices.

The selection of ligands for affinity chromatography is primarily determined

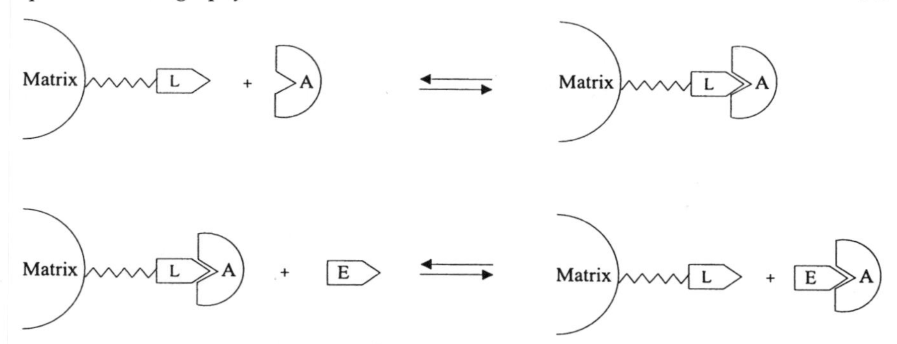

Figure 6.13 *Principle of affinity chromatography with a ligand (L) covalently bound to the matrix through a 'spacer arm'. The analyte (A) has an active site complementary to the ligand and to eluent molecules (E)*

by the ligand's ability to exhibit specific and reversible binding to the analytes (Section 6.7). Furthermore, it is necessary that the ligands contain functional groups that allow the ligand to be covalently bound to the matrix, *e.g.* through a spacer arm. Spacer arms may be needed to reduce unwanted interference between the matrix and ligand/analytes, and to facilitate effective binding, but too long a spacer arm may reduce the selectivity of the separation. Table 6.7 shows some selected examples of ligands and their specificity.

Applications and Practical Procedures

A great variation of ligands has been used for affinity chromatography; a few examples are shown in Table 6.7. The technique is limited only by the

Table 6.7 *Affinity chromatography ligands and their specificity*

Ligand	*Specificity*
Con A (concanavalin A)	Terminal α-D-glucopyranosyl, α-D-mannopyranosyl, or sterically similar residues
Soybean lectin	Glycoproteins containing *N*-acetyl-α- or β-galactopyranosyl residues
Phenylboronate	Glycoproteins
Helix pomatia lectin	*N*-acetyl-α-D-galactosaminyl residues
Lentil lectin	Similar to that of Con A, but lower binding affinity for simple sugars
Protein A	Fc region of IgG and related molecules
Protein G	Similar to that of protein A, but different affinities for IgGs from different species
Heparin	Lipoproteins, lipases, coagulation factors, steroid receptor proteins
Avidin	Biotin-containing enzymes
Cibacron blue	Broad range of enzymes which have NAD^+, $NADP^+$, nucleotide cofactors; serum albumin, *etc.*

availability of immobilized ligands and a great collection of coupling gels are commercially available together with descriptions of coupling procedures.

Practical procedures for efficient utilization of affinity chromatography require knowledge of binding equilibria (Chapter 3) and techniques such as those used in other forms of LC (above and Section 7.8). Once the samples have been applied to the columns, unbound compounds are eluted with buffer, and then either specific or non-specific elution can be used. Non-specific elution may be achieved by changes of pH and/or ionic-strength of the eluents. These changes can be sufficient to change the structure around the binding sites, in such a way that it gives a shift in the equilibrium towards the release of analytes from the ligand. Specific affinity elution requires eluents that are able to bind either to the ligand or to the active site of the analyte bound to the ligand. Examples of such practical procedures are described in Chapter 12 in connection with procedures used for the purification of myrosinases and α-galactosidase, and in the section on page 377 in connection with purification of antibodies.

8 Hydrophobic Interaction Chromatography

Hydrophobic interaction chromatography (HIC) is a LC technique that utilizes an immobile matrix with attached hydrophobic ligands which are able to bind to hydrophobic areas of proteins. The matrices most often used are polysaccharides such as Bio-Gel, Sepharose or Superose (Table 6.5) and the immobilized ligands may be alkyl or aryl groups (Figure 6.14).

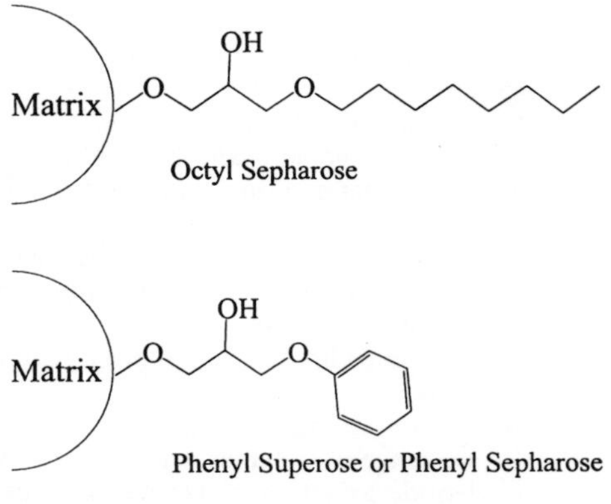

Matrix
OH
O
O
Butyl Sepharose

Matrix
OH
O
O
Octyl Sepharose

Matrix
OH
O
O
Phenyl Superose or Phenyl Sepharose

Figure 6.14 *Selected examples of ligands used in HIC*

The basis for HIC is hydrophobic interactions between the stationary phase and analytes in the mobile phase, as is the case for reversed-phase chromatography (RPC). The differences between HIC and RPC are mainly a question of the degree of substitution on the ligands of the stationary phase; RPC materials have a much higher degree of substitution than HIC media. This implies that proteins with hydrophobic areas on their surface are much more strongly bound to RPC than to HIC column materials. Elution of such strongly bound proteins from RPC columns requires often appreciable concentrations of organic solvents in the eluents, which may lead to denaturation. RPC is, therefore, normally not the method of choice for proteins of high-M_r, but is often chosen for amino acids, peptides and proteins of low-M_r.

In HIC, the type of immobilized ligand and the degree of substitution are the main parameters which primarily determine the protein adsorption or binding to the stationary phase. Also important for the binding of proteins are parameters which affect the protein structure, *e.g.* pH, temperature, additives, and especially type and concentration of salt. The 'salting-out' type of salts like $(NH_4)_2SO_4$ (page 81) promotes the protein–ligand binding in HIC. When bound to the HIC column, elution of the proteins is performed with a gradient over which the concentration of $(NH_4)_2SO_4$ decreases. HIC is, therefore, suitable as an early purification step for proteins which can be precipitated at moderate to high concentrations of $(NH_4)_2SO_4$, if the proteins are able to renaturate when the $(NH_4)_2SO_4$ concentration is reduced. HIC requires a minimum of sample preparation work and can be used as the purification step following $(NH_4)_2SO_4$ precipitation and before gel filtration.

9 Solid Phase Extractions and Reversed Phase Chromatography

Solid phase extraction (SPE) is an efficient and popular LC method, which appears to be of great value in purification procedures (page 74) used prior to high-resolution techniques such as HPLC, FPLC (Chapter 8), and HPCE (Chapter 10). The separation mechanisms in SPE are identical to those known from RPC, and in fact the same type of materials are used in SPE and in RPC. A RPC/SPE matrix consists of hydrophobic ligands grafted on to a porous, insoluble, beaded material. The matrix must both be chemically and mechanically stable and it is generally composed of silica or synthetic organic polymers, *e.g.* polystyrene or other products, as shown in Tables 6.1 and 6.2.

The selectivity of RPC/SPE is predominantly a function of the ligands attached to the gel matrix, and among the more popular types of ligands are n-alkyl groups ($C_2, \ldots, C_8, C_{10}, C_{12}, \ldots, C_{18}$). The hydrophobic C-18 phase has a hydrophobic barrier with a low affinity for hydrophilic compounds. However, an appropriate choice among organic solvents from the elutrope series (Section 2.8) as modifiers for the mobile phase gives possibilities for stationary phases with quite different characteristics (Figure 6.15) that match the required properties for purification of the analytes of interest.[11,12]

When the hydrophobic C-18 phase is solvated with the mobile phase contain-

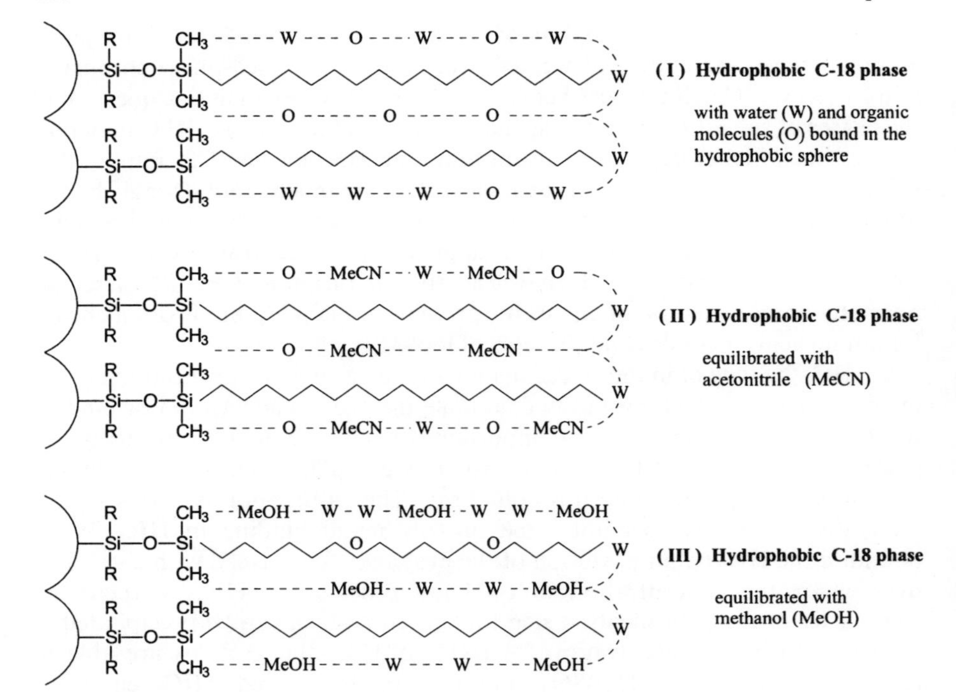

Figure 6.15 *Illustration of three different conditions in a C-18 RPC/SPE stationary phase, which result in quite different properties of the matrix*

ing organic solvents as modifiers, this will give a stationary phase with increased ability to dissolve various analytes according to their hydrophilic–lipophilic properties. If relatively lipophilic analytes are retained in the solvated stationary phase, they can then be released or eluted by increasing the concentration of organic modifiers in the mobile phase. Replacement of mobile phase organic modifiers which have a high dielectric constant (ε, Section 2.8) or high polarity, *e.g.* methanol, with those from the eluotrope series with lower ε, *e.g.* acetonitrile, will in general facilitate the release or elution of lipophilic compounds from SPE column materials. This will, correspondingly, give reduced retention times for analytes separated by RPC.

10 Selected General and Specific Literature

1. A.J.P. Martin and R.L.M. Synge, *Biochem. J.*, 1941, **35**, 1358.
2. E. Stahl, 'Dünnschicht Chromatographie. Ein Laboratoriumshandbuch', 2nd Edition, Springer-Verlag, Berlin, 1967.
3. H. Jork, W. Funk, W. Ficher, and H. Wimmer, 'Thin-Layer Chromatography, Reagents and Detection Methods' WCH Verlagsgesellshaft, Weinheim, 1990.
4. H. Jork, W. Funk, W. Ficher, and H. Wimmer, 'Thin-Layer Chromatography, Reagents and Detection Methods' WCH Verlagsgesellshaft, Weinheim, 1994.

[5] B.O. Eggum and H. Sørensen, in 'Absorption and Utilization of Amino Acids', ed. M. Friedman, CRC Press, Boca Raton, FL, 1989, vol. 3, ch. 17, p. 265.
[6] B.O. Eggum, N.E. Hansen, and H. Sørensen, in 'Absorption and Utilization of Amino Acids', ed. M. Friedman, CRC Press, Boca Raton, FL, 1989, vol. 3, ch. 5, p. 67.
[7] O. Olsen and H. Sørensen, *J. Am. Oil. Chem. Soc.*, 1981, **58**, 857.
[8] L.M. Larsen, O. Olsen, A. Plöger, and H. Sørensen, *Phytochemistry*, 1983, **22**, 219.
[9] J.K. Nielsen, O. Olsen, L.H. Pedersen, and H. Sørensen, *Phytochemistry*, 1984, **23**, 1741.
[10] L.M. Larsen, J.K. Nielsen, and H. Sørensen, *Phytochemistry*, 1982, **21**, 1029.
[11] B. Bjerg and H. Sørensen, in 'World Crops: Production, Utilization, Description. Glucosinolates in Rapeseeds: Analytical Aspects', ed. J.-P. Wathelet, Martinus Nijhoff Publishers, Boston, 1987, Ch. 13, p. 59; 125.
[12] J. Frias, K.R. Price, G.R. Fenwick, C.L. Hedley, H. Sørensen, and C.J. Vidal-Valverde, *J. Chromatogr.*, 1996, **719**, 213.
[13] 'Ion Exchange Chromatography. Principles and Methods', Pharmacia Biotech AB, Uppsala, Sweden, 1995, ISBN 91970490-3-4.
[14] 'Hydrophobic Interaction Chromatography. Principles and Methods', Pharmacia Biotech, 1993, ISBN 91970490-4-2.
[15] 'Affinity Chromatography. Principles and Methods', Pharmacia Biotech, 18-1022-29, 1993.
[16] 'Reversed Phase Chromatography. Principles and Methods', Pharmacia Biotech, 18-1112-93, 1996.
[17] 'Gel filtration. Principles and Methods', Pharmacia Biotech, 7th Edition, 1997, ISBN 91-97-0409-2-6.

CHAPTER 7

Ion-exchange Chromatography

1 Introduction

Ion-exchange chromatography is the liquid chromatographic separation of charged molecules based on equilibria between an immobile and a mobile ion of opposite charges. It requires thus an initial binding of anions to cations in an equilibrium process and finally the possibility of separation of these ions from each other. Most often one of these ions is covalently bound to an insoluble polymeric material used as the stationary phase in columns or bound to membranes. If the immobilized ion is an anion, it requires a positive exchangeable counterion and it is then a cation exchanger. Anion exchangers have, correspondingly, immobile cations and exchangeable anions (Figure 7.1).

The exchangeable ions initially bound to the exchangers may be released in two fundamentally different ways, either by competition with other ions of the

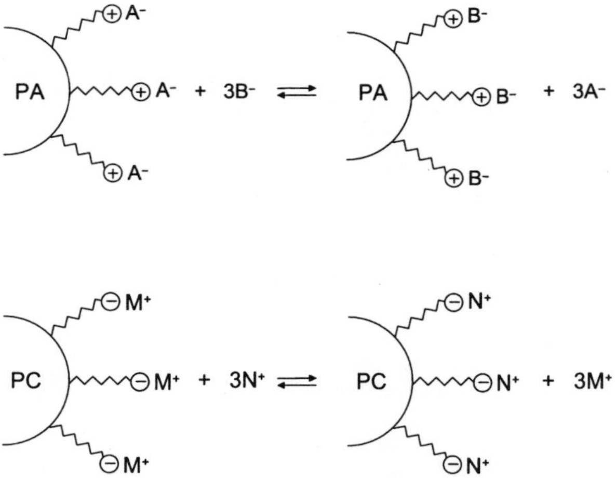

Figure 7.1 *Anion exchanger (PA) and cation exchanger (PC) in equilibrium with exchangeable anions (A^- and B^-) and exchangeable cations (M^+ and N^+), respectively*

same type of charge or by removal/change of the net charge on either the exchanger or the exchangeable ions. The release of the exchangeable ions by competition corresponds to the traditional ion-exchange techniques. In these methods, the binding and release of different exchangeable ions depend on their binding constants (Section 2.7 and page 45), which thereby also determine the sequence of their release and the possibilities for their separation in different fractions. The methods based on the release of the exchangeable ions by removal or change of their net charge or removal of the charges on the exchanger are used in several valuable techniques. These comprise group separations, flash chromatography techniques, and techniques based on the separation of ampholytes such as amino acids, peptides, and proteins according to their pI (Section 2.4), including chromatofocusing.

Efficient utilization of the possibilities in ion-exchange chromatography requires a knowledge of pK_a'-values of both exchangers and analytes as well as of the possibilities for changing pK_a' in the desired direction (Chapter 2). Organic solvents used as modifiers affect the dielectric constant of the solvents and thereby the pK_a'-values (Section 2.8) of protolytically active groups. In addition, organic solvents may be important for the solubility of the analytes, for the properties of the exchangers, and for the non-ionic binding between analytes and exchangers. Adsorption problems can in some cases be appreciable, and it is necessary to realize that it is important to select exchangers and conditions/solvents which are appropriate with respect to the type of analytes and impurities in the extracts subjected to separation in the selected ion-exchange technique.

The optimal choice of ion-exchange technique for high-M_r analytes will most often be different from the exchangers and conditions optimal to ion-exchange separations of low-M_r analytes. In extracts from various biological materials, including food, various amounts of low- and high-M_r compounds will often co-occur, and this needs to be considered in connection with selection of ion-exchange techniques.

2 Ion-exchanger Materials

Ion-exchangers may consist of insoluble matrices of particles, most often porous, or their matrices can alternatively be bound to membranes. Charged groups of the ion-exchanger are covalently bound to the matrix, and electrically balanced with associated exchangeable counterions (Figure 7.1). The matrix may be composed of materials such as silica glass, synthetic resins (polystyrene, polyether resins), polysaccharides (cellulose, dextranes, agarose), and polyacrylate. The charged functional groups bound to the matrix comprise strongly acidic groups such as sulfonic acid ($R\text{-}SO_2\text{-}O^-$, H^+) or strongly basic groups such as quaternary ammonium groups ($R\text{-}N^+(R')_3$, HO^-) and both of these types will be totally ionized in all of the normally used eluents. Examples of weakly acidic and basic functional groups are $HOSi(OR)_3$, RCO_2H, primary, secondary, and tertiary amines ($RN(R')_2$), which will have different pK_a'-values

Table 7.1 *Selected examples of commonly used ion-exchangers*

Type	*Matrix*	*Functional groups*	*Commercial product*
Weakly acidic (cation exchangers)	Polyacrylic acid	$—COO^-$	Amberlite IRC 50
			Bio-Rex 70
			Zeocarb 226
	Polystyrene*	$—CH_2COO^-$	Poros CM
	Cellulose	$—CH_2COO^-$	Cellex CM
	Dextran	$—CH_2COO^-$	CM-Sephadex®
	Agarose	$—CH_2COO^-$	CM-Sepharose®
			Sepharose CL-6B
Strongly acidic (cation exchangers)	Polystyrene	$—SO_3^-$	Amberlite IR 120
			Bio-Rad AG 50
			Dowex 50
			Zeocarb 225
	Dextran	$—CH_2—CH_2—CH_2—SO_3^-$	SP-Sephadex
	Polystyrene*	$—CH_2—CH_2—SO_3^-$	Poros S
	Polystyrene*	$—CH_2—CH_2—CH_2—SO_3^-$	Poros SP
Weakly basic (anion exchangers)	Polystyrene	$—CH_2NH^+R_2$	Amberlite IR 45
			Bio-Rad AG 3
			Dowex WGR
	Cellulose	$—CH_2CH_2NH^+(CH_2CH_3)_2$	DEAE Sephacryl®
	Dextran	$—CH_2CH_2NH^+(CH_2CH_3)_2$	DEAE Sephadex®
	Agarose	$—CH_2CH_2NH^+(CH_2CH_3)_2$	DEAE Sepharose®
	Polystyrene*	Polyethyleneimine	Poros PI
Strongly basic (anion exchangers)	Polystyrene	$—CH_2N^+(CH_3)_3$	Amberlite IRA 401
			Bio-Rad AG 1
			Dowex 1
			Amberlite IRA 410
		$—CH_2N^+(CH_3)_2$ with CH_2CH_2OH on N	Bio-Rad AG 2
			Dowex 2
	Cellulose	$—CH_2CH_2N^+(CH_2CH_3)_3$	Cellex T
	Dextran		QAE-Sephadex
		$—CH_2CH_2N^+(CH_2CH_3)_2$ with $CH_2CH(OH)CH_3$ on N	
	Polystyrene*	Quaternized polyethyleneimine	Poros QE
	Coated silica	$—C(O)NH(CH_2)_3N^+(CH_3)_3$	QMA

* Particles of 10–15 μm with pores through the beads

depending on the R-group, the dielectric constant, and modifiers in the eluent (Section 2.8). Selected examples of often used exchangers are given in Table 7.1.

Recent developments in the technology have also resulted in many new commercial products used in connection with HPLC, FPLC, flash chromatography (FC), and ion-exchange chromatography columns, *e.g.* as Mono Beads (Table 7.2) and on special FC membranes. These media offer the possibilities of fast, high capacity, high-resolution ion-exchange chromatography on an analytical and preparative scale, or in pilot plant/industrial upstream and downstream techniques. Table 7.2 gives specific data for some selected ion-exchange materials that are usable in, for example, group separation techniques.

3 Choice of Ion-exchanger

Both anion- and cation-exchange chromatography can be carried out by use of a variety of ion-exchangers based on different matrices and with functional groups, but successful utilization of ion-exchange techniques depends very much on the choice of ion-exchanger. First, it is necessary to consider the type of compounds to be isolated or purified as well as the presence of other compounds in the solution applied to the ion-exchanger. The capacity (in equivalents, μmol or mg of exchangeable ions per mg or ml of ion-exchanger) and other properties which relate to the functional groups (strongly to weakly acidic or basic groups) are also important, as is the flow properties through the columns or membrane systems. These are defined by the properties of the matrix, including particle size, its rigidity, pore size, swelling properties in different solvents, and chemical and physical stability.

The type of analytes, the effect of pH on their charge and stability, as well as their solubility in the applied eluents, and the adsorption of various compounds to the matrix determine whether anionic/cationic and strong/weak exchangers should preferably be chosen. For low-M_r compounds, the p$K_a{}'$-values of the exchangers' functional groups (Table 7.2) are important when the exchangers are used for group separations requiring removal of charges on the exchanger. For ampholytes (Section 2.4), especially proteins, the pI-values will be important in selecting the ion-exchanger (Figure 7.2). Although it is possible to obtain both positive and negative net charges on the proteins, depending on the selected buffer pH in relation to pI of the proteins, the stability of the proteins can be limited to a relatively narrow pH range, thus limiting the acceptable pH-range for the ion-exchange.

The choice of specific anion- and cation-exchanger material is also primarily dependent on the type of compounds present in the solution to be separated.

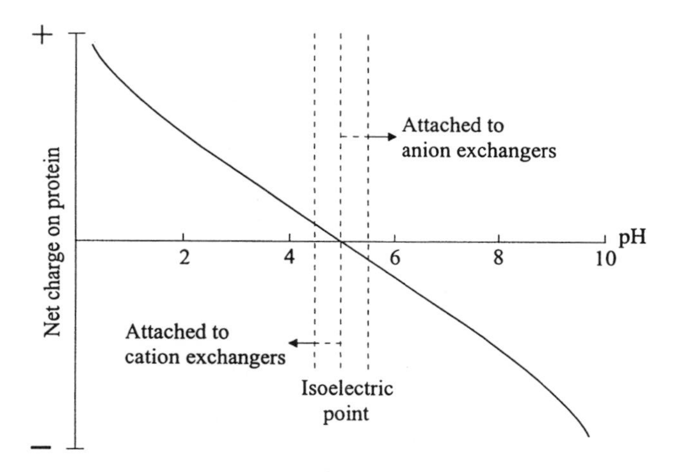

Figure 7.2 *Net charge of ampholytes/proteins as a function of pH and pI*

Table 7.2 *Selected data for some commercially available ion-exchangers*

Exchanger	*Functional groups*	*Capacity* Ion (meq/g dry gel)	*Capacity* Haemoglobin (mg/ml wet gel)	pK'_a
Strongly basic (Table 7.1)	Quaternary ammonium	*ca.* 4		>12
DEAE-cellulose	Diethylaminoethyl, —O—CH₂CH₂—N:(C₂H₅)₂	0.1–1.1		*ca.* 9.5
AE-cellulose	Aminoethyl, —O—CH₂CH₂—N:	0.3–1.0		*ca.* 9.2
DEAE-Sepharose Fast Flow	Diethylaminoethyl		110[a]	9–10
DEAE-Sephadex A-25	Diethylaminoethyl	3.5	30[a]	
DEAE-Sephadex A-50	Diethylaminoethyl	3.5	110[a]	
Ecteola-cellulose	Triethanolamine on cellulose through glyceryl and polyglyceryl chains. Mixed groups	0.1–0.1		6–8
PAB-cellulose	*p*-Aminobenzyl —O—CH_2-φ-NH_2	0.2–0.5		3–5
QAE-Sephadex A-25		3.0	10[a]	always cation
QAE-Sephadex A-50	Diethyl-(2-hydroxypropyl)-aminoethyl —O—CH₂CH₂—N⁺(C₂H₅)₂—CH₂CH(OH)CH₃	3.0	10	always cation

Q-Sepharose Fast Flow	Quaternary aminomethyl, $—OCH_2N^+$—	0.2	120[a]	always cation
Mono Q; hydrophilic polyether resins (Mono Beads)	Quaternary aminomethyl	0.3	65[a]	always cation
CM-Sepharose Fast Flow	Carboxymethyl, $—O—CH_2—CO_2H$		30	3–4
CM-Sephadex C-25	Carboxymethyl	4.5	70	3–4
CM-Sephadex C-50	Carboxymethyl	4.5	140	3–4
P-cellulose	Phosphate, $—O—PO_3H_2$	0.7–7.4		*ca.* 2 & *ca.* 7
SE-cellulose	Sulfoethyl, $—O—(CH_2)_2SO_3^-H^+$	0.2–0.3		<1
SP-Sephadex C-25	Sulfopropyl, $—O—(CH_2)_3SO_3^-H^+$	2.3		<1
SP-Sephadex C-50	Sulfopropyl	2.3	110	<1
SP-Sepharose Fast Flow	Methylsulfonate, $—CH_2—SO_3^-H^+$	0.2	50	<1
Mono S; hydrophilic polyether resins (Mono Beads)	2-Hydroxypropylsulfonate	0.2	70	<1
Mono P; hydrophilic polyether resins (Mono Beads)	Polybuffer Exchanger PBE 94; for chromatofocusing			
Strongly acidic (Table 7.1)	Sulfonic acid		*ca.* 4	<1

[a] HSA, Pharm. Ion Exc. Chrom. 1995.

Crude extracts require generally rigid matrices with good flow properties, *e.g.* matrices of synthetic resins or silica glass with relatively large particle sizes (Table 7.1) or the use of matrices bound on special membranes. Columns with exchangers such as Amberlite, Dowex, Sephadex®, and Bio-Rad, are recommended for purifications of crude extracts, as they have particle sizes, often expressed as mesh values, which allow acceptable flow for even very impure samples.

Ion-exchangers based on synthetic resins and strongly acidic, or basic, functional groups bound to the matrix have in general several advantages in connection with techniques designed to separate inorganic ions and various types of other low-M_r compounds where adsorption is a minor problem. The matrices consist of hydrophobic polymers that are highly substituted with functional groups and have, therefore, a high capacity for ions and generally good flow properties. In general, low-M_r compounds are most often well separated on matrices with small pores (high degree of cross-linking) and relatively small particles (increasing surface area), *e.g.* Dowex 50 W, 200–400 mesh (H^+) cation exchanger or Dowex 1 × 8, 200–400 mesh (AcO^-), anion exchanger. In addition to the value of their relatively large available capacity (*ca.* 2–4 meq g^{-1}), these exchangers also have advantages in group separation techniques (Section 7.6) owing to their ability to rehydrate, thus avoiding drying out of the column material. The high charge density and the hydrophobic matrix tend, however, to denature labile biomolecules such as proteins, and these properties limit their suitability for purifying extracts rich in high-M_r compounds.

Polysaccharide-based ion-exchangers are generally gentler toward various proteins, including enzymes and other labile biological materials, *e.g.* exchangers based on dextran (Sephadex), agarose (Sepharose CL-6B), and cross-linked cellulose (DEAE Sephacryl). These products are often used owing to their relatively well-defined spherical form, high porosity, and acceptable flow properties and capacities. The polysaccharide-based matrices can, however, result in adsorption of some glycoproteins and other glycosides, and many of them do not possess the same good flow properties characteristic of rigid glass and polystyrene matrices.

Sephadex and similar Bio-gel materials (matrix as for gel filtration; page 142) are advantageous as exchangers to labile high-M_r molecules/proteins, and their large pore size may give increased capacity, although the relationship between M_r and pore size is not simple (Table 7.3).

The higher degree of cross-linking in Sephadex C-25 and A-25 compared with Sephadex C-50 and A-50 results in less swelling or changes in bead size as a function of reduced ionic strength and increased pH. Separation of analytes by ion-exchange chromatography will, however, often require elution with buffers of varying ionic strength and/or pH. Matrices which change considerably in their degree of swelling (page 144) will thus very often create flow problems. To overcome such problems it can be necessary to choose exchangers based on Fast Flow, Mono Beads, or silica glass matrices (Table 7.2).

Columns such as Mono Q and Mono S are advantageous in HPLC and

Table 7.3 *Capacity of Sephadex ion-exchangers in relation to pore size*

	Sephadex ion-exchangers (Table 7.2)	
M_r *of proteins*	*With max. capacity*	*Pore size*
<30	C-25 and A-25	Medium
30–200	C-50 and A-50	Large
>200	C-25 and A-25	Medium

FPLC for the high-resolution separation of pre-purified protein samples that are free of particles and undissolved materials which otherwise may destroy these relatively expensive columns.

4 Regeneration of Ion-exchangers

When appropriate ion-exchangers and eluents have been selected, it is essential that the exchangers are correctly regenerated to remove possible impurities, and the exchangers should be brought to equilibrium with the required counterion or starting buffer immediately before use. The procedures needed for sufficient and correct regeneration vary with the type of exchangers, the type of impurities to be removed, and the counterions or buffer required. Regeneration procedures are therefore often recommended and available from the commercial firms producing the exchangers, and readers are advised to follow these instructions. In addition, some examples of regeneration procedures are described in Box 7.1.

5 Packing of Ion-exchange Columns and Sample Application

Packing of the columns and application of the samples are the most critical steps. A poorly packed column gives rise to poor and uneven flow, zone broadening, and loss of resolution. The swollen ion-exchange material, mixed as a slurry with the starting buffer and degassed, is gently transferred to the column down the side of the tube, which contains a small amount of starting buffer at the bottom; all of the material is transferred in one go. The column is then mounted vertically in the required system (Figure 7.3), the outlet is opened and the gel is allowed to settle with only a small amount of starting buffer left at the top.

The amount of sample which can be applied to the column depends on the capacity of the exchanger (Table 7.2), the type of buffer and analytes, and the degree of resolution (R_s) required (page 135). To obtain an acceptable R_s, it is normally not advisable to use more than 10–20% of the exchanger's capacity, and the highest R_s obtainable will only be seen when small, concentrated sample volumes are applied in narrow bands to the top of the column. Information on the available capacity of an ion-exchanger can be determined by a batch test-

Box 7.1 *Selected examples of regeneration procedures for some often used ion-exchange materials*

The regeneration procedures shown are applicable for used ion-exchange material. New ion-exchange material requires further treatment. Consult the manufacturer for precise instructions.

Weakly acidic cation-exchanger

CM-Sephadex C-25 (H^+)

1. Wash the ion-exchange material with 2 M CH_3CO_2H, at least 10 times the column volume
2. Wash the ion-exchange material with deionized water until pH in the effluent is as in the applied water

Strongly acidic cation-exchanger

Amberlite IR-120 (H^+); Dowex 50, 200–400 mesh (H^+)

1. Wash the ion-exchange material with 1 M HCl, at least 15 times the column volume
2. Wash the ion-exchange material with deionized water until pH in the effluent is as in the applied water

Weakly basic anion-exchanger

DEAE-Sephadex A-25 (AcO^-)

1. Wash the ion-exchange material with 2 M CH_3CO_2Na, at least 10 times the column volume
2. Wash the ion-exchange material with 2 M CH_3CO_2H – 96% ethanol (1:1), at least 10 times the column volume
3. Wash the ion-exchange material with deionized water until pH in the effluent is as in the applied water

Strongly basic anion-exchanger

Dowex 1 × 8, 200–400 mesh (AcO^-)

1. Wash the ion-exchange material with 2 M CH_3CO_2Na, at least 15 times the column volume
2. Wash the ion-exchange material with 2 M CH_3CO_2H, at least 10 times the column volume
3. Wash the ion-exchange material with deionized water until pH in the effluent is as in the applied water

QMA-Sep-Pak® (Waters Ass.)

The QMA-Sep-Pak columns should be regenerated with the bottom of the column pointing upwards to remove possible impurities

1. Wash the ion-exchange material with 10 ml 1 M HCl
2. Wash the ion-exchange material with 5 ml 96% EtOH
3. Wash the ion-exchange material with 5 ml 1 M NaOH
4. Wash the ion-exchange material with 5 ml 2 M CH_3CO_2Na
5. Wash the ion-exchange material with 5 ml 2 M CH_3CO_2H – 96% EtOH (1:1)
6. Wash the ion-exchange material with 2 × 5 ml deionized water

tube method similar to that described in Section 7.8 (Box 7.2). Furthermore, the optimal separation conditions based on elution with an appropriate system, pH or salt gradient, can often be obtained from column chromatographic, analytical HPLC, or FPLC experiments.

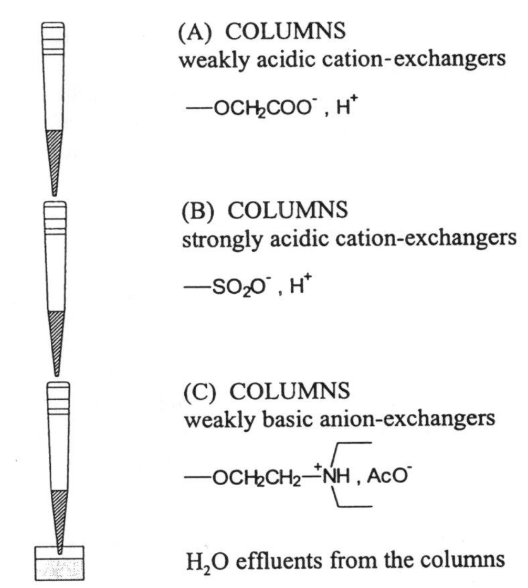

Figure 7.3 *Sequential arrangement of ion-exchange columns:* (A) *column = CM-Sephadex* (H^+) *or CM-Sepharose Fast Flow* (H^+) *(Table 7.2),* (B) *column = Dowex 50 W × 8, 200–400 mesh* (H^+) *(Table 7.1),* (C) *column = DEAE-Sepharose Fast Flow* (AcO^-) *or DEAE-Sephadex* (AcO^-) *(Table 7.2)*

6 Group Separation of low-M_r Compounds

Separation of protolytically active low-M_r compounds by ion-exchange chromatography is an essential technique in group separation and purification of crude extracts (Boxes 4.3 and 4.4) prior to liquid chromatographic methods of analyses such as HPLC and HPCE.

Choice of Techniques and Columns

Group separation can be performed in small-scale procedures with a column arrangement in plastic pipette tips as shown in Figure 7.3. A small lump of glass wool is placed at the bottom of each column and, with appropriate regenerated exchangers (Section 7.4), they are mounted with, for example, sticky tape on a glass plate. The columns could also be small, commercially available columns arranged in the same sequence, *e.g.* in a Supelco vacuum system. Thereby, the speed of the procedure will be increased, so that the total procedure, including wash and elution of the columns, only takes a few minutes.

Depending on the type of low-M_r analytes to be studied, it may be advisable to change column materials as mentioned in Section 7.3. With extracts containing phenolics/aromatic compounds, it could be advantageous to change column (B) (Figure 7.3) from a polystyrene-based matrix (Table 7.1) to columns based on strongly acidic polysaccharide matrices or Mono Beads (Table 7.2). In some

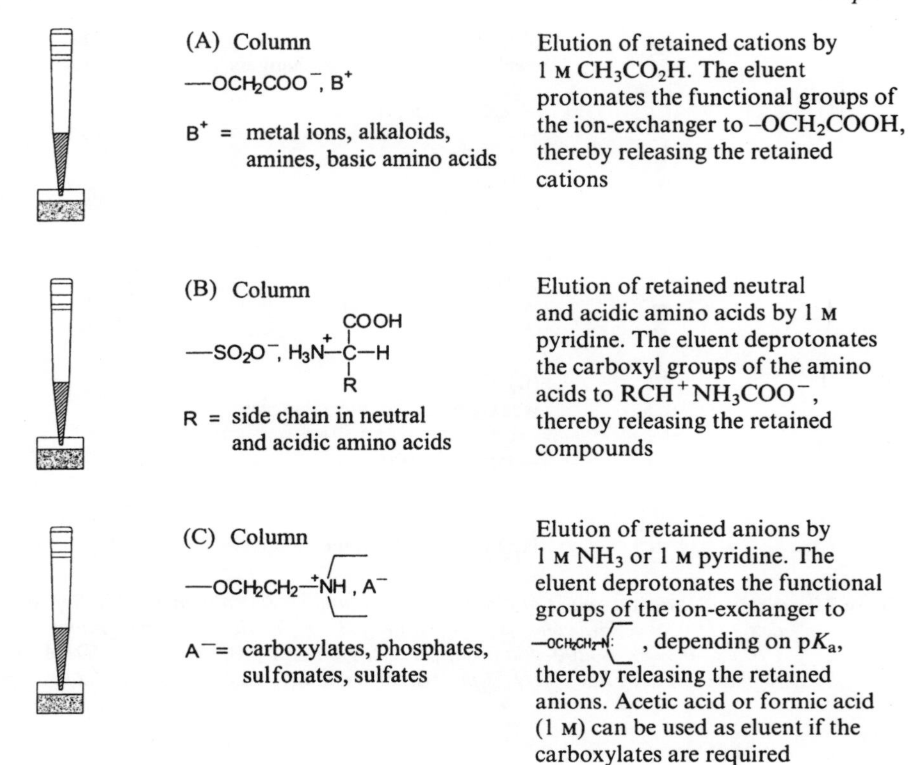

Figure 7.4 *Distribution of compounds with protolytically active groups on columns binding the compounds according to their net charge. Finally, the columns are eluted individually with volatile eluents*

cases, where the analytes of interest are carboxylic acids, saponins, or other compounds with weakly acidic functional groups, column (C) could with advantage be changed to a Dowex 1 × 8, 200–400 mesh (AcO^-). For analytes with strongly acidic functional groups such as sulfate or sulfonate, the column (C) could be an anion exchanger with weakly basic functional groups (Table 7.2), as their charges can be removed by the use of weakly alkaline eluents.

When the extracts have been applied to the column system (Figure 7.3), the subsequent washing of the system with water will result in transport of the analytes to a column where they are bound according to their net charge. The next step will then be elution with an appropriate volatile eluent (Next subsection; Figure 7.4). Tests for the correct function of the ion-exchange columns and evaluation of the amount and type of low-M_r analytes in the extracts considered can be performed in simple systems such as chromatography (Chapter 6) or by HVE (Box 9.1 and Figure 9.4).

Choice of Eluents, Buffers, and pH

The choice of eluents may vary depending on a number of different demands caused by the structure and properties of the analytes and the aim of the ion-exchange technique. In cases where the analytes of interest are low-M_r compounds intended for further studies after concentration, *e.g.* to dryness, it is advantageous to use volatile eluents which leave the eluted compounds as salts or non-volatile compounds (Figure 7.4).

If large volumes of eluent are required, it needs to be relatively cheap and non-toxic. For group separations based on removing or changing the net charge on the exchangeable ions (Figure 7.1) or the ion-exchangers charges (Figure 7.4), it can be advantageous to change the dielectric constant of the eluent, and thereby the pKa' values (Section 2.8), by use of organic modifiers. A change in the dielectric constant may in addition be valuable with respect to reducing adsorption problems and to increasing the solubility of otherwise slightly soluble analytes. Strongly acidic and basic eluents should normally be avoided, especially with more reactive analytes, which otherwise may be transformed into artefacts.

7 Column Chromatographic Separation of low-M_r Compounds

Separation of individual low-M_r compounds with identical net charge is usually achieved on polystyrene-based ion-exchangers in columns of appropriate size. For this purpose, column materials with small particles, small diameters, or high mesh values will have the best separation efficiency or performance with a high number of theoretical plates (N) and good resolution (R_s). The elution of the columns will then require pumps, *e.g.* peristaltic pumps, and is performed as either isocratic or gradient elution (Figure 7.5). The eluents may contain competing ions or elution may be based on pH changes with gradient elution altering the charges of analytes or functional groups on the column material as for group separation (Section 7.6). If the purpose is the preparative isolation of appreciable amounts of analytes, it will be advantageous to use volatile eluents, as is the case for group separations or purifications prior to analytical procedures sensitive to high salt concentrations.

Amino acid analysis based on commercially available amino acid analyser instruments has been one of the most often used column chromatographic ion-exchange techniques.[1] This method is based on the use of a strongly acidic cation-exchanger and a starting pH below 2, thus ensuring that most amino acids will have a positive net charge. Gradient elution using increasing pH results in release of acidic and neutral amino acids in a sequence mainly determined by their pK_{a1}' values, which are close to 2 (Section 2.4). However, their elution sequence is also affected by adsorption, especially for the aromatic amino acids. The basic amino acids are eluted last, by use of a salt gradient. Detection of the amino acids eluted is based on, for example, post-column ninhydrin reaction and colour determination is performed by the amino acid

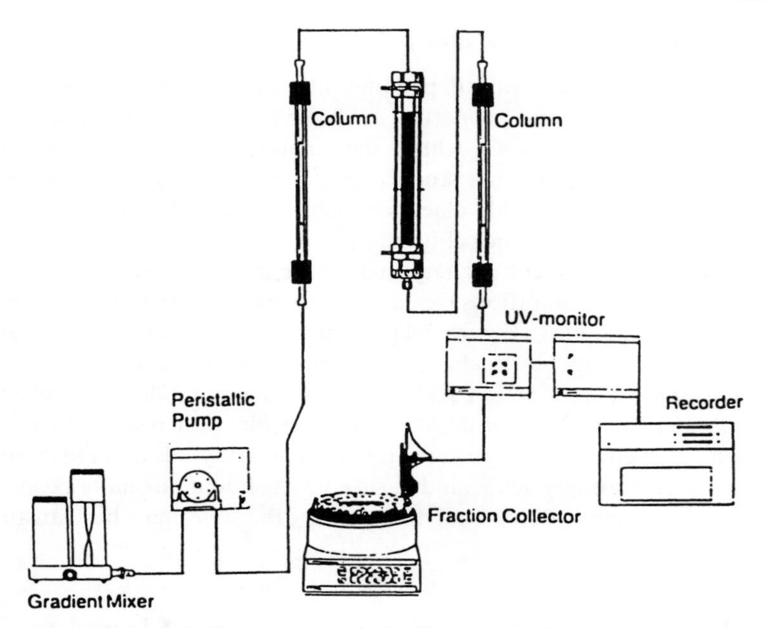

Figure 7.5 *Illustration of instrumentation for column chromatography where the application of samples and washing are performed with the columns in series. The columns are separated and eluted individually either isocratically or with use of a gradient*

analyser. Isolation and separation of appreciable amounts of structurally closely related amino acids can be performed in a similar manner by using one or more larger columns (Figure 7.5) connected in series, corresponding to the sequence in Figure 7.3, and with the use of volatile eluents.[1,2]

Isolation and separation of structurally closely related weakly acidic compounds can be performed by use of the corresponding principles with anion-exchange columns, *e.g.* Dowex 1 × 8, 200–400 mesh (AcO^-), utilizing the variations in pK_a'-values, around 3–5, for carboxylic acids.[1,3]

Column chromatographic isolation and separation of low-M_r compounds such as cations with $pK_a' > 12$ and anions with $pK_a' < 2$ can be performed by a combination of the group separation principles and separations based on, for example, adsorption chromatography.[4,5]

8 Column Chromatographic Separation of High-M_r Compounds

For low-M_r compounds, the group separation technique described in Section 7.6 (Figures 7.3 and 7.4) gives information on ion-exchangers, eluents, and analytical tests required for selection of appropriate methods for the various types of low-M_r compounds. Selection of appropriate ion-exchange methods for ampholytes/proteins comprise evaluation both of the protein stability and of the possibilities for binding and release of the proteins from the ion-exchanger. The procedure described in Box 7.2 can be helpful in this connection.

Box 7.2 *Procedure for evaluation of protein stability and possibilities for binding and release of the proteins from the ion-exchanger*

1. Equilibrate and swell a series of different ion-exchange materials with varying buffer types and ionic strengths (50 mg in 0.5 ml buffer) and place in test tubes
2. Add a known constant amount of a protein solution (50 μl) to each tube
3. Mix the contents for about 5 min and transfer the solution or mixture to pipette tips with a small lump of glass wool at the bottom, arranged as in Figure 7.3
4. Assay the effluent for the substance of interest (assay or spot-test as described in Chapter 12 for selected enzymes)
5. Columns with bound protein are eluted with buffers of 1–2 pH-units higher than the binding buffer (in the case of cation-exchange chromatography) or lower (in the case of anion-exchange chromatography). Alternatively, use buffers with up to 0.5 M NaCl
6. Assay as described in 4.

Ion-exchange chromatography of high-M_r compounds, *e.g.* proteins including labile enzymes, will give the best results when using appropriate buffers (Sections 2.5–2.7), correct pH, and ionic strength (Section 2.3). As buffers themselves consist of ions, they can also participate in the ion-exchange equilibria (Figure 7.6), and as a general rule it will be advantageous to use anionic buffers (*e.g.* phosphate or acetate) for binding of proteins to cation exchangers, and cationic buffers (*e.g.* TRIS) with anion exchangers.

The required concentration of buffer may vary depending on the structure and properties of the polymer/protein, but it should be as low as possible (10 mM will often be sufficient).

The elution of ampholytes/proteins can be performed by use of a pH gradient, with the release of the proteins when the buffer pH passes the proteins' pI (Figure 7.2). The release of proteins from the exchanger can alternatively be performed by use of a salt gradient. This will then require desalting if the next

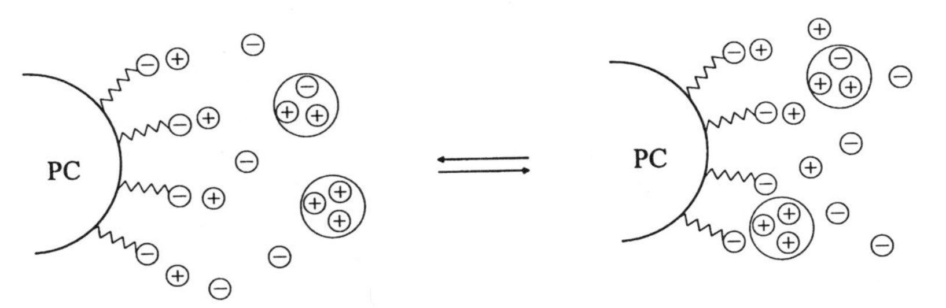

Figure 7.6 *Ion-exchange equilibrium between a cation exchanger (PC), the buffer ions, ⊖ and ⊕, and charged polymers,* (⊖⊕⊕), (⊕⊕⊕)

step implies concentration before, for example, electrophoresis, HPCE, or other procedures which may be sensitive to high salt concentrations. However, an appropriate pH, salt or buffer concentration may play an important role in stabilizing protein structures in solution, and this, the type of ions, ionic strength, and modifiers (Sections 2.8 and 4.3), are important in protecting the proteins against denaturation and/or precipitation.

When the column material and eluents for ion-exchange chromatography of high-M_r compounds have been selected, columns of the correct size have to be filled with regenerated ion-exchange material and placed in a system such as that shown in Figure 7.5.

9 Flash Chromatography

Chromatography with a potential possibility for a very high flow rate compared with other types of column chromatography is known as 'flash chromatography' (FC). The ion-exchange materials are, for example, prepared as disks on special membranes where each layer is thin and several membranes are placed in layers. Several FC systems are commercially available, *e.g.* with disks produced in different sizes and as laboratory scale ion-exchange chromatography devices. These FC techniques combine a solid, high capacity matrix with special high flow rate design for rapid and easy chromatographic separations. The column material on the disks can be anion exchangers such as DEAE (diethylaminoethyl) or QAE (quaternary aminoethyl) (Table 7.2). Specific examples describing the use of FC techniques are presented in Chapter 12.

10 Chromatofocusing

Chromatofocusing is a column chromatographic method separating proteins and other ampholytes according to their pI in a pH-gradient formed in an ion-exchanger column by a buffer consisting of a mixture of different ampholytes (polybuffers). The commercially available polybuffers (Pharmacia Biotech) are designed to cover different pH ranges *e.g.* 9–6 (Polybuffer 96), 7–4 (Polybuffer 74), or 11–8 (Pharmalyte pH 8–10.5).[8]

The ion-exchange matrices produced by Pharmacia Biotech for chromatofocusing columns are special anionic polybuffer exchangers (PBE 94 and PBE 118) based on Sepharose 6B and with a capacity of about 2–4 meq/100 ml gel, but varying with pH. It is recommended that PBE 118 is used together with the Pharmalyte pH 8–10.5 buffer and PBE 94 should be used together with Polybuffer 96 or Polybuffer 74 depending on the required pH-ranges.

Chromatofocusing gives best results when relatively long, narrow columns are used, *e.g.* i.d. 0.9 × 60 cm in height. As for other techniques, it is necessary to have well-packed columns as described in Section 7.5. Otherwise, it is an experimentally simple technique.

To perform chromatofocusing, the anion exchanger, *e.g.* PBE 94, is equilibrated with the start buffer which has to be adjusted to a pH slightly above the upper limit of the pH-gradient. The eluent buffer, *e.g.* Polybuffer 96, is adjusted to the pH selected to be the lower limit of the pH-gradient. The protein sample dissolved in the eluent buffer or start buffer is applied to the column which then is eluted with the eluent Polybuffer. The sample should not contain large amounts of salts ($I < 0.05$ M).

The proteins initially in the start buffer are negatively charged, whereas proteins dissolved in the eluent buffer, with pH < pI, will obtain a positive net charge. When the proteins are applied at high pH they will, however, obtain a negative net charge and bind to the anion exchanger. As more eluent buffer with a lower pH moves down the column, a pH-gradient will develop, from low pH at the top of the column to high pH at the bottom. The proteins with high pI will as a result of effects from the eluent buffer be transformed into first uncharged (pH = pI) and then positively charged molecules, which forces them to move down through the column until they again pass a pH > pI. This situation will continue until the compound leaves the column. The same will happen with the other proteins, in sequence, according to their pI, resulting in fractionations so that they leave the column when their pI is equal to the pH in the buffer mixture leaving the column. Thereby, proteins with lowest pI will be the last eluted proteins.

Chromatofocusing thus separates proteins according to their pI, as the case is with isoelectric focusing (IEF; page 190) and preparative isoelectric membrane electrophoresis (PrIME; page 204). It is a very efficient column chromatographic technique for the separation of, for example, isoenzymes of myrosinases and aromatic choline esterase,[6,7] and polybuffers will normally not interfere with enzyme assays (Chapter 12) or amino acid analysis. However, in some other types of analysis, the polybuffer ampholytes can create problems, and the ampholytes can be difficult to separate from proteins with $M_r < 15 \times 10^3$. With respect to the pI determined by chromatofocusing, these values may differ from that determined by IEF[7] owing to differences in binding of different buffer ions and ampholytes (Section 2.4).

11 Selected General and Specific Literature

1. B.O. Eggum and H. Sørensen, in 'Absorption and Utilization of Amino Acids', ed. M. Friedman, CRC Press, Boca Raton, FL, 1989, vol. 3, ch. 17, p. 265.
2. I. Kristensen, P.O. Larsen, and H. Sørensen, *Phytochemistry*, 1974, **13**, 2803.
3. E.P. Kristensen, L.M. Larsen, O. Olsen, and H. Sørensen, *Acta Chem. Scand., Ser. B.*, 1980, **34**, 497.
4. B. Bjerg, O. Olsen, K.W. Rasmussen, and H. Sørensen, *J. Liq. Chromatogr.*, 1984, **7**, 691.
5. B. Bjerg and H. Sørensen, in 'World Crops: Production, Utilization, Description. Glucosinolates in Rapeseed: Analytical Aspects', ed. J.-P. Wathelet, Martinus Nijhoff Publishers, Boston, 1987, ch. 13, p. 59.
6. L. Buchwaldt, L.M. Larsen, A. Plöger, and H. Sørensen, *J. Chromatogr.*, 1986, **363**, 71.
7. L.M. Larsen, T.H. Nielsen, A. Plöger, and H. Sørensen, *J. Chromatogr.*, 1988, **450**, 121.
8. 'Chromatofocusing with Polybuffer™ and PBE™', Pharmacia Fine Chemicals, Uppsala, 1994.

CHAPTER 8

High-performance Liquid Chromatography and Fast Polymer Liquid Chromatography

1 Introduction

The resolution and efficiency of liquid chromatography (LC) are closely associated to the active surface area of the materials used as stationary phase. This is caused by the fact that increasing surface area will increase the number of possible equilibria, the number of theoretical plates, and thereby the possibility of separating structurally closely related analytes (Chapter 6). Decreasing the particle size of the chromatographic materials and/or use of porous materials is thus recommended. Column chromatography using stationary phases consisting of small particles gives, however, increased resistance to flow of the mobile phase through the column. The result of this will be a need for pumps giving a relatively high pressure, and therefore column materials of high mechanical stability are needed to obtain an acceptable flow rate. The term HPLC (high-performance liquid chromatography) was coined to cover all LC developments based on these principles, originally with P in HPLC meaning pressure, but now meaning performance, as the pressure required for HPLC is not especially high, generally 5000–10 000 psi = 3.4–6.8 × 10^8 dyn cm^{-2} = 3.4–6.8 × 10^7 Pa = 340–680 bar. During the 1970s HPLC was developed as an efficient and popular technique for separation of various types of low-M_r compounds. With the introduction of reversed-phase and ion-pairing techniques it became a technique which was useful for biomolecules otherwise difficult to handle.[1] Since then, HPLC has developed dramatically with respect to available column materials and the wide range of instruments commercially available. This has also been the case for the technique known as fast polymer liquid chromatography (FPLC). This technique is based on the same ideas and principles as HPLC, with the main differences being in the type of column materials, lower pressure, and partly in the instrumental design, which for FPLC are specially adapted to work with more or less labile high-M_r compounds or polymers as proteins, nucleic acids, and polysaccharides.[2–4]

2 Instrumentation

A HPLC/FPLC system requires a solvent delivery system in which the mobile phase is pumped through the column, preferably with possibility of using different types of gradients: linear, convex, or concave. The columns used are generally made of stainless steel or other materials which can withstand the applied pressures. The compounds are eluted from the columns packed with the selected materials (Section 8.3), and they are commonly detected by UV-vis, fluorescence, electrochemical, RI, or some other special detectors, including diode-array (Section 5.6). A fraction collector is needed for preparative techniques, and a computer system can with advantage be used for both control of the system and treatment of the analytical data. In Figures 8.1 and 8.2 are shown typical HPLC and FPLC systems, respectively.

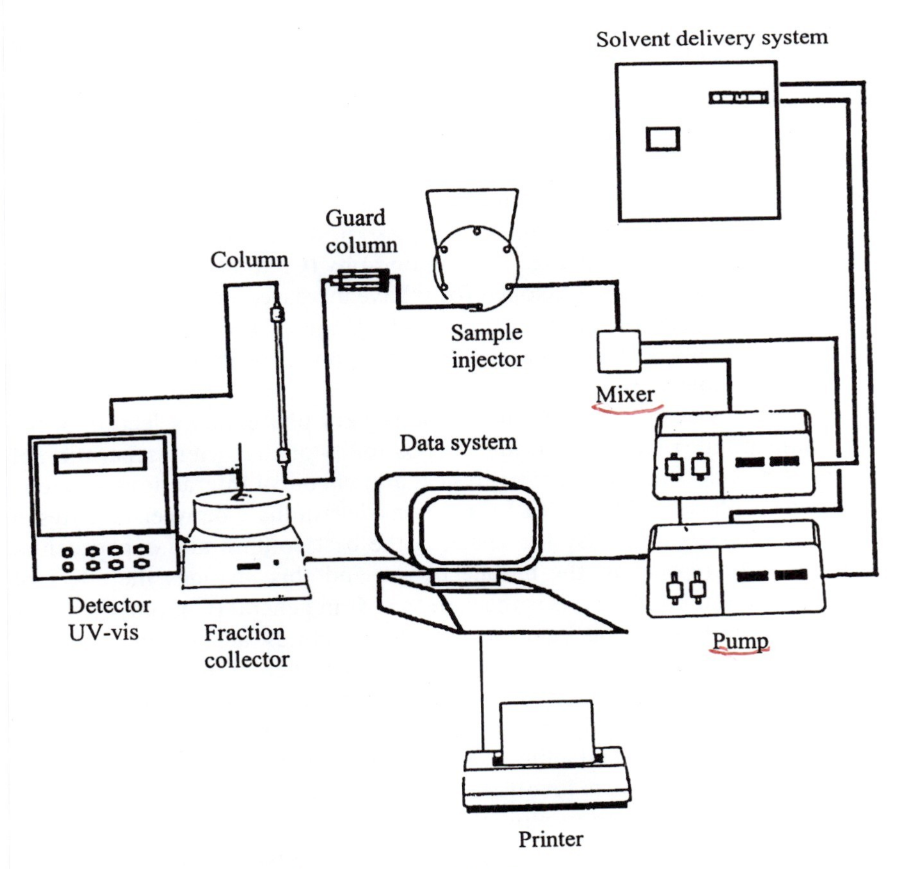

Figure 8.1 *Schematized illustration of an HPLC system*

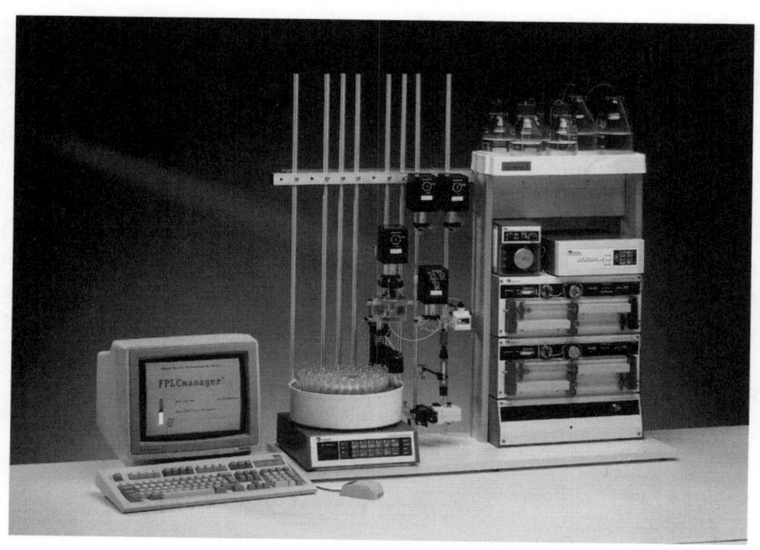

Figure 8.2 *Illustration of a FPLC system purchased from Pharmacia Biotech, Uppsala, Sweden*

3 Columns and Column Materials

A wide range of columns and column materials are available, allowing HPLC and FPLC to be used for all of the techniques known from traditional column LC (Chapter 6) and ion-exchange chromatography (Chapter 7). HPLC and the columns/column materials developed for it have increased the interest in its use for different methods of analyses for a wide range of low-M_r biomolecules. FPLC is for various reasons generally the most popular technique for analyses of high-M_r biomolecules.

The properties of the analytes and the stationary phase, and selection of the appropriate mobile phases, determine the possibility of an efficient separation of analytes. An understanding of the physical and chemical interactions whereby the stationary phases in HPLC/FPLC columns discriminate between analytes in the mobile phase is necessary for selecting the best column and experimental conditions with respect to the mobile phase, and this includes the effect of temperature and modifiers; this is the case for LC in general (Chapter 6). There are thus a number of mechanisms by which solutes are retained by the stationary phase, which are determined by the type of column material chosen from the large number available, such as pre-packed columns that discriminate between biochemically important analytes on the basis of their size and shape, charges, polarity, and special affinities or binding (Chapters 11–14). Most work on hydrophilic biomolecules involves ion-exchange chromatography (Chapter 7), size-exclusion chromatography (Section 6.6), normal and reversed-phase chromatography (RPC) (Section 6.9), hydrophobic interaction chromatography (HIC) (Section 6.8), affinity chromatography (Section 6.7) and ion-pair chromatography.[1,5] The basic types of column materials are either: (1) silica-based materials, which can be either totally porous particles, partially porous, or pellicular particles, *i.e.* porous particles coated on an inert solid core, *e.g.* glass beads, or (2) polymeric materials, *e.g.* those in Tables 7.1 and 7.2 and in Chapter 6.

In normal phase LC, the stationary phase is polar, as for paper and thin-layer chromatography (PC and TLC; Section 6.5), and the mobile phase is a relatively polar solvent, *e.g.* water–acetonitrile or water–methanol. RPC and HIC employ column materials that have been chemically modified to render them hydrophobic, *e.g.* with C_8 or C_{18} hydrocarbon chains, which again have modified properties caused by the mobile phase, as is the case for solid phase extraction (SPE; Section 6.9). RPC with column materials such as Bondapak-C_{18}, Corasil, LiChrosorp, Ultrasil, Ultrapore, Bio-Sil, Hi-Pore, and Ultrapak, as well as several other types of materials, is useful for various types of hydrophilic and amphiphilic low-M_r compounds (Chapter 11). However, RPC on such materials can result in denaturation of proteins and other high-M_r compounds, but special column materials of intermediate polarity have found some applications, along with the use of solvent gradients, in the RPC of proteins. The stationary phases for ion-exchange chromatography comprise a great number of special HPLC/FPLC materials, as is the case for other types of ion-exchange chromatography (Chapter 7, Tables 7.1 and 7.2). The stationary phases used for HPLC/FPLC based on size-exclusion separations are available in a range of pore sizes, as gels or particles with more or less rigid structures corresponding to the materials mentioned in Section 6.6.

HPLC/FPLC columns are manufactured so that they can withstand pressures up to about 700 bar (*vide supra*). Straight columns of 15–50 cm length and 2–4 mm i.d. are generally used, although smaller microbore columns, which may lead to higher resolution, are available. Microbore columns typically have an i.d. of 1–2 mm and are generally about 25 cm long. They can sustain flow rates of 0.05–0.2 $cm^3 min^{-1}$ as opposed to about 1–2 $cm^3 min^{-1}$ normally used with traditional HPLC/FPLC columns. Preparative columns are also commercially available with, for example, an i.d. of 2.5 cm and with possibilities of flow rates of 50–150 $cm^3 min^{-1}$. The temperature of the columns may be kept constant by use of column ovens, where the temperature can be adjusted as required.

4 Liquid Chromatography Principles including Ion-pair Reversed-phase Chromatography and Chiral Chromatography

HPLC/FPLC is in principle only a special utilization of LC and is therefore based on the same general chromatographic theories, including distribution coefficients (K_d) of the analytes between the stationary and mobile phases (Chapter 6).

Normal-phase Liquid Chromatography

Hydrophilic column materials saturated with solvent/water constitute the stationary phase in normal-phase LC. The mobile phase is generally a mixture of polar organic solvents with a composition defined in accordance with the properties of the analytes to be separated. The analytes need to be soluble both in the mobile phase and in the hydrophilic solvent/water saturated with mobile phase bound to the stationary phase. If the analytes are nearly insoluble in the mobile phase they will remain at the top of the column (Figure 6.7), which in TLC and PC (Section 6.5) corresponds to $R_f = 0$ (page 132). If the analytes only are soluble/dissolved in the mobile phase they will then have a retention time of $t_R = t_s = 0$ (Figure 6.9) or $R_f = 1$.

The mechanism of separation in normal-phase LC exploits the ability of analytes to displace molecules of the mobile phase bound in the stationary hydrophilic phase. This polar but mainly uncharged surface corresponds thus in many respects to the immobile layer to be discussed in connection with charged surfaces (Figure 9.2; Section 9.2).

The main use and advantages of normal-phase LC are in separations based on PC and TLC, and in HPLC separations of analytes with low solubility in water and systems using reversed-phase LC (RPC).

Reversed-phase Liquid Chromatography

The stationary phases in systems used for reversed-phase LC (RPC) are non-polar materials, *e.g.* produced by covalent binding of hydrocarbon chains such as $—(CH_2)_nCH_3$ to the hydroxy groups of silica, and chain lengths often used are those with $n = 7$ (C_8), and $n = 17$ (C_{18}) (Section 6.9). The mobile phases are relatively polar organic-solvents–water mixtures which solvate the stationary phase, as discussed for SPE (Section 6.9) and illustrated in Figure 6.15. The properties of the stationary phase will thus be both a function of the column material selected and the composition of the mobile phase.

Although the majority of column materials commercially available from various companies are based on C_8 and C_{18} alkylsilane groups attached to silica, they often exhibit great variations in their properties depending on the product type, which may vary both as a result of the silica preparation and the degree of alkylsilanylation.

The mobile phase should be selected in accordance with the properties of the analytes, allowing them to be soluble/dissolved both in the stationary and the mobile phase with an appropriate K_d (Section 6.2) and t_R, as also discussed for normal-phase LC (previous subsection). The organic solvents in the mobile phase are referred to as modifiers. They need to be soluble in the mobile phase, and their effects or properties follow more or less their position in the eluotrope series (Section 2.8).

RPC is mainly based on the distribution of analytes between the solutions on the stationary phase and the more hydrophilic mobile phase, involving hydrophobic interactions or the analytes equilibrium distribution between these two phases (Section 6.2). This means that the obtainable separation of analytes is principally determined by the characteristics of the mobile phase. Understanding the parameters defining the properties of solutions (Chapter 2) and binding, as well as the relationship between structure and properties of analytes (Chapter 3), are therefore crucial for the development and use of optimal HPLC/FPLC methods of analyses. These factors are also the basis for the attraction of RPC to the analyst, as small changes in the mobile phase composition or in the samples applied to the system profoundly affect the separation characteristics of HPLC/FPLC systems. Important changes to consider are:

- concentration of salts;
- concentration of impurities;
- pH;
- amount of organic modifier;
- type of organic modifier;
- temperature.

Appropriate selection of column material and an optimized composition of the mobile phase to match the properties of the analytes under consideration, give the basis for efficient HPLC/FPLC systems measured as theoretical plates per metre (N m^{-1}), height equivalent to a theoretical plate (HETP), or resolution (R_s) (page 135) and allow analytes closely related in structure/properties to be separated.

The order of elution in RPC is generally opposite to that of analytes in normal-phase LC. This means that polar analytes elute first in RPC, whereas lipophilic analytes may be strongly associated with the stationary phase, and high concentrations of organic modifiers in the mobile phase may be needed for their elution. Separation of analytes with appreciable variation in polarity may thus require gradient systems using increasing proportions of organic modifiers in the eluent.

RPC is one of the most widely used forms of LC, owing to its potential for relatively high resolutions. However, HPCE techniques (Chapter 10) with their even greater potential for high resolution are now an efficient supplement or substitute for several HPLC methods. For separations of lipophilic compounds, *e.g.* lipids, RPC requires the use of non-aqueous mobile phases, and for the separation of high-M_r compounds/proteins it is generally advisable to use FPLC systems specially adapted to these purposes.

Ion-pair Reversed-phase Chromatography

Ions require an ion of opposite charge, a counterion, to form salts, which results in hydrophilic properties. The possibilities of extraction of ions from an aqueous solution to an organic solution by use of counterions with lipophilic side chains is, however, an old organic chemistry technique. Extraction of anions as sulfonates or sulfates to an organic phase thus requires a counter-cation with a lipophilic side chain. Correspondingly, the extraction of cations to

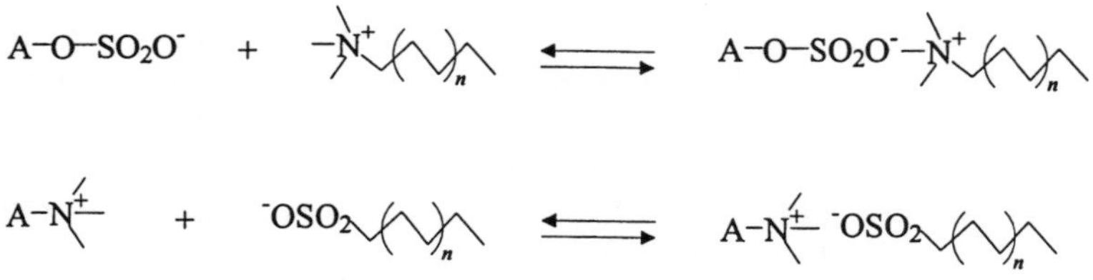

$A{-}O{-}SO_2O^- \quad + \quad {-}N^+{}_n \quad \rightleftharpoons \quad A{-}O{-}SO_2O^-{-}N^+{}_n$

$A{-}N^+ \quad + \quad {}^-OSO_2{}_n \quad \rightleftharpoons \quad A{-}N^+\ {}^-OSO_2{}_n$

Figure 8.3 *Illustration of ion-pairing equilibria for anion analytes (A-OSO_2O^-) and cation analytes (A^+NR_3). Increasing the length of the hydrocarbon side chain in the counterions facilitates formation of ion-pairs that are extractable to organic solvents*

an organic phase will be facilitated by use of a counter-anion with a lipophilic side chain, as illustrated in Figure 8.3.

Use of the ion-pairing technique in combination with RPC gives possibilities for efficient HPLC separation of highly polar/hydrophilic compounds as sulfate esters, *e.g.* glucosinolates.[1] These principles have also been used for successful HPLC separation of choline esters by use of anionic counterions. Success with ion-pair reversed-phase chromatography (I-P RPC) depends on the appropriate selection of counterions, their concentration, the use of organic modifiers in the mobile phase, and an appropriate stationary phase.

Ion-exchange Liquid Chromatography and Chromatofocusing

The principal feature underlying ion-exchange LC is described in Chapter 7. A great number of ion-exchange materials are commercially available (Section 7.2), including exchangers specifically developed for HPLC and FPLC techniques. Among these, amino acid analyses based on ion-exchange techniques are among the most often used column chromatographic LC methods (Section 7.8). Chromatofocusing, which is a special ion-exchange technique that was also developed for FPLC analyses of proteins,[2,3] gives in several cases efficient methods of analyses, as described in Section 7.10. Special ion-exchange materials such as Mono Q and Mono S (Section 7.2) have advantages when used in various FPLC ion-exchange methods, *e.g.* for protein separations and isolation, as illustrated by specific examples included in Chapter 12. The advantages of HPLC and FPLC ion-exchange techniques are the high efficiency obtainable, the easy change of gradient systems, which can be based on salt and pH-gradients, and their simplicity and speed.

Chiral Chromatography

Knowledge of the stereochemistry of analytes in biological materials, food, and feed is in several cases an important part of the information required from analytical chemistry in relation to food analyses. This is because enzymes/proteins are composed of asymmetric molecules containing several chiral centres. Enzymes have therefore the potential ability of exhibiting specificity towards different stereoisomers, including enantiomers and even compounds with differences in pro-chiral centres (Section 3.3).

Stereoisomers with physico-chemical differences in their properties, *e.g.* geometric or *cis-trans* isomers as well as diastereoisomers, can be separated by efficient conventional LC methods without the use of chiral reagents or phases. This is, however, not the case with enantiomers/optical antipodes, which require derivatization with chiral reagents for production of diastereoisomers. Alternatively, enantiomers can be separated by use of chiral stationary or mobile phases.

The chiral selectors often used in mobile phases for HPCE are described in the section on chiral selectors (page 256). Chiral stationary phases used in HPLC/FPLC also include various cyclodextrins bound to silica. Collectively, cyclodextrins are sometimes referred to as chiral cavity phases because their capacity to separate enantiomers relies on the ability of the enantiomers to enter the three-dimensional cyclodextrin cage. This gives possibilities for interactions which may be different from optical antipodes. Descriptions of the various other chiral stationary phases can be obtained from the commercial companies representing these HPLC columns or products.

Size-exclusion, Hydrophobic Interaction Chromatography, and Affinity Chromatography

Purification and analysis of proteins by use of FPLC/HPLC can be easily automated and optimized by selecting the appropriate techniques. Size-exclusion or gel filtration (Section 6.6) thus give a gentle technique, which can be used both for separation of molecules according to size/shape and for desalting or buffer exchange. It is a technique which is not especially sensitive to the presence of relatively high salt concentrations in the samples. Therefore, it can be used in multi-dimensional purification schemes after steps that result in a high ionic strength (Section 12.4; Table 12.4). As a desalting method, gel filtration can be used before techniques which require a low ionic strength, *e.g.* ion-exchange techniques (flash chromatography and chromatofocusing; Sections 7.9 and 7.10, respectively), electrophoresis (Chapters 9 and 10), and freeze-drying. The gel filtration media (page 142) need to be selected according to the purpose of using the technique, *e.g.* the type and size of molecules to be separated; media with a high degree of cross-linking (small G or P numbers) should be selected for desalting purposes. Determination of M_r and/or evaluation of separations of high-M_r compounds according to their M_r, size, or shape require the use of suitable reference compounds and selectivity curves (Figure 8.4). Significant deviations from the selectivity curves may result from variations in the hydrophobicity, shape, and binding of proteins to the column materials. This will, however, only be realized, if, for example, the M_r of the proteins of interest is known or obtained by other techniques.

Affinity chromatography, described in Section 6.7, and hydrophobic interaction chromatography (HIC, Section 6.8) are also techniques which, like gel filtration, can be used when the samples have relatively high I. Both techniques

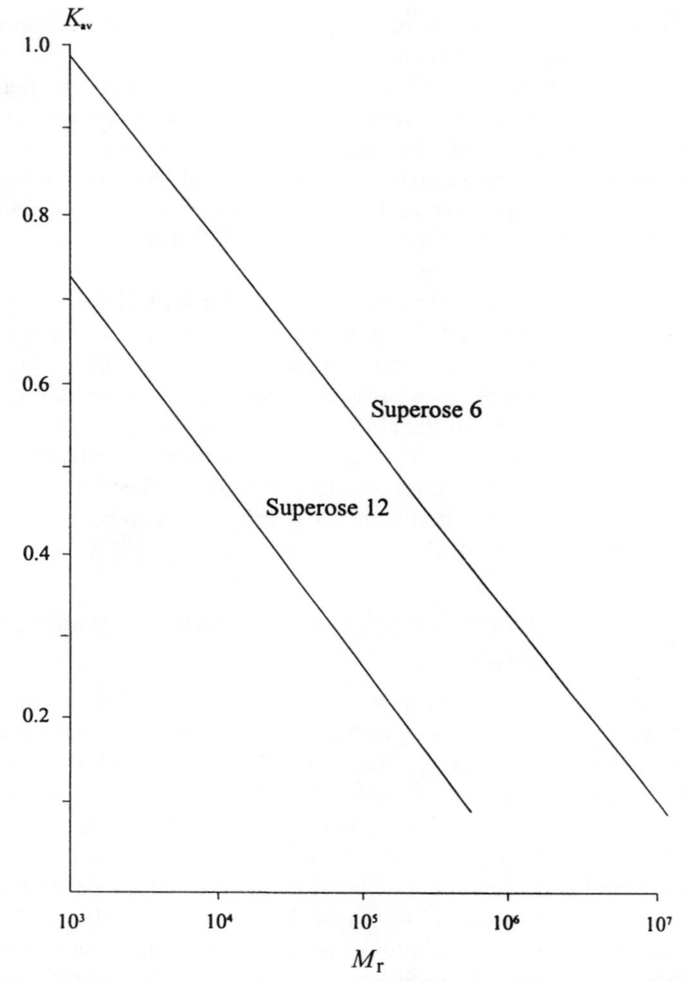

Figure 8.4 *Selectivity curves for Superose 6 and Superose 12, using globular proteins as reference compounds*

can, with advantages, be used as HPLC/FPLC methods of analysis. Affinity chromatography is a highly biospecific technique, and HIC offers also a highly effective technique, providing the compounds or analytes are stable at the relatively high salt concentrations required.

5 Applications of HPLC/FPLC

HPLC/FPLC methods have found wide applicability owing to the obtainable speed, efficiency, high resolution, high recovery, and sensitivity. Column materials and techniques are available for the analysis of nearly all types of biological molecules and constituents in food and feed. HPLC/FPLC employs column packings with small particle diameters (2–10 μm); such small beads and porous materials have a high surface area as well as reduced diffusion pathways for the analytes, which gives these techniques their advantages. However, these column materials also have high requirements for efficient group separations and purifications of the extracts prior to HPLC/FPLC to avoid destruction or changes to them (Chapters 4 and 7).

6 Selected General and Specific Literature

1. P. Helboe, O. Olsen, and H. Sørensen, *J. Chromatogr.*, 1980, **197**, 199.
2. L. Buchwaldt, L.M. Larsen, A. Plöger, and H. Sørensen, *J. Chromatogr.*, 1986, **363**, 71.
3. L.M. Larsen, T.H. Nielsen, A. Plöger, and H. Sørensen, *J. Chromatogr.*, 1988, **450**, 121.
4. T. Børresen, N.K. Klausen, L.M. Larsen, and H. Sørensen, *Biochim. Biophys. Acta*, 1989, **993**, 108.
5. S. Clausen, O. Olsen, and H. Sørensen, *J. Chromatogr.*, 1983, **260**, 193.
6. B. Bjerg, O. Olsen, K.W. Rasmussen, and H. Sørensen, *J. Liq. Chromatogr.*, 1984, **7**, 691.
7. B. Bjerg and H. Sørensen, in 'World Crops: Production, Utilization, Description. Glucosinolates in Rapeseeds: Analytical Aspects', ed. J.-P. Wathelet, Martinus Nijhoff Publishers, Boston, 1987, p. 125.
8. B.O. Eggum and H. Sørensen, in 'Absorption and Utilization of Amino Acids', ed. M. Friedman, CRC Press, Boca Raton, FL, 1989, vol. 3, ch. 17, p. 265.
9. B.O. Eggum, N.E. Hansen, and H. Sørensen, in 'Absorption and Utilization of Amino Acids', ed. M. Friedman, CRC Press, Boca Raton, FL, 1989, vol. 3, ch. 5, p. 67.
10. J.S. Fritz, D.T. Gjerde, and C. Pohlandt, in 'Ion Chromatography', Huthig, Amsterdam, 1982.
11. R.L. Grob, 'Modern Practice of Gas Chromatography', Wiley-Interscience, New York, 1985.
12. E. Heftmann, 'Chromatography–Fundamentals and Applications of Chromatographic and Electrophoretic Methods', Journal of Chromatography Library, Elsevier, New York, 1983, vols 22A and 22B.
13. W.H. Scouten, 'Affinity Chromatography–Bioselective Adsorption on Inert Matrices', Wiley-Interscience, New York, 1981, vol. 59.
14. H. Sørensen and P.M.B. Wonsbek, in 'Advances in the Production and Utilization of Cruciferous Crops', ed. H. Sørensen, Martinus Nijhoff/Dr. W. Junk Publ., Boston, 1985, p. 127.
15. H. Sørensen, in 'Rapeseed/Canola: Production, Chemistry, Nutrition, and Processing Technology', ed. F. Shahidi, Van Nostrand Reinhold, New York, 1990, ch. 9, p. 149.
16. W.W. Yau, J.J. Kirkland, and D.D. Bly, 'Modern Size-Exclusion Liquid Chromatography', Wiley-Interscience, New York, 1979.

CHAPTER 9

Electrophoresis

1 Introduction

Electrophoresis is the transport of ions or molecules in solutions under the influence of an applied electric field. The solutions containing charged compounds need to be included in a support media, which can be paper, silica, various types of gels, or 'free solutions' included in tubes or capillaries. The support media is placed in contact with electrodes (Figure 9.1), and an applied electric field will then force the charged compounds to move; anions toward the anode and cations toward the cathode.

Different types of instrumentation and electrophoretic techniques can be used. The choice may be determined mainly by the type of compounds to be analysed, the purpose of the study, and the apparatus and resources available. HPCE, considered in Chapter 10, comprises a series of techniques that exhibit great efficiency in studies of nearly all types of compounds. The various types of other electrophoretic techniques considered in this chapter each have their own limited area of application, which means that they have to be selected according to the properties of the compounds considered and the required information. Appropriate techniques for low-M_r compounds are generally different from techniques recommendable for high-M_r compounds. For some compounds on the borderline between low- and high-M_r, HPCE is often the only acceptable technique, as illustrated and explained in connection with the methods of analyses recommended for selected peptides/small proteins (Chapter 12).

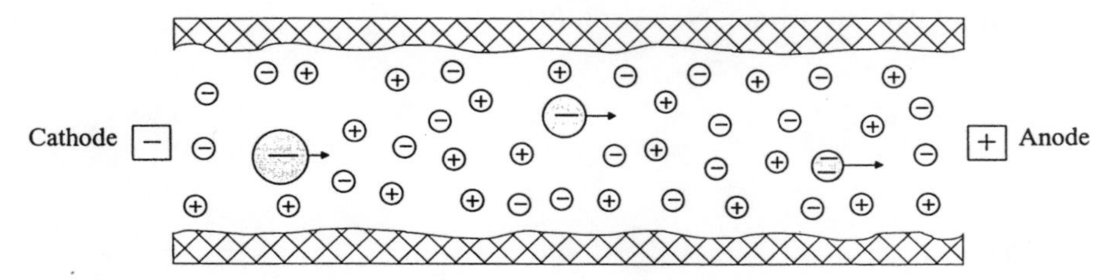

Figure 9.1 *Electrophoretic transport of charged compounds in a polar medium; anion analytes and buffer anions (−) move toward the anode; buffer cations (+) move toward the cathode*

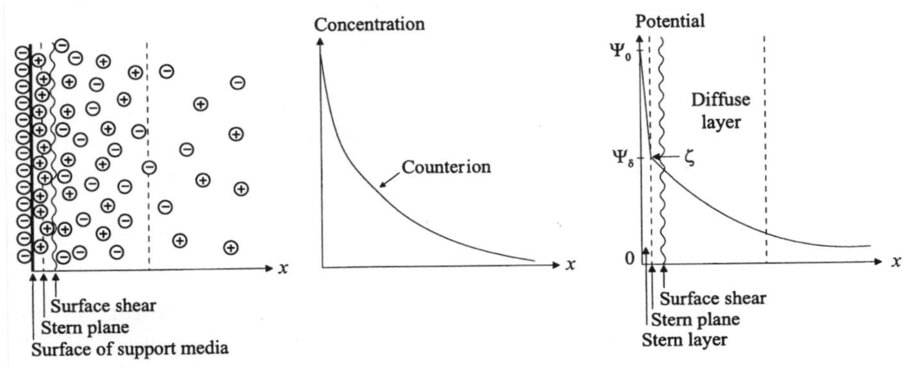

Figure 9.2 *Illustration of the accumulation of ions at different distances (x) from a charged surface of a support media, with an indication of the Stern plane (centre of ions directly adsorbed to the support media) and the potential changes from ψ_0 at the surface to ψ_δ in the Stern plane, approaching zero outside the diffuse layer*

The selection of technique determines whether the type of support media influences will be more or less pronounced and important. Solid support media, paper or gels, may give limitations and variations in obtainable results owing to variations in pore size and as a result of adsorption problems. The narrow tubes as those used in HPCE give the support media (silica wall) a high surface area in relation to the volume of the solution. Thereby, it generates electroosmotic flow, the Stern layer and zeta (ζ) potential (Section 9.2, Figure 9.2), which are more important factors in HPCE than generally found for paper- or gel-electrophoresis techniques, where the support media most often have a limited amount of charged groups bound to the stationary support media.

If we consider the transport in polar solutions of charged compounds in an electric field [E ($\mathrm{V\,cm^{-1}}$)] (Figure 9.1), the compounds will at constant velocity (v) obtain an electrophoretic mobility (μ_e). The size of μ_e depends on the properties of the solution [viscosity (η); pages 126 and 210], the M_r of the compounds, the size of net charges (z), and the radius of equivalent hydrodynamic radius (R) for globular proteins (high-M_r compounds). With other shapes for the compounds, *e.g.* prolate, oblate, and long rods, R shall be substituted with another function (Figure 6.5). With $e = 4.8 \times 10^{-10}$ esu $= 14.4 \times 10^{-14}$ V, the force from the electric field (F^*) can be calculated from:

$$F^* = Eze \tag{9.1}$$

At constant velocity, this will be equal to the friction (F') between molecules and the solution as defined by Stokes' equation (page 126):

$$F' = 6\pi\eta Rv \tag{9.2}$$

With $F^* = F'$, the result of this will then be:

$$\mu_e = \frac{v}{E} = \frac{ze}{6\pi\eta R} \tag{9.3}$$

where μ_e is given in $cm^2 V^{-1} s^{-1}$. With the assumption of $M_r = \frac{4}{3}\pi R^3 N_0 \bar{v}^{-1}$, where N_0 = Avogadro constant and $\bar{v}$ = partial specific volume ($cm^3 g^{-1}$), it gives:[1]

$$\mu = \frac{ze}{6\pi\eta}\left(\frac{3M_r\bar{v}}{4\pi N_0}\right)^{-1/3} = kzM_r^{-1/3} \tag{9.4}$$

The logarithmic form of this equation corresponds to a linear relation between μ_e and M_r, when the parameters included in k are kept constant. The analytes will have μ_e that is strongly dependent on the net charge (z), and in cases with appreciable electro-osmotic flow, this should be included as correction to μ_e. Reproducible results from electrophoresis require well-defined buffers (Chapter 2), and an efficient cooling system to remove heat generated during electrophoresis.

With the analytical potential in electrophoresis, a great number of principles and techniques has been developed, originally with zone electrophoresis as the main one. Selected examples of zone electrophoresis principles and techniques will be presented in the following sections, with focus on special selected techniques still considered as valuable methods for some purposes. A brief explanation of 'free solution' electrophoresis will be presented, including the moving boundary electrophoresis which has historical interest, as it was developed with the purpose of overcoming various problems met in zone electrophoresis. Moving boundary electrophoresis gave, however, not the expected results, but this technique can also be considered as the start of techniques as isotachophoresis and the newest and most popular technique, HPCE. With HPCE, the separations depend to a great extend on effects from the capillary wall, and it will therefore only in part be 'free solution' electrophoresis.

2 Electro-osmosis, Stern Layer, and Zeta Potential

Electrophoresis in an applied electric field results in transport of all 'free' and charged compounds in the solution. These compounds will in an aqueous buffer be surrounded by associated water and the amount of water included in this solvation sphere (Section 2.2) is different for different compounds. All types of electrophoresis will therefore create a flow of solvent/water either toward the anode or toward the cathode (Figure 9.1), depending on the types of ions having the most pronounced solvation. In the majority of techniques used for zone electrophoresis, the solvent flow, electro-osmosis, or electro-endo-osmosis, will normally be of limited importance compared to the transport of charged analytes by the electric field. However, in some cases it can be an important factor, which should be included in calculations of the electrophoretic mobility for charged analytes, and it can also be of importance in connection with determination of isoelectric points *e.g.* by isoelectric focusing. In the cases where the support media contains ions bound on the surface, a corresponding amount of counterions is required. This represents an appreciable amount when the surface area of the support media is large compared to the volume of solvent included in, for example, capillaries, and electro-osmosis will therefore be a dominating factor in various HPCE techniques (page 209).

The concentration of counterions in aqueous solutions will be high close to the charged surface of the support media and then gradually decrease with increasing distance to the surface. This will also be the case for the potential as illustrated in Figure 9.2, and the solvent close to the charged surface will therefore be less mobile. The solvent layer has for various theoretical considerations arbitrarily been divided by the Stern plane into a double layer: the Stern layer and the diffuse layer.[2] The adsorption of ions in this double layer was assumed to follow the Langmuir-type adsorption isotherm as the case is for other types of binding and chromatographic considerations (page 41).

Use of buffers with surface active counterions, as cationic surfactants (Table 2.6), in capillaries with negatively charged surfaces (Figure 9.2) will give a reversal of the charges due to adsorbed ions forming a double layer and thereby also reversal of the electro-osmotic flow (pages 233–234). Treating the Stern layer as a molecular condenser of thickness δ and with a permittivity $\varepsilon^* = \varepsilon_p \varepsilon_o \pi$($\varepsilon_p$ = dielectric constant of the solution; Section 2.8) gives a charge density σ_o at the surface of the shear (Figure 9.2):

$$\sigma_o = \frac{\varepsilon^*}{4\pi\delta}(\psi_o - \psi_\delta) \tag{9.5}$$

where ψ_o = potential at the surface of the support media and ψ_δ = potential at the Stern plane.

This equation is closely related to the equation for electro-osmotic mobility (μ_{eo}; equation 9.7) with:

$$\delta\sigma_o = \mu_{eo}\eta \quad \text{and} \quad \psi_o - \psi_\delta = \zeta \text{ (zeta potential)} \tag{9.6}$$

where η = viscosity.

The electrokinetic behaviour depends thus on the electrokinetic or zeta potential at the surface shear between the charged surface and the electrolyte solution. The exact location of the shear plane is, however, an unknown feature of the electric double layer, but it is a region that changes with the type of solution and its viscosity as indicated by the above mentioned equations. Depending on the size of the double layer, the equations for the electro-osmotic mobility (μ_{eo}) can be written as:[3,4]

$$\mu_{eo} = \frac{\varepsilon^*\zeta}{6\pi\eta} \quad \text{or} \quad \mu_{eo} = \frac{\varepsilon^*\zeta}{4\pi\eta} \tag{9.7}$$

For the more applied aspects considered in this chapter, it is sufficient to consider electroosmosis as a result of solvation of charged compounds transported in an electric field and thereby causing a solvent flow. The main factor in electroosmosis is owing to formation of an electric double layer at the surface of the support medium, involving cations from the solutions in the cases with negatively charged groups on the support media. In HPCE, where electroosmosis is much more important than in the zone electrophoresis techniques considered in this chapter, it is also easy to change the direction of the electroosmotic flow, as mentioned above (pages 233–234). Otherwise, zone electrophoresis will only have limited electroosmotic flow as a result of a negatively charged surface caused by, for example, carboxylate, phosphate, or sulfate groups in the polysaccharides of the solid support. Application of an electric field results in movement of cations in the mobile layer (Figure 9.2) toward the cathode, with the cations pulling solution with them and thereby creating a net electroosmotic flow in this direction. An electro-osmotic flow directed towards the anode may be obtained for support media having a positively charged surface or by use of some buffer solutions (Figure 9.4, *vide infra*). The main conclusions of this section are that the electro-osmotic flow usually has limited disturbing effects in zone electrophoresis, it can create a problem in isoelectric focusing and the HPCE techniques take advantage of EOF in most instances.

3 Zone Electrophoresis

The technique of zone electrophoresis involves the migration of charged particles in a support media to minimize convectional disturbances. Various forms of support material, solid or gel framework materials, are generally used, *e.g.*:

- paper
- cellulose acetate
- silica
- starch gel
- agar/agarose gels
- polyacrylamide gel

A general feature of the supporting medium is that it should be relatively inert. In fact, factors as adsorption, electroosmosis (Section 9.2), and molecular sieving may influence the migration of analytes, depending on the medium chosen.

Adsorption is the retention of analytes by the supporting material. Adsorption may cause tailing of the sample leading to non-distinct bands, decreased electrophoretic mobility, and reduced resolution.

Molecular sieving is a characteristic of gels (Figure 6.2) by which the movement of large molecules is increasingly hindered by decreasing pore size, as the molecules have to travel through the pores. This feature is seen in paper and in polyacrylamide type gels. Conversely, Sephadex gels used in chromatographic techniques (Chapter 6), have particles with small pores that exclude large molecules, which therefore only travel outside the particles and with the highest velocity. The pore size of the various gels may be regulated during preparation, and electrophoresis in gels has formed the basis for many specialized techniques (Section 9.5).

Paper electrophoresis – HVE

Low-voltage paper electrophoresis is seldom used nowadays owing to long analysis times, and a considerable diffusion of, especially, small molecules (pages 127–130). These problems are, however, diminished by use of high-voltage paper electrophoresis (HVE). A typical experimental arrangement for this technique is shown in Figure 9.3.

The samples are spotted on the paper, which then is saturated with buffer and the contact with the power supply is made possible by use of wetted wicks immersed in the two buffer reservoirs. An efficient cooling system is necessary owing to the high currents produced.

HVE is an experimentally simple technique which may be carried out by use of different buffer systems and corresponding fixed time intervals, *e.g.* as described in Box 9.1.

Amino acids are positively charged at pH 1.9 even if they have $\mathrm{p}K_a'$ values close to 2. At pH 6.5, the net charges will be zero for neutral amino acids, positive for basic amino acids, and negative for acidic amino acids (Section 2.4). HVE at pH 6.5 gives, in accordance with the net charge of the compounds, mobilities (μ_e) for basic and acidic amino acids which are a function of their size or M_r (Equation 9.4). This is also the case for relatively small peptides. HVE at pH 3.6 gives for the acidic amino acids appreciable differences in μ_e following the variations in their $\mathrm{p}K_a'$-values and thereby in their net charges. HVE under these conditions has also proved over many years to be an efficient qualitative method of analysis of amino acids and peptides,[5] with possibilities for

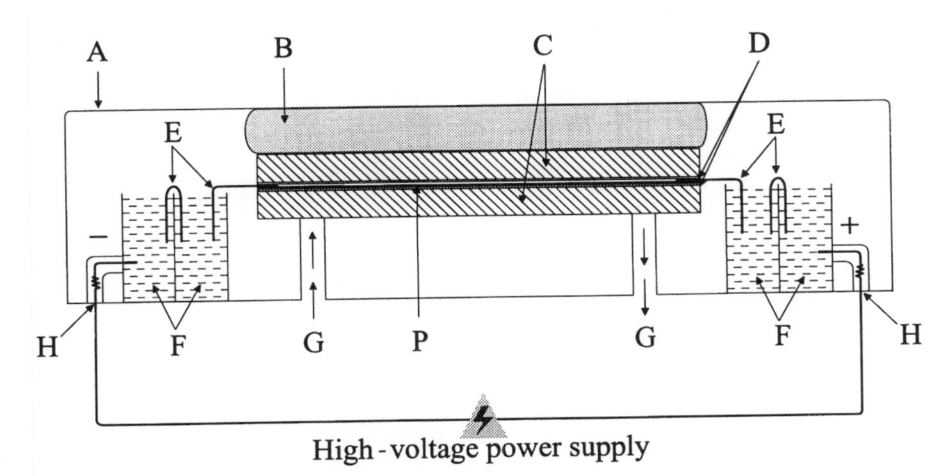

Figure 9.3 *Schematic illustration of an horizontal electrophoresis unit for high-voltage electrophoresis (HVE). A = cover, B = pad, C = cooling plates, D = insulating plastic sheet, E = paper wicks saturated with buffer, F = separated buffer compartments, G = cooling water, H = electrodes connected to 10 kV, 100 mA high-voltage power supply unit, P = paper sheet with the analyte and buffer solution*

discriminating between diastereoisomeric 3,4-substituted acidic amino acids as well as α- and γ-glutamyl derivatives.

Separations of macromolecules such as proteins by HVE are poor owing to severe problems with adsorption and molecular sieving in the paper capillaries which are of different sizes. Large peptides or small proteins will generally create

Box 9.1 *HVE of low-*M_r *compounds*

Electrophoresis paper sheet (P in Figure 9.3) and paper wicks (E in Figure 9.3) are prepared from Whatman No. 3 MM paper (sizes are determined by the actual instrument). For the electrophoresis, one of the following buffer systems are chosen, depending on the compounds to be separated. The run time and applied voltage (electrophoresis paper sheet: 23 × 46 cm) are stated for various compounds. The resulting current will vary depending on the actual instrument; however, *ca.* 90 mA will in general be appropriate

Buffer, pH 1.9

This buffer is prepared from glacial acetic acid (80 ml), formic acid (20 ml), and 900 ml water with pH adjusted to 1.9.

Acidic amino acids: run time, 2 h, 3 kV*
Neutral amino acids: run time, 1 h, 3 kV*
Basic amino acids, alkaloids, biogenic amines, anthocyanins: run time, 0.5 h, 3 kV*
Glucosinolates, sulfate/phosphate esters: run time, 2 h, 3 kV†

* Use a starting line, 15 cm from the anode, † Use a starting line, 15 cm from the cathode

Buffer, pH 3.6

This buffer is prepared from pyridine (5 ml), glacial acetic acid (50 ml), and 945 ml water with pH adjusted to 3.6.

Acidic amino acids and peptides: run time, 2 h, 3.2 kV†
Carboxylic acids, glucosinolates, sulfate/phosphate esters: run time, 1–2 h, 3.2 kV†
Alkaloids, biogenic amines, anthocyanins: run time, 1–2 h, 3.2 kV†
Basic amino acids and peptides: run time, 0.5–1 h, 3.2 kV*

* Use a starting line, 15 cm from the anode, † Use a starting line, 15 cm from the cathode

Buffer, pH 6.5

This buffer is prepared from pyridine (50 ml), glacial acetic acid (2 ml), and 948 ml water with pH adjusted to 6.5.

The majority of acidic amino acids and peptides: run time, 50 min, 5 kV†
The majority of carboxylates, glucosinolates: run time, 50 min, 5 kV†
The majority of sulfate/phosphate esters: run time, 50 min, 5 kV†
Alkaloids, biogenic amines, anthocyanins: run time, 0.5 h, 5 kV*
Basic amino acids, peptides and other cationic compounds: run time, 0.5 h, 5 kV*

* Use a starting line, 15 cm from the anode, † Use a starting line, 15 cm from the cathode

After analysis, the paper is removed, dried, and the sample components located, *e.g.* by their colour or fluorescence (pages 102–114) or by staining with various dyes (Table 5.6) as used for paper chromatography (Chapter 6).

the same problems in HVE. Furthermore, as they are not denatured as required in connection with protein staining used for other traditional types of zone electrophoresis (*vide infra*), these techniques will be unable to solve the analytical problems posed by with such compounds. The new technique based on HPCE are, however, very promising for such proteins and peptides.

HVE techniques have also been found to be well suited for a wide range of low-M_r compounds other than amino acids and peptides, *e.g.* anthocyanins (page 104). Carboxylic acids, glucosinolates,[6] biogenic amines,[7] and aromatic choline esters are other examples of compounds where HVE has found successful application. An illustration of such results is shown in Figure 9.4.

The compounds are identified after HVE by their characteristic colour in long wavelength UV (Figure 9.4, where 4-H = 4-hydroxybenzoylcholine, *c*-I and *t*-I are *cis*- and *trans*-isoferuloylcholine, *c*-C and *t*-C are *cis*- and

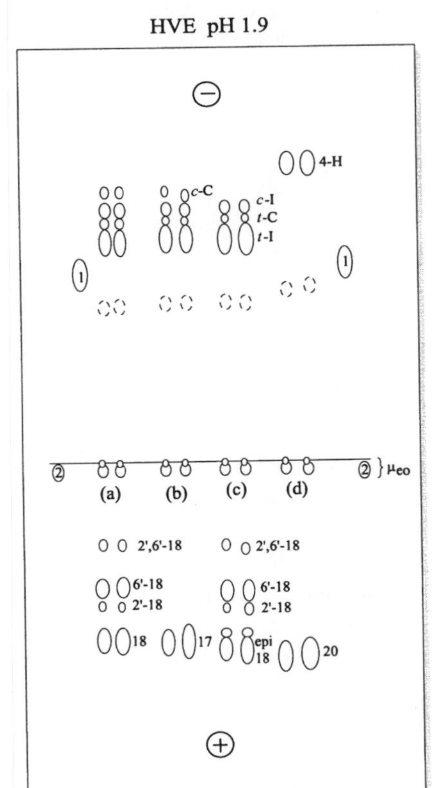

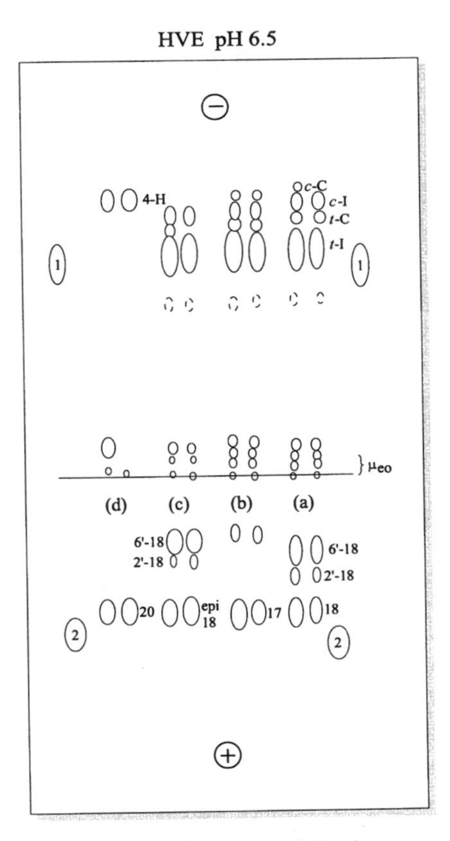

Figure 9.4 *Results obtained by HVE at pH 1.9 and 6.5 as described in Box 9.1. of extracts from seeds of (a)* Barbarea vulgaris *R. Br., (b)* B. verna *(Mill.) Asch., (c)* B. intermedia *Bor., and (d)* Lepidium campestre *(L.) R. Br.; sinapine (1) and sinapic acid (2) are included as references. See text for identification of compounds*

trans-coumaroylcholine, 18 = glucobarbarin, 17 = gluconasturtiin, epi-18 = glucosibarin, 20 = glucosinalbin, 2′-18, 6′-18, and 2′,6′-18 are the glucosinolates 18 and epi-18 with isoferuloyl as ester on the 2′ and 6′ carbons in the thioglucose part). It is seen that HVE allows an efficient separation of aromatic choline esters from anions (glucosinolates), which also separate according to their size (M_r and *cis-trans* steroisomers).[6] The electroosmotic flow is of different size, and against the cathode at HVE pH 6.5 and against the anode for HVE pH 1.9.

Cellulose Acetate

Cellulose acetate is a modified form of paper in which the hydroxy groups of cellulose have been acetylated. This results in a less hydrophilic material than untreated paper, and care should be taken to prevent drying out by evaporation during electrophoresis. One of the advantages of cellulose acetate is that very little adsorption occurs, leading to improved resolution even at low voltage. The material is also transparent, facilitating spectrophotometric determination of the analytes on the electropherogram. Electrophoresis based on cellulose acetate will in general separate the same range of compounds as paper electrophoresis.

Silica

Different materials, such as silica, kieselguhr, cellulose, *etc.*, can be prepared on glass plates, as for thin layer chromatography, and used for electrophoresis in a horizontal system, as described for paper electrophoresis (Figure 9.3). Thin-layer electrophoresis (TLE) constitutes a convenient system to combine chromatography and electrophoresis in two dimensions, and has been used successfully for the separation of protein and nucleic acid hydrolysates.

Starch Gel

Starch has been used as supporting material in the earliest types of gel electrophoresis systems. Starch gels can be prepared with high porosity [<2% (w/v) starch] or low porosity [8 to 15% (w/v) starch] by heating and then cooling a mixture of partially hydrolysed starch. The pore size of starch gels is, however, difficult to reproduce exactly, and the actual pore size can not be determined. Starch gels have in particular been used for the separation of isoenzymes. The general use of starch gels has now primarily been superseded by agar/agarose gels or polyacrylamide gels (following two subsections).

Agar/Agarose

Agar is a mixture of polysaccharides isolated from seaweed. The dominant polysaccharides are two galactose-based polymers, agarose and agaropectin. A 1% (w/v) agar solution forms a gel with large pore size and low frictional resistance, which is well suited for the separation of large molecules.

Agarose, which is now mainly preferred to agar, is a linear polysaccharide made up of repeating units of galactose and 3,6-anhydrogalactose. Substitution of the sugar residues with ionic groups, especially sulfate, may occur to various degrees, introducing charges to the otherwise neutral polymer. This is not desirable, as electroosmosis as well as ionic interaction between the gel and sample molecules may occur. Agarose is available in different purity grades, based on sulfate concentration, and a high grade should generally be preferred. Agarose gels are easily prepared from a dry powder (Boxes 9.4 and 9.9).

Usually, a concentration of agarose between 1 and 3% (w/v) is suitable. As for agar, the pore sizes of the agarose gel are relatively large, and this material is therefore often used for techniques where molecular sieving is unwanted, *e.g.* immunoelectrophoresis (page 380) and isoelectric focusing (page 190), and a range of different gels for this purpose are commercially available. Agarose is also widely used in nucleic acid work.

Polyacrylamide Gels

Electrophoresis in polyacrylamide gels is often referred to as PAGE, an abbreviation of *p*oly*a*crylamide *g*el *e*lectrophoresis. Polyacrylamide gels are made by polymerization of acrylamide monomers in the presence of a cross-linking agent as well as catalysts (Figure 9.5).

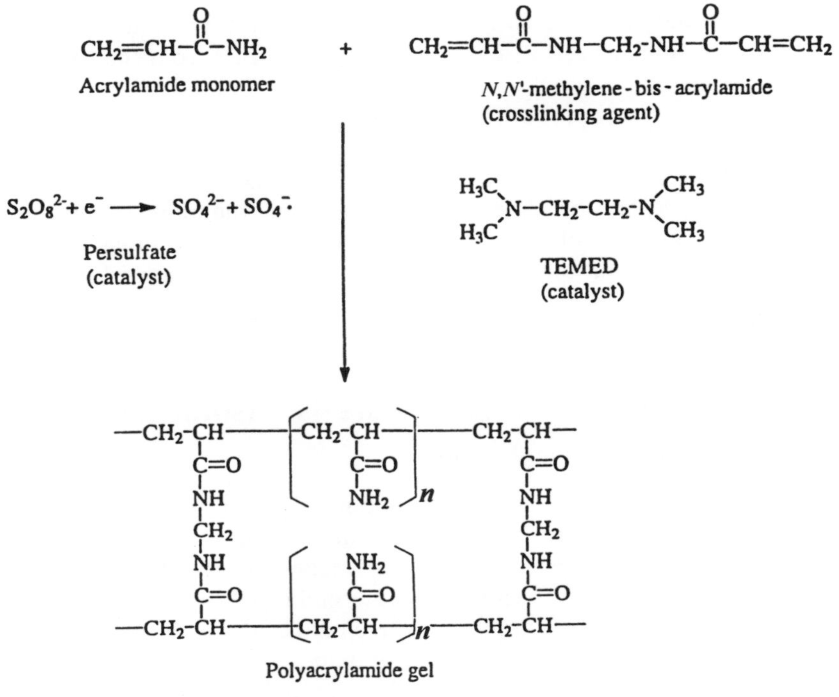

$$CH_2{=}CH{-}C({=}O){-}NH_2 \quad + \quad CH_2{=}CH{-}C({=}O){-}NH{-}CH_2{-}NH{-}C({=}O){-}CH{=}CH_2$$

Acrylamide monomer

N,N'-methylene-bis-acrylamide (crosslinking agent)

$$S_2O_8^{2-} + e^- \longrightarrow SO_4^{2-} + SO_4^{-\cdot}$$

Persulfate (catalyst)

$$(H_3C)_2N{-}CH_2{-}CH_2{-}N(CH_3)_2$$

TEMED (catalyst)

Polyacrylamide gel

Figure 9.5 *The formation of polyacrylamide gels*

The polymerization is based on free radicals with N,N,N',N'-tetramethylethylenediamine (TEMED) initiating the reaction by donation of an electron to persulfate. As oxygen inhibits the formation of free radicals, gel solutions are normally degassed prior to reaction. The occasional introduction of the cross-linking agent N,N'-methylene-bis-acrylamide (Bis) during polymerization results in formation of a complex network, the porosity of the gels being determined by the relative proportion of the acrylamide monomer and the cross-linking agent. Procedures for preparation of polyacrylamide gels are described in Boxes 9.2 and 9.7.

In PAGE it is common to use T (total acrylamide, %) and C values (cross-linking, %) to characterize the gel. The T and C values are calculated as:

$$T(\%) = \frac{\text{acrylamide (g)} + \text{Bis (g)}}{\text{volume (ml)}} \times 100 \tag{9.8}$$

$$C(\%) = \frac{\text{Bis (g)}}{\text{acrylamide (g)} + \text{Bis (g)}} \times 100 \tag{9.9}$$

At low C values (*ca.* 2%), the gel is homogeneous, whereas with increasing C, the gel becomes more irregular; C is often held at about 5%, whereas the T value varies between 3 and 30%, the low percentage gels having the largest pore sizes. Owing to the high reproducibility of gel formation, the porosity of gels can be chosen to match specific separation tasks, *e.g.* analysis of ionic compounds of similar charge but different size and shape. It is also possible to prepare polyacrylamide gradient gels in which the pore size varies from the top to the bottom of the gel.

Polyacrylamide gels show very limited adsorption, as well as electroosmosis, and, unlike agarose gels, they cover a wide range of pore sizes. A drawback is that some of the reagents for polyacrylamide gels are hazardous, and care has to be taken during preparation. Alternatively, to minimize handling of hazardous reagents, many suppliers of electrophoresis devices also sell a range of casted gels and acrylamide solutions. Polyacrylamide gels are well suited for various purposes, and are especially used for analysis of proteins and other macromolecules.

4 Moving Boundary Electrophoresis – Isotachophoresis

Moving Boundary Electrophoresis

Moving boundary electrophoresis is electrophoresis performed in free solution, and is a technique developed by A. Tiselius in Sweden about 60 years ago.[8] Electrophoresis is performed in a Tiselius apparatus, essentially a U-tube consisting of large diameter ascending and descending electrophoresis tubes to which a potential difference is applied. Charged molecules with similar electrophoretic mobility tend to move together as bands, with boundaries formed between analytes of slightly different mobility. Special scanning optics (Schlieren) are used to detect changes in refractive index at separated boundary interfaces. The technique suffers, however, from serious problems caused by tubes that are too big and the evolution of temperature gradients, and thereby density gradients, within the electrophoretic medium. The resulting formation of convection currents easily disrupts separation by mixing of zones. For these and several other reasons, this technique is only of historical interest; it can be considered as the start of new developments in electrophoresis, resulting in isotachophoresis and the much more important HPCE technique.

Isotachophoresis

Isotachophoresis is a form of moving boundary electrophoresis, in which separation of the ionic analytes is achieved by stacking them into discrete zones as a function of their mobilities. The sample is inserted between two electrolyte solutions, a leading electrolyte having a higher mobility than the sample ions, and a terminating electrolyte, having a lower mobility (Figure 9.6).

When the electric field is applied, potential gradients evolve resulting eventually in the same velocity for the various ions in the sample solution. The reason for this phenomenon is that the electric field will be stronger in regions with low mobility ions, causing them to move at the same velocity as more mobile ions, which are affected by a lower electric field. Once equilibrium is achieved, the sample bands will migrate at the same speed in order of mobility, from most to least mobile. Either cations or anions can be separated in this system, but not both at once. The general use of small diameter electrophoresis tubes reduces the problem of temperature and density gradients seen in moving

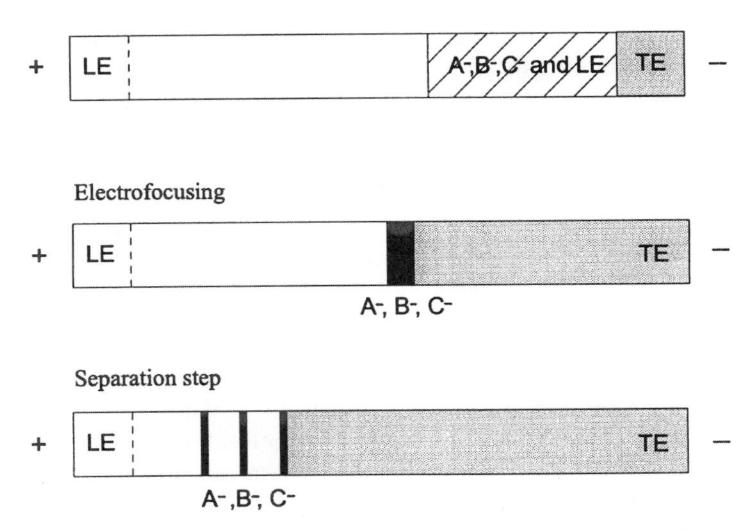

Figure 9.6 *Isotachophoresis; schematic illustration of the separation of anion analytes (A^-, B^-, C^-) using a fast moving leading electrolyte (LE) and a terminating electrolyte (TE)*

boundary electrophoresis. Moreover, a dynamic focusing of bands occurring at zone boundaries makes the resolution of this technique relatively high. This method can be used as an analytical tool for separation of small amounts (μg) of analytes. It can also be used as a preparative technique for the separation of relatively large quantities of samples, especially when columns containing polyacrylamide are used.

5 Specialized Electrophoretic Techniques

In addition to HVE designed for the separation of low-M_r compounds on paper (pages 183–186), slab gels of agarose (Box 9.9), or polyacrylamide gels without Ampholine® or without SDS prepared on glass plates, can be used for free zone electrophoresis of native proteins. However, the more specialized electrophoretic techniques described in the following sections have found more general applications in the analysis of proteins.

Isoelectric Focusing

Isoelectric focusing (IEF)[12] is a high-resolution technique, where ampholytes are separated according to their isoelectric point (pI), *i.e.* the pH at which the net charge is zero (Section 2.4). The technique is mainly used for proteins and is especially well suited for separating isoenzymes. Determination of pI of particular components is also widely used, but the exact value may vary owing to binding of various buffer constituents (Section 2.4). With IEF it is possible to separate compounds differing only by 0.01 pH units in pI. Separation is achieved by use of a pH-gradient, established by a mixture of low-M_r carrier ampholytes as amino acids and peptides with M_r 600–900, with closely spaced pI and individual pI which cover a wide pH-range. When an electric field is applied, all ions, including the sample ions, will travel according to their actual charge. The carrier ampholytes will create a pH-gradient and the analytes will proceed into the pH-gradient, where their charge will gradually approach zero when they reach the pH corresponding to their isoelectric point. The separation principle is illustrated in Figure 9.7.

Application of sample may be performed wherever it is appropriate, as the sample molecules will travel according to their actual charge at the specific position in the pH-gradient. This also means, that even a diffuse sample application will end up with sharp band focusing. The pH-gradient chosen may be narrow or wide depending on the actual sample composition (pI-values) as well as the resolution needed. To obtain this, it is possible to purchase commercial ampholyte mixtures covering a wide range of different pH-intervals. As control, the use of a sample of pI-markers in both sides of the gel to indicate the success of the established pH-gradient is recommended (Table 9.1). This is also required if IEF is used for determination of pI of a protein. Reproducibility of such data requires that the proteins are handled with great care to avoid any modifications of the protein structure and composition during extraction (Chapter 4) or caused by binding of buffer ions or other molecules (Chapter 3).

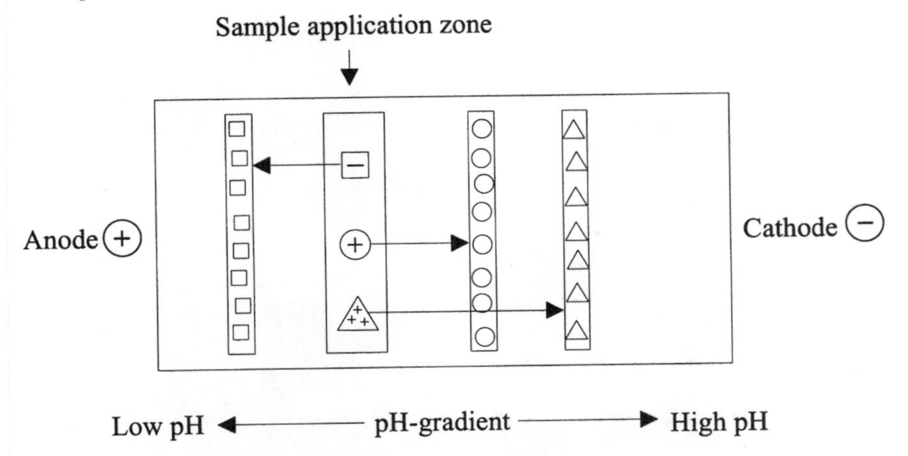

Figure 9.7 *The principle of isoelectric focusing, illustrating the transport of ampholytes in a pH-gradient until they loose their net charge. Subsequently, the focused bands will only move as a result of electro-osmosis*

Proteins are thus often associated with lipids and/or carbohydrates (DF; Chapter 14), which will result in shifted pI or streaks in the gel.

IEF may be performed in gels as well as in free solution, which has gained renewed attention in connection with HPCE and preparative isoelectric focusing (*vide infra*). The systems used in the more traditional slab gel zone electrophoresis can be horizontal and vertical systems for both analytical and preparative separation. For free solution IEF, the use of a sucrose gradient in vertical column systems is possible. Various commercial instruments are available, both vertical and horizontal systems (*e.g.* Bio-Rad® Mini-Protein apparatus, Easy 4-III Camlab apparatus®, and Pharmacia Gel Electrophoresis apparatus, including the Phast Gel Electrophoretic System™) (Figure 9.8).

Table 9.1 *Example of pI-marker mixtures available for a pH-gradient from pH 3.5 to 9.5*

pI-Marker	*pI-Value*
Amyloglucosidase	3.50
Methyl red	3.75
Soybean trypsin inhibitor	4.55
β-Lactoglobulin A	5.20
Bovine carbonic anhydrase B	5.85
Human carbonic anhydrase B	6.55
Horse myoglobin	6.85; 7.35
Lentil lectin	8.15; 8.45; 8.65
Trypsinogen	9.30

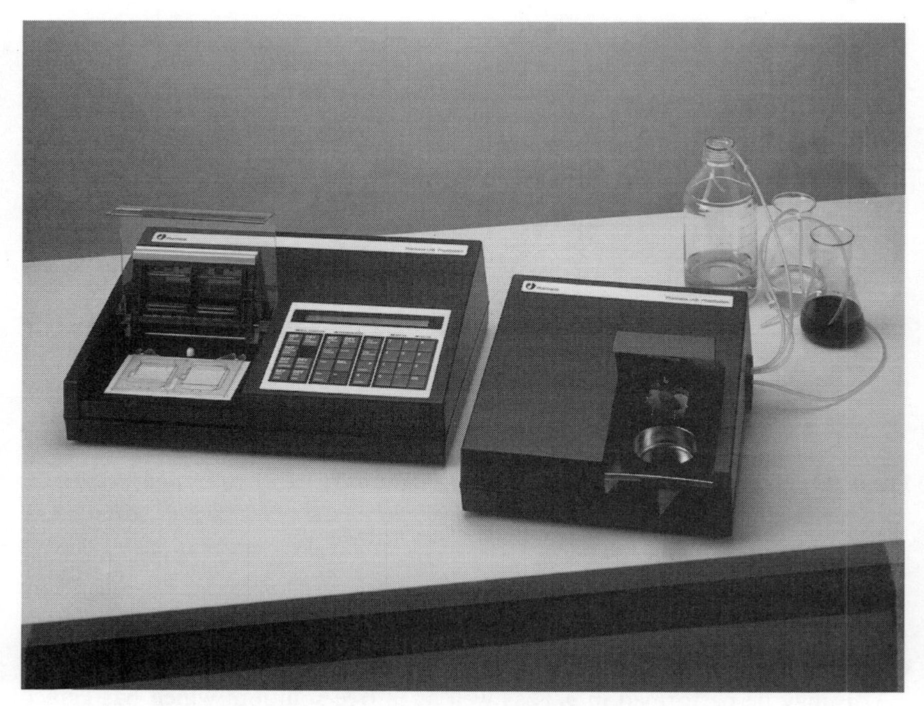

Figure 9.8 *Pharmacia Phast Gel Electrophoretic System™; separation unit (left) and the development unit (right), purchased from Pharmacia Biotech, Uppsala, Sweden*

The components of both vertical and horizontal systems serve the same functions, although they may look rather different. In instruments based on the vertical system, it is important that the upper buffer compartment is well sealed against leaks by appropriate mounting of the glass plates (Figure 9.9A).

IEF gels are usually either agarose gels or highly porous polyacrylamide gels [2–4% or, for example, 5% *T* and 3.3% *C* (page 188)] as molecular sieving should preferably be avoided in this technique. As the cost of the ampholyte mixture is rather high, ultrathin gels of about 0.2 mm thickness are preferred. The dimensions of the gel depend on the actual electrophoretic system, but apparatus exists (*e.g.* the Bio-Rad® Mini-Protein apparatus and the Pharmacia Phast Gel Electrophoretic System™) in which the maximum gel size of 5 × 5 cm is sufficient. For most analytical applications, the mini-slab gel systems have gained considerable popularity owing to the reduced amount of time and materials required for electrophoresis. The possibility of using two gel plates in the Phast Gel Electrophoretic System™, *e.g.* one for protein staining and one for enzyme staining, and the automatized staining/destaining system are also part of the reason for the popularity of the systems. IEF separations are performed at relatively high voltages (up to 2000 V), and a cooling system, as in other zone electrophoresis systems, including HVE, is

(A)

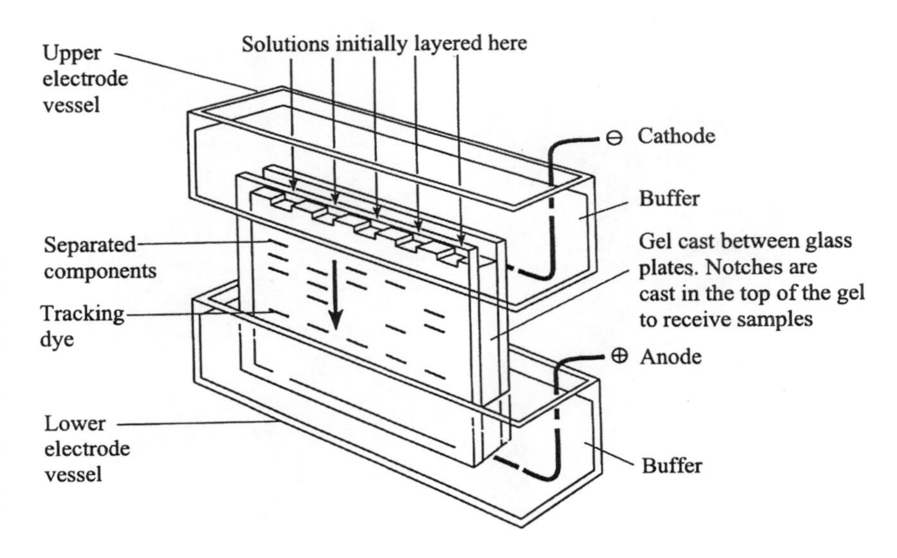

(B)

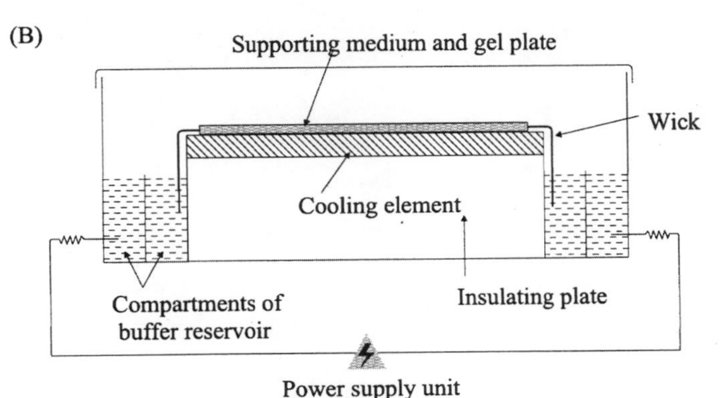

Figure 9.9 (A) *Slab gel between glass plates separated by spacers (1 mm thick) for use in a vertical apparatus and with wells produced by the use of a well-forming template.* (B) *A schematic illustration of a horizontal apparatus where sample applications are made on the slab gel plate at an appropriate distance from the end and not in wells as for vertical systems*

therefore required. With small, ultrathin gels and high voltages, separation is usually complete within a few hours. An example of the procedure used for IEF (Pharmacia Phast Gel Electrophoretic System^{TM}) of dietary-fibre-associated proteins from rapeseed is described in Boxes 9.2 and 9.3 and the results from the analysis are shown in Figure 9.10.

IEF may also be performed using agarose gels for the Pharmacia Phast Gel Electrophoresis System^{TM} (Boxes 9.4 and 9.5).

In IEF, proteins are often better resolved when a high-voltage gradient is

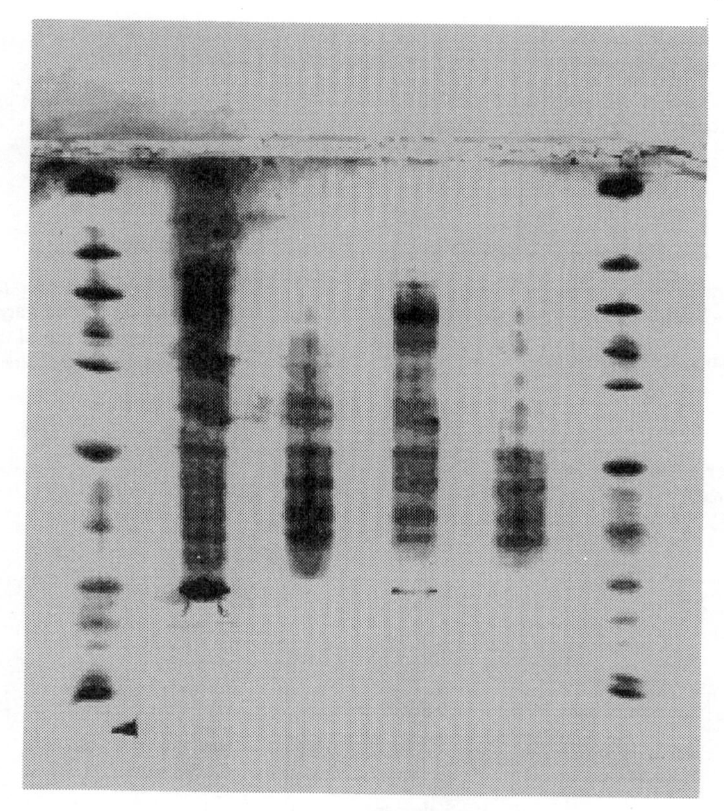

Figure 9.10 *Separation of dietary-fibre-associated proteins from rapeseed by isoelectric focusing (pH 3.5–9.5) on the Pharmacia Phast Gel Electrophoretic System™ using a polyacrylamide gel. The bands on the left and right of the gel originate from a pI-marker sample*

established across the gel, and this gradient can only be maintained if efficient cooling is achieved. To obtain efficient heat transfer from the gel to its surroundings it is advantageous in IEF, as well as other types of zone electrophoresis, to use slab gels rather than disc gel electrophoresis (page 196). Slab gels have therefore become more widely used than disc gel electrophoresis. Furthermore, the use of slab gels allows easy comparison of several protein samples run on the same gel and provides uniformity during polymerization of the gel, running conditions, staining, and destaining. However, as mentioned in the subsection below, disc gel electrophoresis still has advantages in connection with studies of some native and biologically active proteins/enzymes. The staining for enzyme activity in gels from disc electrophoretic separations is thus less sensitive to the time used in the required reactions and it usually gives a more sensitive detection.[9]

Box 9.2 *Preparation of polyacrylamide gels for isoelectric focusing*

This example comprises the preparation of 10 small gels for use in the Pharmacia Phast Gel Electrophoretic System™. The gels are intended for isoelectric focusing, which explains the addition of Ampholine® to the gel mixture.

Reagents for preparation of 10 gels (4.5 × 5 cm)

3.33 ml polyacrylamide stock solution*
1.00 ml Ampholine® (*e.g.* pH gradient 3.5–9.5)
9.3 μl TEMED (*N,N,N′,N′*-tetramethylethylenediamine)
15.6 ml deionized water
39.9 μl ammonium persulfate solution†

*Polyacrylamide stock solution: 30% acrylamide + 0.93% *N,N′*-methylene-bis-acrylamide. Store in darkness. Use within 2 weeks of preparation
†40.00 mg ammonium persulfate/100 μl; should be freshly prepared

Excess reagent is collected for waste disposal

Procedure for preparation of gels

1. Prepare a mould of two glass plates (125 × 260 mm): the hydrophobic side of a GelBond PAG film is adhered to one of the glass plates by few drops of water and the surface of the other glass plate is pretreated with Repel-silane. The glass plates are mounted with 0.5 mm rubber spacings between them and are kept together by clips
2. Mix the listed reagents (except ammonium persulfate) and degas the solution
3. Add the ammonium persulfate to the mixture and mix by gentle agitation
4. Transfer the mixture to the mould (1.) and leave to polymerize for at least 1 h at room temperature
5. Disassemble the mould and cut the gel stuck to the gel bond film into *ca.* 4.5 × 5 cm pieces
6. Store gels in a moist chamber at 4 °C and use them within 2 weeks

Box 9.3 *Procedure for IEF (polyacrylamide gels, pH 3.5–9.5) on the Pharmacia Phast Gel Electrophoretic System™*

Gels are prepared as described in Box 9.2.

Preparation of samples

The samples are to be solubilized in deionized water or a low concentration buffer. The protein concentration must be *ca.* 1 mg ml^{-1}

Prefocusing and focusing conditions

Prefocusing	2000 V	2.5 mA	1.8 W	15 °C	75 Vh
Sample application	200 V	2.5 mA	1.8 W	15 °C	15 Vh
Focusing	2000 V	2.5 mA	1.8 W	15 °C	210 Vh

Silver staining of polyacrylamide gels

Reagent list

Prepare the following reagents in a fume cupboard. The volumes stated indicate the minimum amount of the reagent used per run.

(*continued overleaf*)

1. 20% trichloroacetic acid (TCA) (200 g TCA/1000 ml deionized water), 80 ml
2. 50% EtOH + 10% glacial acetic acid (500 ml EtOH + 100 ml glacial acetic acid + 400 ml deionized water), 80 ml
3. 10% EtOH + 5% glacial acetic acid (100 ml EtOH + 50 ml glacial acetic acid + 850 ml deionized water), 400 ml
4. 8.3% glutardialdehyde (27 ml glutardialdehyde + 54 ml deionized water), 80 ml; should be freshly prepared
5. Deionized water, 400 ml
6. 0.5% silver nitrate (2.5 g silver nitrate/ 500 ml deionized water), 80 ml
7. 'Developer' (32 ml 12.5% Na_2CO_3* + 128 ml deionized water + 64 μl formalin), 160 ml; should be freshly prepared
8. 5% acetic acid (50 ml glacial acetic acid + 950 ml deionized water), 80 ml

*125 g Na_2CO_3/1000 ml deionized water or 337.5 g $Na_2CO_3 \cdot 10\ H_2O$/1000 ml deionized water

Staining and destaining procedure based on the Pharmacia Phast Gel Developing Device

Step No.	Reagent No.	In*	Out*	Time (min)	Temp. (°C)
1	1	1	5	5	20
2	2	9	2	2	50
3	3	3	2	2	50
4	3	3	2	4	50
5	4	4	4	6	50
6	3	3	2	3	50
7	3	3	2	5	50
8	5	5	0	2	50
9	5	5	0	2	50
10	6	6	6	10	40
11	5	5	0	0.5	30
12	5	5	0	0.5	30
13	7	7	2	0.5	30
14	7	7	2	4	30
15	8	8	2	5	50

*No. of the rubber tube used for taking the actual solution in or out

Staining and destaining reagents are collected for waste disposal

Disc Gel Electrophoresis

Disc gel electrophoresis is an old zone-electrophoretic technique performed in glass tubes, where the separated and stained compounds show up as thin discs. This technique is still used with advantages for the separation of some native enzymes or other biologically active proteins where staining for biological activity is required.[9] The gels used in the glass tubes are most often polyacrylamide (page 187) and for free zone electrophoresis, the buffer used in the gel and buffer reservoirs is usually TRIS, adjusted to the required pH, *e.g.* 8.7. Thereby most proteins will be negatively charged and if this is the protein of interest, the anode should then be placed in the lower reservoir (Figure 9.11).

A selected procedure for disc gel electrophoresis of protein-type proteinase inhibitors is shown in Box 9.6, but the technique could also be used as slab gel isoelectric focusing (previous subsection).

Box 9.4 *Preparation of agarose gels for isoelectric focusing*

Reagents for preparation of 2 gels (4.5 × 5 cm)

3.5 mg IsoGel agarose (FMC bioproducts)
3.5 ml deionized water
0.15 ml Pharmalyte® (Pharmacia)

Procedure for preparation of gels

1. IsoGel agarose and water are boiled until the agarose is dissolved and then cooled to 60 °C
2. 3 ml of 1. is mixed with Pharmalyte®
3. Prepare a mould of two glass plates (110 × 100 mm). The hydrophobic side of a GelBond film (FMC bioproducts) is stuck to one of the glass plates by a few drops of water. The glass plates are mounted with 0.5 mm rubber spacings and held together with clips
4. The glass plates are heated at 60 °C for 5 min
5. The agarose solution (60 °C) is poured between the plates
6. The gel is kept at 4 °C overnight
7. Disassemble the mould and cut the gel stuck to the gel bond film into *ca.* 4.5 × 5 cm pieces

Box 9.5 *Procedure for IEF (agarose gels) on the Pharmacia Phast Gel Electrophoresis System*™

Gels are prepared as described in Box 9.4.

Prefocusing and focusing conditions

Prefocusing	2000 V	2.0 mA	3.5 W	18 °C	75 Vh
Sample application	200 V	2.0 mA	3.5 W	18 °C	15 Vh
Focusing	2000 V	5.0 mA	3.5 W	18 °C	510 Vh

Coomassie staining of agarose gels

Reagent list

Fixation solution: 10% TCA
Colour solution I: 5 g Coomassie Brilliant Blue R 250 in 450 ml EtOH
Colour solution II: 100 ml glacial acetic acid and 450 ml deionized water
Colour solution: colour solution I is dissolved in colour solution II by heating to maximum 60 °C. The solution is filtered. It may be reused until the colour starts to precipitate
Destaining solution; 450 ml EtOH, 100 ml glacial acetic acid, and 450 ml deionized water

Procedure

1. Immediately after electrophoresis the gels are transferred to the fixation solution and placed for 15 min on a shaking table
2. The gel is washed quickly but carefully with deionized water and placed with the gel upward
3. A piece of nylon cloth is placed on the gel, then a piece of chromatography paper, *ca.* 1 cm of adsorbent tissue, a glass plate, and a weight. Press for minimum of 15 min.
4. Repeat 1–3

(continued overleaf)

5. The gel is dried with warm air (40 cm distance) until dry and clear
6. The gel is stained immediately after drying by incubation in the colour solution for 15 min on a shaking table
7. The gel is destained until a clear background is achieved
8. The gel is washed with deionized water and dried

Silver staining of agarose gels

Reagent list

Fixation solution; 30 g TCA, 10 g $ZnSO_4 \cdot 7H_2O$, 10 ml glycerol to 200 ml with deionized water
Silver stain I; 25 g Na_2CO_3 to 500 ml with deionized water
Silver stain II; 1.0 g NH_4NO_3, 1.0 g $AgNO_3$, 5.0 g wolframkiesel acid, and 7.0 ml formaldehyde (37%) to 500 ml with deionized water
Stop solution; 5 ml glacial acetic acid to 500 ml with deionized water
Staining and destaining reagents are collected for waste disposal

Procedure

1. The agarose gel is immersed in fixation solution for 15 min immediately after focusing
2. The gel is washed quickly but carefully with deionized water and placed with the gel upward
3. A piece of nylon cloth is placed on the gel, then a piece of chromatography paper, *ca.* 1 cm adsorbent tissue, a glass plate, and a weight. Press for minimum 15 min
4. Repeat 1–3.
5. The gel is dried with warm air (40 cm distance) until dry and clear
6. The dry gel is immersed in a mixture of 32 ml silver stain I and 68 ml silver stain II, mixed immediately prior to use, until the desired colour intensity is achieved (4–5 min)
7. Staining is stopped by transfer of the gel to stop solution for 5 min
8. The gel is washed with deionized water for 10 min
9. The gel is dried

Box 9.6 *Disc gel electrophoresis for isoelectric focusing of protein-type proteinase inhibitors*

With IEF in gel rods, inhibitors of chymotrypsin or trypsin may be identified by specific staining. The gels are incubated with the enzymes which will not cleave the substrate in the areas where enzyme-inhibitor complexes are formed. Therefore unstained areas will appear in the presence of inhibitors. Glass cylinders, 90 mm with i.d. 2.7 mm are used (J.K. Chavan and J. Hejgaard, *J. Sci. Food Agric.*, 1981, **332**, 857). The apparatus follows the principles shown in Figure 9.11.

Reagents for preparation of gels

1.75 ml polyacrylamide stock solution*
3.125 ml deionized water
250 μl Pharmalyte® (pH 3–10, Pharmacia)
100 μl ammonium persulfate solution†

*Polyacrylamide stock solution: 22.2% acrylamide + 0.6% *N,N'*-methylene-bis-acrylamide. Store in darkness. Use within 2 weeks from preparation
†2% w/v in deionized water. Shall be freshly prepared

Excess reagents are collected for waste disposal (*continued opposite*)

Procedure for preparation of gels

1. Eight glass cylinders are mounted with parafilm at one end and filled up to 1 cm from the edge with the mixture (above)
2. The gel surface is covered with n-butanol and the gels left for polymerization for 1 h
3. After polymerization, the n-butanol is removed, the gel surfaces are rinsed with deionized water and samples diluted 1:1 in sucrose (60% in deionized water) are applied to the top of the gels and then overlaid with a solution of 2% Pharmalyte® in 10% sucrose

Electrophoresis

1. Electrophoresis is performed for 3 h at 175 V, ambient temperature, with 1% (v/v) orthophosphoric acid as anode solution and 10 mM NaOH as cathode solution
2. After electrophoresis, the gel rods are removed from the glass cylinders by injecting water with a syringe between the glass and the gel. For colouring, the gels are transferred to individual reagent glasses with lids to allow incubation in water bath

Staining for inhibitors

1. The gels are incubated 5 min with enzyme at 37 °C (bovine α-chymotrypsin (Sigma, 0.05 mg ml^{-1}) or bovine trypsin (Sigma, 0.2 mg ml^{-1}) in 0.1 M phosphate buffer, pH 7.6, with the enzyme kept on ice until used
2. The gels are washed twice with deionized water
3. The gels are incubated for *ca.* 5 min at 37 °C with freshly prepared colour solution [6 mg *N*-acetyl-DL-phenylalanine-β-naphthyl-ester (Bachem, Switzerland) dissolved in 2.5 ml dimethylformamide and then mixed with 22 ml 0.1 M phosphate, pH 7.6, followed by addition of 12 mg Echtblausalz B (Fluka AG, Switzerland)]
4. After colouring, the gels are washed once with deionized water
5. The gels are kept in 7.5% acetic acid

Staining with Coomassie G-250

For a general protein detection, the gels may be stained with Coomassie G-250.

1. The gels are fixed for minimum 1 h in fixing solution (147 g TCA, 44 g sulfosalicylic acid and 910 ml deionized water)
2. The gels are incubated with staining solution (100 ml Coomassie G-250 (0.41 mg ml^{-1} in deionized water) is mixed with 3 ml perchloric acid for 1 h and filtered) for 1–2 h
3. The gels are destained to appropriate colour in destaining solution [5% (v/v) 99% acetic acid, 7% (v/v) 99% EtOH and 10% (v/v) ethyl acetate in deionized water]
4. The gels are kept in 7.5% acetic acid

Staining and destaining reagents are collected for waste disposal

SDS–PAGE

Sodium dodecyl sulfate – polyacrylamide-gel electrophoresis (SDS–PAGE) is a unique technique for the separation of proteins according to their subunit size. SDS is an anionic detergent, binding strongly to proteins, which then become denaturated with degradation of the quaternary, tertiary, and secondary protein structure. The proteins are saturated with detergent, resulting in an overall negative charge, masking the charge of the protein itself. Sample preparation also includes the use of 2-mercaptoethanol or dithiothreitol, which are reducing

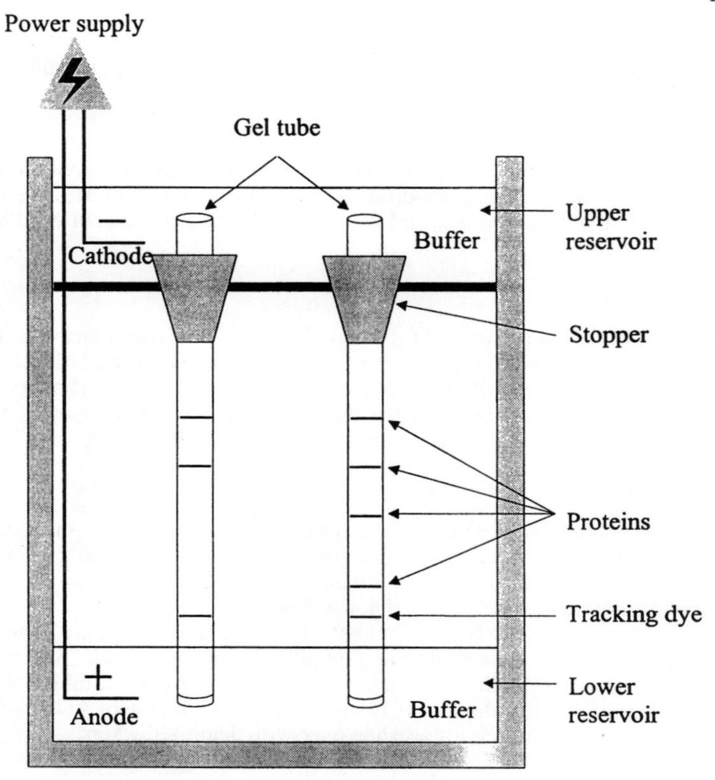

Figure 9.11 *Schematic diagram of a disc gel electrophoresis apparatus*

agents that are included to break possible disulfide bridges, leaving the proteins as rodlike monomers and so allowing a more uniform SDS binding. When adding SDS in excess to a protein mixture, about 1.2 g of the detergent will bind to 1 g of proteins or one SDS molecule for every two amino acid residues. This creates a negative charge on the proteins that is determined by the SDS molecules. During electrophoresis, the protein–SDS complexes will travel with different velocities toward the anode as a result of their negative charge and M_r, combined with the molecular sieving properties of the gel (Section 9.3). The instrumentation required corresponds to the IEF technique (Figure 9.8) and an example of SDS–PAGE based on use of the Pharmacia Phast Gel Electrophoretic System™ is described in Boxes 9.7 and 9.8.

Relative molecular mass standards with proteins of known M_r may be included to determine the M_r of the sample proteins. It should be noted that the relative molecular mass standard is to be reduced and generally treated as the sample. A plot of the distance migrated against log M_r results in a straight line, as illustrated in Figure 9.12. The M_r determination of glycoproteins using SDS may lead to incorrect estimates.

Box 9.7 *Preparation of polyacrylamide gels for SDS–PAGE*

This example involves the preparation of 10 gels for use in the Pharmacia Phast Gel Electrophoretic System™. The gels are intended for SDS–PAGE, which explains the addition of SDS to the gel mixture.

Reagents for preparation of 10 gels (4.5 × 5 cm)

7.20 ml polyacrylamide stock solution*
8.00 ml 1.88 M TRIS, pH 8.8
0.20 ml SDS 10%
20.0 μl TEMED
4.60 ml deionized water
200 μl ammonium persulfate solution†

*Polyacrylamide stock solution: 30% acrylamide + 0.93% *N,N'*-methylene-bis-acrylamide. Store in darkness. Use within 2 weeks from preparation
†20.0 mg ammonium persulfate/200 μl; should be freshly prepared

Excess reagent is collected for waste disposal
The exact procedure for preparation of the gels is as described in Box 9.2.

Box 9.8 *Procedure for SDS–PAGE on the Pharmacia Phast Gel Electrophoretic System™*

Preparation of samples

The samples shall be denatured with SDS. At a protein concentration in the sample around 5–10 mg protein/ml, use 10–50 μl sample and mix with 200 μl denaturation buffer (see composition below). Boil this mixture for 5 min. Let cool and the denatured sample is ready for use.

Denaturation buffer:

0.2 g dithiothreitol
2.0 ml SDS (10%)
1.0 ml glycerol (87%)
0.2 ml bromphenol blue (0.1%)
2.0 ml TRIS (0.02 M, pH 6.8)
Add deionized water to a total volume at 10 ml

Preparation of buffer strips

In contrast to IEF, SDS–PAGE requires the use of buffer strips. These can either be bought commercially or prepared in the laboratory as described below:

1. Prepare 100 ml solution of the following composition (pH 8.1):
   0.2 M tricine
   0.2 M TRIS
   0.55% SDS
   2.0% agarose
2. Heat this solution under magnetic stirring, and boil for 5–6 min
3. Let cool by placing the solution in a 56 °C waterbath
4. After the solution has reached 56 °C, transfer it to the buffer strips template placed on a heated, horizontal surface
5. When the gel is rigid, the buffer strips are ready for use

(*continued overleaf*)

Separation conditions

250 V	10.0 mA	3.0 W	15 °C	60 Vh
50 V	0.1 mA	0.5 W	15 °C	0 Vh*

*This step is optional. Its purpose is to reduce the risk of proteins migrating off the gel should the user miss the alarm that marks the end of the method [after step 1 (60 Vh)]

Silver staining of polyacrylamide gels

Reagent list

Prepare the following reagents in a fume cupboard. The volume stated indicates the minimum amount of the reagent used per run.

1. 10% acetic acid + 10% glycerol (50 ml glacial acetic acid + 50 ml glycerol + 400 ml deionized water), 80 ml
2. 50% EtOH + 10% acetic acid (250 ml EtOH + 25 ml glacial acetic acid + 225 ml deionized water), 80 ml
3. 10% EtOH + 5% acetic acid (50 ml EtOH + 25 ml glacial acetic acid + 425 ml deionized water), 400 ml
4. 8.3% glutardialdehyde (27 ml glutardialdehyde + 54 ml deionized water), 80 ml; should be freshly prepared
5. Deionized water, 400 ml
6. 0.5% silver nitrate (2.5 g silver nitrate/500 ml deionized water), 80 ml
7. 'Developer' (32 ml 12.5% Na_2CO_3* + 128 ml deionized water + 64 μl formalin), 160 ml; should be freshly prepared
8. 5% acetic acid (50 ml glacial acetic acid + 950 ml deionized water), 80 ml

*125 g Na_2CO_3/1000 ml deionized water or 337.5 g $Na_2CO_3 \cdot 10\ H_2O$/1000 ml deionized water

Staining and destaining procedure based on the Pharmacia Phast Gel Developing Device

Step No.	Reagent No.	In*	Out*	Time (min)	Temp. (°C)
1	2	9	2	2	50
2	3	3	2	2	50
3	3	3	2	4	50
4	4	4	4	6	50
5	3	3	2	3	50
6	3	3	2	5	50
7	5	5	0	2	50
8	5	5	0	2	50
9	6	6	6	6.5	40
10	5	5	0	0.5	30
11	5	5	0	0.5	30
12	7	7	2	1	30
13	7	7	2	5	30
14	8	8	0	2	50
15	1	1	2	5	50

*No. of the rubber tube used for taking the actual solution in or out

Staining and destaining reagents are collected for waste disposal

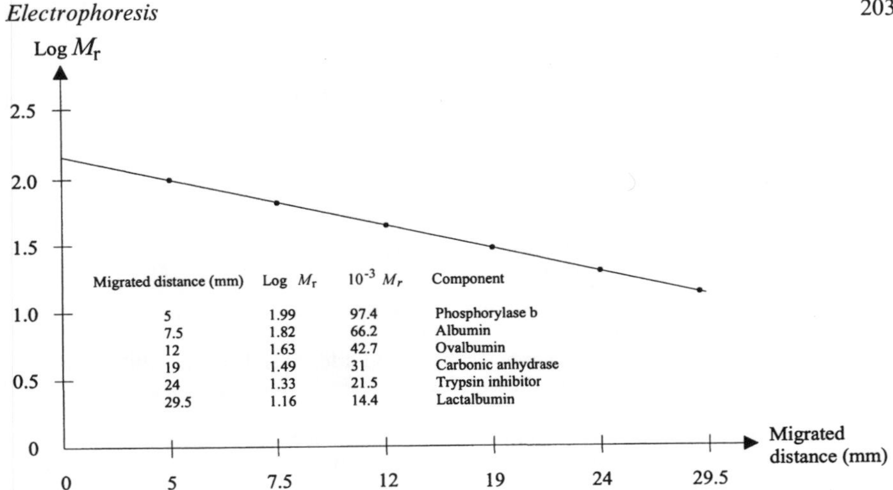

Figure 9.12 *Results from SDS–PAGE of a standard mixture of proteins. The migration distance for a protein in the gel leads to a direct calculation of the* M_r *of the protein*

To improve resolution, the polyacrylamide gel used for SDS–PAGE may be divided into a main separation gel, in which separation takes part, and, prior to this, a stacking gel, to concentrate the protein sample into sharp bands. The phenomenon of isotachophoresis (page 189), obtained by differences in pH and ionic strength between the electrophoresis buffer and the stacking gel, is widely used for this purpose.

In addition to M_r determination, SDS–PAGE is widely used to assess the purity of a protein sample, *e.g.* during a purification procedure.

Native Gel Electrophoresis

In native gel electrophoresis, separation is based on different electrophoretic mobilities owing to the native charge, size, and shape of the sample molecules, and on the properties of the gel in which electrophoresis is performed. The PAGE technique can be performed using homogeneous or gradient gels, the gradient gels (varying pore size) being useful for analysing complex samples. Native gel electrophoresis (Box 9.10) will generally be of special interest for samples to be detected on the basis of their biological activity (Section 9.6), *e.g.* enzymes. The preparation of agarose gels with uniform pore size is described in Box 9.9.

Immunoelectrophoretic Techniques

When antibodies with specificity for the investigated proteins are available, the immunoprecipitin reactions can be employed for the characterization and identification of the sample proteins. These investigations may include protein

Box 9.9 *Preparation of agarose gels*

This example shows the preparation of one 10 × 10 cm agarose gel (1% w/v).

1. Prepare a TRIS/tricine stock solution from:

   11.30 g tricine
   22.15 g TRIS
   0.50 g NaN_3

   Add deionized water to a total volume of 500 ml
   Adjust pH to 8.6 with 6.0 M HCl
2. Make a 1:4 dilution of the TRIS/tricine stock solution, *e.g.* 200 ml stock solution + 800 ml deionized water
3. Weigh 2.0 g agarose to an Erlenmeyer flask and add 200 ml of the diluted TRIS/tricine stock solution. Dissolve the agarose by heating and magnetic stirring. Allow to boil for 5–6 min
4. Add 16 ml of the solution to a centrifuge tube and place it in a waterbath (56 °C)
5. After the solution has reached 56 °C, transfer it to a 10 × 10 cm glass plate placed on a heated, horizontal surface
6. When the gel is rigid, use a gel punch to make the required number of wells in the gel, which is now ready for use

Excess reagent is collected for waste disposal

identity investigations, quantitative immunoelectrophoresis (rocket immunoelectrophoresis), and two-dimensional electrophoresis. These techniques require a knowledge of antibody–antigen reactions, and will be described in Chapter 13.

Preparative Isoelectric Membrane Electrophoresis (IsoPrime™)

A method for preparative isoelectric focusing has been developed by Righetti and co-workers[10,12] and commercialized by Pharmacia (Uppsala, Sweden). The system encompasses a separation module with up to seven membranes and eight separation chambers. A set of buffered polyacrylamide membranes with incorporated, covalently linked ampholytes separate the chambers through which protein samples circulate. Each chamber represents a pH area defined by the partitioning membranes. During electrophoresis a protein migrates between the chambers until it reaches a chamber in which the pH is equal to the pI of the protein. After electrophoresis, the separated protein samples are collected from the chambers. This method differs from slab gel IEF and chromatofocusing owing to the covalently bound ampholines in the acrylamide membranes, thus resulting in isolation of proteins without contaminating ampholines.

6 Detection and Recovery

A wide range of detection methods exist, their relevance depending on the actual type of electrophoresis performed and, of course, the type and amount of components to be detected. Whereas the various possibilities for detection in

Box 9.10 *Native electrophoresis in agarose gels*

Electrophoresis

1. The samples are applied to the wells, and 2 μl of 0.1% bromophenol blue is added as migration marker to one of the wells
2. The gel plate is mounted on the cooling plate of the electrophoresis apparatus using a few drops of water between the plate and the apparatus
3. Wicks (5 × 10 cm) are placed with one end overlapping (by 5 mm) the ends of the gel and the other end immersed in buffer (TRIS-tricine buffer, pH 8.6, see Box 9.9)
4. A voltage of 20 V cm^{-1} is applied, and the electrophoresis is run until the marker has run 6 cm (*ca.* 30 min)
5. The protein bands are developed immediately

Staining with Coomassie Brilliant Blue R-250

Reagent list

Fixation solution: 12% TCA
Coomassie solution: 0.25 g Coomassie Brilliant Blue R-250, 45 ml MeOH, 10 ml glacial acetic acid is mixed under stirring. When the coloured powder is nearly solubilized, 45 ml deionized water is added and the solution is filtered
Destaining solution: 45 ml MeOH, 45 ml deionized water, and 10 ml glacial acetic acid

Procedure

1. The gel plate is immersed in fixation solution for *ca.* 10 min
2. A piece of filter paper (10 × 10 cm) wetted with deionized water is placed on the gel followed by adsorbent tissue, a glass plate, and a weight of *ca.* 1 kg. The gel is pressed for a minimum of 20 min with appropriate adjustments of the adsorbent tissue
3. The filter paper is removed and the gel is dried under warm air (40 cm distance)
4. The gel is placed in the Coomassie solution for 3–5 min
5. The gel is transferred to the destaining solution for 5–10 min until the background is appropriately destained. For repeated destaining steps, use destaining solutions with decreasing acetic acid content

Reagents for fixation, staining, and destaining are collected for waste disposal

free zone electrophoresis are limited, zone electrophoresis, using a supporting media such as paper or gels, opens up several alternative procedures. Staining techniques are commonly used, the actual procedure depending on the sample components considered. Some reagents often used for protein staining are:

- Coomassie Brilliant Blue R-250 (Boxes 9.5, 9.6, 9.10);
- Amido black 10B;
- Lissamine green in aqueous acetic acid;
- Bromophenol blue in acetic acid and zinc sulfate;
- Procion blue;
- Nigrosine in acetic acid or trichloroacetic acid;
- Procion S;
- Silver (Boxes 9.3, 9.5, 9.8);
- Gold (Box 13.9);
- PAS-stain.

Amido black staining is more time consuming than staining with Coomassie Brilliant Blue, but may be used whenever Coomassie Brilliant Blue is less suitable (*e.g.* for cellulose acetate, to which Coomassie Brilliant Blue binds strongly). In this way, the large spectrum of staining methods offers an alternative for most purposes. Staining with silver is a way of achieving very high sensitivity (detection down to 1 ng protein). In this technique, silver ions are reduced to metallic silver on the protein, leaving a grey to black band. Silver staining may be used in combination with other, less sensitive, staining methods. The PAS-staining (periodic acid-Schiff stain) is a method used for the detection of glycoproteins, resulting in often weak red to pink bands. The method is not very sensitive.

Natural ultraviolet absorbance of the sample components gives another possibility for detection, especially if the molecule contains an efficient chromophore group (Chapter 5), but it is normally not an efficient technique and requires a supporting material that has no, or limited, absorption at the relevant wavelength. Proteins can thus be detected by their absorbance at 280 nm. A more sensitive detection of various proteins is possible by initial complexing with compounds that induce fluorescence under UV light, *e.g.* ANS (1-anilinonaphthalen-8-sulfonate), fluorescamine, or dansyl chloride.

Detection of enzymes may be performed by traditional protein detection methods, however, *in situ* determination of the enzymatic activity is a widely used alternative. In Refs. 11 and 13, a range of different methods is listed for detecting enzymatic activity, and specific procedures are described elsewhere (Chapter 12). Autochrome methods provide formation of an (insoluble) coloured product, or a product which can be followed by UV or fluorescence, from a specific substrate. Problems often met with in detecting products of enzyme-catalysed reactions can in some cases be solved by the use of reactions coupled with a third compound, leading to a more highly coloured product (simultan-capture reaction). In some situations, *e.g.* when the substrate is too large to penetrate the electrophoresis gel, a sandwich technique or simply an overlayered gel technique, in which the substrate is incorporated into a support medium (*e.g.* a gel), may be suitable. The gel containing the substrate is placed on top of a second gel or between the split electrophoresis gel containing the enzyme, as in a sandwich, and incubation is performed. Detection of enzyme activity may be as described above. In SDS–PAGE, none of the methods described for *in situ* determination of enzymatic activity is possible owing to the denaturation of proteins prior to separation.

Electrophoresis is basically an analytical technique; however, further examination of separated compounds may sometimes be of interest. For this purpose, the blotting technique will be the first step to recover the actual components. Analogous to southern blotting (used for DNA), western blotting has been developed for the recovery of proteins. The transfer of proteins may be performed mechanically (Figure 9.13) or by electroelution, in which a current is passed at right angles to the gel, causing the proteins to travel out of the gel.

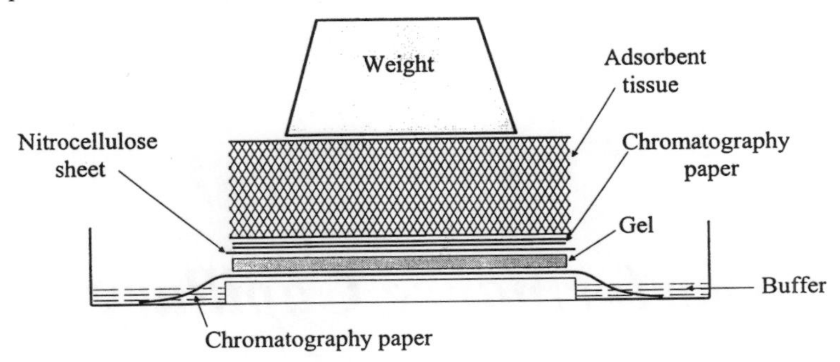

Figure 9.13 *Principle of western blotting*

As illustrated, capillary forces will cause transference of the proteins from the gel onto the nitrocellulose sheet covering the gel. Once transferred, the proteins can be examined further.

7 Selected General and Specific Literature

1. E.T. McGuiness, *J. Chem. Educ.*, 1973, **50**, 826.
2. O. Stern, *Z. Electrochem.*, 1924, **30**, 508.
3. E. Hückel, *Phys. Z.*, 1924, **25**, 204.
4. M. Smoluchowski, in 'Handbuch der Elektrizitat und des Magnetismus', ed. B. Graetz, Leipzig, 1914, vol. 2, p. 366.
5. B.O. Eggum and H. Sørensen, in 'Absorption and Utilization of Amino Acids', ed. M. Friedman, CRC Press, Boca Raton, FL, 1989, vol. 3, ch. 17, p. 265.
6. O. Olsen and H. Sørensen, *J. Am. Oil. Chem.*, 1981, **58**, 857.
7. B.O. Eggum, N.E. Hansen, and H. Sørensen, in 'Absorption and Utilization of Amino Acids', ed. M. Friedman, CRC Press, Boca Raton, FL, 1989, vol. 3, ch. 5, p. 67.
8. A. Tiselius, *Trans. Faraday. Soc.*, 1937, **33**, 524.
9. A. Arentoft, H. Frøkiær, H. Sørensen, and S. Sørensen, *Acta Agric. Scand. B*, 1994, **44**, 234.
10. P.G. Righetti, E. Wenisch, and M. Faupel, *J. Chromatogr.*, 1989, **475**, 293.
11. G. Othmar and D.M. Gersten, in 'Enzyme assays – a practical approach', ed. R. Eisenthal and M.J. Danson, IRL Press, Oxford University Press, Oxford, 1992, p. 217.
12. P.G. Righetti, 'Isoelectric Focusing: Theory, Methodology, and Applications', Elsevier Biomedical Press, Amsterdam, 1983.
13. G.H. Rothe, 'Electrophoresis of Enzymes. Laboratory Methods', Springer Verlag, Berlin, 1994.

CHAPTER 10

High-performance Capillary Electrophoresis

1 Introduction

High-performance capillary electrophoresis (HPCE) represents a modern analytical technique derived principally from traditional electrophoresis, but with references to chromatography as well. This combination provide us with possibilities for many modes of operation, and HPCE has been shown applicable to a wide variety of charged and uncharged compounds of high- and low-M_r.

The ideas behind HPCE can be traced back a long way, but free-zone capillary electrophoresis (FZCE) in relatively small diameter capillaries was first performed successfully in the late fifties and sixties.[1] The possibilities of the technique were, however, not fully recognized until 1981, when Jorgenson and Lukacs,[2] demonstrated high separation efficiencies, using a high field strength in narrow capillaries (<100 μm). Since then, there has been a period of rapid growth within HPCE, one of the major events being the introduction of micellar electrokinetic capillary electrophoresis (MECC).[3] Moreover, the development of commercial instruments, instead of the home-made prototypes, has made the technique accessible to a larger group of analysts, mainly those interested in the practical application of the technique.

This chapter will include an introduction to the basic principles of HPCE together with an overview of the instrumental and operational aspects, including parameters affecting the performance of HPCE. The many possibilities of the technique in food analysis are demonstrated by selected examples in the Chapters 10 to 14.

2 Basic Principles

In HPCE, a buffer-filled capillary is placed between two buffer reservoirs, and an electric field is applied across the capillary (Figure 10.1).

This results in an acceleration of ions in the capillary, eventually reaching a constant velocity proportional to the electric field applied. Both size and shape

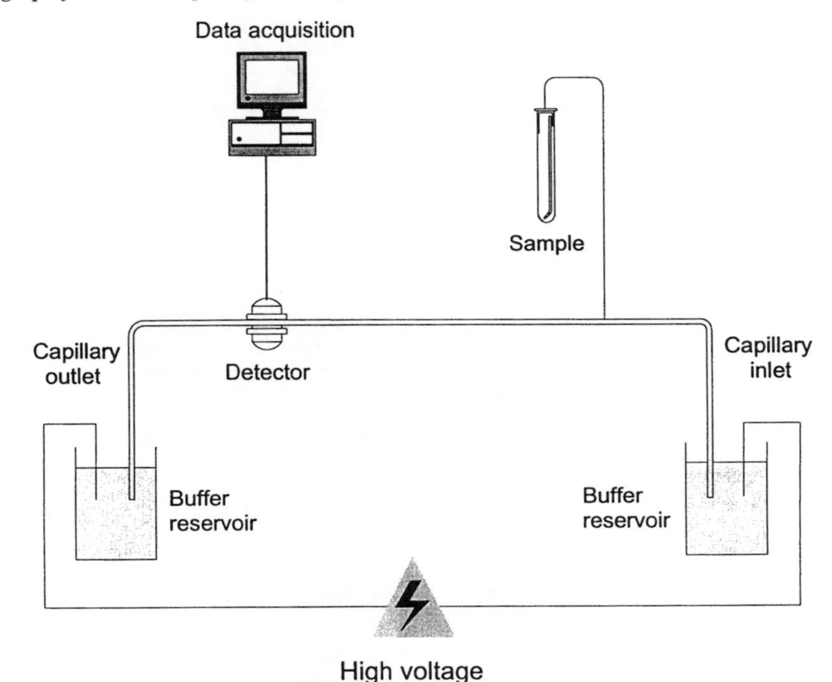

Figure 10.1 *Diagrammatic presentation of a HPCE system*

of the analytes affect the electrophoretic mobility, as for traditional electrophoresis (Chapter 9). The migration time (MT) for a compound migrating from injection to detection point (effective length) can theoretically be determined as:

$$MT = \frac{lL}{\mu_e V} \tag{10.1}$$

where l = effective length of capillary (cm), L = total length of capillary (cm), μ_e = electrophoretic mobility of the compound ($cm^2\ V^{-1}\ s^{-1}$), and V = applied voltage (V).

Electro-osmosis

The EOF, described in Chapter 9, is a very important feature in HPCE, having a great impact on the actual MT obtained. The uncoated fused silica capillaries most often used have a negatively charged surface owing to ionized silanol groups (SiO^-), the degree of charge depending on the pH of the solution in the capillary. The exact pI of the silanol groups in the fused silica is difficult to interpret and will be dependent on buffer modifiers or dielectric constants of the buffer (Section 2.8), but at pH > 3 the size of the electro-osmotic flow becomes significant (Figure 10.30). During electrophoresis, the electro-osmotic flow is

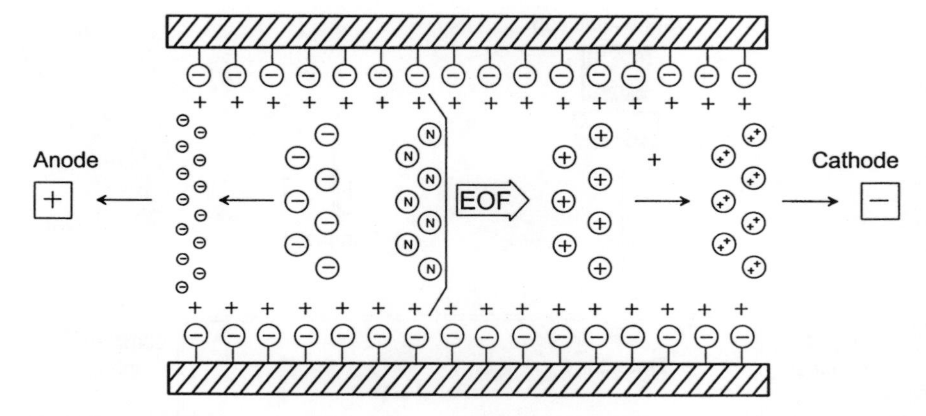

Figure 10.2 *The electroosmotic flow (μ_{eo}) and sample migration in a fused silica capillary. The velocity of the electro-osmotic flow is* $v_{eo} = \mu_{eo} V$ *($cm^2 s^{-1}$)*

directed against the cathode, and since this flow is relatively fast, most sample molecules, regardless of charge/weight ratio (pages 179–180), will be carried along to the detector (Figure 10.2).

The electroosmotic flow thus creates an apparent mobility of the analytes, μ_{eap}, which is μ_e (electrophoretic mobility of a given analyte) with μ_{eo} (electro-osmotic mobility) added vectorially; units of $cm^2 V^{-1} s^{-1}$ (Equation 10.2). The size of μ_{eo} can be determined as shown in Equation 10.3.

$$\mu_{eap} = \mu_e + \mu_{eo}; \qquad (10.2)$$

$$\mu_{eo} = \frac{\varepsilon^* \zeta}{4\pi\eta} \qquad (10.3)$$

where $\varepsilon^* = \varepsilon_o \varepsilon_r 4\pi$ = permittivity of the medium ($F\ m^{-1} = C\ V^{-1}\ m^{-1}$), ε_o = permittivity in vacuum (a constant of $8.854 \times 10^{-12}\ C^2\ J^{-1}\ m^{-1} = 8.854 \times 10^{-12}\ F\ m^{-1}$), $F = C\ V^{-1}$ = 'Farad', ε_r = dielectric constant of the solution (dimensionless figure; Section 2.8), ζ = zeta potential (V), and η = viscosity of the medium (Pa s = dyn s cm^{-2}). In the presence of electro-osmosis, μ_e in Equation 10.1 should thus be replaced by μ_{eap}. In practice, positively charged ions reach the cathode first, since their electrophoretic mobility and electro-osmotic flow are in the same direction, whereas anions are retarded by their electrical attraction towards the anode. Neutral molecules travel with a velocity determined by the electroosmotic flow. Separation of anions, cations, and uncharged analytes can thus be obtained in a single analysis.

It is possible to determine the size of the electroosmotic flow by using a neutral marker with a detectable signal. The marker has to be inert to the capillary wall, the sample molecules, and the buffer components. Often used markers are mesityl oxide, methanol, or acetone. Measurement of the electro-osmotic flow as a decrease in detector signal is also possible, if the introduced sample solution contains a lower buffer concentration than the capillary separation buffer, and if the buffer components are detectable (*e.g.* indirect UV detection). It should be stated that the electroosmotic flow changes with the state of the capillary. In practical laboratory work, it is normally not necessary to know the actual size of the electro-osmotic flow; however, it may well be necessary to control it to improve separation. Monitoring electro-osmotic flow (and current) can also be used as a routine check of the whole electrophoretic system, including buffer, capillary, and instrument settings to avoid conclusions based on incorrectly performed analyses.

Band Broadening and Dispersive Effects

Separation in electrophoresis, traditional as well as in HPCE, is based on differences in the electrophoretic mobility of analytes. During analysis, the analytes are focused into zones, and the difference in distance necessary to resolve two zones depends on the length of these zones. Dispersion, spreading of the analyte zone, increases zone length (band broadening) and should be controlled. In this respect, it is important to be aware of the factors affecting band broadening in HPCE.

Diffusion and Electromigration Dispersion

Diffusion of analytes along the capillary (longitudinal diffusion) results in a Gaussian concentration profile, which can be characterized by its variance, which again is related to the diffusion coefficient, *D*, of the analyte (pages 127–130). The contribution of diffusion to band broadening is generally small for large compounds having low diffusion coefficients, *e.g.* proteins and DNA, whereas diffusion for small molecules, under otherwise optimized HPCE conditions, can be a considerable factor, controlling zone length. The flat flow profile obtained in HPCE (Figure 10.3) is beneficial in this respect, as it reduces the importance of diffusion, and radial diffusion (across the capillary) in particular should be unimportant.

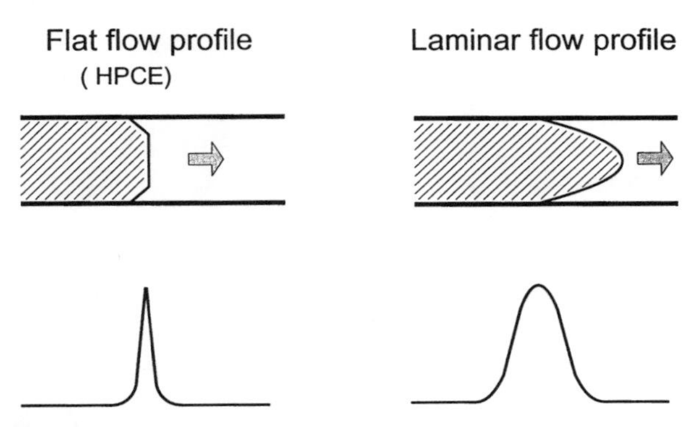

Figure 10.3 *Plug flow profile in HPCE compared with laminar flow and an illustration of the resulting analyte zones (peaks) as they appear in the electropherogram*

Lowering diffusion, *e.g.* by increasing viscosity of the media, also decreases the mobility of sample molecules, which is not desirable in HPCE. Diffusion can be said to define the fundamental limit of efficiency in HPCE.

Peaks in electropherograms from HPCE are often asymmetric, a fact which can be related to the phenomena of electromigration dispersion, *i.e.* differences in sample zone and running buffer conductivities. If the mobility in the sample zone is higher than that of the buffer ions, a peak with fronting will appear (Figure 10.4). The explanation for this is that sample ions which diffuse into the rear buffer zone, are accelerated owing to the higher electric field there until they reach their own zone again; however, the diffuse front appears when sample ions diffuse into the frontal buffer zone, where they are accelerated. Note that the concentration profile of the sample ions will be reversed compared with the peak

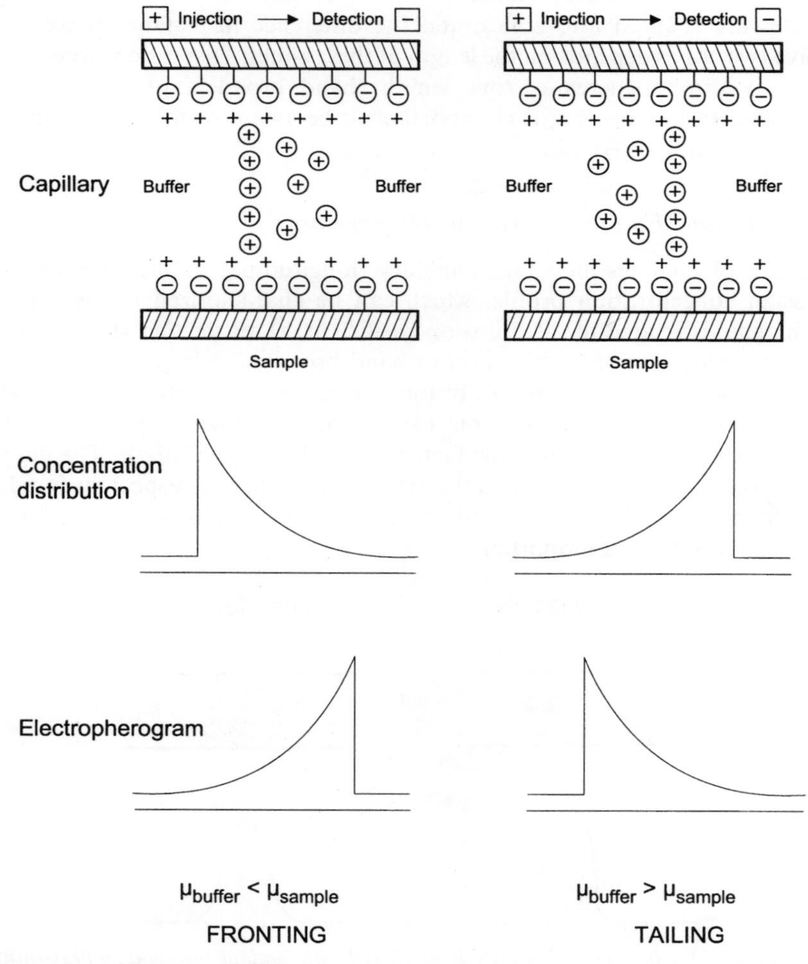

Figure 10.4 *Fronting and tailing of peaks as a consequence of electromigration dispersion*

appearing in the electropherogram. Similar to fronting, sample ions migrating slower than buffer ions may cause peak tailing. Only when sample and buffer ions migrate with the same velocity can peak symmetry be obtained. Neutral sample components will not be affected by mobility differences.

The contribution from electromigration dispersion to band broadening can be compared with the contribution from diffusion, when the concentration of the sample is more than 100 times lower than the concentration of the buffer. A high concentration of sample ions increases the significance of electromigration dispersion.

An important issue in HPCE concerning peak shape, which is not directly related to either diffusion or electromigration dispersion, is the fact that different compounds pass the detector zone at different velocities. This results in different residence times in the detector region, and thereby different peak widths. The importance of this phenomena in relation to quantification is discussed in the section on page 219.

Heat Development and Temperature Gradients

One of the advantages of HPCE over traditional electrophoresis is the effective heat dissipation obtained owing to the large surface-to-volume ratio, especially in small bore capillaries. The high field strengths applied may, however, result in power generation exceeding heat dissipation, and in such cases the temperature in the capillary will rise. The temperature increase in itself is no problem, except of course for thermally labile components and if the buffer reaches boiling point; however, generation of a temperature gradient across the capillary can be detrimental to separation.

Temperature gradients in the capillary arise because heat is produced in the whole of the capillary but only conducted away by the capillary wall. The temperature in the middle of the capillary thus exceed the temperature of the solution near the capillary wall. Temperature differences across the capillary create inhomogeneity in viscosity and thereby mobility, resulting in band broadening. Excessive heat generation and formation of temperature gradients can be recognized in several ways. Reduction of separation efficiency or a disproportionate increase in current (Ohm's law) or solute mobility at increased voltage are phenomena indicating this type of problem.

Different attempts to limit problems with heat development are:

- Reduce capillary diameter;
- reduce electric field;
- change buffer composition;
- use efficient temperature control systems.

A reduced capillary diameter increases the surface-to-volume ratio, thereby increasing the dissipation of heat. The capillary diameter can only be decreased down to a certain limit, if problems with detection (sensitivity), sample loading, and capillary clogging are to be avoided. Internal capillary diameters of 25 to 50 μm are often preferred.

Reduction of the electric field gives a proportional decrease in heat generated but also a concomitant reduction in separation efficiency, which is not desirable.

The concentration of buffers used affect the current in the capillary and thereby the heat developed. A lower buffer concentration thus reduces current; however, buffer capacity is also decreased and analyte–wall interactions may be increased. A possibility for high concentration buffers is to use large buffer ions having a low charge (TRIS, borate, *etc.*), as the conductivity of such buffers will be relatively low.

Temperature control systems applied in HPCE comprise forced air convection, Peltier thermoelectric devices, and flowing liquid thermostatting. The temperature should be controlled over the whole of the capillary, including capillary ends and the detector region. In some systems, sample vials and buffers can be cooled as well. Theoretically, liquid-based cooling systems should be more effective than air-based ones; however, the efficiency of the different heat dissipation systems depends also on the design of the equipment.

Analyte–Capillary Wall Interactions

The large surface-to-volume area, advantageous for heat dissipation, leads to a well-known problem in HPCE: the analyte–capillary wall interaction. Local turbulence in the capillary can be caused by the adsorption of analytes to the capillary wall, resulting in band broadening, severe peak tailing, irreproducible analyses (altered electroosmotic flow), non-quantitative analyses, and reduced separation efficiency. The adsorption may be reversible or irreversible, and may include hysteresis (page 217); irreversibility and hysteresis result in a permanent change of the state of the capillary. Adsorption is primarily caused by electrostatic interaction between positively charged analytes and the negatively charged capillary wall with its immobile and mobile surface regions[4,5] or by hydrophobic interactions. In particular, proteins and large peptides having small diffusion coefficients and numerous positive charges (especially at $pH < pI$) may display significant adsorption.

Various approaches to reduce the analyte–capillary wall interactions have been proposed. Selected examples include the use of:

- Buffers with extreme pH.
- Buffers with high concentration/ionic strength.
- Buffers with additives [*e.g.* zwitterions, surfactants, alkali salts, divalent cationic amines, charge reversal reagents, zeta potential repressors (ethylene glycol, cellulose derivatives), *etc.*].
- Coated capillaries (static coatings).

Most of the procedures also cause reductions in the electro-osmotic flow. The success of the various approaches depends strongly on the actual separation conditions, and on the size and net charge of the compounds to be analysed. In general, large molecules with intermediate pI-values are the most difficult to analyse. Furthermore, unwanted changes in analyte properties may occur when using some of the above-mentioned approaches, *e.g.* unfolding of proteins or aggregation of proteins, which may lead to worse separations. Variation in buffer composition and possible modifications of capillaries are further treated in Section 10.5.

The number of coils on the capillary and the diameter of these coils have been reported to give band broadening. Ideally, the capillary should be straight; however, this is seldom possible in practice.

Injection Plug Length and Effects from the Detector Zone

Injection of sample may cause band broadening. Injected sample volumes should preferably be small (nl), although the precision of the injected volume increases with the volume injected. If the injection plug length is longer than the dispersion caused by diffusion, the separation efficiency will be reduced.[6] In order to diminish the band broadening effect of injection, sample stacking is most often used (page 264).

The length of the detector zone should be as short as possible to avoid effects on measured sample zone width. The detector cell volume will thus be small, resulting in a dilemma between separation efficiency and sensitivity, as is also the case discussing injection plug length.

Separation Efficiency

The efficiency of separation can be expressed as the number of theoretical plates, N (page 134):[2]

$$N = \frac{(\mu_e + \mu_{eo})Vl}{2DL} \tag{10.4}$$

where μ_e and μ_{eo} are as in Equation 10.2, V = applied voltage (V), l = effective length of capillary (cm), D = molecular diffusion coefficient of compound in the zone (cm^2 s^{-1}), and L = total length of capillary (cm). D and μ_e are constant for a given analyte, whereas the field strength can easily be increased resulting in higher N as long as the heat produced can be effectively removed. From Equation 10.4, it seems that an increase in the electro-osmotic flow (in the same direction as μ_e) improves separation efficiency; however, this is only partly true, as there is a concomitant reduction in the resolution, R_s (page 135):

$$R_s = 0.177(\mu_{e1} - \mu_{e2})\sqrt{\frac{V}{D(\mu_{av} + \mu_{eo})}} \tag{10.5}$$

where μ_{e1}, μ_{e2}, and μ_{av} = electrophoretic mobility of compounds 1 and 2, and their average mobility, respectively ($cm^2\ V^{-1}\ s^{-1}$), V = applied voltage (V), and D = mean molecular diffusion coefficient of compounds in the zones ($cm^2\ s^{-1}$). An increase in the electroosmotic flow will thus result in shorter time of analysis and sharper but more poorly resolved peaks. Large differences in electrophoretic mobilities, high voltages, and low diffusion coefficients will increase R_s. It is possible to manipulate the difference in migration times by various methods, *e.g.* change of buffer pH (acid–base equilibria of analytes and constituents of the mobile phase as well as the electroosmotic flow), formation of complexes between analytes and electrolytes in the buffer (*e.g.* borate complexation with carbohydrates, and complexation of some inorganic cations with tartaric acid), interaction with neutral molecules (*e.g.* crown ethers or cyclodextrins), and ion-pair formation or hydrophobic interaction (*e.g.* charged and neutral analytes with micelles). High-M_r compounds as proteins have low diffusion coefficients (pages 127–130) and efficient separation can in general be obtained, provided that adsorption problems can be overcome.

In practice, determination of N and R_s can be performed directly from the electropherogram using the equations (page 134):

$$N = 16\left(\frac{MT}{w}\right)^2 = 5.54\left(\frac{MT}{w_{1/2}}\right)^2 \tag{10.6}$$

where w = peak width at base line (time units), $w_{1/2}$ = peak width at half-height (time units), and $w = 1.699w_{1/2}$, and

$$R_s = \frac{1.18(MT_2 - MT_1)}{w_1 + w_2} = \frac{2(MT_2 - MT_1)}{1.699(w_{1/2}(1) + w_{1/2}(2))} \tag{10.7}$$

where MT_1 and MT_2 = migration time of compounds 1 and 2, respectively ($MT_2 \geqslant MT_1$), w_1 and w_2 = peak width at base line (time units) of compounds 1 and 2, respectively, and $w_{1/2}(1)$ and $w_{1/2}(2)$ = peak width at half height (time units) of compounds 1 and 2, respectively.

The equations using peak width at half height are in general to be preferred if the peaks of interest show tailing or fronting. The number of theoretical plates is often expressed per meter of capillary to facilitate the comparison between different systems. N/m may exceed a million for open tube capillaries, and may be even higher for gel-filled ones. Theoretical calculation of N in general end up with a higher value than the figure calculated from the actual analysis, as the theoretical equation accounts only for band broadening caused by longitudinal diffusion. As discussed above (page 211), other dispersive effects are often present. The magnitude of R_s should exceed 1 (page 135); in practice a resolution of 1.25 is sufficient even for quantitative analysis.

Repeatability and Identification

The migration time of a compound may vary from analysis to analysis, without concomitant changes in separation conditions. This variation in migration time may be caused by different parameters, including among others:

- Capillary wall effects;
- changes in buffer composition;
- ionic strength in sample;
- instrumental factors.

The state of the capillary wall affects the size of the electro-osmotic flow, which determines the absolute migration time of a compound. Capillary wall effects may be caused by capillary ageing, previous capillary treatment, stripping off wall coating in covalently coated capillary, changes in double layer of surfactant at capillary wall (cationic surfactants in MECC, pages 233–236), or adsorption of analyte and/or impurities to the capillary wall. A preconditioning of the capillary prior to each run greatly improves the repeatability of migration time. An example of such washing procedure is shown in Box 10.1. Sufficient equilibration time is necessary to avoid pH hysteresis, as the electroosmotic flow has been shown to depend on whether the capillary is preconditioned at alkaline or acidic pH values.[7] Hysteresis loops are thus often met in connection with the adsorption of molecules or ions to the layers at the capillary wall, and so affect the zeta potential and electroosmotic flow (pages 179–182 and 209). This means that the adsorption isotherms corresponding to Langmuir's theory (page 41) are not always reversible but show hysteresis loops in which it takes considerable time to obtain equilibrium and total saturation of the capillary wall.

Changes in buffer composition during analyses may be caused by buffer ion depletion. Buffer ion depletion is seen when buffer vials are small, not replenished with fresh buffer for each run, and if numerous runs are performed with the same buffer. Evaporation of buffer is another problem in some

Box 10.1 *Preconditioning of capillary*

1. 2–10 min:	0.1 M NaOH or 1.0 M NaOH
2. 2 min:	H_2O
3. 5 min:	1.0 M HCl–EtOH (1:9)
4. 2 min:	1.0 M NaOH
5. 2 min:	H_2O
6. 4–8 min:	Separation buffer

Washing media and rinsing times may be varied depending on the state of the capillary and the actual mode of HPCE operation used. Preparation of a new capillary thus requires at least 20 min washing with 1.0 M NaOH, 5 min with 0.1 M NaOH, and 5–10 min with separation buffer, preferably at elevated temperature (*e.g.* 60 °C), whereas the step using the acid–alcohol mixture may be omitted.

instruments, which may be solved by the caption of buffer vials combined with cooling of the autosampler.

The ionic strength of the sample compared with the separation buffer has a great impact on sample stacking and thereby separation efficiency (page 264). Furthermore, analyses of samples having very high ionic strengths compared with the buffer may also show very large variations in migration times from run to run (up to more than 100%). Simple dilution of the sample will solve this problem.

Instrumental factors, including, for example, variation in temperature or applied voltage, is not always user controllable. In some commercial instruments, the analysis is automatically interrupted if a variable deviates more than allowed during the analysis sequence.

Inclusion of an internal standard in the analysis makes it possible to calculate relative migration times, *RMT*, of sample components:

$$RMT = \frac{MT_1}{MT_2} \tag{10.8}$$

where MT_1 = migration time of the actual analyte and MT_2 = migration time of the internal standard (reference peak). RMT = 1 for the internal standard. The use of relative migration times has been shown to give much better repeatabilities than absolute values (Table 10.1).

Relative migration times can be used for identification purposes, if the assumption is made that changes in migration time of the reference peak parallel changes in the migration time of unknown peaks. To verify an identification of peaks based on *RMT*, spiking, in which the expected pure compound is injected together with the sample, can also be performed. The resulting peak in the electropherogram should be higher than the original peak,

Table 10.1 *Relative standard deviation (RSD) of migration time and relative migration time for glucosinolates analysed by MECC**

	RSD (%)	
Peak no.	*MT*	*RMT*
1	2.38	0.28
2	2.51	0.18
3	2.50	0.10
4	2.51	1.00
5	2.54	0.08
6	2.50	0.11
7	2.46	0.19

*S. Michaelsen, P. Møller, and H. Sørensen, *J. Chromatogr.*, 1992, **608**, 363

and be without any indication of a shoulder. Spectral analysis of peaks by use of a diode array detector (Chapter 5) is another possible way of identification. Comparing spectra obtained at the beginning, middle, and end of the analyte peak with corresponding spectra of the expected pure compound may be very useful for identification purposes. If another compound is hidden in the analyte peak, such comparisons of spectra will often reveal this.

Quantification

In HPCE, peak areas used for quantification have to be corrected for differences in analyte mobility. This is because different residence times in the detector region affect the magnitude of the peak area. Analytes moving with high velocities thus spend less time in the detector window, leading to decreased peak area and *vice versa*. Theoretically, the peak area should be divided by the analyte velocity; however, by using the same length of capillary from injection to detection window, it is easy to correct for this phenomena by simply dividing integrated peak area (A) of a given analyte by MT of the analyte:

$$NA = \frac{A}{MT} \tag{10.9}$$

where NA = normalized peak area. The reproducibility of the normalized peak areas are critical for the quantitative analyses. The major factors affecting this area are:

- Precision of injection.
- Precision of integration.
- Changes in sample concentration/composition/viscosity.

Variation in injection volume is one of the main reasons for changed peak areas during consecutive runs. Nowadays, many commercial instruments correct the injection time if the specified injection force varies from the programmed value. This leads to higher precision of injection. However, the use of very short injection times (1–2 s) is still not recommended. The implementation of a rinsing step after sample injection by dipping the capillary end into water may also lead to improved precision of injected volume. This is owing to small amounts of sample present near and just outside the capillary opening when the capillary has been dipped into the sample. Furthermore, the capillary end should be cut in a clean cut to avoid disturbance owing to a jagged capillary end. The precision of injection may also be affected negatively by variations in temperature, resulting in changed viscosity and thereby altered injection volume. Another factor is the sudden application of high voltages at the start of analysis, which may lead to thermal expansion of the buffer with expulsion of sample as a result. Gradual increment of separation voltage during the first min of separation may be a solution. Finally, hydrodynamic injection has proved to give considerably better reproducibility than electrokinetic injection, especially with short injection times. Injection modes are further treated below.

The precision of peak area integration is affected by the sample concentration, a low signal-to-noise ratio leading to integration errors. Optimization of the integration parameters combined with increased sample concentration may limit this problem and save time otherwise taken up in reprocessing data.

A problem in some of the commercially available HPCE instruments is evaporation from the samples, leading to increased peak areas and thus incorrect concentration determination. In many newer instruments this problem has, however, been solved using special evaporation caps combined with cooling of the autosampler during analyses. Adsorption of sample components to the capillary wall may affect peak areas seriously, the actual degree depending on the state of capillary. If problems with sample adsorption arise, the separation buffer should be revised as a first step, including change of pH, concentration, additives, *etc*. Finally, when analysing series of samples obtained from, for example, different time intervals during a processing step of food, it is necessary to be aware of differences in sample viscosities, as this will influence injection volumes. If possible, sufficient dilution should be made.

The use of an internal standard may compensate for some of the problems discussed above and is recommended to obtain higher repeatability in quantification. Relative normalized peak areas, RNA, are calculated as:

$$RNA = \frac{NA_1}{NA_2} \quad (10.10)$$

where NA_1 = normalized peak area of the actual analyte, and NA_2 = normalized peak area of the internal standard (reference peak). RNA = 1 for the internal standard. Response factors for the different analytes have to be determined for the particular HPCE system and method to get correct quantification from the use of an internal standard. The internal standard chosen must be absent from the original sample, and migrate at a velocity different from the sample components. Moreover, it should be pure and well defined, unaffected by other compounds present in the sample, and with electrophoretic and detection behaviour similar to the analytes. It may be advantageous to use two internal standards with well-defined concentrations and a known relationship between their concentrations. Thereby, it will be possible to reveal the presence of absorbing peaks below the internal standards, as it is unlikely that the same amount of interfering compounds will occur at the position of two different internal standards. The response factors of the internal standards, which are important for quantitative methods of analyses based on HPLC and HPCE, need not be identical, even with the same UV–Vis-detection wavelength used, and with peak areas such as RNA for HPCE. This is especially the case for MECC, where compounds with large MT generally are more strongly bound to the micelles than compounds with small MT. Compounds with large MT thus give hypochromic effects (page 103), resulting in relative low RNA.

3 Instrumentation

The commercial HPCE systems available have no well-defined standard instrument specifications, and the apparatus from different companies may be quite different from each other with respect to technical data. A typical commercial HPCE system consists, however, basically of a high-voltage power supply (max 30 kV), two buffer reservoirs, electrodes, capillary tube, capillary temperature controller, capillary filling system, detector, and data acquisition equipment. Enhanced features are, for example, multiple detectors, temperature-controlled autosampler, replenishment system for buffer reservoirs, and fraction collector. It is not the intention here to give an overview of the present commercial instruments and their features as such an overview will very soon be outdated owing to continuous introduction of new and updated instruments. The following comprises a discussion of more general parameters such as capillary type and dimensions together with a presentation of selected detection and injection modes in HPCE.

Capillary Type and Dimension

One of the most important components in a HPCE system is the capillary itself. The capillary has to be of a high quality tubing material and fused silica has up to now been the material of choice. Fused silica has good thermal and optical properties (UV–Vis transparent) and is available in the very small dimensions necessary in HPCE, with internal diameters below 100 μm. It is inexpensive and chemically inert, except for the hydroxy groups at the silica surface, which are responsible for the electro-osmotic flow and can interact with charged molecules in the sample/separation buffer. Fused-silica capillaries are coated at their outer surface with a protective polyimide layer, making them relatively flexible, robust, and thus easy to handle. A capillary with uncoated detection window is fragile and should be handled with care. Other possible materials for capillaries are Teflon and quartz, whose performance and cost, however, is less favourable than fused silica.

The dimensions of HPCE fused silica capillaries varies over a wide range, with typical internal diameters in the 25–75 μm range, corresponding to an outer diameter of around 350–400 μm, whereas the effective length of the capillary in general is 50 to 75 cm, and the total length is about 5 to 20 cm longer. Changing capillary internal diameter and length will affect many parameters in HPCE, such as:

- migration time;
- heat generation and dissipation;
- buffer composition range;
- adsorption;
- detection limits;
- injection volumes;
- separation efficiency.

The capillary length should in general be as short as possible, to minimize time of analysis; however, other factors should also be considered in this respect. The shorter the capillary, the larger the injection volume as the drop in injection force per cm of capillary increases (Equation 10.11). Minimizing capillary length (or increasing capillary inner diameter) results in lower electrical resistance, higher currents, and thereby increased heat generation, but the surface area for dissipation of heat is decreased. Capillary tubes with small inner diameters should be preferred whenever short capillaries are required, as heat dissipation increases with decreasing diameter. This also gives the possibility to select a separation buffer with a relatively high ionic strength. However, some drawbacks of small diameter capillaries also exist, *e.g.* increased adsorption due to a higher surface area to volume ratio. The detection sensitivity, especially when optical detection is used, is also decreased by a smaller inner capillary diameter, a direct consequence of the short light path-length (Lambert–Beers law; page 98), and this is further emphasized by the small injection volume in HPCE. The limits for detection are also affected negatively by the distortion and scatter of light caused by the rounded capillary wall. Possible ways to increase detection sensitivity are discussed below in the subsection detection modes.

Injection Modes

Precise and reproducible sample injection is crucial in HPCE when working with quite small injection volumes (nl). Quantitative sample injection can be performed in various modes, the most commonly used being direct on-column methods as hydrodynamic and electrokinetic injection.

Hydrodynamic Injection

Hydrodynamic injection is usually accomplished using either:

- Vacuum
- Pressure
- Siphoning action

All three injection modes are common in commercial HPCE systems. A vacuum is applied at the detector end of the capillary, pressure at the injection end, and the siphoning action is obtained by an elevation of the injection reservoir relative to the reservoir at the detection end of the capillary. To avoid unwanted sample injection by siphoning during vacuum or pressure injection, it is thus important to hold the liquid in sample and buffer reservoirs at the same levels. Siphoning is typically used in HPCE systems that are without pressure or vacuum injection capabilities.

The hydrodynamic injection volume (ml) can be quantified using Poiseuille's law for liquid flow through a circular tube:

$$\text{Volume} = \frac{\Delta P \pi r^4 t}{8\eta L} \tag{10.11}$$

where ΔP = pressure difference across the capillary (dyn cm^{-2}), r = inner radius of the capillary (cm), t = injection time (s), η = viscosity of buffer (dyn s cm^{-2}), and L = total length of the capillary (cm). The injection time and pressure difference are the main factors used for controlling the injected sample volume. For vacuum and pressure injection, typical injection times vary from 1 to 10 s, whereas the pressure difference is instrument specific and monitored during injection. See Box 10.2 for a calculation example of injection volume.

ΔP for siphoning injection is calculated from the equation:

$$\Delta P = \rho g \Delta h \tag{10.12}$$

where ρ = sample density (kg m^{-3}), g = gravitational acceleration constant (9.82 N kg^{-1}), and Δh = height difference between sample and buffer reservoirs (m). The unit for ΔP is N m^{-2} (= Pa = 10 dyn cm^{-2}). The height difference is typically 5 to 10 cm, held for 10 to 30 s.

Calculation of the injection volume is important in determining the detection limits (Box 10.3). The sample plug-length in the capillary can be calculated from the injection volume by dividing with πr^2 (see Box 10.2). To prevent sample overloading, the plug length should not exceed 2% of the total capillary length

Box 10.2 *Calculation of injection volume and sample plug length using hydrodynamic injection in HPCE*

Injection volume

Pressure difference across the capillary (ΔP): 127 mmHg (5″)
1 mmHg = 1.333×10^2 Pa ⇔ 127 mmHg = 16929.1 Pa
1 dyn cm^{-2} = 0.1 Pa ⇔ 16929.1 Pa = 16.93×10^4 dyn cm^{-2}
Capillary diameter = 50.0×10^{-4} cm ⇔ capillary radius (r) = 25.0×10^{-4} cm
Injection time (t) = 1.0 s
Viscosity (η) = 0.798×10^{-2} dyn s cm^{-2} (water at 30 °C)
Total capillary length (L) = 76.0 cm

$$\text{Volume} = \frac{16.93 \times 10^{-4}\ \text{dyn cm}^{-2} \times 3.1416(25.0 \times 10^{-4}\ \text{cm})^4 1.0\ \text{s}}{8 \cdot 0.798 \times 10^{-2}\ \text{dyn s cm}^{-2} \times 76.0\ \text{cm}} = 4.28 \times 10^{-6}\ \text{ml}$$

Sample plug length

Injection volume: 4.28×10^{-6} ml
Capillary radius (r) = 25.0×10^{-4} cm

$$\text{Sample plug length} = \frac{4.28 \times 10^{-6}\ \text{ml}}{3.1416(25.0 \times 10^{-4}\ \text{cm})^2} = 0.218\ \text{cm}$$

in ordinary HPCE modes. A precise injection implies a well functioning temperature control, as the viscosity is temperature dependent and may vary with 2–3% per °C. Inclusion of an internal standard (page 73) can eliminate the problem. In contrast to electrokinetic injection (*vide infra*), the injection volume for hydrodynamic injection is virtually unaffected by differences in the electrophoretic mobilities of the sample components.

A special variant of injection is that two (or more) individual injections may be performed one after the other followed by a single separation step. This procedure can be of interest for several reasons, *e.g.* to avoid mixing of samples or if a marker of electroosmotic flow, unwanted in the sample but important to the analysis, is to be included. Kuhn and Hoffstetter-Kuhn[4] note that for small injection volumes (< 5 nl) the systemic error of this method should be within the experimental error of separation.

Electrokinetic Injection

Electrokinetic injection is performed by moving the electrode from the buffer reservoir to the sample vial and applying a voltage, typically lower than the separation voltage, for a short period of time (normally 5 to 30 s). Sample ions will then enter the capillary as a function of their electrophoretic mobility combined with sample flow owing to electro-osmosis. Electrokinetic injection thus discriminates among the different sample ions, the more mobile analytes being injected in higher amounts than those of less mobility (Figure 10.5)

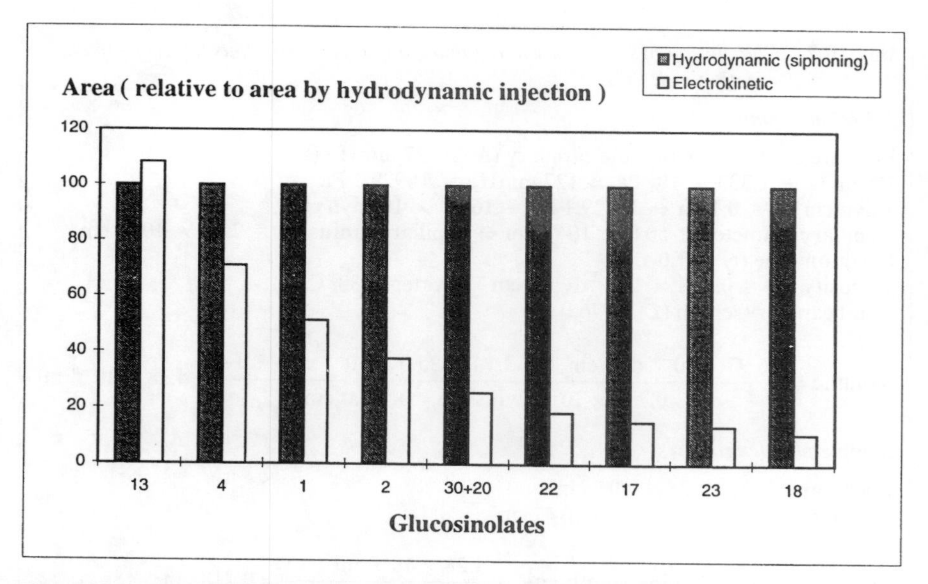

Figure 10.5 *Peak areas for glucosinolates analysed by MECC using hydrodynamic (siphoning) and electrokinetic injection*

Theoretically, the actual quantity [Q (mol)] of an ion introduced into the capillary can be calculated as:

$$Q = \frac{(\mu_{eo} + \mu_e)\pi r^2 VCt}{L} \quad (10.13)$$

where μ_{eo} = electro-osmotic mobility ($cm^2\ V^{-1}\ s^{-1}$), μ_e = electrophoretic mobility of the analyte ($cm^2\ V^{-1}\ s^{-1}$), r = inner radius of the capillary (cm), V = applied voltage (V), C = concentration of the actual analyte ($mol\ ml^{-1}$), t = injection time (s), and L = total length of the capillary (cm). The concentration of the sample is, however, seldom known. The sample volume (ml) introduced can be calculated as:

$$\text{Volume} = (v_{eo} + v_e)\pi r^2 t \quad (10.14)$$

where v_{eo} = the electro-osmotic flow velocity ($cm\ s^{-1}$), v_e = the migration velocity of analyte ($cm\ s^{-1}$), r = inner radius of the capillary (cm), and t = injection time (s); v_e, which will vary depending on the actual analyte, is small compared with v_{eo}. This is also the reason that reversed charged ions are not repelled from migrating into the capillary, as the electro-osmotic flow will carry them towards the capillary despite a v_e in the opposite direction.

The electro-osmotic and electrophoretic mobilities are dependent on the composition of the sample solutions, and electrokinetic injection is in general not as reproducible as hydrodynamic injection, which is the usually preferred injection mode. Electrokinetic injection can, however, be advantageous in certain situations, *e.g.* for highly viscous samples or in capillary gel electrophoresis, when hydrodynamic injection is not feasible. This is also the case when working with wide-diameter capillaries (>100 μm i.d.), as the injection volume with hydrodynamic injection (1 s) in general will be too high.

Detection Modes

Many of the detection methods in HPCE are similar to those used in traditional liquid column chromatography, *e.g.* HPLC. Sensitive detection with minimum zone dispersion is indispensable owing to the minute injection volumes characteristic for the HPCE technique, and on-column detection, whenever possible, is far the most common. Optical on-column detection modes imply that a small part of the polyimide coating of traditionally fused silica capillaries is removed to obtain a detection window. This is most easily done by placing the capillary window site in a flame from a match for about 1–2 s followed by thorough cleaning of the now uncoated part using paper or tissue with methanol or ethanol. Optical on-column detections are complicated in certain HPCE modes, *e.g.* in capillary electrochromatography using packed columns. Some selected detection methods applicable in HPCE are:

- UV–Vis absorption (inclusive diode array detection)
- conductivity

- fluorescence
- laser-induced fluorescence
- refractive index
- thermal lens
- amperometry
- potentiometry
- radioactivity
- mass spectrometry

UV–Vis absorption and fluorescence detection are the most commonly used techniques in commercial HPCE apparatus, although attempts are currently made to include other techniques as standards as well.

Detection may be direct or indirect, the latter being interesting because analytes showing limited or no UV–Vis absorbance, fluorescence, electrochemical activity, *etc*. can be detected in this way. The principle is a displacement of mobile phase background electrolyte (buffer ions or additives) by analytes as illustrated in Figure 10.6. The detector signal obtained is a decrease in signal, because of the lower concentration of buffer molecules or additives in the analyte peaks compared with the baseline signal. The background electrolyte chosen should basically give a high signal to obtain detectable peaks from analytes in low concentration. It is also important that the background electrolyte has a mobility close to those of the sample components to prevent peak fronting/tailing owing to electromigration dispersion. Examples of suitable background electrolytes are given in Ref. 8. Pre- or post-column derivati-

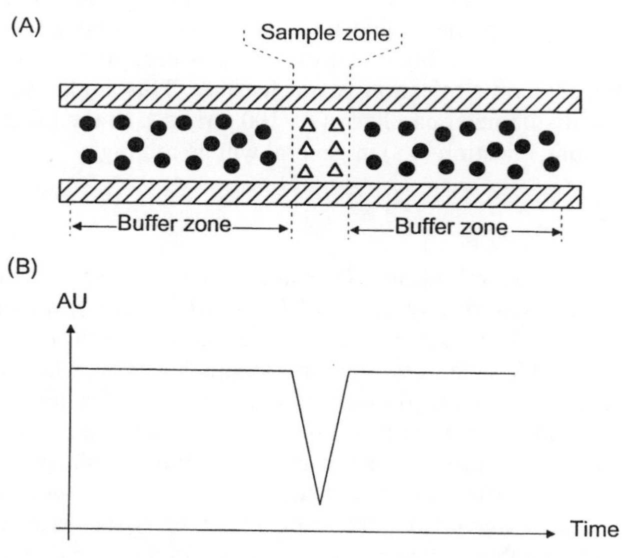

Figure 10.6 *Schematic presentation of the principle used in indirect detection (UV–Vis).* (A) *Capillary with sample molecules and UV–Vis absorbing background electrolyte and* (B) *corresponding electropherogram. Indirect detection can be performed using the same instrumentation as used for the corresponding direct detection methods*

Box 10.3 *Calculation of detection limit in HPCE*

Calculation of detection limits in HPCE demands knowledge of:

injection volume
concentration of analytes in the sample
M_r of analytes in the sample

In general, an acceptable signal-to-noise ratio is 3:1

Example

Sinapine analysed by MECC (sodium cholate), UV-detection (235 nm)

Concentration detection limit for sinapine

0.325 mM sinapine in sample
20 times dilution of sample gives lowest detectable amount (1 s injection):
0.325 mM/20 = 0.01625 mM = 16.25×10^{-6} M

Mass detection limit for sinapine

Injection volume:
4.28 nl (Box 10.2)

Injected moles:
$4.28 \times 10^{-9}\ \mathrm{L} \times 16.25 \times 10^{-6}\ \mathrm{mol\ L^{-1}} = 6.96 \times 10^{-14}\ \mathrm{mol} = 69.6\ \mathrm{fmol}$
M_r, sinapine: 369.4 g mol^{-1}

Injected grams:
$6.96 \times 10^{-14}\ \mathrm{mol} \times 369.4\ \mathrm{g\ mol^{-1}} = 2.57 \times 10^{-11}\ \mathrm{g} = 25.7\ \mathrm{pg}$

zation of sample components with chromophores or fluorophores is another way to enhance detection (and separation) in HPCE. There have been successful approaches to this, *e.g.* for amino acids and carbohydrates (Chapter 5).

Problems in detection sensitivity for HPCE are in general mainly owing to the short optical path-length giving relatively high concentration detection limits (g L^{-1} or M), although the mass detection limit (g or mol) usually are very low depending on the analyte (Box 10.3).

In general, UV–Vis absorption detectors have a concentration detection limit of around 10^{-6} M, whereas laser-induced fluorescence improves the sensitivity to about 10^{-12} M or even better. Several attempts have been made to extend the optical path-length for UV detection in HPCE, including the use of:

- rectangular or quadratic capillaries;
- Z-shaped flow cell;
- bubble-shaped flow cell;
- axial illumination;
- multireflection flow cell.

An example of the new capillary types with rectangular or quadratic appearance, in which detection may be performed across the short and long axis, respectively, is shown in Figure 10.7.

The rectangular capillaries, which in addition to the increased light path-

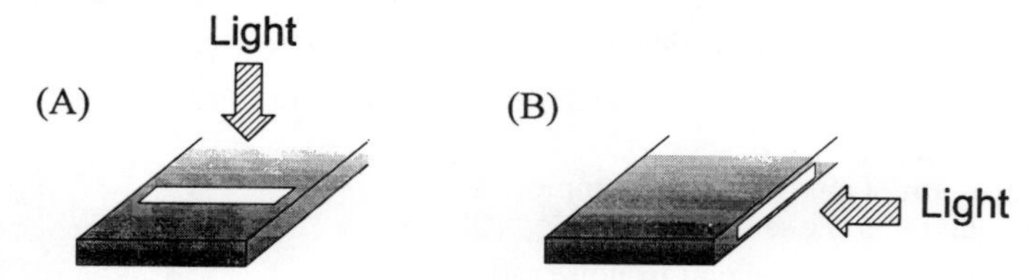

Figure 10.7 *Rectangular capillary with arrangements for detection* (A) *across the short axis and* (B) *across the long axis of the capillary*

length also have a good heat dissipation, are only available in borosilicate glass with no protective coating, making them more fragile than the traditionally used fused silica capillaries. The rectangular capillaries are now commercially available, however, not all instruments can be equipped with these.

The principles of the Z- and bubble-shaped flow cells are illustrated in Figure 10.8. The bending of the capillary in the Z-shaped flow cell results in a considerable increase in optical path-length (*e.g.* 40 times, 75 μm capillary, 3 mm Z-cell); however, the improvement in sensitivity may be limited by an inadequate focusing of light. Lambert–Beers law is thus not directly applicable here, and this is in fact also true for ordinary unbent capillaries, where the curved circular cross-section leads to scattering of light (page 98) and a non-linear relationship between absorbance and concentration. The gained sensitivity obtained with the Z-cell is in general accompanied by a decrease in efficiency.

The optical path-length in the bubble-shaped flow cell is typically less than

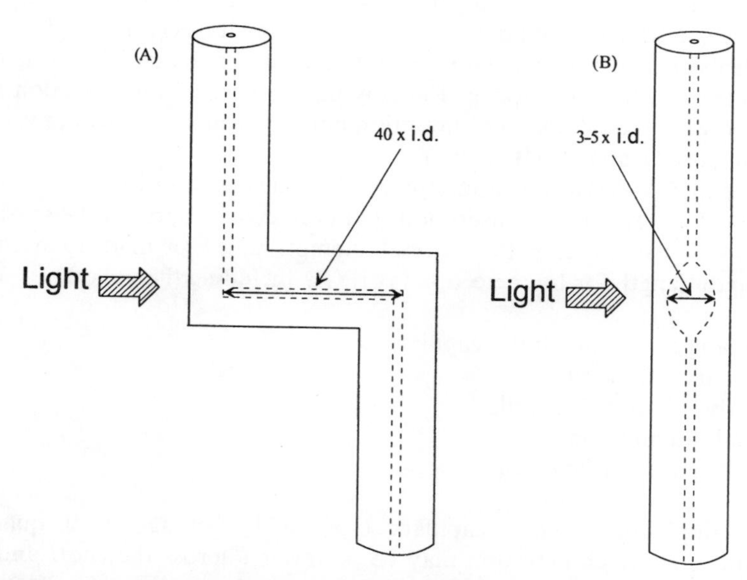

Figure 10.8 *Design of* (A) *Z-shaped flow cell and* (B) *bubble-shaped flow cell. Notice the increased optical path-length as indicated by* $\leftrightarrow$

obtained for the Z-cell (*e.g.* 3–5 times capillary i.d.). One advantage of the bubble-cell is, however, a concentration (stacking) of the sample zone as it enters the bubble, being caused by a decreased velocity of sample ions owing to increased volume and lowered electrical resistance and thereby reduced electric field in the bubble. The band broadening in the bubble cell is thus limited, despite the increased volume.

For some apparatus both Z- and bubble-shaped flow cells are commercially available, built directly into the capillary; this is of course considerably more expensive than the cost of ordinary capillaries, which are often sold by the metre. A theoretical overview of the less common axial illumination and multireflection flow cell is given in Ref. 9.

4 Selected Modes of HPCE Operation and Examples of Applications

HPCE comprise a big family of related techniques, each differing in the mechanism or combination of mechanisms of separation. It may be somewhat confusing to distinguish between the various modes, as a large number of abbreviations and different designations have been used at random. An overview of selected techniques has been given in Table 10.2.

A short description of the basic separation principles of the various methods are given in the following, whereas specific separation parameters and variation of these are treated separately in Section 10.5. The applicability of HPCE in food analysis is very wide and covers analysis of a heterogeneous group of compounds, *e.g.* amino acids, peptides, proteins, protein/peptide–antigen/inhibitor complexes, carbohydrates (mono-, oligo-, and polysaccharides and their derivatives), lipids, steroids (*e.g.* saponins), vitamins, inorganic ions,

Table 10.2 *Selected modes of HPCE*

Acronym	*HPCE technique, designation*
CAE	Capillary affinity electrophoresis
CEC	Capillary electro-osmotic chromatography*
	Capillary electrochromatography*
CFZE	Capillary free zone electrophoresis†
CGE	Capillary gel electrophoresis
CIEF	Capillary isoelectric focusing
CITP	Capillary isotachophoresis
CZE	Capillary zone electrophoresis†
FSCE	Free solution capillary electrophoresis†
FZCE	Free zone capillary electrophoresis†
HICE	Hydrophobic interaction capillary electrophoresis
MCE	Microemulsion capillary electrophoresis
MECC	Micellar electrokinetic capillary chromatography‡
MEKC	Micellar electrokinetic chromatography‡

*,†,‡Acronym/designation covering the same technique

nucleic acids, nucleotides, oligonucleotides, flavonoids, other phenolics, biogenic amines, carboxylic acids, inositol phosphates, glucosinolates, mycotoxins, food additives and degradation products, and drugs, *etc*. Selected examples of applications of HPCE in food analysis are included in connection with the description of the various techniques, and additional examples are included in Chapter 11 to 14.

Free-zone Capillary Electrophoresis (FZCE)

FZCE, the fundamental and simplest mode of HPCE, has been described in detail in Section 10.2. In contrast to the more sophisticated HPCE modes, separation in FZCE is based solely on differences in electrophoretic mobility of analytes (charge, size), with the electro-osmotic flow making separation of oppositely charged sample components possible in one and the same analysis (Figure 10.2).

Applications of FZCE

Basically, FZCE is not applicable to uncharged compounds; however, this may be evaded by complexing or derivatization procedures that introduce a charge on the compounds. Two examples of carbohydrate analysis by FZCE are given in Figures 10.9 and 10.10.

Negatively charged oligosaccharide–borate complexes are formed as a

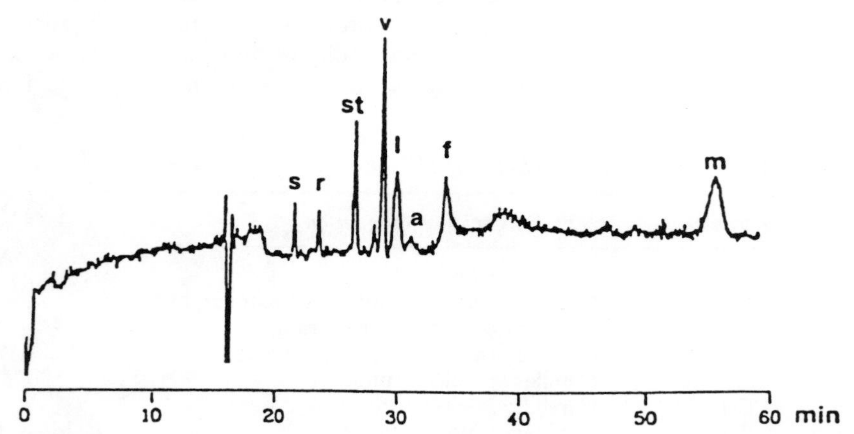

Figure 10.9 *FZCE separation of oligosaccharides present in a pea extract purified by group separation. Peaks: s = sucrose, r = raffinose, st = stachyose, v = verbascose, a = ajugose, l = lactose (internal standard), f = false peak, and m = melibiose (internal standard); buffer: 100 mM disodium tetraborate, pH 9.9; temperature: 50 °C, voltage: 10 kV; injection: hydrodynamic (vacuum) (+ to −); detection: UV, 195 nm; capillary: 50 μm i.d., 72 cm total length, 50 cm effective length*

(Reprinted from A.M. Arentoft, S. Michaelsen, and H. Sørensen, *J. Chromatogr. A*, **652**, 517 (1993), Elsevier Science, Amsterdam)

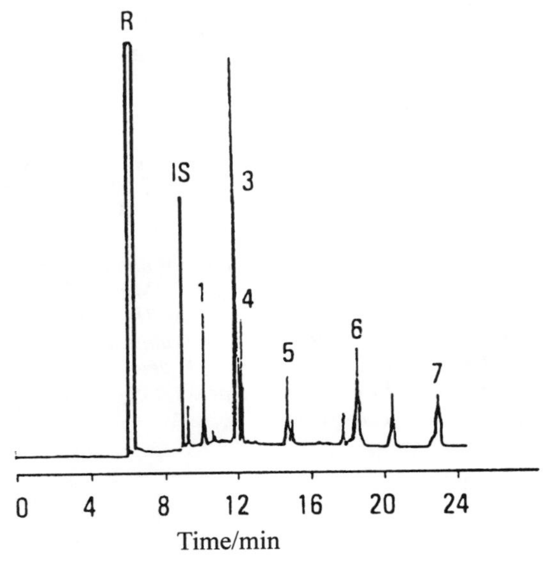

Figure 10.10 *FZCE separation of ABEE-derivatized monosaccharides in polysaccharides extracted from* Radix althaeae. *Peaks: R = reagent, IS = 2-deoxy-D-ribose, 1 = rhamnose, 3 = glucose, 4 = arabinose, 5 = galactose, 6 = glucuronic acid, and 7 = galacturonic acid; buffer: 200 mM boric acid, pH 10.5; temperature: 30 °C; voltage: 25 kV; injection: hydrodynamic; detection: bubble cell, UV, 306 nm; capillary: 50 μm i.d., 72 cm total length, 50 cm effective length*
(Reproduced by permission from C. Woodward and R. Weinberger, Application Note, Hewlett-Packard, Bioscience, 1994)

consequence of the buffer used resulting in both better separation and improved UV detection (195 nm) compared with the uncomplexed carbohydrates. Derivatization of sugars is another method of obtaining improved detection limits, here using the ethyl ester of 4-aminobenzoic acid (ABEE) (Figure 10.10).

FZCE is also used for the analysis of free inorganic cations and anions as well as organic acids *e.g.* in fermentation broth, to obtain information on any surplus or depletion of substrates as well as on the ongoing metabolism (Figures 10.11 and 10.12).

The analysis of some inorganic cations may be complicated by their complexation with various anions such as phosphate, phytate, and nicotinamide. Complexation is further discussed below (page 254).

Finally, FZCE is also widely used for the analysis of peptides and proteins. Different approaches have in this connection to be used to limit interactions with the capillary wall (Chapters 11 to 13).

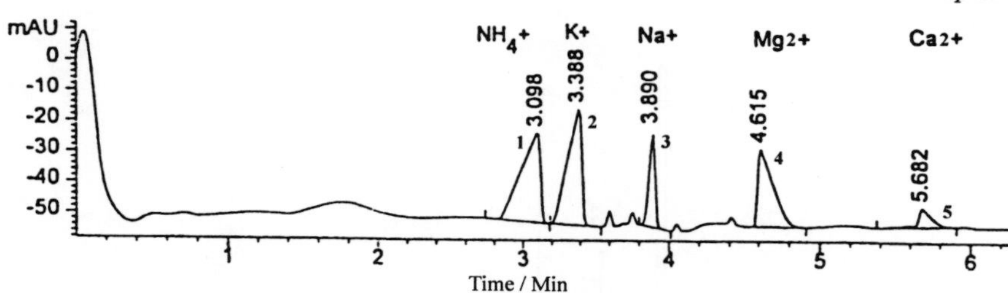

Figure 10.11 *FZCE separation of free inorganic cations in fermentation broth. Peaks: 1 = NH_4^+, 2 = K^+, 3 = Na^+, 4 = Mg^{2+}, 5 = Ca^{2+}; buffer: 10 mM imidazole, 5 mM tartaric acid, 3 mM 18-crown-6-ether, pH 4.5 (HCl); temperature: 30 °C; voltage: 30 kV (ramped over the first min); injection: hydrodynamic (pressure) (+ to −); detection: UV (indirect), signal 235 nm, reference 210 nm (positive peaks); capillary: 75 µm i.d., 64 cm total length, 56 cm effective length*

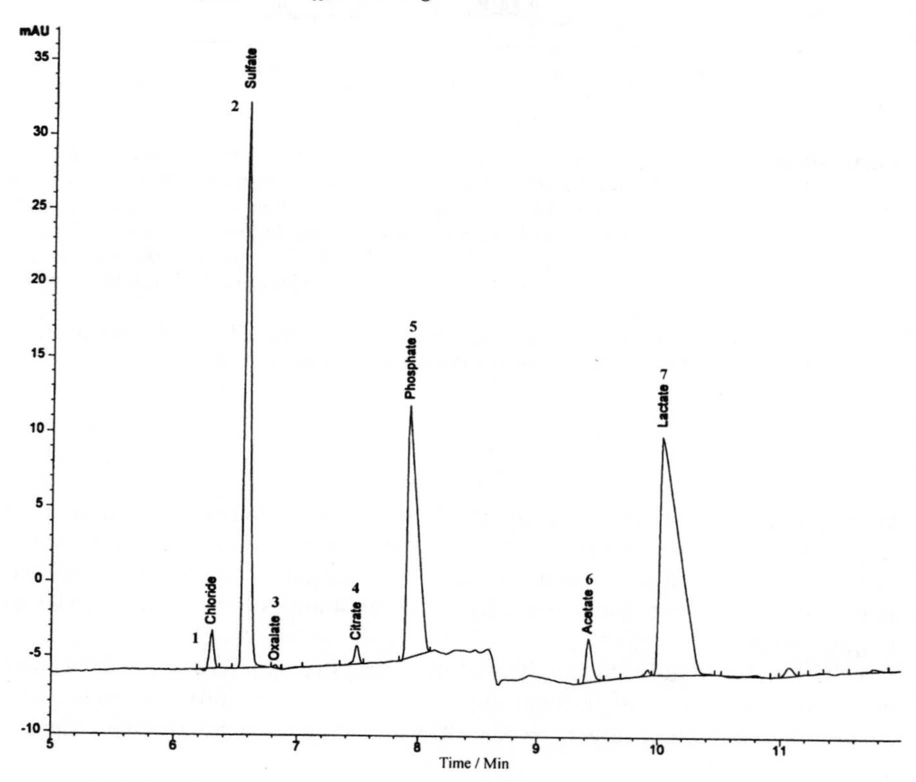

Figure 10.12 *FZCE separation of free inorganic anions and organic acids in fermentation broth. Peaks: 1 = Cl^-, 2 = SO_4^{2-}, 3 = oxalate, 4 = citrate, 5 = PO_4^{3-}, 6 = acetate, 7 = lactate; buffer: 6 mM potassium dichromate, 0.1 mM hexadecyltrimethyl ammonium bromide (CTAB), pH 9.1 (NH_4OH); temperature: 30 °C; voltage: 15 kV (ramped over the first min); injection: hydrodynamic (pressure) (− to +); detection: UV (indirect), signal 310 nm, reference 380 nm (positive peaks); capillary: 75 µm i.d., 72 cm total length, 64 cm effective length*

Micellar Electrokinetic Capillary Chromatography (MECC)

MECC differs basically from FZCE by the inclusion of surfactants in the buffer phase. Surfactants are amphophilic molecules, containing a hydrophobic and a hydrophilic part. Surfactants present in low concentration will exist primarily as their monomeric form, whereas an increase in concentration above the critical micellar concentration (c.m.c.) leads to formation of micelles. A more detailed description of micelle formation is given in Chapter 2.

The separation principle in MECC is based on a distribution of analytes between the aqueous buffer phase and the so-called pseudostationary micellar phase (Figure 10.13). The interaction of analytes with the micelles are typical of hydrophobic character, but may also be based on ion-pairing.

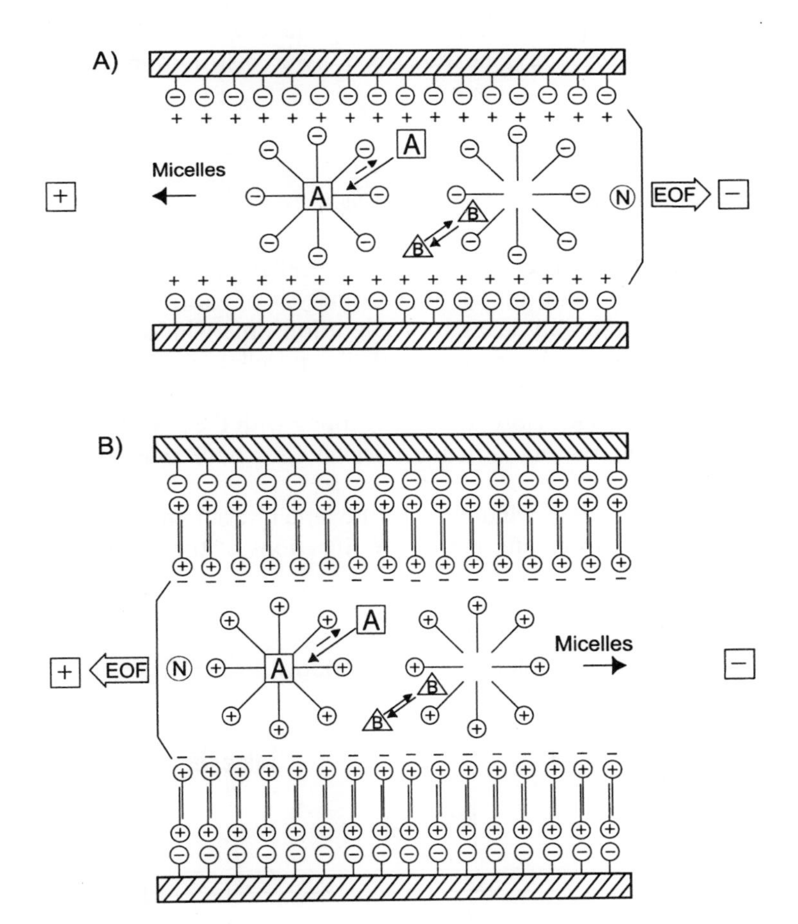

Figure 10.13 *Schematic illustration of the separation principle in MECC using* (A) *anionic and* (B) *cationic surfactants. N = neutral analytes with no interaction with the micelles, A = analytes fully solubilized in the micelles, and B = analytes partly solubilized in the micelles*

The use of cationic surfactants reverses the direction of the electro-osmotic flow compared to FZCE (without additives) (Figure 10.2) and MECC using neutral or anionic surfactants [Figure 10.13(A)]. This is because of the formation of a double layer of surfactant at the capillary wall, changing the net charge of the wall from negative to positive [Figure 10.13(B)]. The change in direction of the electroosmotic flow has no direct effect on the separation principle in MECC, however, it is usually necessary to switch the polarity of the electrodes (cathode instead of anode at the injection site).

As illustrated in Figure 10.13, the micellar and electroosmotic flow, the latter directed towards the detector site, are opposite each other, no matter the charge of the surfactants. Compounds interacting with the micelles will thus be retarded in their migration, however, their net flow will be in the direction of the electroosmotic flow, because the strength hereof in general exceeds the micellar flow. The interaction of sample components with the micelles can, as in traditional chromatography, be described by a distribution constant, K_D, defined as:

$$K_D = \frac{[S]_m}{[S]_b} \tag{10.15}$$

where $[S]_m$ = solute concentration in the micellar phase and $[S]_b$ = solute concentration in the buffer phase. MECC analysis of neutral compounds with no electrophoretic mobility leads to a migration time window (Figure 10.14) with analytes having no interaction with the micellar phase (K_D close to 0) moving along with the electro-osmotic flow (t_0), whereas analytes fully solubilized in the micelles (K_D close to 1) move along with the micelles (t_m). Sample molecules partly interacting with the micelles thus have migration times (t_s) between t_0 and t_m.

A wide migration time window will altogether result in better separation possibilities, and it can be affected by modifications of the size of the electro-

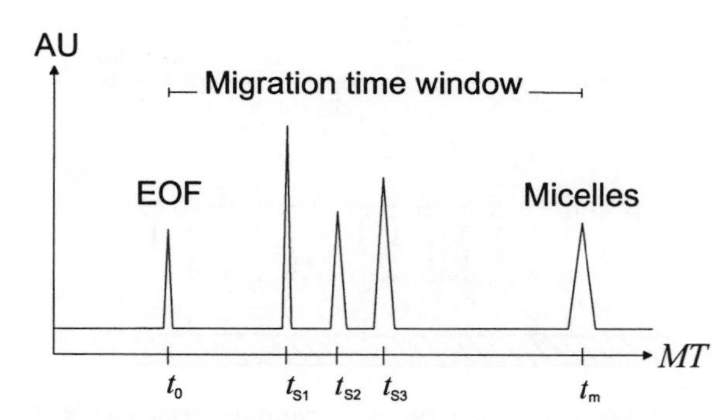

Figure 10.14 *Migration time window in MECC.* t_{s1}, t_{s2}, *and* t_{s3} *indicates migration times for sample components showing increasing interaction with the micellar phase*

osmotic flow and micellar flow, respectively (Section 10.5). Actually, ionic compounds analysed in MECC can have migration times from less than for the electro-osmotic flow to longer than for the micelles, depending on charge and interaction with the micelles.

Applications of MECC

The MECC technique is especially interesting for selective separation of neutral compounds and ionic compounds which have limited difference in electrophoretic mobility. There are often several possibilities for surfactant choice within an analysis, and one has to take care choosing the type resulting in the best separation. It is often possible to improve the separation obtainable by FZCE by the inclusion of surfactants in the buffer phase, although exceptions of course exist. The wide applicability of MECC in food analysis, using various forms of surfactants, is illustrated by selected examples.

The analysis method for amino acids shown in Figure 10.15 is only one among many, where different derivatizations and surfactants have been used.

The applicability of MECC in ion analysis is exemplified in Figure 10.16, which shows the separation of ions in milk using a cationic surfactant. Nitrate and nitrite accumulate often to unwanted high concentrations in various vegetables and occurrence of these ions in food are of interest owing to their possible involvement in nitrosamine formation. Nitrite do not occur in milk owing to its reaction with milk proteins. The thiocyanate ion occurs in many cruciferous vegetables and other products used as food, especially in insufficiently processed materials. This ion competes with iodide needed for correct function of the thyroid gland.

MECC separation of indoles, physiologically very active compounds obtained as one of the degradation product groups derived from glucosinolates, is shown in Figure 10.17. Various types of indolyls in food give an unpleasant smell and taste, even at very low concentrations. However, degradation products from some indolylglucosinolates, which occur in relatively high concentrations in cruciferous vegetables, have got special attention in relation to cancer and in search for more healthy foods.

Glucosinolates can be effectively separated by MECC using CTAB as surfactant (Figure 10.18). These compounds occur in all plants belonging to the plant order Capparales and only in some few other plants. The glucosinolates and their degradation products are of great importance for the quality of cruciferous crops and efficient methods of analyses are thus of great importance in the quality control and studies of the physiological effects these compounds may have. The possibility to separate structural epimers by MECC is seen in Figure 10.18 for progoitrin (No. 4) and epiprogoitrin (No. 5) differing only in the position of the hydroxy group on C-3 in the side chain.

Phenolics in both free and especially bound forms (see also Chapter 14) are important constituents of all plants, and phenolics are often considered in relation to the quality of food and feed stuffs. Analysis of structural isomers is

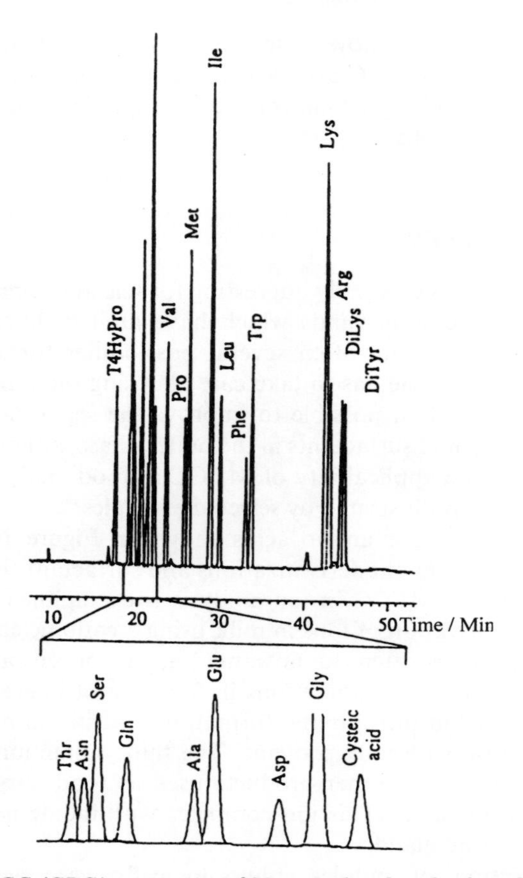

Figure 10.15 *MECC (SDS) separation of 21 neutral and acidic dansylated (Dns) amino acids. Peaks: abbreviations describes the dansylated amino acids. Special abbreviations: T4HyPro =* N-*Dns*-trans-*4-Hy-*L*-Pro, DiLys = di-Dns-*L*-Lys, DiTyr =* N,O-*di-Dns-*L*-Tyr; buffer: 100 mM boric acid, 150 mM SDS, pH 8.3; temperature: 27 °C, voltage: 15 kV; injection: hydrodynamic (vacuum) (+ to −); detection: UV, 216 nm; capillary: 50 μm i.d., 72 cm total length, 52 cm effective length*
(Reprinted from S. Michaelsen, P. Møller and H. Sørensen, *J. Chromatogr. A*, **680**, 299 (1994), Elsevier Science, Amsterdam)

here demonstrated by the separation of ferulic/isoferulic acid and 4-hydroxybenzoic/salicylic acid (Figure 10.19). In addition, *cis*-forms of sinapic, ferulic, and coumaric acid could be effectively separated from the dominant *trans*-forms, as indicated by arrows in Figure 10.19.

Bile acid surfactants, differing considerably in structure and micellar composition compared with the more commonly used alkyl-chain surfactants such as SDS and CTAB (page 262), are suitable for the analysis of various compounds, illustrated here by use in the separation of phospholipids, which is of great interest in lipid research (Figure 10.20).

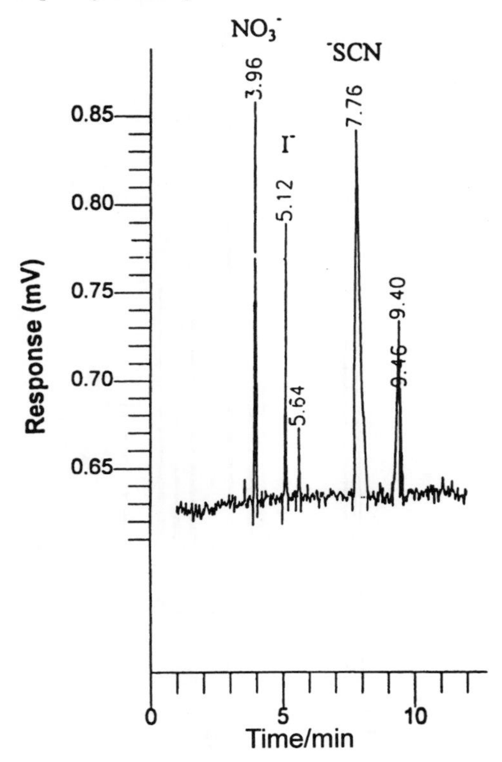

Figure 10.16 *MECC (DTAB) separation of nitrate, iodide, and thiocyanate ion in milk using ion-exchange for isolation. Buffer: 18 mM disodium tetraborate, 30 mM Na_2HPO_4, 50 mM decyltrimethylammonium bromide (DTAB), 10% propan-2-ol, pH 7.0; temperature: 60 °C, voltage: 20 kV; injection: hydrodynamic (vacuum) (− to +); detection: UV, 235 nm; capillary: 50 μm i.d., 76 cm total length, 53 cm effective length*
(Reprinted from C. Bjergegaard, P. Møller, and H. Sørensen, *J. Chromatogr. A*, **717**, 409 (1995), Elsevier Science, Amsterdam)

Hydrophobic Interaction Capillary Electrophoresis (HICE)

In HICE, surfactants are included in the buffer phase at concentrations below c.m.c., or in presence of high concentrations of organic modifier (*e.g.* acetonitrile), thus preventing the formation of micelles for surfactant concentrations above c.m.c.. The presence of surfactant affects the analyte mobility, depending on the charge of the surfactant and analyte. If the charge sign is the same, hydrophobic interactions will increase the net charge of the analyte and thereby also the analyte mobility, whereas the analyte mobility will be decreased if the charge is of opposite sign to that of the surfactant molecules. Furthermore, uncharged analytes can become charged by hydrophobic interaction with the surfactant thereby making separation possible. Addition of

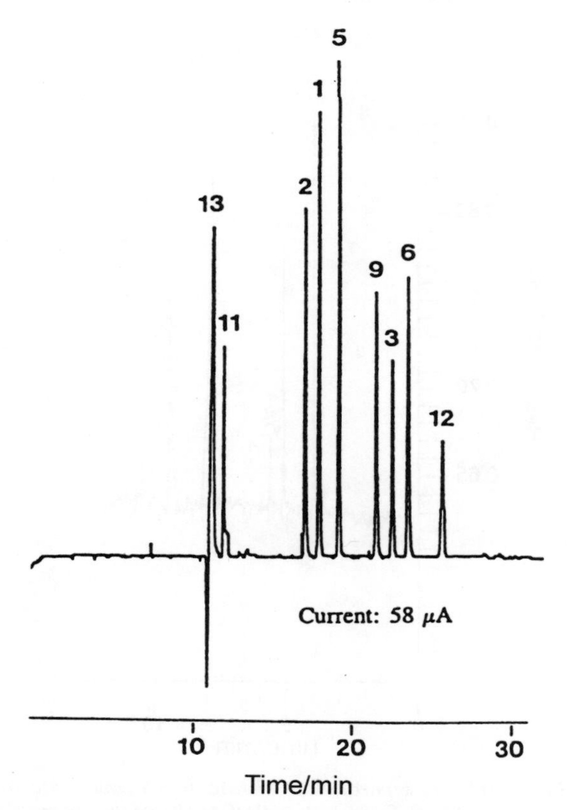

Figure 10.17 *MECC (DTAB) separation of indoles. Peaks: 1 = indol-3-ylcarboxylic acid, 2 = indol-3-ylacetic acid, 3 = indol-3-ylacetonitrile, 5 = indol-3-ylaldehyde, 6 = 3-acetoxyindole, 9 = 3-(indol-3-yl)-propionic acid, 11 = tryptophan, 12 = tryptamine, 13 = 5-hydroxytryptophan; buffer: 18 mM disodium tetraborate, 30 mM Na_2HPO_4, 50 mM dodecyltrimethylammonium bromide (DTAB), 10% propan-2-ol, pH 7.0; temperature: 30 °C, voltage: 20 kV; injection: hydrodynamic (vacuum) (− to +); detection: UV, 235 nm; capillary: 50 μm i.d., 76 cm total length, 53 cm effective length* (Reproduced from C. Feldl, P. Møller, J. Otte, and H. Sørensen, *Anal. Biochem.*, 1994, **217**, 62)

lyotropic or antichaotropic salts (page 81) together with the surfactant will shift the equilibrium towards greater interaction and thereby change analyte mobility even further.

Applications of HICE

Very hydrophobic neutral analytes can with advantage be analysed using HICE. An example of separation of the alkyl aryl ketone homologues tetradecanophenone (C_{20}), hexadecanophenone (C_{22}), and octadecanophenone

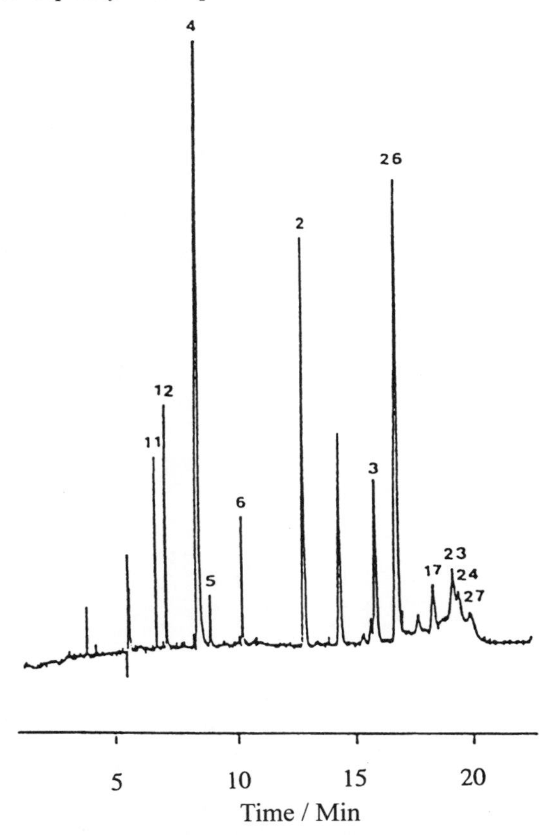

Figure 10.18 *MECC (CTAB) separation of glucosinolates in QMA extract of rapeseeds. Peaks: 11 = glucoraphanin, 12 = glucoalyssin, 4 = progoitrin, 5 = epiprogoitrin, 6 = napoleiferin, 2 = gluconapin, 26 = 4-hydroxyglucobrassicin, 3 = glucobrassicanapin, 17 = gluconastutiin, 23 = glucobrassicin, 24 = neoglucobrassicin, 27 = 4-methoxyglucobrassicin; buffer: 18 mM disodium tetraborate, 30 mM disodium hydrogenphosphate, 50 mM cetyltrimethylammonium bromide (CTAB), pH 7.0; temperature: 30 °C, voltage: 20 kV; injection: hydrodynamic (vacuum) (− to +); detection: UV, 235 nm; capillary: 50 μm i.d., 72 cm total length, 50 cm effective length* (Reproduced by permission from S. Michaelsen, P. Møller, and H. Sørensen, *GCIRC-Bull.*, 1991, **7**, 97)

(C_{24}) is shown in Figure 10.21. Note the reversal of migration order when changing from anionic to cationic surfactants.

Capillary Electrochromatography (CEC)

CEC is performed using capillaries packed with chromatographic material. The separation is based on electrophoresis and on distribution of the analytes

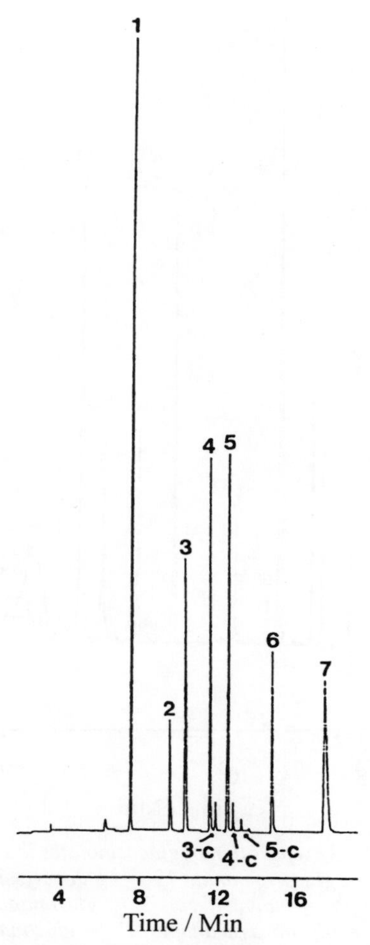

Figure 10.19 *MECC (CTAB) separation of phenolic carboxylic acids. Peaks: 1 = 4-hydroxybenzoic acid, 2 = isovanillic acid, 3 = sinapic acid, 4 = ferulic acid, 5 = coumaric acid, 6 = isoferulic acid, 7 = salicylic acid; buffer: 18 mM disodium tetraborate, 30 mM Na_2HPO_4, 50 mM cetyltrimethylammonium bromide (CTAB), pH 7.0; temperature: 40 °C, voltage: 20 kV; injection: hydrodynamic (vacuum) (− to +); detection: UV, 280 nm; capillary: 50 μm i.d., 66 cm total length, 44.5 cm effective length. Arrows show the cis forms of sinapic, ferulic and coumaric acid, respectively*
(Reprinted from C. Bjergegaard, S. Michaelsen, and H. Sørensen, *J. Chromatogr. A*, **608**, 403 (1992), Elsevier Science, Amsterdam)

between the packing particles and the aqueous buffer phase (Figure 10.22), similar to traditional chromatography and in part to MECC. Main differences between CEC and MECC are though, that the particles in CEC makes up a stationary phase (held in the capillary by frits in both capillary ends), whereas the micelles in MECC contribute further to the separation by moving at a

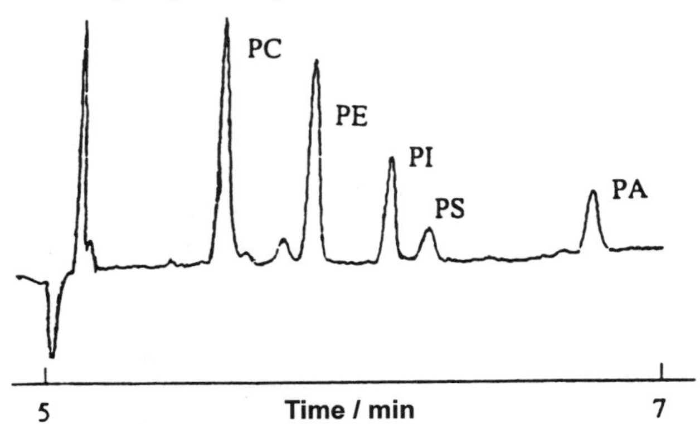

Figure 10.20 *MECC (sodium deoxycholate) separation of phospholipids from soybean lecithin. Peaks: PC = phosphatidylcholine, PE = phosphatidylethanoleamine, PI = phosphatidylinositol, PS = phosphatidylserine, PA = phosphatidic acid. Buffer: 6 mM borax, 10 mM phosphate, 35 mM sodium deoxycholate, 30%* n-*propanol, pH 8.5 (HCl); temperature: 50 °C, voltage: 30 kV; injection: hydrodynamic (pressure) (+ to −); detection: UV, 200 nm; capillary: 50 μm i.d., 57 cm total length, 52 cm effective length*
(Reproduced by permission from K. Verleysen, R. Szücs, and P. Sandra, in 'Proceedings of Eighteenth International Symposium on Capillary Chromatography', Italy, 20–24 May, 1996)

certain speed depending on their actual charge (pseudostationary phase). Moreover, the interaction of the analytes and the packing material take place on the surface of the particles, whereas the analytes in MECC are partly solubilized in the micelle. Semiporous packing material will, however, allow passage of sample molecules depending on their size, introducing an additional separation parameter to CEC.

The fixed charges on the capillary wall and, if charged, on the packing material result in creation of the electro-osmotic flow by movement of solvated counter-ions, similar to what is seen in, for example, FZCE and MECC. In the case of anionic packing material, the electro-osmotic flow will be in the direction of the cathode (Figure 10.22), whereas the use of positively charged packing material will reverse the electro-osmotic flow, owing to the high particle surface area compared with the capillary wall. In contrast to the laminar flow profile obtained in traditional pressure-pumped liquid chromatography systems, CEC results in a plug-like flow profile (Figure 10.3), which is only slightly disturbed by the presence of packing material, and the separation efficiency therefore tends to be higher in CEC. However, the high surface area in CEC increases the risk of problems caused by severe adsorption, but with pure reference samples CEC seems to be a valuable technique, although it may have problems with co-occurring impurities in practical samples. An evaluation of some of the parameters determining the performance of capillary electrochromatography in packed columns are given in Ref. 10.

(A)

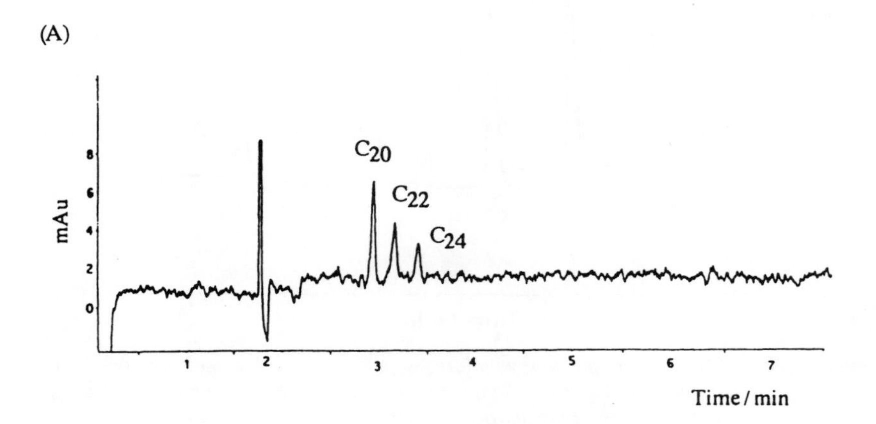

(B)

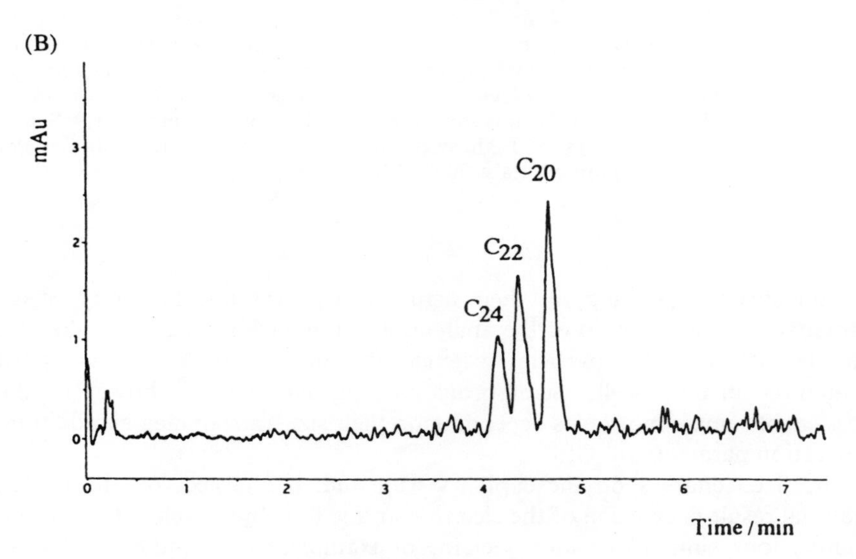

Figure 10.21 *Separation of tetradecanophenone (C_{20}), hexadecanophenone (C_{22}), and octadecanophenone (C_{24}) using HICE.* (A) *Buffer: 5 mM NaH_2PO_4, 50 mM SDS, 50% acetonitrile, pH 7.0; temperature: 25 °C, voltage: 30 kV; injection: hydrodynamic (+ to −); detection: UV, 214 or 254 nm; capillary: 50 μm i.d., 37.5/42.5 cm total length, 30/35 cm effective length.* (B) *Buffer: 5 mM phosphate, 20 mM CTAB, 50% acetonitrile, pH 2.8; temperature: 25 °C, voltage: 30 kV; injection: hydrodynamic (vacuum) (− to +); detection: UV, 214 or 254 nm; capillary: 50 μm i.d., 37.5/42.5 cm total length, 30/35 cm effective length*

(Reprinted from E.S. Ahuja and J.P. Foley, *J. Chromatogr. A*, **680**, 73 (1994), Elsevier Science, Amsterdam)

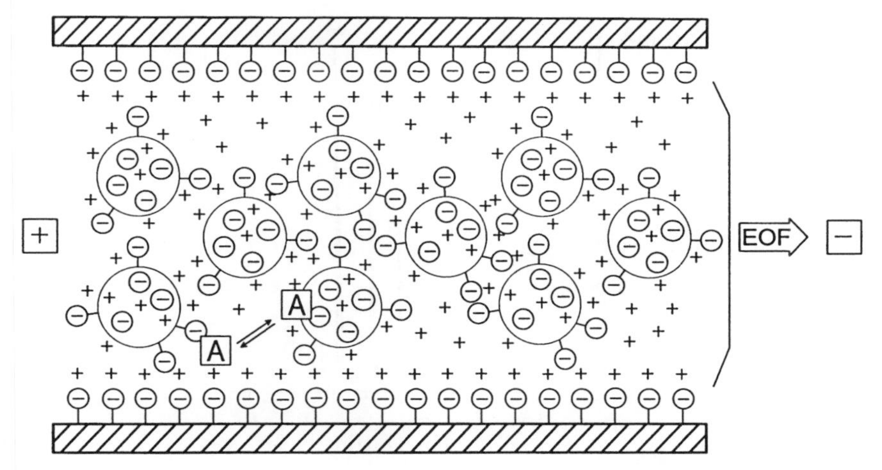

Figure 10.22 *Example of capillary electrochromatography, using a negatively charged packing material (ion-exchange chromatography). Positively charged packing material and uncharged hydrophobic material (reverse phase chromatography) are also theoretical possibilities*

Applications of CEC

The applications of CEC are comparable to FZCE and MECC; however, for large-molecule separation, the technique is probably not very useful. The adsorption problem is increased by the presence of chromatographic material in the column, and analytes known to suffer from such problems, *e.g.* macromolecules such as proteins, should not be analysed by CEC.

CEC is not a commonly used HPCE technique, mainly because of the difficulties in preparing the packed capillaries. If the technique is to be used routinely in food and feed analyses, packed capillaries need to be commercially available, and this is now possible (*e.g.* Electropak^{TM}, Unimicro Technologies, Inc.). Figure 10.23 shows an example of application of CEC.

Capillary Gel Electrophoresis (CGE)

The separation principle in CGE, using capillaries filled with preferably inert gels or polymer networks, is identical to that in traditional gel electrophoresis, a molecular sieving mechanism with large molecules being retarded more than small molecules. In contrast to CEC, in which analytes interact with a stationary phase, separation in CGE is thus principally based on differences in size. In cases where analytes (*e.g.* glycoproteins) interact with the gels or polymer networks, this technique will be more like affinity chromatography. The materials used for CGE are typically:

- polyacrylamide (cross-linked or linear not-cross-linked);
- agarose;

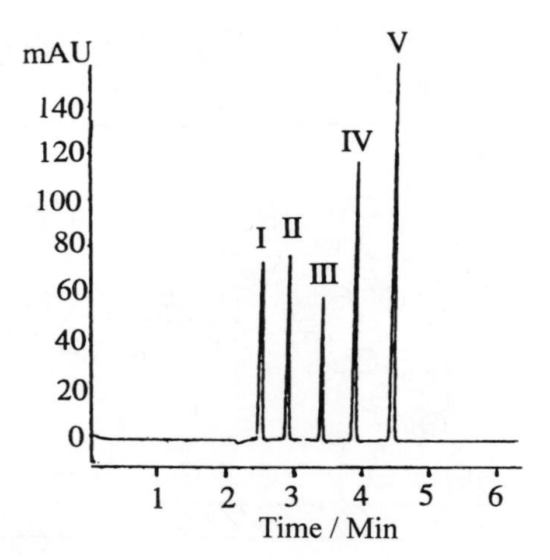

Figure 10.23 *CEC separation of a neutral test mixture. Peaks: I = thiourea, II = benzamide, III = anisole, IV = benzophenone, V = biphenyl; buffer: Acetonitrile/TRIS (0.05 M, pH 7.8) buffer 80/20 v/v mobile phase; temperature: 15 °C, voltage: 30 kV; injection: electrokinetic; detection: UV, 210 nm; capillary: Spherisorb ODS1 packed, 50 μm i.d., 25 cm total length*
(Reproduced by permission from S.M. Shariff, M.M. Robson, S. Roulin, P. Myers, K.D. Bartle, M.R. Euerby, and C.M. Johnson, in 'Proceedings of Eighteenth International Symposium on Capillary Chromatography', Italy, 20–24 May, 1996)

- methylcellulose, various derivatives;
- dextrans;
- polyethylene glycol.

In contrast to traditional slab-gel electrophoresis, CGE does not imply rigidity of the gels, but allows the use of various low viscosity polymer network solutions as well. Some characteristics of CGE using different kinds of capillary filling materials are collected in Table 10.3.

Table 10.3 *Characteristics, advantages, and disadvantages of CGE using different separation materials. In general, the same type of buffers as used in FZCE are applicable*

Gels	*Low viscosity polymer network solutions*
Complex preparation (polyacrylamide gels)	Easy to prepare
	Easy to fill into capillary (by pump)
Risk of gel failure during analysis	Several runs possible
Can not be replaced	Easy to replace
Only electrokinetic injection	Both electrokinetic and hydrodynamic injection

Polymerization of acrylamide in the presence of a cross-linking agent (page 187) results in a chemical gel with a pore size determined by covalent bondings, whereas physical interactions are responsible for the pore size in the so-called physical gels (linear not cross-linked polyacrylamide/agarose). Contrary to polyacrylamide, agarose requires no polymerization procedure, and it is also possible to work with low viscous polymer network solutions of agarose at temperatures above 35–40 °C (gelling temperature). The pore size in low viscosity solutions based on *e.g.* cellulose derivatives, dextrans, or polyethylene glycol is concentration dependent and based on polymer entanglement. In general, the polymer concentration necessary is inversely proportional to the size of the sample components.

It is possible to purchase commercially prepared gels as well as polymer network solutions, however, this is often an expensive solution. An experimental description of gel-preparation can be found in Ref. 9. Bubble formation and shrinkage are some of the difficulties connected with user-prepared gels, and this, combined with problems with the integrity of gels during the high field strength separation in CGE, means that polymer network solutions are often preferred. Also advantageous is the low UV-absorbance compared with polyacrylamide gels, which allows more sensitive optical detection. The very high separation efficiencies possible with CGE ($10–30 \times 10^6$ range) has, however, only been obtained using gels.

Unlike the above-mentioned HPCE modes, the electroosmotic flow in CGE is a parameter which should be suppressed, as separation according to size may otherwise be disturbed. In CGE using cross-linked polyacrylamide gels and pretreated capillaries, the electro-osmotic flow is very small or even absent. Elimination of the electro-osmotic flow working with low viscous polymer network solutions demands the use of, for example, coated capillaries (page 265).

Applications of CGE

The lack of electro-osmosis in CGE implies that the technique is only applicable for charged analytes, typically with a similar charge to mass ratio making separation in FZCE insufficient. CGE has especially proved its high efficiency in connection with the analysis of oligo- and polynucleotides, and is also suitable for the analysis of, for example, restriction digests, DNA sequencing reaction products, and charged oligo- and polysaccharides. CGE analysis of proteins is often a M_r determination performed using polyacrylamide added SDS (capillary SDS-PAGE) (Figure 10.24).

Promising results from the separation of bovine caseins, including hydrolysis products, in cheese by CGE at low pH with a linear polymer is shown in Figure 10.25.

Capillary Isoelectric Focusing (CIEF)

CIEF is based on the same separation principles as traditional isoelectric

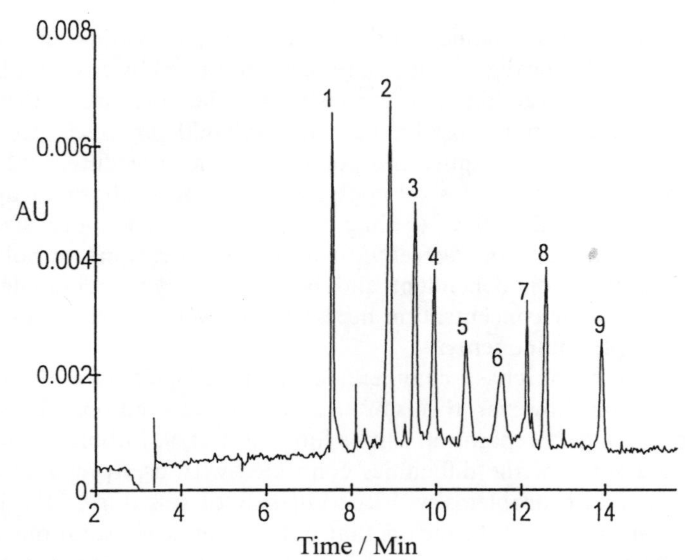

Figure 10.24 *Apparent* M_r *determination of proteins using capillary SDS-PAGE. Peaks* (10^{-3} M_r *in parenthesis): 1 = Orange G (marker), 2 = α-lactalbumin (bovine milk) (14.2), 3 = soybean trypsin inhibitor (Type IS) (20.1), 4 = carbonic anhydrase (bovine erythrocyte) (29), 5 = ovalbumin (chicken egg) (45), 6 = bovine serum albumin (66), 7 = Phosphorylase B (97.4), 8 = β-galactosidase (116), and 9 = myosin (205); buffer: Dionex Corporation Proprietary Sieving Buffer; temperature: 25 °C, voltage: 20 kV; injection: electrokinetic (10 kV) (− to +); detection: UV, 210 nm; capillary: 75 μm i.d., 50 cm total length (Dionex Corporation Proprietary Coated Fused Capillary)*
(Reproduced by permission from G.M. McLaughlin, R.M. McCormick, D.C. Siu, W.A. Ausserer, K. Srinivasan, and K.W. Anderson, in 'Proceedings of The Fifth Annual Frederick Conference on Capillary Electrophoresis', Frederick, MD, USA, October, 1994)

focusing (page 190), which are separation of analytes in a pH-gradient due to differences in their pI-values. Whereas traditional isoelectric focusing is usually performed in gels with only semiquantitative detection (protein staining), CIEF can be run in free solution with the possibility of quantitative on-column detection. As in CGE, CIEF usually demands elimination of the electroosmotic flow (by dynamic or covalent coating) for efficient separation.

CIEF is basically a three step procedure, although modifications to this concept exist. The three main steps are:

- loading (injection)
- focusing
- mobilization

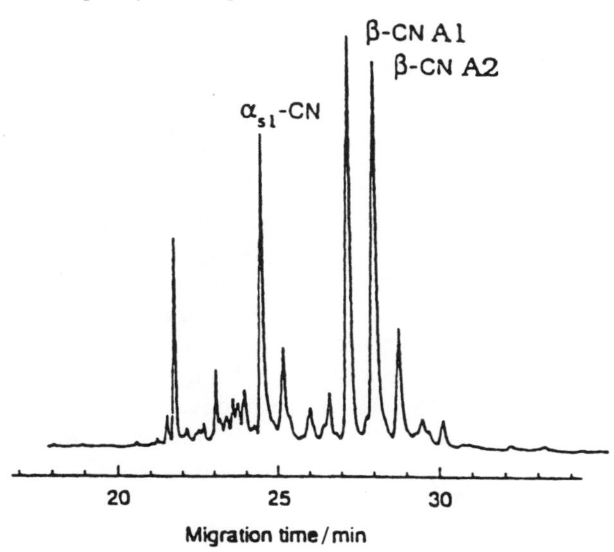

Figure 10.25 *CGE separation of caseins from Feta cheese (six weeks old). Buffer: 10 mM NaH_2PO_4 with 0.02% hydroxymethylpropyl cellulose, 6 M urea, pH 2.5; temperature: ambient, voltage: 14 kV; injection: hydrodynamic (+ to −); detection: UV, 214 nm; capillary: CElect P150 (Supelco), 50 μm i.d., 43 cm total length, 36.5 cm efficient length*
(Reproduced by permission from J. Otte, K.R. Kristiansen, M. Zakora, and K.B. Quist, *Le Lait*, 1997, **77**, 241)

Desalting of the sample is important in CIEF, as the formation of the pH-gradient (see below) can be disturbed by high ionic strengths. The sample may be injected at the cathode or anode end of the capillary, either separately or along with the ampholyte solution. In the latter case, where the sample is predissolved in the ampholyte, it is possible to fill the capillary completely with sample/ampholyte, giving a lower limit of detection. The amount of sample loaded should, however, not be too high, as precipitation of sample components, *e.g.* proteins, may occur during focusing of zones at the pI of the proteins (page 82). The risk of aggregation and precipitation can be reduced by the addition of dispersive agents, *e.g.* ethylene glycol to the buffer.

Diluted versions of buffers used in traditional IEF can be used in CIEF, and unlike other HPCE modes, different buffers are used at the cathode and anode end of the capillary. The pH of the cathodic buffer should in general be higher than that of the most basic ampholyte, whereas the pH in the anodic buffer should be lower than that of the most acidic ampholyte. This is to prevent contact between the ampholytes and the electrodes, which could result in oxidation/reduction processes that alter the composition of the buffer solution. The ampholyte solution used can be purchased commercially, and typically covers a pH-range from about 3 to 10, although narrower gradients are also obtainable. Ampholytes usually absorb UV light below 250 nm, making sensitive detection difficult, especially because the composition in the capillary is continually changing during focusing and mobilization. On-column detection at 280 nm is therefore the most common in CIEF although other types of detection modes have also been applied.

Establishment of a pH-gradient and focusing of sample components are performed simultaneously when voltage is applied to the capillary. The ampholytes, being uniformly distributed throughout the capillary at the start, begins to separate with the positively charged components migrating towards the cathode, and the negatively charged migrating towards the anode. This movement of ampholytes results in an increase in pH at the cathode, whereas the pH declines at the anode. Whenever the ampholytes reach a pH corresponding to their pI-value, they stop migrating. Sample components present in the capillary also migrate according to their respective pI-values. Eventually, ampholytes and sample components have ceased to move, and the current decline to near zero, an indication of ended focusing (Figure 10.26).

Mobilization of the focused sample zones beyond the detector can be performed by different techniques:

- electrophoretic mobilization
- hydrodynamic mobilization
- electro-osmotic mobilization

Electrophoretic mobilization implies the replacement of acid with base at the anode, base with acid at the cathode, or addition of salt/zwitterions to one of the buffer reservoirs followed by application of an electric field. The requirement of electroneutrality then forces ampholytes and sample molecules to migrate through the capillary, the direction depending on which reservoir the additives were added to. Using hydrodynamic mobilization, the entire capillary content, including the sample zones, is moved past the detector by vacuum or pressure. Application of an electric field (as during focusing) reduces band broadening. A special problem in CIEF mobilization arises when sample components focus behind the detector zone. As indicated in Figure 10.26, such sample zones will remain undetected. An extension of the pH-gradient range, so that the most basic/acidic analytes in the sample will focus prior to the detector, can be a solution. Introduction of such a 'blinded end' can be performed by using an ampholyte with a wider pH-range or by addition of a strong base/strong acid to the ampholyte solution used.

Electro-osmotic mobilization implies controlling rather than eliminating the electroosmotic flow, and is a specialized form of CIEF in which simultaneous focusing and mobilization of analytes occurs. The electro-osmotic flow is reduced by additives in the ampholyte solution (*e.g.* 0.1% methyl cellulose), and a low electro-osmotic flow allows focusing of the sample zones before passing through the detector. A disadvantage of electro-osmotic mobilization is

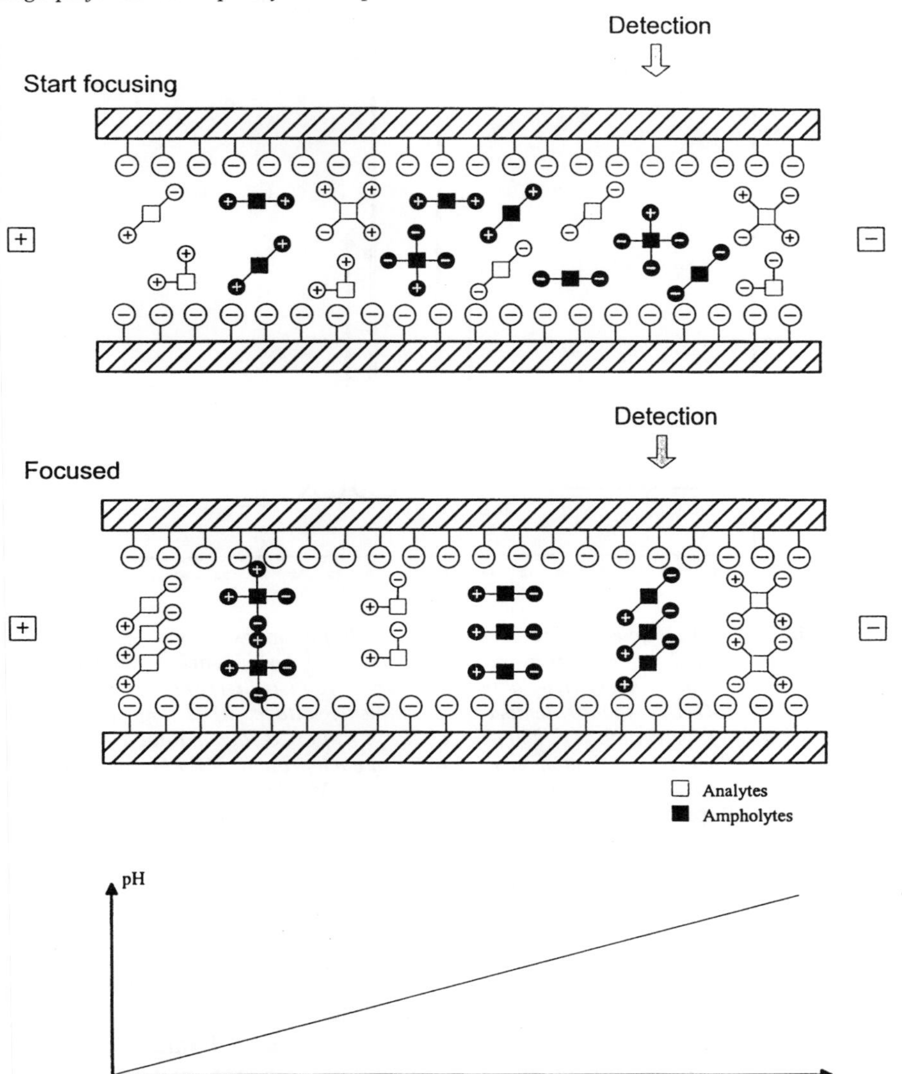

Figure 10.26 *Focusing of sample zones in CIEF*

that the linearity of the pH gradient may be disturbed, and it may be more difficult to obtain reproducible separations.

CIEF is a high resolution technique with the ability to separate sample components which differ by less than 0.05 pH-units in pI. Parameters with a beneficial effect on resolution are, among others, a high electric field, low diffusion coefficients, and a narrow pH-gradient.

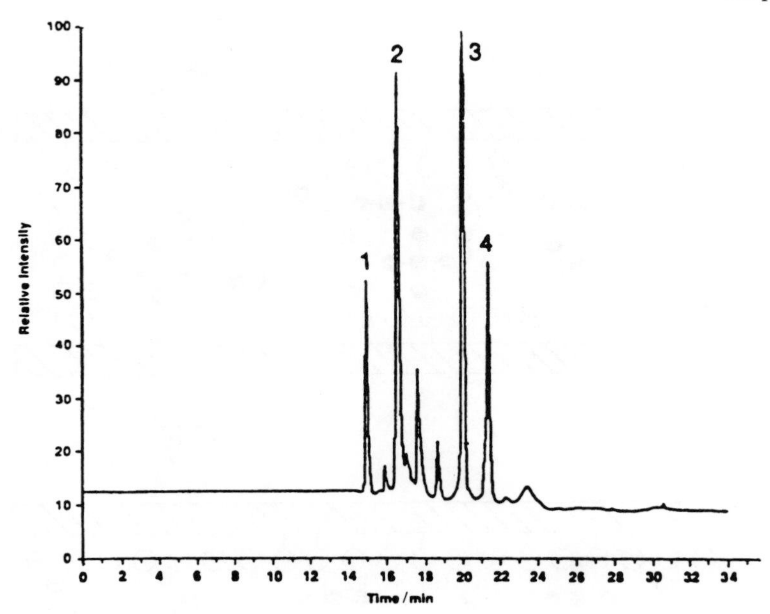

Figure 10.27 *CIEF of selected proteins. Peaks (pI in paranthesis): 1 = cytochrome* c (9.6), *2 = chymotrypsinogen A (9.0), 3 = horse heart myoglobin (7.2), 4 = horse heart myoglobin (6.8); buffer: 10 mM* Na_2HPO_4 *(anode), 20 mM NaOH (cathode); temperature: ambient; voltage: 30 kV; injection: (+ to −); detection: UV, 280 nm; capillary: 75 μm i.d., 60 cm total length; 40 cm efficient length; sample: 1 mg* ml^{-1} *of each protein, 5% Pharmalyte® (pH 3–10), 0.1% methylcellulose, and 1% TEMED*
(Reprinted with permission from *Anal. Chem.*, 1991, **63**, 2852. Copyright (1991) American Chemical Society)

Applications of CIEF

The applicability of CIEF is, as for traditional IEF, directed towards the separation of zwitterionic compounds with different pI-values, *e.g.* proteins and peptides (Figure 10.27). Determination of exact pI-values is also a possibility with CIEF, provided that mobilization is not performed using electro-osmosis.

Other Techniques

Several other HPCE techniques exist, including among others capillary isotachophoresis (CITP), which has been widely used in food analysis for a long time, microemulsion capillary electrophoresis (MCE), capillary affinity electrophoresis (CAE) and other less widespread HPCE techniques, which will not be presented in details here. Miniaturized HPCE systems applicable for single molecule analysis is under current development, but is of limited interest in food analysis.

5 Parameters of Analysis Affecting Separation in HPCE

Characteristic for HPCE is the many possibilities to control and change separation. Different parameters exist depending on the actual separation mechanisms, however, there will be some overlap within the various HPCE techniques. The variation in analysis parameters often also results in changes in the electro-osmotic flow as well, but this basically does not affect selectivity, whereas the migration time and thus resolution of the compounds will be altered. In the following a description of some of the variables will be given.

Temperature and Voltage

Temperature and voltage are easily changeable parameters whose primary effect will be on the electro-osmotic flow, the viscosity of the capillary solution, and thereby on the migration time of sample components. Theoretically, an increase in temperature or field strength should result in improved separation efficiency, however, this is provided that heat dissipation is effective. An example of temperature influence in MECC is given in Figure 10.28, which shows how very small changes in temperature can result in large changes in the separation of selected amino acids.

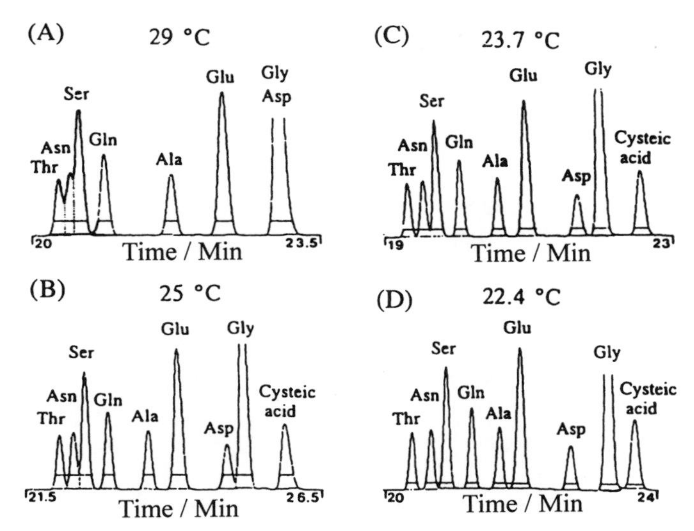

Figure 10.28 *Influence of temperature on the separation of 9 amino acids.* (A) *29,* (B) *25,* (C) *23.7, and* (D) *22.4 °C. Peaks: Thr = threonine, Asn = asparagine, Ser = serine, Gln = glutamine, Ala = alanine, Glu = glutamic acid, Asp = aspartic acid, Gly = glycine; buffer: 100 mM boric acid, 150 mM SDS, pH 8.3; voltage: 13 kV; injection: hydrodynamic (vacuum) (+ to −); detection: UV, 216 nm; capillary: 50 μm i.d., 72 cm total length, 52 cm effective length*
(Reprinted from S. Michaelsen, P. Møller, and H. Sørensen, *J. Chromatogr. A*, **680**, 299 (1994), Elsevier Science, Amsterdam)

Another interesting aspect of temperature influenced separation is given in Figure 10.29 showing the analysis of borate complexed monosaccharides.

The increased temperature alters the reaction rates of the different dynamic equilibria in the borate/analytes solution and thus affects the electrophoretic separation in a positive manner. That temperature influences complex forma-

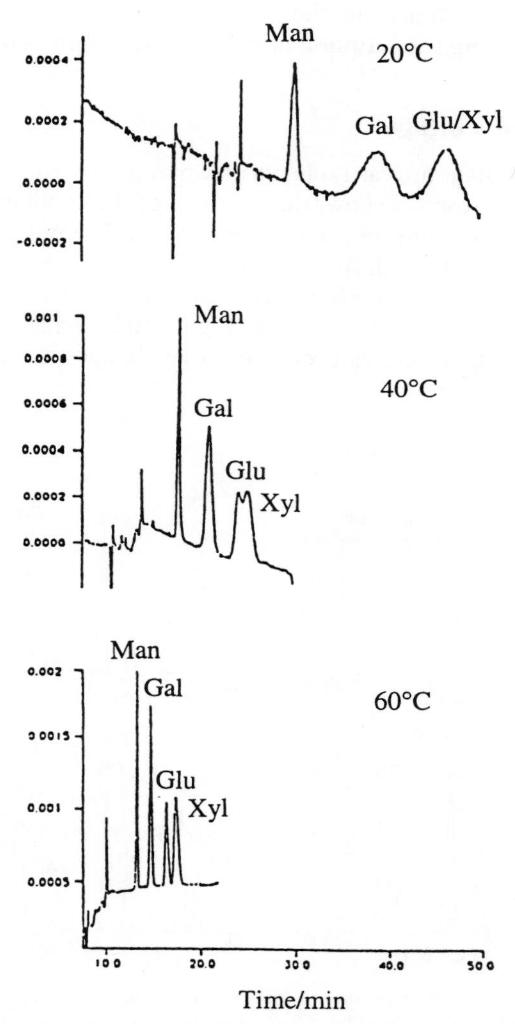

Figure 10.29 *FZCE separation of monosaccharides at different temperatures.* (A) *20,* (B) *40, and* (C) *60 °C. Peaks: Man = mannose, Gal = galactose, Glu = glucose, and Xyl = xylose. Buffer: 50 mM disodium tetraborate, pH 9.3; voltage: 20 kV; injection: hydrodynamic (+ to −); detection: UV, 195 nm; capillary: 75 µm i.d., 94 cm total length, 87 cm efficient length*
(Reprinted with permission from *Anal. Chem.*, 1991, **63**, 1541. Copyright (1991) American Chemical Society)

tion/dissociation and chemical equilibria in general is well known, and it may also affect the partitioning of analytes between the micellar and aqueous buffer phase, resulting in changes of selectivity, in for example, MECC. Temperature-induced structural changes of analytes, *e.g.* proteins, may also affect selectivity.

In some commercial HPCE apparatus, it is possible to run voltage (current) or temperature gradients. This may, for example, be useful for reducing the time of analysis whenever there is a substantial difference in migration time between analytes. A low voltage at the start of the analysis may also prevent problems with rapid heating and thermal expansion of the buffer, which may lead to expulsion of sample. Reduction of the temperature below ambient may be advantageous in some situations, *e.g.* when using high ionic strength buffers.

Buffer Electrolyte

The buffer (Chapter 2) is very important for the performance of HPCE as it may affect migration times (current, electro-osmotic flow, generated heat, analyte–wall interactions) and resolution (analyte–wall interactions, possible complexation). Therefore, careful consideration of composition, concentration, as well as pH of the chosen electrolyte system is very important. This means that we need information both on the type and concentration of electrolytes and their counterions (ionic strength, Chapter 2), in addition to the traditional information on buffers (type, pH, molarity). All buffers should be passed through a 0.45 μm filter before use.

Composition and Concentration

Table 10.4 shows some commonly used buffers in HPCE together with their approximate pK'_a-values. The choice of the electrolyte involves consideration of

Table 10.4 *Selected buffers for HPCE, and approximate* pK'_a*-values*

Electrolyte	pK'_a	*Electrolyte*	pK'_a
Phosphate	2.12, 7.21, 12.32	TES	7.50
Citrate	3.13, 4.76, 6.34	HEPES	7.55
Formate	3.74	HEPPS	8.00
Succinate	4.19, 5.57	Tricine	8.15
Acetate	4.74	Glycine amide, HCl	8.20
MES	6.15	TRIS	8.30
ADA	6.60	Bicine	8.35
PIPES	6.80	Glycylglycine	8.40
ACES	6.90	Morpholine	8.49
MOPSO	6.90	Borate	9.24
MOPS	7.20	CHES	9.50
		CAPS	10.40

several factors, but basically knowledge of the desired pH-range will reduce the possibilities, as good buffer capacity is limited to ±1 pH-unit of the pK'_a-value.

Electrolytes with strong UV-absorption should in general be avoided in experiments based on UV–Vis detection (pages 94 and 225); however, such electrolytes could be advantageous in HPCE methods based on indirect detection. The molarity of the analytes in the detection zone should be lower than the molarity of the chromophore electrolyte, otherwise there will be a decrease in the linearity of the analysis. Common electrolytes for indirect detection are shown in Table 10.5.

The analytes should be soluble and stable in the chosen buffer (most often aqueous), and sometimes it will be necessary to include an organic modifier (Section 2.8) to obtain this. On certain occasions, buffers are chosen owing to their complexing properties. This is, for example, true for borate in monosaccharide separation and tartaric acid in inorganic cation separation. The complexing properties may, however, also be disadvantageous owing to unwanted interaction between the analytes and constituents in the buffer or sample (page 27). In MECC, when using anionic detergents, the buffer cation should preferably be the same as the counterion of the surfactant, and *vice versa*

Table 10.5 *Selected electrolytes for indirect UV-detection in HPCE and their approximate pK'_a, λ_{max} and selected ε-values. The actual wavelength used will be determined by the analytes and the actual system. The electrolytes are arranged with approximate decreasing electrophoretic mobility*

Electrolyte	pK'_a	λ_{max} *(nm)*	ε *(M^{-1} cm^{-1})*
Cations			
Histamine	5.9, 9.8	211	5800
4-Aminopyridine	9.1	260	18000
2-Aminopyridine	6.8	300	6100
Pyridine	5.3	255	2700
Imidazole	6.9	211	5800
Benzylamine	9.6	206, 257	9300, 195
4-Methylbenzylamine	9.4	206, 257	9300, 195
2-Aminobenzimidazole	7.5	283	7800
Anions			
Chromate	0.7	310	–
Pyromellitic acid (PMA)	1.9, 2.9, 4.5, 5.6	298	2500
Trimellitic acid (TMA)	2.5, 3.8, 5.2	293	1500
Phthalate (PHA)	2.9, 5.5	281	1600
Benzoate	4.2	271	760
p-Hydroxybenzoate	4.6, 9.4	251	12300

for cationic detergents. This is to prevent adverse effects owing to possible exchange of buffer and counterions. The buffer counterions have also been shown to influence the electro-osmotic flow and, to a lesser extent, the electrophoretic mobility of analytes.[11] Matching the buffer ion mobilities to those of the analytes can be used to obtain symmetric peaks; however, this is normally not a problem.

The different electrolyte systems vary in conductivity and temperature dependence, and this should be considered especially using high ionic strength buffers, as current generation and thus heat development otherwise may be too high. Large, minimally charged ions are, in principal, preferable in this connection, *e.g.* the zwitterionic buffers such as MES, tricine, TRIS, and bicine (Table 2.4), whose conductivity at pHs around pI is low. High ionic strength buffers (Section 2.3) are most often used as a tool to limit adsorption problems, *e.g.* in protein analysis, taking advantage of the competition between buffer ions and analytes for the charged sites at the capillary wall. As the electroosmosis and electrophoretic mobility of ions in general decreases with increasing ionic strength of the separation buffer, the time of analysis may be prolonged.

pH

Theoretically, a pH-interval of 2–12 is possible in FZCE owing to the high stability of uncoated fused-silica capillaries. The stability of the analytes will, however, usually limit the actual pH-range allowed, and the chosen pH should moreover match the buffer capacity of the electrolyte chosen. It is important to maintain a stable pH during and between analyses, as pH affects the analyte's electrophoretic mobility as well as the electroosmotic flow (Figure 10.30).

It is the ionization level of silanol groups which determines the size of the electro-osmotic flow; however, the pH-hysteresis effect (page 217) may cause some variation. The effect of pH on the electrophoretic mobility is owing to an effect on analyte net charge, shape, and possible interaction with surfactants and other buffer constituents. This is evident, especially for zwitterionic compounds such as proteins bearing a positive net charge at pH below pI and a negative net charge at pH above pI. If the protein stability allows it, extreme pH buffers can be chosen to limit the interaction of proteins with the capillary wall, *e.g.* a strongly acidic for suppressing the dissociation of silanol groups at the capillary wall, or a strongly alkaline one to change the net charge of the protein to negative. In micellar systems, one has to be careful that the chosen pH is not in conflict with the properties of the surfactant used. An example is the buffer pH in MECC using sodium cholate, which should exceed 6.8 to avoid precipitation of the surfactant (pK'_a-value = 6.4).[12]

Buffer Additives

Buffer additives constitute an efficient tool for changing the selectivity in especially FZCE and MECC. Buffer additives are easy to use as they are simply added to the electrolyte systems, and then removed by flushing of the

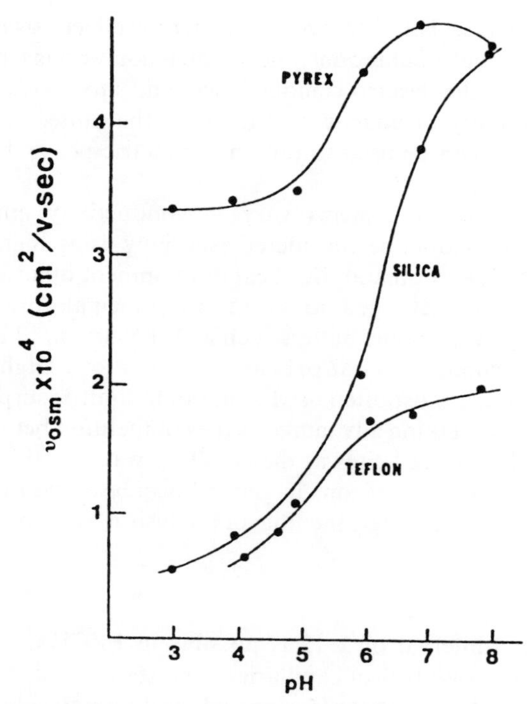

Figure 10.30 *Schematic illustration of the effect of pH on electroosmosis at constant ionic strength* (I = *0.06*)
(Reproduced by permission from K.D. Lukacs and J.W. Jorgenson, *J. High Resolut. Chromatogr. Chromatogr. Commun.*, 1985, **8**, 407)

capillary with new buffer or water when no longer needed. The following sections give a brief introduction to some of the most widespread buffer additives in HPCE.

Chiral Selectors

Different chiral selectors applicable to HPCE exist, the cyclodextrins and crown ethers being the best known. The native cyclodextrins are cyclic oligosaccharides, divided into the α-, β-, and γ-forms, differing in their number of glucose units (Figure 10.31). A comparison of different cyclodextrin characteristics are given in Ref. 13.

The aqueous solubility of β-cyclodextrins is 8 to 12 times lower than for the α- and γ-forms, limiting the concentration range applicable, however, this may be overcomed using derivatized forms of the native cyclodextrin. Organic modifier added to the buffer in combination with the cyclodextrins is another possibility. The cyclodextrins form a truncated cone, with an internal cavity of hydrophobic character varying in diameter depending on the actual cyclodextrin. The selectivity is brought about by stereospecific inclusion of the analyte into the

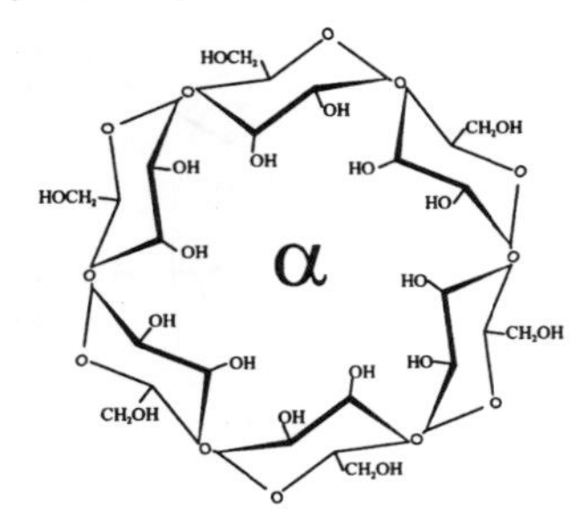

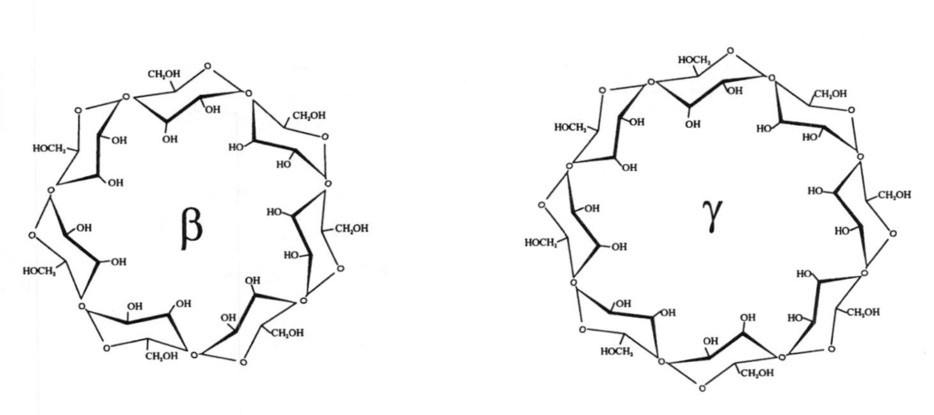

Figure 10.31 *Chemical structure of the native cyclodextrins α, β, and γ*

cavity of cyclodextrins, the complex formation primary being stabilized by hydrophobic interactions (Figure 10.32).

Cyclodextrins are not limited to FZCE, but have also proved their applicability for *e.g.* MECC, CEC, and CGE. In MECC, using SDS and γ-cyclodextrins, a second chiral additive (*e.g.* *d*-camphor-10-sulfonate or 1-methoxyacetic acid) may improve resolution of enantiomers.[14]

Crown ethers are known to form stable inclusion complexes with some inorganic and organic cations, *e.g.* metals and organic amines, and are, as the cyclodextrins, successful in enantiomer separations.

Organic Modifiers

Organic modifiers (Section 2.8) may be useful for improving the solubility of analytes hardly soluble in the aqueous buffers normally used in HPCE, and may also affect the distribution of analytes between various phases *e.g.* in MECC and chiral separations, using inclusion complexes. Organic modifiers applicable in HPCE comprise a long list of mainly alcohols, some selected examples are methanol, ethanol, propanol, butanol, pentanol, glycerol, ethylene glycol, 1,2-butanediol, PEG (*e.g.* 200 and 400), acetonitrile, acetone, and dimethyl sulfoxide.

(A)

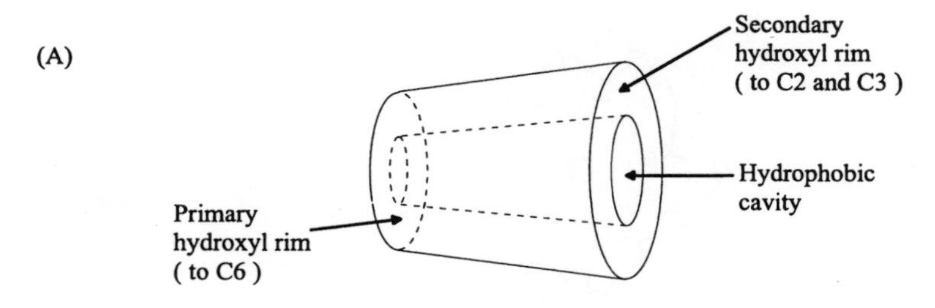

(B)

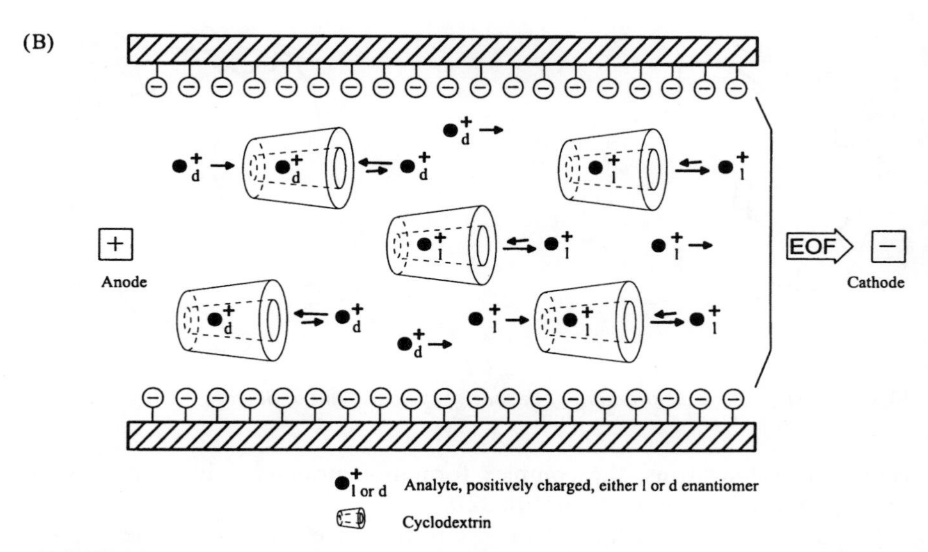

Figure 10.32 (A) *Schematic structure of cyclodextrins and* (B) *principle of chiral separation for positively charged analytes using cyclodextrins as additives*

The electro-osmotic flow is in general lowered by addition of organic modifiers. The alcohols with the longest chain lengths have the greatest effect on electro-osmotic flow, and this effect is mainly due to a decreased dielectric constant and an increased viscosity of the buffer solution in combination with interactions between the alcohol and the capillary wall. The change in dielectric constant of the buffer phase may also shift the pK'_a-value of the capillary wall silanol groups towards a higher value (Section 2.8), altering the size of the electro-osmotic flow this way. Diols, as *e.g.* glycerol and ethylene glycol, increases viscosity (Chapter 6) more effectively than alcohols with only one hydroxy group. Any effect on viscosity also affects the electrophoretic mobility of analytes. Contrary to most of the organic modifiers used, acetonitrile has only limited effect on the electro-osmotic flow, probably due to a limited interaction with the capillary wall. Still, with protolytic active compounds, a changed dielectric constant (Section 2.8) will increase the pK'_a-values and thereby the mobility of the electrolyte.

The choice of an organic modifier can be difficult, and some may replace each other with only limited change in performance of analysis. Working at high temperatures, it can be important to check the boiling point, *e.g.* propan-1-ol (b.p. 97.4 °C) may be preferred to propan-2-ol (b.p. 82.4 °C). Generally, the concentration of organic modifier included varies from 1–30% (v/v) or even higher as separations using HPCE have been performed in non-aqueous buffers.[15]

Amine Modifiers

Amine modifiers function by their interaction with the silanol groups on the capillary wall, reversing/reducing the electro-osmotic flow. Moreover, ion pairing with negatively charged analytes, surfactants, *etc.* may also be possible. Inclusion of diamines in buffers for protein separation may result in reduced adsorption to the capillary wall due to ionpairing with the negatively charged groups on the capillary wall. Examples of amine modifiers are:

- 1,3-diaminopropane
- 1,4-diaminobutane (putrescine)
- 1,5-diaminopentane (cadaverine)

A drawback of these primary alkyldiamines is their strong odour. Tertiary alkylamines provide another alternative. Amine modifiers are primarily used in protein separations.

Zwitterions

Zwitterions are usually included in the buffer to reduce the interaction of positively charged analytes with the negatively charged capillary wall, but may also contribute to the buffer effect in itself. They increase ionic strength without increasing conductivity and may therefore be included in high concentrations (500 mM or higher). Some zwitterions used in HPCE (*e.g.* FZCE, MECC) are shown in Figure 10.33.

$$H_3C-{}^{+}N(CH_3)_2-CH_2-COO^- \qquad NH_3^+-CH_2-CH_2-SO_3^-$$

Betaine   Taurine

Figure 10.33 *Taurine and betaine, zwitterionic additives in HPCE*

Surfactants

The use of surfactants in MECC and HICE is described in Section 10.4.2 and 10.4.3, and details on the micelle formation can be found in Chapter 2. The present Section aims to give an overview of some of the typical surfactant systems used in HPCE (Table 10.6).

The list of surfactants is much longer (see, for example, Chapter 2), and only a few have been tested in MECC and HICE. The basic demands for a surfactant suitable for MECC are high enough solubility in the buffer to form micelles, relatively low viscosity of the micellar solution, and UV transparency at the wavelengths of interest.

It may be very difficult to predict the most suitable surfactant for a given analysis. Some authors mention, that the polar group of the surfactant affects selectivity more than the non-polar group. Surfactants with different polar groups should therefore be tested if large changes in selectivity are wanted, whereas smaller adjustments may be achieved by changing the non-polar group. Non-ionic surfactants are often tested when other surfactants have failed. These surfactants have shown to give good results if, for example, peptides with small differences in hydrophobicity are to be separated.[16] In some cases, the interaction of the analyte with the chosen surfactant is too strong, resulting in long analysis times and broad, poorly defined peaks. One has thus to consider the possible interaction between analyte and surfactant. There may be strong ion binding which could be reduced/eliminated by including ion-pairing reagents or changing to a neutral or oppositely charged surfactant. Ion-pairing reagents may also be used to increase the interaction of analytes with micelles, *e.g.* tetraalkylammonium in SDS-based MECC of anionic analytes. Increased migration times are thus obtained owing to ion-pairing of the anionic analytes, with the cationic ion leading to increased incorporation into the anionic SDS micelle. For cationic analytes, migration times will decrease due to competition between the cationic ion-pairing reagent and the cationic analytes.

A further possibility is the use of surfactants forming so-called inverse or reversed micelles, having the hydrophilic parts turned inward and the hydrophobic part situated on the surface. Reverse micelles are formed by the group of bile salt surfactants, which may also act as chiral selectors. Furthermore, other special surfactants for enantiomeric separations exist, *e.g.* sodium-*N*-dodecanoyl-L-valinate. Too strong a hydrophobic interaction between the analytes and surfactants may *e.g.* be overcome by the addition of an organic modifier to the buffer phase. In this connection, one should be aware that even small amounts of organic solvent may produce marked changes in the c.m.c. and aggregation number of the micelles. In high concentrations (typically above 20%) the micelles structure may be broken down (Figure 10.34).

Reverse micelles are in general more tolerant to organic modifiers in the buffer, owing to their special structure. Changing the surfactant type and including a low concentration of organic modifier may have very positive effects on the separation of analytes as exemplified in Figure 10.35.

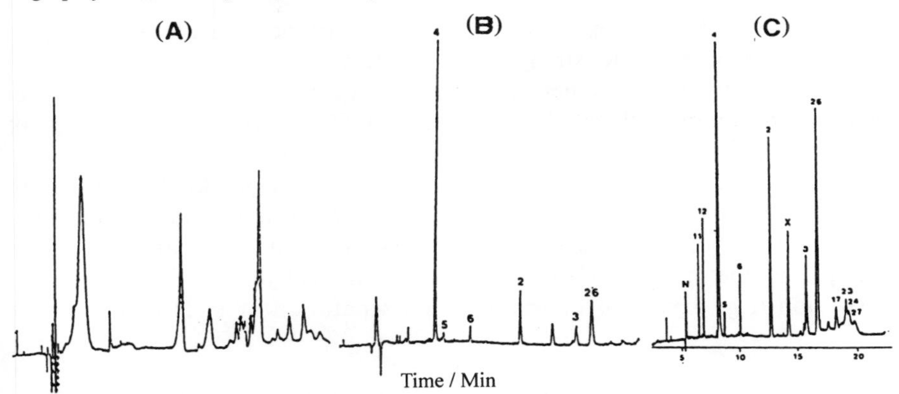

Figure 10.34 *The influence of high concentrations of organic solvent on separation of glucosinolates in rapeseed using MECC (CTAB).* (A) *Eluate (50% acetonitrile, high salt concentration),* (B) *eluate 4 times diluted with water (12.5% acetonitrile, reduced salt concentration), and* (C) *eluate evaporated and redissolved in water (no acetonitrile, high salt concentration)*

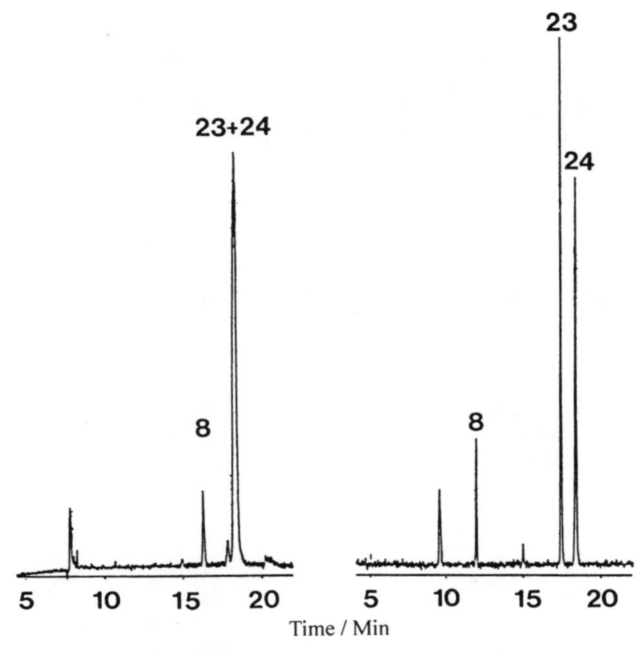

Figure 10.35 *Effect of changing surfactant and including organic modifier in the separation buffer for glucosinolates. Separation conditions as in Figure 10.18 except for 10% propan-2-ol and 50 mM DTAB instead of 50 mM CTAB*

In optimizing MECC separations, it is also important to take into account the effect of increased ionic strength/salt concentration on the properties of micelles. In general, the higher salt concentration the lower c.m.c. and higher aggregation number, which will influence MECC separation. Also the type of counterion may influence the size of c.m.c. (Table 10.6).

The surfactant systems are usually simple, with only one kind of surfactant involved. Alternatively, mixed micelles, in which different surfactants or a surfactant and some non surfactant additives form a micelle, can be used. An example is the mixing of SDS with optically active additives for chiral separation (*e.g.* digitonin for enantiomeric separation of PTH amino acids).

Table 10.6 *Some characteristics of selected surfactants applicable in MECC and HICE**†‡

Surfactant	*Solvent*	*Temp. (°C)*	*c.m.c. (mM)*
Anionic			
Sodium dodecyl sulfate (SDS) $CH_3(CH_2)_{11}OSO_3^-Na^+$	H_2O	25	8.2
Sodium tetradecyl sulfate (STS) $CH_3(CH_2)_{13}OSO_3^-Na^+$	H_2O	40	2.2
Cationic			
Dodecyltrimethylammonium chloride/bromide (DTAC/DTAB) ($CH_3(CH_2)_{11}N^+(CH_3)_3Cl^-/Br^-$)	H_2O	25	20.0/16.0
Cetyltrimethylammonium chloride/bromide (CTAC/CTAB) ($CH_3(CH_2)_{15}N^+(CH_3)_3Cl^-/Br^-$)	H_2O	30/25	1.3/0.9
Zwitterionic			
3-[3-(Chloroamidopropyl) dimethylammonio]-1-propanesulfonate (CHAPS)	H_2O	–	8.0
3-[3-(Chloroamidopropyl) dimethylammonio]-2-hydroxy-1-propanesulfone (CHAPSO)	H_2O	–	8.0
N-Dodecyl betaine	–	–	0.8
Bile salts			
Sodium cholate (pK'_a = 6.4)	H_2O	25	14.0
Sodium deoxycholate (pK'_a = 6.6)	H_2O	25	5.0
Sodium taurodeoxycholate (pK'_a = 1.9)	H_2O	25	3.0
Non-ionic			
Triton X-100	–	–	0.24
n-Octyl β-D-glucoside	H_2O	25	25.0
n-Dodecyl β-D-glucoside	H_2O	25	0.19

*M.J. Rosen, in 'Surfactants and Interfacial Phenomena', Wiley, New York, 1978, ch. 3, p. 83. †A. Helenius, D.R. McCaslin, E. Fries, and C. Tanford, in 'Methods in Enzymology', 1979, vol. LVI, p. 734. ‡Notice how different counterions may affect c.m.c. In charged surfactant systems with high c.m.c., thermal problems may arise owing to the high resultant concentration of surfactant

Other additives can also be used to improve selectivity in MECC, *e.g.* urea. Urea affects micelle formation owing to an influence on the water structure (page 81). In practice, when moving between different surfactants, the capillary should be changed as well. A used capillary can easily be kept unused for several months provided a thoroughly final washing (NaOH or other solvents able to rinse out residual surfactant followed by water and air flush).

With the addition of surfactant in HICE at concentrations below c.m.c., a possibility for control of the electroosmotic flow is obtained. Depending on the charge and concentration of surfactant, the electroosmotic flow can either be increased, reduced, or reversed owing to the hydration of surfactants and association of cationic surfactants to the capillary wall (Figure 10.36).

Control of the electroosmotic flow is an important parameter in most HPCE techniques. Other possibilities for modification of the electroosmotic flow are described in Section 10.5. To determine the electroosmotic mobility in MECC, a neutral compound which is not soluble in the micelles must be chosen, *e.g.* methanol. To determine the electrophoretic mobility of micelles, a neutral

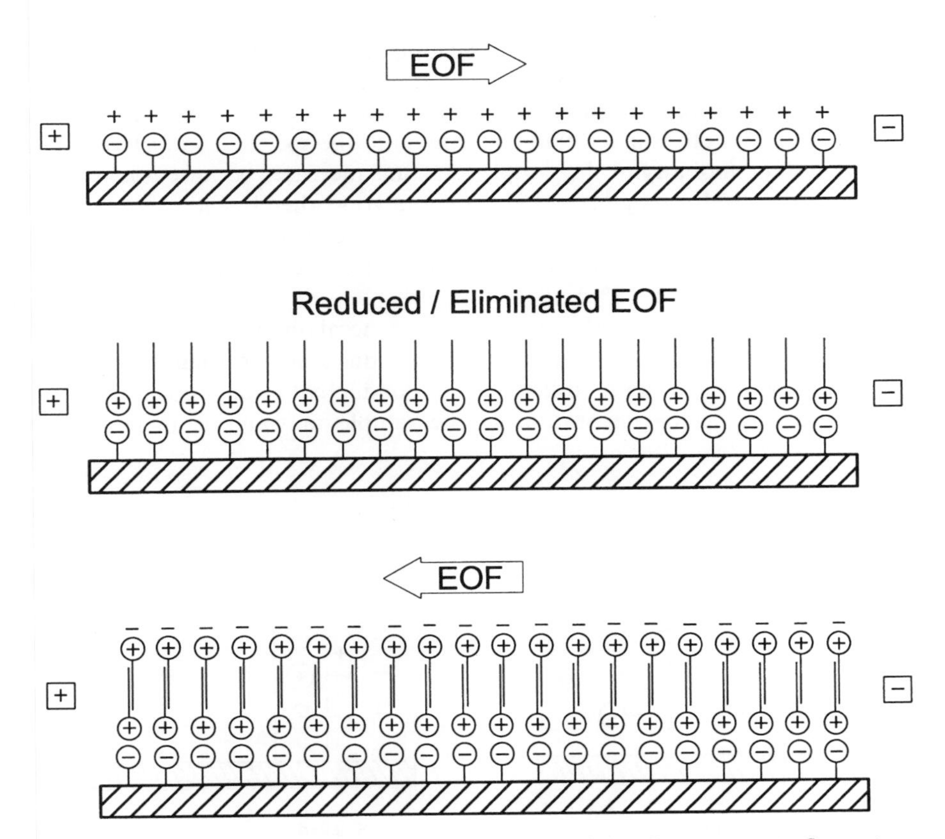

Figure 10.36 *Reduction (elimination) and reversal of the electro-osmotic flow using a cationic detergent. An increased electro-osmotic flow is obtained using anionic detergents*

compounds which is fully solubilized in the micelles must be chosen, *e.g.* Sudan III or IV.

A more detailed discussion on MECC principles and applications can be found in Ref. 17.

Others

Many possibilities for buffer additives exist in addition to those mentioned above. Among many others are linear polymers (as in CGE), inorganic salts (NaCl, KCl, Na_2SO_4 *etc.*), urea, and complexing agents such as metal ions (*e.g.* Cu^{2+}, Zn^{2+}, Mg^{2+} with oligonucleotides) and borate, which actually may serve as buffer electrolyte itself.

Sample

An important factor affecting the performance and sensitivity of HPCE is the possibility of sample stacking. Sample stacking or concentration of the sample occurs, when voltage is applied over the capillary containing a sample plug with a lower specific conductivity than the buffer (Figure 10.37). The field strength will be higher along the sample plug compared with the buffer, because the electric field strength is inversely proportional to the specific conductivity. The electrophoretic mobility of the sample ions thus increases because it is proportional to the field strength, and the sample zone thereby narrows. Sample stacking is possible with both hydrodynamic and electrokinetic injection modes. The injected sample plug should preferably be small compared with the total length of the capillary (<2%).

Good conditions for sample stacking are in general obtained by preparing the sample in a buffer concentration about 10 times lower than that of the separation buffer. A large difference in conductivity may counteract the stacking effect,[18] and preparation of the sample in water is therefore not

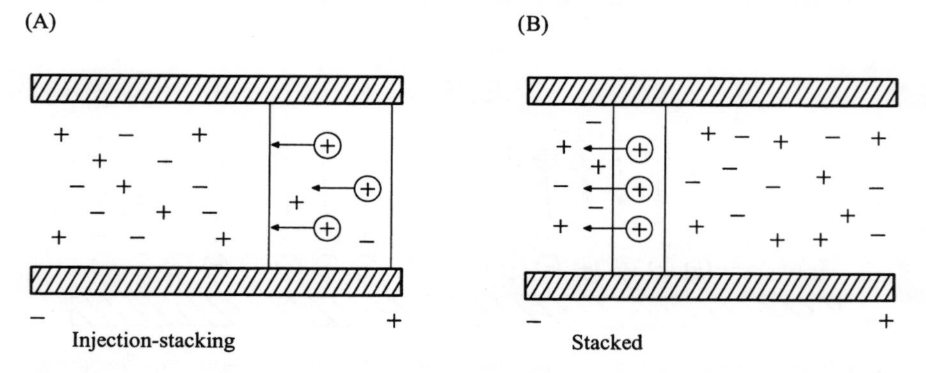

Figure 10.37 *Sample stacking.* (A) *Time of injection.* (B) *After sample stacking has occurred*

always beneficial. If the conductivity of the sample and buffer are in the same range, stacking may be obtained by injection of a short plug of water prior to the sample injection. One problem with sample stacking is the risk for thermal degradation of analytes in the sample zone. The high electric field may lead to severe increases in temperature, and care should therefore be taken, especially with thermolabile compounds. Prestacking conditions, such as using a low voltage when the sample is in the stacking zone, can reduce the problem.

Capillary

The importance of the type and dimensions of a capillary for the performance of HPCE have been described in the subsection beginning on page 221, whereas the present section is concentrated on capillary coating. Capillary coating is a tool for limiting analyte–capillary wall interactions, and may even be a necessity in some techniques (*e.g.* CGE and CIEF) in which the electroosmotic flow should be reduced or eliminated. Two different approaches to capillary coating exist:

- permanent coating (covalently bonded or physically adhered phases);
- dynamic coating.

Silyl coupling of different reagents to the capillary wall (Si–O–Si–X) is the main form used for covalent binding of coatings. The reagents may be of various types: polyacrylamide, polyethylene glycol, maltose, *etc.*, and are in general relatively simple to prepare. A direct coupling of polyacrylamide to the silanol groups (Si–C) is another approach which is more stable but less easy to prepare. A detailed description of some important coating procedures can be found in Ref. 4. Various permanently coated capillaries are also commercially available. Polymers for physically adhered phases include, among others, cellulose derivatives, poly(ethylene glycol), poly(vinyl alcohol), and polyethyleneimine, the latter being cross-linked after adsorption. Moderate or highly hydrophilic/hydrophobic coatings of the kind found in liquid chromatography are also used in HPCE, and gas chromatography type coatings have been tested as well.

Dynamic coating is obtained by addition of an additive to the buffer phase which may interact with the capillary wall and thus reduce the adsorption of analytes. Among the buffer additives used are:

- surfactants
- hydrophilic polymers
- zwitterions
- quaternary amines

These have been described in the previous sections. In addition, high ionic strength buffers (competition with analytes) or buffers with extreme pH (reduced electroosmotic flow at low pH or repulsion of anionic analytes at high pH) may be useful in this respect.

Permanently coated capillaries may suffer from a gradual loss of surface coverage limiting their long-term stability, and complicating the repeatability of results. However, some recently available permanently coated capillaries show considerably better long term stability. The dynamic coating approach will result in a continuous 'regeneration' of the coating owing to the presence of additive in the buffer phase. Moreover, its simplicity and lower cost may be advantageous.

6 Development of a HPCE Method of Analysis

Several factors have to be considered, developing a separation procedure in HPCE. The approaches are many, but depend basically on the analytes to be analysed. The optimization of the method is usually done using a standard mixture that includes the compounds of interest and the final validation of the method should, in addition, include several real samples differing in composition to ensure applicability for actual samples that may contain potentially interfering compounds.

Summary of the Influence of Separation Parameters in HPCE

The following lists give an overview of the separation parameters to be used in HPCE and some of the possible effects observed. Most of the parameters have been discussed in more detail in previous sections. The aim here is to give a good starting point when an actual separation has to be performed.

Temperature

- Higher temperature lowers the viscosity of buffer electrolytes and increases electro-osmotic flow, leading to shorter migration times.
- Higher temperature alters the equilibria when using MECC or complex formation for separation, and thus affects the selectivity and migration times.
- Higher temperature gives a higher volume of sample injected.

Voltage

- Higher voltages give higher electro-osmotic flow and increased analyte velocity, leading to shorter migration times.
- Too high a voltage may result in such a high current that the buffer reaches its boiling point, leading to worse separations and/or degradation of analytes.

Buffer Electrolyte Composition

- High conductivity and temperature dependence may lead to heat generation, which has a negative influence on separations.

- Choose the electrolyte for the correct pH interval to obtain high buffer capacity.
- In certain analyses, an electrolyte with complexing properties should be considered.
- Matching of the buffer ions with the counterions of surfactants in MECC may be advantageous.

Buffer Electrolyte Concentration

- Typically 5–100 mM.
- High concentrations and ionic strength of electrolyte give high currents, reduced electro-osmotic flow, and reduced peak tailing, leading to longer migration times and better separations.
- High ionic strength buffers may be advantageous to limit adsorption problems.

pH

- High pHs give an increased number of charges on the capillary wall, leading to increased electro-osmotic flow and shorter migration times.
- The selectivity of analytes that change net charge with pH will be greatly affected by changes in pH.
- Be aware of the risks of analyte–capillary wall interactions, especially when analysing proteins. Use of extreme pH buffers is a possibility.
- Be aware of the effects of pH on analytes and buffer additives (*e.g.* net charges, solubility, and stability).

Buffer Additives

Chiral selectors

- Changed migrations/separation of analytes if complexation/interaction occurs.
- Often gives higher viscosity, leading to decreased electro-osmotic flow and longer migration times.
- Mixing chiral additives may improve resolution.

Organic modifiers

- Lead to reduced dielectric constants and effects on pK'_a.
- Lead to increased solubility of hydrophobic analytes.
- Lead to decreased electro-osmotic flow and longer migration times, except when a changed distribution is involved.
- Lead to a changed distribution of analytes when included in, for example, MECC and chiral separations.
- Possible to perform separations in non-aqueous buffer.

Amine modifiers

- Lead to decreased/reversed electro-osmotic flow, resulting in longer migration times/change in polarity and changes in migration order.
- Ion-pairing with analytes leads to a higher net positive charge, resulting in changed migration times and separation.
- Ion-pairing of diamines with the negatively charged groups on the capillary wall leads to reduced adsorption of positively charged analytes, *e.g.* proteins, to the capillary wall.

Zwitterions

- Lead to reduced interaction of positively charged analytes with the capillary wall.
- Some zwitterions can also function as buffers.
- Without effect on conductivity, meaning that the ionic strength/concentration of the buffer may be increased.

Surfactants

- Even though SDS may often be the first detergent of choice when optimizing MECC for all kinds of analytes, it will not always give the best results. Other anionic detergents, *e.g.* the cholate system, have various advantages, especially in studies of biologically active or native proteins. Including cationic detergents for, especially, neutral or anionic analytes may prove very effective.
- Increased concentration gives an increased number of micelles/monomers, leading to a decrease in electro-osmotic flow and changed distribution of analytes which changes the migration times and separations.
- Increased salt concentrations will lead to decreased c.m.c. and increased aggregation number, changing the distribution of analytes and thereby the separations.
- The temperature also influences the distribution of analytes.
- Ion-pairing of analytes will either increase (equally charged analytes and surfactants) or decrease (oppositely charged analytes and surfactants) hydrophobic interactions between analytes and micelles, which in turn will influence separations.
- Strong interaction with micelles can be reduced by addition of organic modifiers.
- Zwitterionic and uncharged surfactants will give a low current and thereby less problems with heat generation.
- The counterion of the ionic surfactant may influence the c.m.c. and thereby separations.
- Mixed micelles may be used to improve selectivity.

Ion-pairing reagents

- Reduced interaction of positively charged analytes with the capillary wall.
- See also amine modifiers, zwitterions, and surfactants.

Sample

- Conductivity preferably about 10 times lower than conductivity in buffer to obtain sufficient stacking.
- Be aware of thermal degradation if conductivity in sample is very low.
- The sample must not contain compounds that will interfere with the chosen HPCE separation and detection mode.
- Presence of high concentrations of organic solvent may be detrimental to separations in, for example, MECC.

Capillary Dimension/Type

- 5–250 μm i.d. Fused-silica capillaries, typically 25–75 mm i.d.
- Increased diameter results in higher current, increased heat generation, and reduced surface area for heat dissipation.
- Increased diameter also results in decreased adsorption problems (lower surface area to volume ratio) and increased sensitivity.
- Consider special capillaries to increase sensitivity (*e.g.* Z or bubble cells).

Capillary Coating

- Permanent/static or dynamic coatings can be used.
- Should be able to reduce analyte–capillary wall interactions for, for example, protein separations.
- Suitable to reduce/eliminate electro-osmotic flow in some HPCE modes, *e.g.* CGE and CIEF.

Injection

- The injected volume has to be relatively low, using normal injection modes to obtain good separations (max. 2% of the capillary length).

Others

- Linear polymers (*e.g.* in CGE) can be used to reduce electro-osmotic flow and reduce diffusion of analytes as well as decrease interaction with the capillary wall (dynamic coating).
- Other compounds can be used to increase viscosity and reduce diffusion and electro-osmotic flow.
- Urea or ethylene glycols can be included to solubilize analytes and reduce analyte–capillary wall interactions.
- Complexing agents can be included to improve separation.

In Figure 10.38, the basic considerations made in using various modes of HPCE (FZCE, MECC, and HICE) have been schematized. Other selected modes such as CGE, CIEF, and CEC are commented on separately.

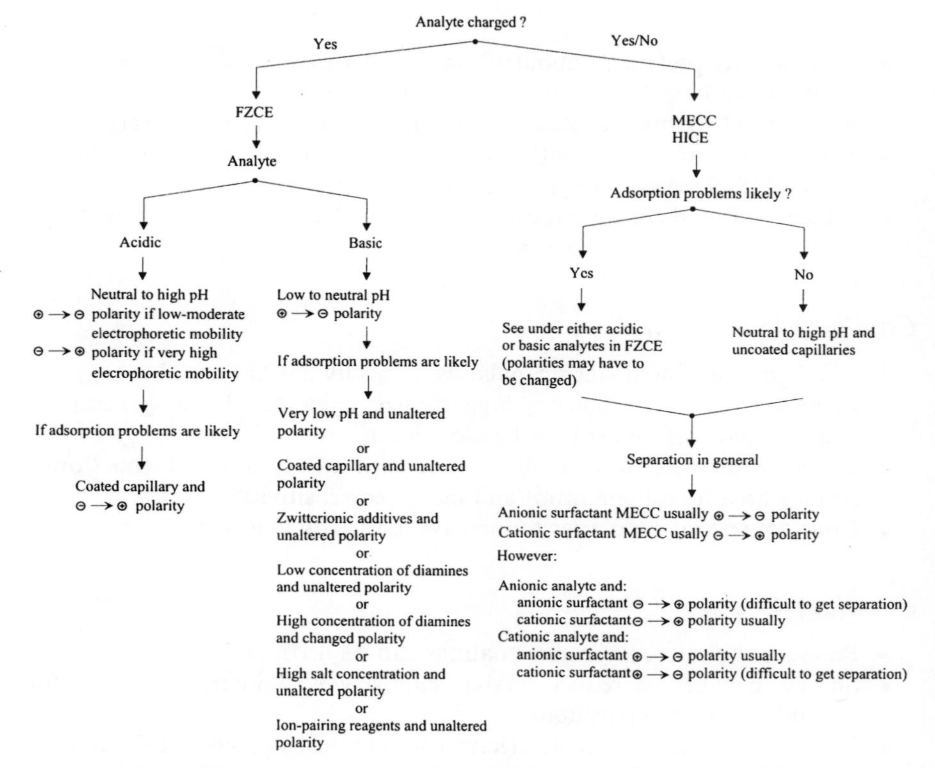

Figure 10.38 *Basic proposal for choice of various modes of HPCE (FZCE, MECC, and HICE)*

CGE

- Size-dependent molecular sieving mechanism.
- No or very low electro-osmotic flow.
- Analytes have to be charged.
- Especially used for oligo- and polynucleotides, restriction digests, DNA sequencing reaction products, oligo- and polysaccharides, and proteins (SDS–proteins).

CIEF

- Separations based on differences in analyte pI-values in a pH-gradient.
- Usually no electro-osmotic flow.
- Especially used for peptides and proteins.

CEC

- New technique with applications comparable to FZCE and MECC.
- Risk of destruction of 'expensive' capillaries by impurities.
- Electro-osmotic flow in either direction, depending on the packing material.
- Mostly for small molecules in pure well-defined groups and therefore not likely to give adsorption problems on particle surfaces.

Suggested Optimization Procedure

Based on a knowledge of the effects of separation parameters in HPCE, a sensible development and optimization of a MECC method may be performed by considering the procedure described in the following. In general, the influence of the various parameter changes is followed by changes in migration time, normalized areas, resolution, and number of theoretical plates. Some comments on the actual strategy for optimization of the separation of aromatic choline esters are included as an example. Moreover, as a supplement to these more general guidelines, detailed specific guidelines described in the literature could be worthwhile consulting. In Ref. 19, for example, is described a strategy for optimizing the resolution in SDS-MECC, whereas another more detailed description of the separation and optimization can be found in Ref. 20. It is hoped that this can provide the reader with some general ideas for method development.

Literature Search

Many excellent methods are described. Use these if possible and save time. However, many of the methods described have only been developed using standards, which often makes them unsuited for real samples. Knowledge of extraction, group separation, and clean up procedures (Chapters 4 and 6–9) are crucial for successful work with real samples.

Analyte and Sample Information

- pK'_a-values and number of charged groups.
- UV–Vis spectra/fluorescence information.
- Possible effects of buffer composition and type on spectroscopic properties and on protolytic active groups.
- Solubility in aqueous solutions of various pH and in the presence of surfactants, organic solvents, and other buffer additives.
- Stability at different pHs and temperatures.
- Possible interfering compounds in the samples to be analysed.

Example. Aromatic choline esters are cationic low-M_r compounds. A standard mixture with closely related aromatic choline esters solubilized in water has been prepared and used for the development of a HPCE procedure.

Select the Mode of HPCE Most Likely to Give Success (Figure 10.38)

Usually a simple separation principle such as FZCE is chosen as a first choice if no knowledge of the possible superiority of other modes can be obtained.

Example. Theoretically, analysis of charged low-M_r compounds such as aromatic choline esters in FZCE can be a good solution; however, MECC introduces an additional separation factor, namely the partitioning of analytes between micellar and buffer phase. The choice of MECC leads to a new choice: which surfactant? Cationic surfactants do not allow separation based on ion-pairing, only hydrophobicity; however, adsorption to the negatively charged capillary wall will be eliminated. Anionic surfactants, on the other hand, should be supplemented with an additive to avoid this interaction. The analysis time should be considered, especially when there is a need for a high number of analyses. Basically, cationic analytes elute before the solvent front when using an anionic surfactant, provided that ionic interactions are not too strong. Tests with SDS showed long migration times and poor separation. Instead, negatively charged bile salts (sodium cholate), forming reversed micelles that limit ionic interactions, gave promising results.

Capillary Dimension and Type

- Capillary diameter is a trade-off between sensitivity and the desire for high electrolyte concentrations.
- If there seems to be a special need for high sensitivity, consider the use of special capillaries.
- Is there something indicating a possible need for coated capillaries (decreased/eliminated electroosmotic flow and/or reduced analyte–capillary wall interactions)?
- Determine relevant pre-conditioning conditions.

Select Suitable Electrolyte (Type, Concentration, pH)

- High buffer capacity in the required pH range;
- no interference with detection principle;
- usually low conductivity is preferred.

Example. As the use of sodium cholate limits the pH-range to above 6.8 (pK'_a-value for sodium cholate = 6.4), a phosphate buffer was chosen, although others may have been comparably good.

Optimize Temperature, and Voltage

- Investigate if changes in voltage and temperature may influence separations.
- The variation of temperature usually covers the 20–60 °C range; however, some apparatus have possibilities for temperature regulation below ambient.
- The variation of voltage usually covers the 5–30 kV range.
- Check the influence of voltage on buffer temperature by making a plot of voltage (x) against current (y) (Ohm's plot).

Optimize Additives

- Some additives may have been selected when the HPCE mode was chosen but others may also be relevant.
- Organic modifiers, amine modifiers, and ion-pairing reagents may be especially useful for optimization of separations at this stage.

Example. Late elution and very poor separation of the aromatic choline esters in the standard mixture was obtained without buffer additives; consequently, taurine, a zwitterionic compound, was included to eliminate the interaction of the positively charged aromatic choline esters with the negatively charged capillary wall. Organic modifiers and surfactant concentration are both factors which affect the partitioning of analytes between the aqueous and micellar phase, and different combinations of surfactant and modifier were shown to have a great impact on resolution and efficiency of the separation system. Repeated analyses with small adjustments finally resulted in baseline separation of the six analytes in the standard mixture.

Optimize Detection Wavelength

- Determine optimal signal and reference wavelength if UV–Vis detection.
- Check optimal excitation and emission wavelength if fluorescence.
- In indirect detection, the concentration of the background electrolyte should be low enough to give high sensitivity but high enough to give an acceptable linearity range.

Optimize Sample Concentration and Composition

- Optimize dilution rate with respect to sensitivity and conductivity to obtain stacking.
- Test group separation, ion-exchange, solid phase extraction, or other clean-up procedures if interfering compounds are present and/or conductivity is too high.

Optimize Injection Method

- Select injection mode (usually hydrodynamic to avoid bias with electrophoretic mobility differences).
- Optimize injection time in combination with the above optimization of sample concentration and composition. Preferably, at least 3–5 s injection times should be used.
- Injection times that are too long may lead to run failure owing to problems with conductance in the capillary.

Optimize Precision (includes many of the above-mentioned parameters)

- Use as high a sample concentration and injection volume as possible. This reduces integrator errors and effects of adsorption to the capillary wall.
- Always use normalized peak areas.
- Avoid differences in viscosity in samples and standards.
- Minimize evaporation losses from vials.
- Use internal standards whenever possible.
- Samples should have the same temperature as the autosampler when an analysis is started.
- Ensure that the buffer capacity is high enough.
- Replenish buffers between each analysis.
- Constant/controlled temperature of capillary, sample vials, and buffer reservoirs should be used (not the same temperature!).
- Use a voltage ramp at the beginning of the analysis.
- Use correct preconditioning conditions, including capillary washing and pre-runs of samples.
- Avoid short injection times (less than 1–2 s).
- Use self-regulating hydrodynamic injection if possible.
- Assure equal levels of solvent in inlet and outlet vials.

Optimize Sensitivity (includes many of the above-mentioned parameters)

- Consider capillary dimension (see above). Special capillaries, with, for example, bubble cells or Z-shaped cells may be used.
- Reduce peak tailing in general.
- Derivatization of analytes may be necessary, but this most often results in completely changed separations and the need to re-optimize separations.
- Consider detection wavelength (see above).
- Consider injection time (see above).
- In some cases, electrokinetic injection may improve sensitivity but the injection is biased by the analyte mobilities and the sample matrix.

Validate the Optimized HPCE Method

The optimized HPCE method has to be validated to document the performance and disclose possible problems and limitations. Generally, suitability, linearity and recovery, precision, limit of detection, limit of quantitation, and ruggedness are evaluated during the validation. The criteria for accepting the various parameters depend on the actual application. An analysis of food quality, antinutritional, and physiologically active compounds, additives, or impurities, *e.g.*, in pharmaceuticals, requires documentation of very low detection limits, whereas, for example, food composition or characterization focus on the main components present and very low detection limits are usually not required.

In the following, a short description of the information required and the parameters included in a typical validation protocol are given. The validation report should clearly specify the area of applications under which the method is valid and a very thorough method description should be given, including information on samples, the method of analysis itself, and calculations performed. The suitability tests include determination of whether HPCE is suited for performing the analysis in question and whether the method is able precisely to specifically determine analytes in the presence of other components that might occur in the sample (*i.e.* no interference). Acceptance criteria may be fully resolved analyte peaks and interfering peaks. The linearity range of the method and the recovery of a known amount of analyte has to be determined as well as the sensitivity, which is the ability to differentiate between small differences in amounts of analytes. Acceptance criteria may be a reasonably wide linearity range with close to 100% recovery and as high a sensitivity as possible. Linearity studies are usually performed by simple dilution of a standard mixture (*e.g.* 100, 80, 60, 40, 30, 20, 15, 10, 5, 2.5, 1, *etc.*). Determination of precision (normalized peak areas, relative migration times) includes repeatabilities, which is the precision of the method under identical conditions (same day, same conditions), and reproducibility, which is the precision of the method under different conditions (different days, buffers, capillaries, laboratories, *etc.*). Overlapping 95% confidence intervals for analyte amounts determined on different days can be used as acceptance criteria for reproducibility and as low as possible a relative standard deviation (RSD) of analyte amounts for repeatability. The limit of detection may often be determined from the linearity study results as the analyte amount giving a signal-to-noise ratio of 3:1; however, some analysts accept 2:1 as the lower limit. See Box 10.3 for an example of the calculation of detection limits (concentration/mass). The limit of quantitation is the lowest quantifiable concentration with an acceptable RSD. A signal-to-noise level of 10:1 often gives an acceptable RSD. Finally, the ruggedness of the method is determined by testing the influence of small variations of the most relevant separation parameters and the stability of standards and samples. A Youden experimental design[21] may be used to test the influence of small variations of many parameters with a small number of experiments. In this way, a test of, for example, seven parameters with both a high and a low value can be performed in only eight different experiments (including repetitions). The stability of standard and sample components is determined after different relevant storage conditions (*e.g.* 1 day in an autosampler, a few days in a refrigerator, weeks or months in a freezer). The acceptance criteria for storage conditions may typically be overlapping 95% confidence intervals for the determined concentrations in the samples.

7 Conclusion

Year by year, an increasing number of publications and interest in scientific meetings in the area of HPCE have arisen, and although the technique is still developing there are a variety of established methods and general procedures available. In particular, the high versatility of HPCE makes the technique a promising tool in food analysis, allowing fast, effective, and qualitative separation of a wide range of different compounds, and reliable quantitative analysis can be obtained, providing there is awareness of the possible problems in this connection.

8 Selected General and Specific Literature

[1] S. Hjertén, *Chromatogr. Rev.*, 1967, **9**, 122.
[2] J.W. Jorgenson and K.D. Lukacs, *Anal. Chem.*, 1981, **53**, 1298.
[3] S. Terabe, K. Otsuka, and T. Ando, *Anal. Chem.*, 1985, **57**, 834.
[4] R. Kuhn and S. Hoffstetter-Kuhn, 'Capillary Electrophoresis: Principles and Practice', Springer-Verlag, Berlin, 1993, 375 pp.
[5] S. Michaelsen and H. Sørensen, *Pol. J. Food Nutr. Sci.*, 1994, **3**, 5.
[6] D.N. Heiger, 'High Performance Capillary Electrophoresis – An Introduction', 2nd Edition, Hewlett-Packard Company, France, 1992, ch. 2, p. 31.
[7] W.J. Lambert and D.L. Middleton, *Anal. Chem.*, 1990, **62**, 1585.
[8] R. Weinberger, 'Practical Capillary Electrophoresis', Academic Press, New York, 1993, 312 pp.
[9] S.F.Y. Li, 'Capillary Electrophoresis. Principles, Practice, and Applications', Journal of Chromatography Library, Elsevier, Amsterdam, 1992, vol. 52, 578 pp.
[10] E.G. Behnke and E. Bayer, *J. Chromatogr. A*, 1995, **716**, 207.
[11] I.Z. Atamna, C.J. Metral, G.M. Muschik, and H.J. Issaq, *J. Liq. Chromatogr.*, 1990, **13**, 2517.
[12] C. Bjergegaard, L. Ingvardsen, and H. Sørensen, *J. Chromatogr. A*, 1993, **653**, 99.
[13] K.A. Cobb, V. Dolnik, and M. Novotny, *Anal. Chem.*, 1990, **62**, 2478.
[14] H. Nishi, T. Fukuyama, and S. Terabe, *J. Chromatogr.*, 1991, **553**, 503.
[15] Y. Wahlbroehl and J.W. Jorgenson, *Anal. Chem.*, 1986, **58**, 479.
[16] H.E. Schwarts, R.H. Palmieri, J.A. Nolan, and R. Brosen, 'Separation of Proteins and Peptides by Capillary Electrophoresis – An Introduction', Beckman Instruments, Fullerton, CA, 1992, 69 pp.
[17] G.M. Janini and H.J. Issaq, *J. Liq. Chromatogr.*, 1992, **15**, 927.
[18] A. Vinther and H. Søeberg, *J. Chromatogr.*, 1991, **559**, 3.
[19] S. Terabe, 'Micellar Electrokinetic Chromatography', Beckman Instruments, Fullerton, CA, 1992, 46 pp.

[20] K.D. Altria, 'Capillary Electrophoresis Guidebook. Principles, Operation, and Applications', Methods in Molecular Biology, Humana Press, Totowa, NJ, 1996, vol. 52, 349 pp.
[21] W.J. Youden and E.H. Steiner, 'Statistical Manual of the Association of Official Analytical Chemists', AOAC, 1975.
[22] P.F. Cancalon, *J. AOAC Int.*, 1995, **78**, 12.
[23] J. Lindeberg, *Food Chem.*, 1996, **55**, 73.
[24] C.A. Monnig and R.T. Kenney, *Anal. Chem.*, 1994, **66**, 280R.

CHAPTER 11

Analytical Determination of Low-M$_r$ Compounds

1 Introduction

Natural products, and thereby feed and food, are composed of a great number of different compounds, which often are divided into high relative molecular mass (high-M_r) and low relative molecular mass (low-M_r) compounds. Based on weight amount, high-M_r compounds are the quantitatively dominating constituents of nearly all biological materials, food, and feed dry matter. The analytical methods required for determination of high-M_r compounds (Chapters 12–14) will for various reasons also differ from many of the methods of analyses, which are recommendable for determination of low-M_r compounds.

Based on the number of different compounds, low-M_r compounds occurring in natural products, food and feed are, however, quantitatively dominant compared with high-M_r compounds. The low-M_r compounds including oligomeric compounds, peptides, and bioactive peptides or small proteins call for special attention, especially because of their potential biological, antinutritional, and toxic effects, their nutritional value and thus the both positive and negative effects they may have on food and feed quality.[1,2] Analytical determination of individual amphiphilic, polar, or hydrophilic low-M_r compounds is considered to be of upmost importance for evaluation and analytical control with food quality and it may require determination of both structure (constitution) and stereochemistry (configuration) of the individual compounds.

This chapter focuses on selected liquid chromatographic (LC) methods of analyses not yet covered in the previous chapters, but which are considered to be important tools in determinations of non-volatile low-M_r compounds. The analyses have been arranged in a sequence that corresponds to the described group separations considered to be important for reliable quantitative determinations of individual compounds. High-performance liquid chromatography (HPLC) with the background information treated in Chapter 8 and especially high-performance capillary electrophoresis (HPCE), with the background information treated in Chapter 10, are the methods of choice for these purposes. HPLC and HPCE have thus potential as highly efficient methods of analyses for reliable quantitative determinations of the individual low-M_r compounds found in natural products. However, the possibilities for optimal utilization of both HPLC and HPCE depend on the use of appropriate sample preparation prior to HPLC or HPCE. Therefore, correct procedures are needed for sampling (page 71), extractions (Chapter 4), use of internal standards (page 73), and group separations (Section 7.6) independent of the choice among the various HPLC and HPCE methods of analyses. The problems caused by impurities or interfering compounds may, however, vary. HPLC is thus very sensitive to impurities and insufficient sample preparation, whereas many HPCE methods are less affected. Micellar electrokinetic capillary chromatography (MECC) and free zone capillary electrophoresis (FZCE) can, however, be sensitive to high ionic strength (I; Section 2.3), as well as to type and concentration of various compounds co-occurring with the analytes.

Although HPCE techniques including MECC are relatively new methods of analyses, they are increasingly used with advantage as a substitute or supplement to the traditionally applied HPLC methods of analyses used for determination of food and feed quality and in research for determination of different types of natural products.[2–5] As group separations generally are required prior to both HPLC and HPCE methods of analyses for low-M_r compounds, the following sections are arranged according to this separation theory and knowledge concerning experimental techniques.

2 Inorganic Ions and Elemental Analyses

Elemental analyses of biological materials, feed, and food can be determined by relatively cheap, fast, and simple methods. The results can be of value in combination with LC methods of analyses used for more detailed information on the actual compounds of interest in relation to structure determination and evaluation of feed and food quality.[1,2]

Ash, C, H, N, S, and O Determination

Chemical composition of biological materials comprises inorganic ions often measured gravimetrically as ash by standard procedures of combustion or ashing of the sample.[6] The composition of organic matter based on determination of C, H, N, S, and O are often performed by standard procedures in commercially available instruments utilizing the Dumas principle. Alternatively, determination of N by the Kjeldahl method is often used for calculation of crude protein based on the nitrogen content multiplied by 6.25. Other more or less specific methods for protein determination are described in the subsection on pages 99 and 328.

Cations

Atomic spectroscopy (Section 5.2) is a fast and simple method of analysis for different cations. For isolated cations, *e.g.* in the A-eluate from group separation (Section 7.6), HPCE used as FZCE (page 232) also gives the possibility for determination of individual cations (Figure 10.11).

Anions

Determination of phosphorus can be performed by use of various standard procedures, *e.g.* after ashing followed by colorimetric analysis[7] or by atomic absorption spectrometry determination of phosphorus in oils and fats.[8] HPCE can also be used with advantages for the determination of inorganic anions as chloride, sulfate and phosphate (Figure 10.12) or nitrite, nitrate, iodide, and the thiocyanate ions using their chromophoric properties (Figure 10.16).

3 Low-M_r Organic Compounds with Positive Net Charge, Alkaloids and Biogenic Amines

Alkaloids are basic nitrogen-containing organic compounds. This definition is identical to that for biogenic amines, and no clear borderline exists between these two groups of biomolecules. They are heterogeneous groups of more than 3000 different types of compounds, and can often give appreciable physiological effects.[9] Common to these low-M_r compounds is the possibilities for analytical investigations based on the use of A-eluates from group separations (Section 7.6). Preliminary qualitative detection of the different types of compounds may be performed by thin-layer chromatography (TLC) or paper chromatography (PC) (Section 6.5) or by high-voltage paper electrophoresis (HVE) (page 183). The information obtained thereby on appropriate solvent systems or other specific separation conditions may then provide the basis for selection of appropriate quantitative HPLC or HPCE methods of analyses.

Aromatic Choline Esters

Choline esters are important parts of all living cells, where they occur in phospholipids (Figure 10.20), and in the neurotransmitter acetylcholine. In seeds of especially cruciferous plants, various benzoic and cinnamic acid derivatives (Figure 5.10) are accumulated as aromatic choline esters (Figure 5.18) to levels of often 10–30 μmol/g dry matter. Therefore, they will also be present in various feed and food mixtures. They are easily isolated as A-column eluates (Section 7.6) which can be used both for HPLC and HPCE determination of the individual compounds as shown in Figures 11.1 and 11.2.

Lupine Alkaloids and Guanido Compounds

As the case is for the majority of alkaloids, lupine alkaloids can give antinutritional or toxic effects, when they occur in too high a concentration in feed and

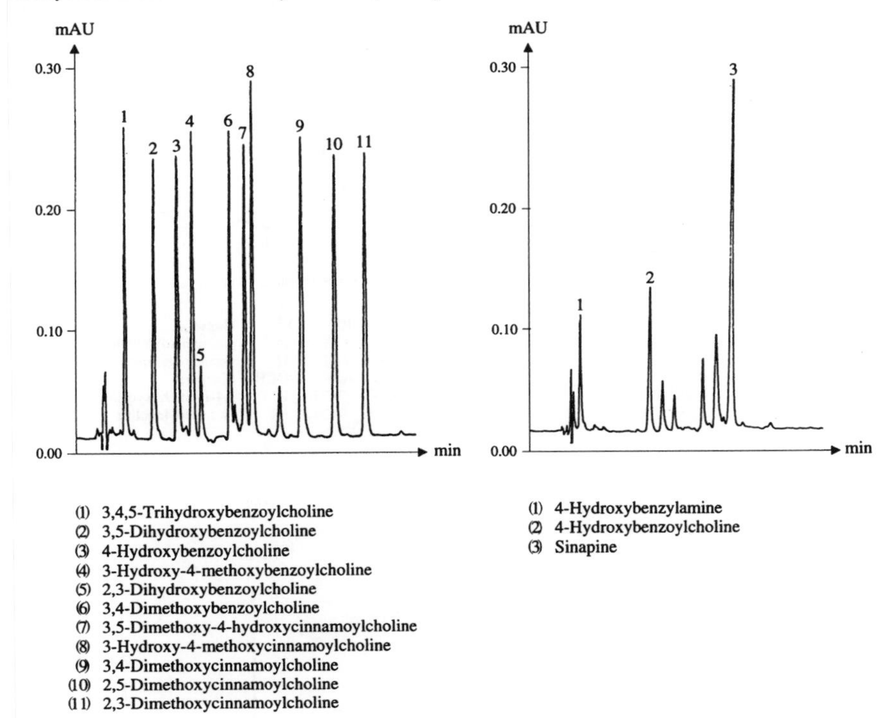

Figure 11.1 *Determination of individual aromatic choline esters by use of reversed phase ion-pairing HPLC. The figure on the right is the HPLC chromatogram of the A-column eluate obtained by group separation of the compounds in an extract from 0.1 g of seeds of* Sinapis alba *cv. Trico. The left hand figure is the HPLC chromatogram of different aromatic choline esters (reference compounds)* (Reprinted from S. Clausen, O. Olsen, and H. Sørensen, *J. Chromatogr.*, **260**, 193 (1983), Elsevier Science, Amsterdam)

food. It is also noteworthy that aromatic amino acids are major precursors of many alkaloids *e.g.* as morphine, codeine, papaverine, colchicine, ricinine (Figure 5.14), and various indolalkaloids. The aromatic groups result in efficient chromophore systems which give the basis for UV–Vis detection of these compounds.

Quinolizidine- or lupine alkaloids and guanido compounds, *e.g.* canavanine, galegin and hydroxygalegin (Figure 11.3) occur in some few plants of the family Leguminosae.

In general, guanido compounds and alkaloids as found in lupine (Figure 11.3) have poor chromophore systems (section 5.3.5) reducing the sensitivity for their detection by direct UV–Vis determination used in HPLC and HPCE instruments. However, MECC based on systems as used for aromatic choline esters (Figure 11.2) results in efficient separations of individual guanido compounds and lupine alkaloids isolated as A-column eluates (Section 7.6) as shown in Figure 11.4.

Benzoic acid derivatives

No.	R^3	R^4	Name
2	-H	-OH	4-Hydroxybenzoylcholine
3	$-OCH_3$	-OH	Vanillylcholine
9	-OH	$-OCH_3$	Isovanillylcholine
10	$-OCH_3$	$-OCH_3$	Hesperaline

Cinnamic acid derivatives

No.	R^3	R^4	R^5	Name
4	-H	-OH	-H	*p*-Coumaroylcholine
5	$-OCH_3$	-OH	-H	Feruloylcholine
6	$-OCH_3$	-OH	$-OCH_3$	Sinapine
7	$-OCH_3$	$-OCH_3$	-H	3,4-Dimethoxycinnamoylcholine
8	-OH	$-OCH_3$	-H	Isoferuloylcholine

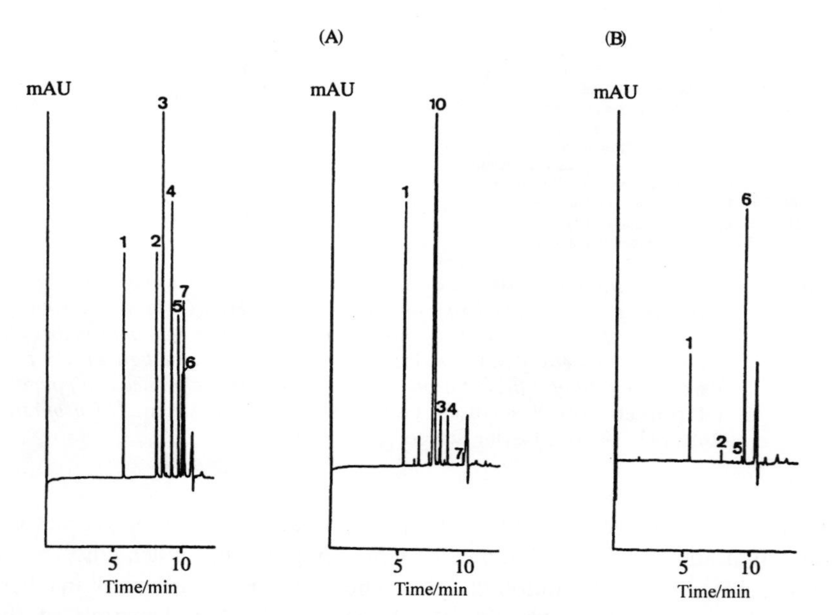

Figure 11.2 *Determination of individual aromatic choline esters by use of MECC, with trigonellinamide (1) as internal standard. The electropherogram on the left is of a mixture of reference compounds. The aromatic choline esters accumulated in seeds of* (A) Hesperis matronalis *L. and* (B) Sinapis alba *L. were obtained after purification by the procedure mentioned in Figure 11.1. Buffer: 100 mM Na_2HPO_4, 500 mM taurine, 35 mM sodium cholate, 2% propan-2-ol, pH 7.3; temperature: 30 °C; voltage: 20 kV; injection: hydrodynamic (vacuum) (+ to −); detection: UV, 235 nm; capillary: 50 μm i.d., 76 cm total length, 53 cm effective length*
(Reprinted from C. Bjergegaard, L. Ingvardsen, and H. Sørensen, *J. Chromatogr. A*, **653**, 99 (1993), Elsevier Science, Amsterdam)

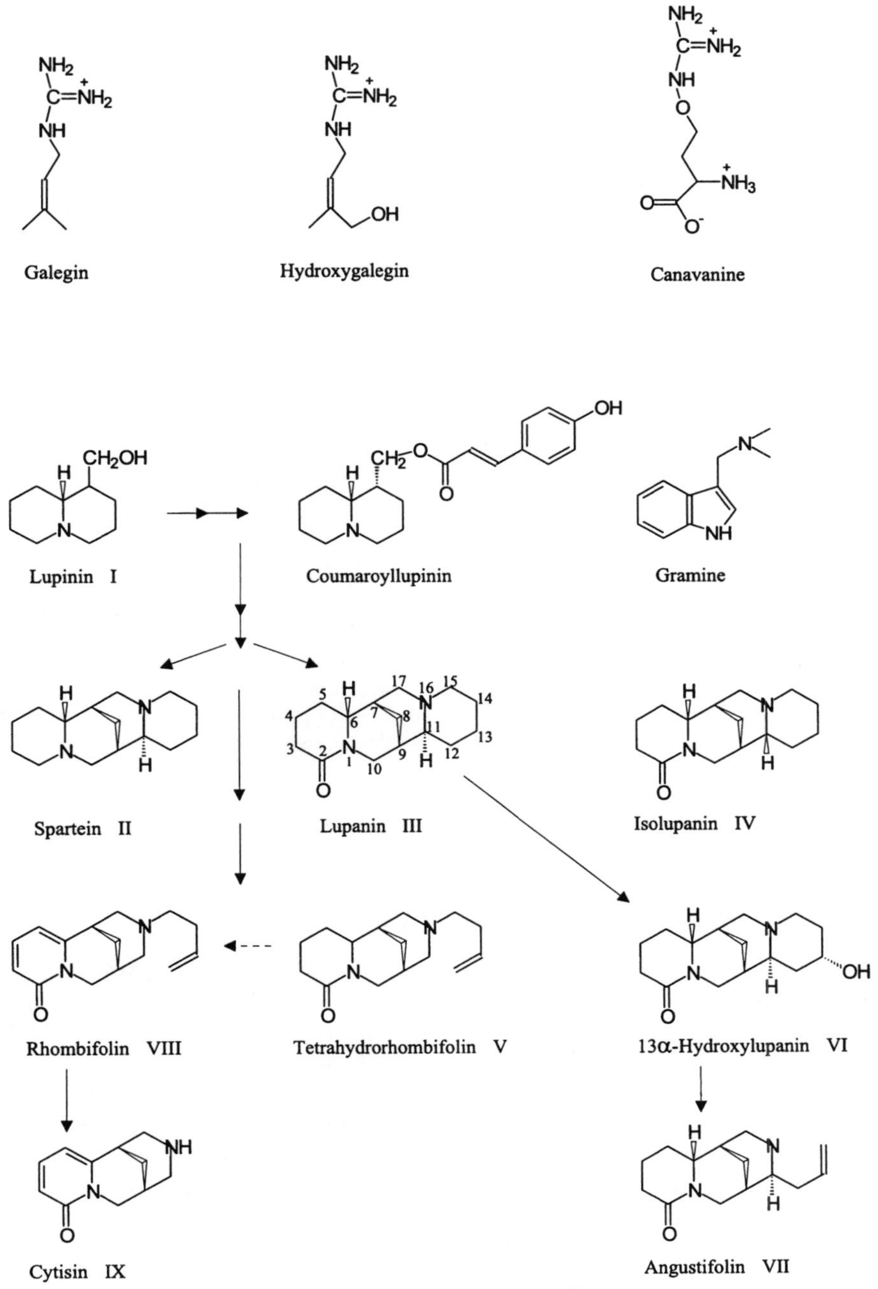

Galegin

Hydroxygalegin

Canavanine

Lupinin I

Coumaroyllupinin

Gramine

Spartein II

Lupanin III

Isolupanin IV

Rhombifolin VIII

Tetrahydrorhombifolin V

13α-Hydroxylupanin VI

Cytisin IX

Angustifolin VII

Figure 11.3 *Structural formula of galegin, hydroxygalegin, and canavanine together with formulae of alkaloids occurring in lupine*

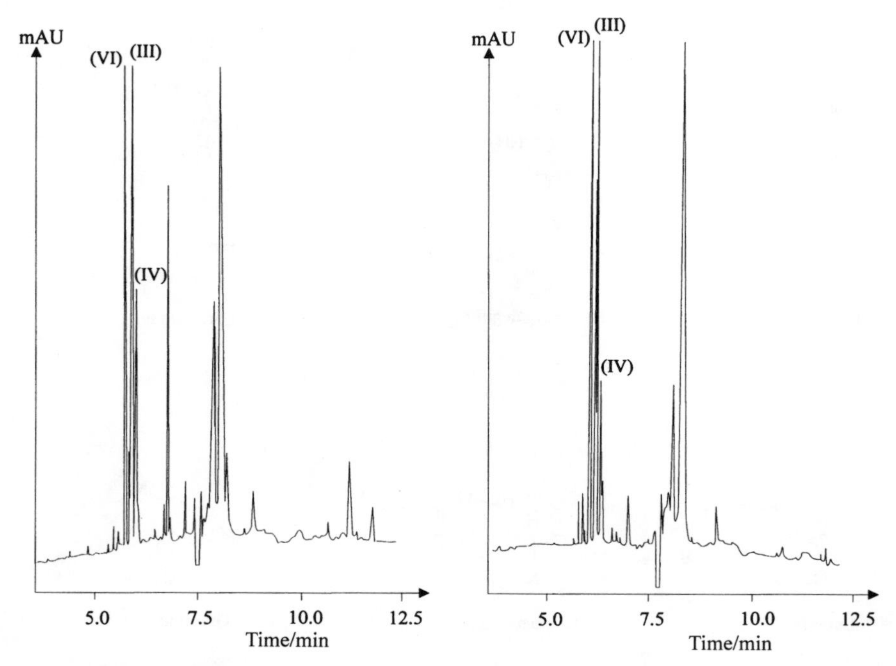

Figure 11.4 *Separation by MECC of lupine alkaloids (Figure 11.3) isolated as A-column eluate from lupine seeds extracts, using the MECC method developed for aromatic choline esters, but with detection at 205 nm (Figure 11.2). The electropherograms show alkaloids from* Lupinus angustifolius *L. cv W-26 (left), and* L. albus *L. cv. Wat alkaloids (right), with compounds numbered as in Figure 11.3*

Heteroaromatics and Basic Amino Acids

Basic amino acids and basic heteroaromatics (Figure 5.14) can be isolated as A-column eluates, whereas other heteroaromatics may occur in subsequent fractions or groups from the column purification (Sections 7.6 and 11.4). Separation and quantitative determination of individual basic protein amino acids can be performed by use of standard amino acid analysers (Figures 10.15 and 10.28). Separation of more complicated mixtures of compounds in this group may require more advanced techniques, and heteroaromatics which generally have good chromophoric systems (pages 108–111) can with advantage be determined by HPLC and HPCE methods of analyses based on direct UV or diode-array detection (Section 5.6). Selected examples of HPLC and HPCE methods used for determination of heteroaromatics and aromatic amino acids are shown in the Figures 11.5–11.7.

Separation and determination of aromatic and heteroaromatic compounds

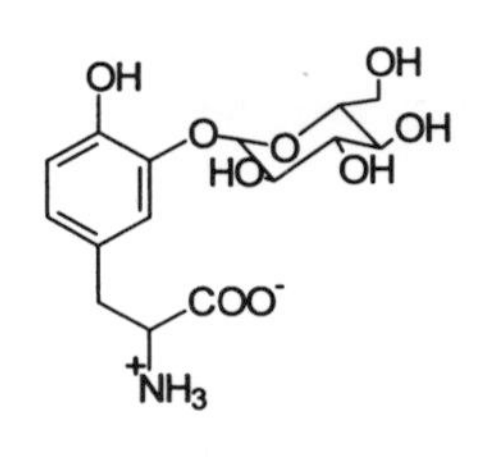

Dopa-glucoside
(2*S*)-3-(3'-β-D-glucopyranosyloxy-
4'-hydroxyphenyl) alanine

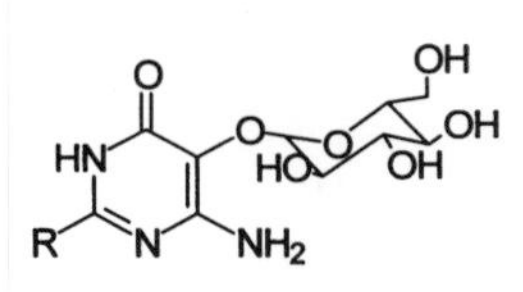

R=NH₂ : Vicine
2,6-Diamino-5-(β-D-gluco-
pyranosyloxy)-4-pyrimidinone

R=OH : Convicine
6-amino-2-hydroxy-5-(β-D-gluco-
pyranosyloxy)-4-pyrimidinone

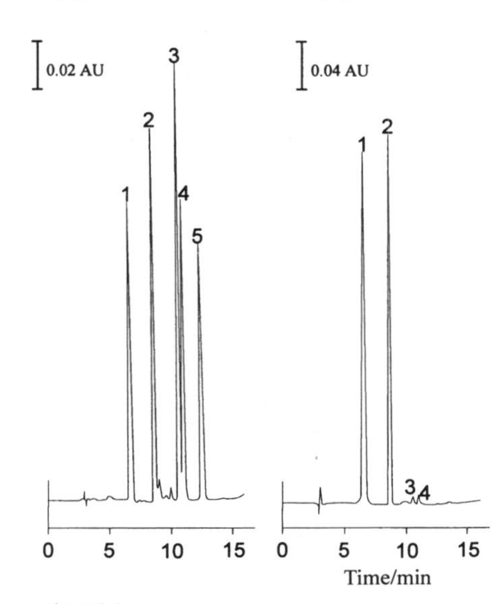

(1) Vicine
(2) Convicine
(3) Dopa
(4) Dopa-glucoside
(5) Tyrosine

Figure 11.5 *HPLC chromatograms of an artificial mixture of reference compounds* (A) *and of an extract from* Vicia faba *seeds* (B). *Support: Nucleosil 5 C-18, 250 × 4.6 mm. Mobile phase, a linear gradient of solvent A-solvent B (1:99) to (99:1) for 10 min with the final conditions maintained for an additional 5 min, flow rate 1 ml min*$^{-1}$. *Solvent A: 12.5 mM phosphate buffer (pH 2.0) modified with 25% MeOH. Solvent B: 10 mM phosphate buffer (pH 2.0). Recorder speed: 5 mm min*$^{-1}$*; detection: UV, 280 nm*
(Reprinted from B. Bjerg, O. Olsen, K.W. Rasmussen, and H. Sørensen, *J. Liq. Chromatogr.*, 1984, **7**, 691, by courtesy of Marcel Dekker Inc.)

by use of HPCE give often advantages compared with the use of HPLC. Different types of detergents can be used in MECC, *e.g.* sodium cholate (NaCh), sodium dodecylsulfate (SDS), and alkyltrimethylammonium compounds with varying chain lengths of the alkyl group (Table 2.6). The alkyltrimethylammonium bromide systems failed to give acceptable separation of the compounds in the standard mixtures of compounds denoted Pisum and Vicia standard mixtures (Figure 11.6), whereas the results obtained by use of NaCh and SDS were promising. Thus all compounds in the Pisum and Vicia standard mixtures were separated within 20 min of analysis (Figures 11.6 and 11.7).

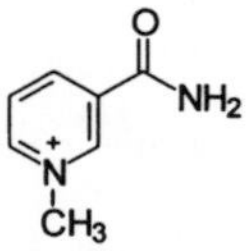

(1) Trigonellinamide

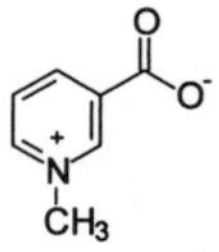

(2) Trigonelline

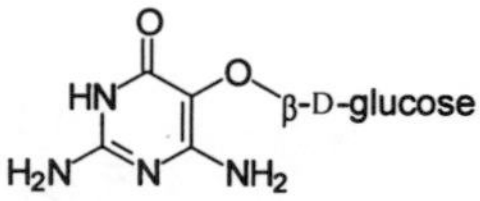

(3) Vicine

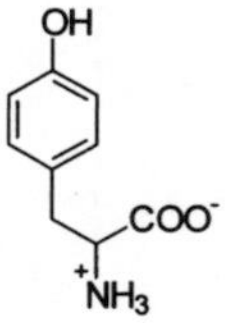

(4) Tyrosine

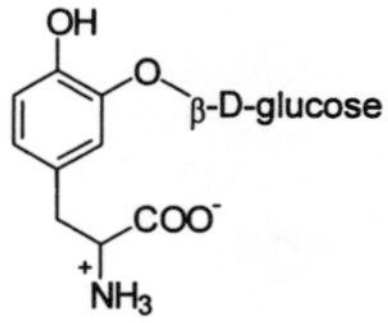

(5) Dopa-glucoside

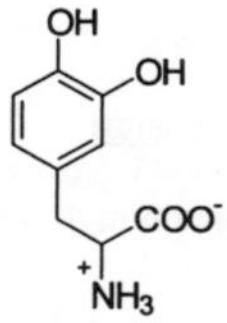

(6) Dopa

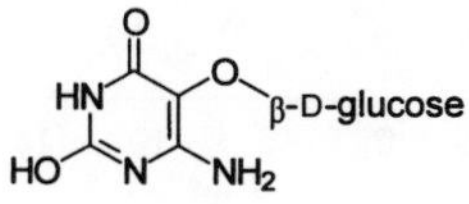

(7) Convicine

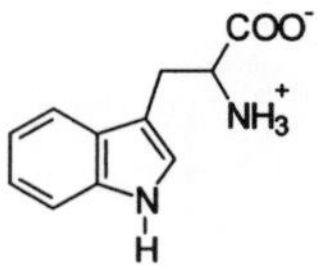

(8) Tryptophan

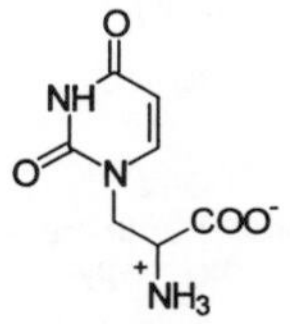

(9) Willardine

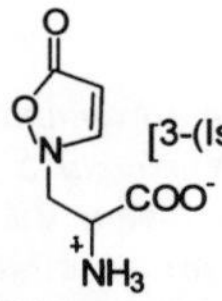

[3-(Isoxazoline-5-one-2-yl)alanine]

(10) Isoxazoline

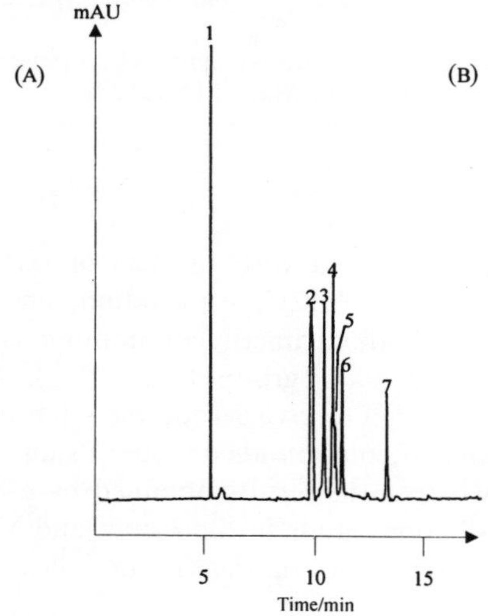

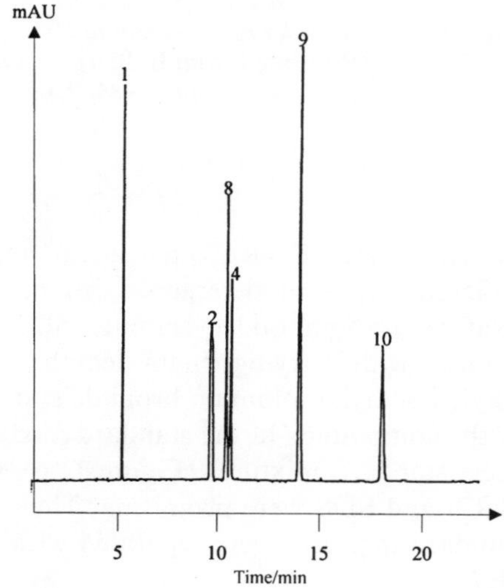

(A)

mAU

Time/min

(B)

mAU

Time/min

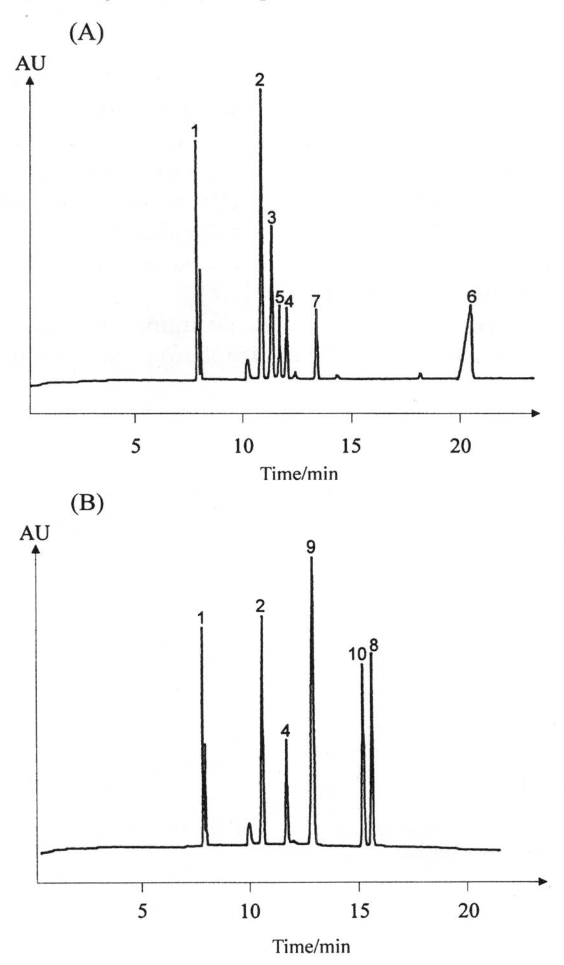

Figure 11.7 *Electropherograms of* (A) *Vicia and* (B) *Pisum standard mixtures separated by MECC (SDS). Peak numbers as in Figure 11.6. Buffer: 30 mM Na_2HPO_4, 18 mM disodium tetraborate, 50 mM SDS, 5% propan-1-ol, pH 7.0; temperature: 40 °C; voltage: 18 kV; injection: hydrodynamic (vacuum) (+ to −); detection: UV, 260 nm; capillary: 50 μm i.d., 76 cm total length, 44.5 cm effective length*
(Reprinted from C. Bjergegaard, H. Simonsen, and H. Sørensen, *J. Chromatogr. A*, **680**, 561 (1994), Elsevier Science, Amsterdam)

Figure 11.6 *Electropherogram of* (A) *Vicia and* (B) *Pisum standard mixtures separated by MECC (NaCh). Buffer: 100 mM Na_2HPO_4, 15 mM sodium cholate, 10% propan-1-ol, pH 7.3; temperature: 50 °C; voltage: 20 kV; injection: hydrodynamic (vacuum) (+ to −); detection: UV, 260 nm; capillary: 50 μm i.d., 76 cm total length, 44.5 cm effective length*
(Reprinted from C. Bjergegaard, H. Simonsen, and H. Sørensen, *J. Chromatogr. A*, **680**, 561 (1994), Elsevier Science, Amsterdam)

Biogenic Amines

Biogenic amines comprising catecholamines are compounds that can give appreciable physiological effects and they will thus often affect the quality of food and feed.[9] Isolation of biogenic amines in A-eluates (Section 7.6) as a group of compounds gives the basis for further analyses and quantification of the individual compounds by use of either HPLC[9] or HPCE. MECC based on the cholate micelles also gives possibilities for evaluating the applied group separation system, as this MECC system separates the analytes in groups corresponding to their net charge (Figure 11.8).

To obtain improved separation of biogenic amines in MECC, the change of temperature and voltage used for the electropherogram in Figure 11.8 will give

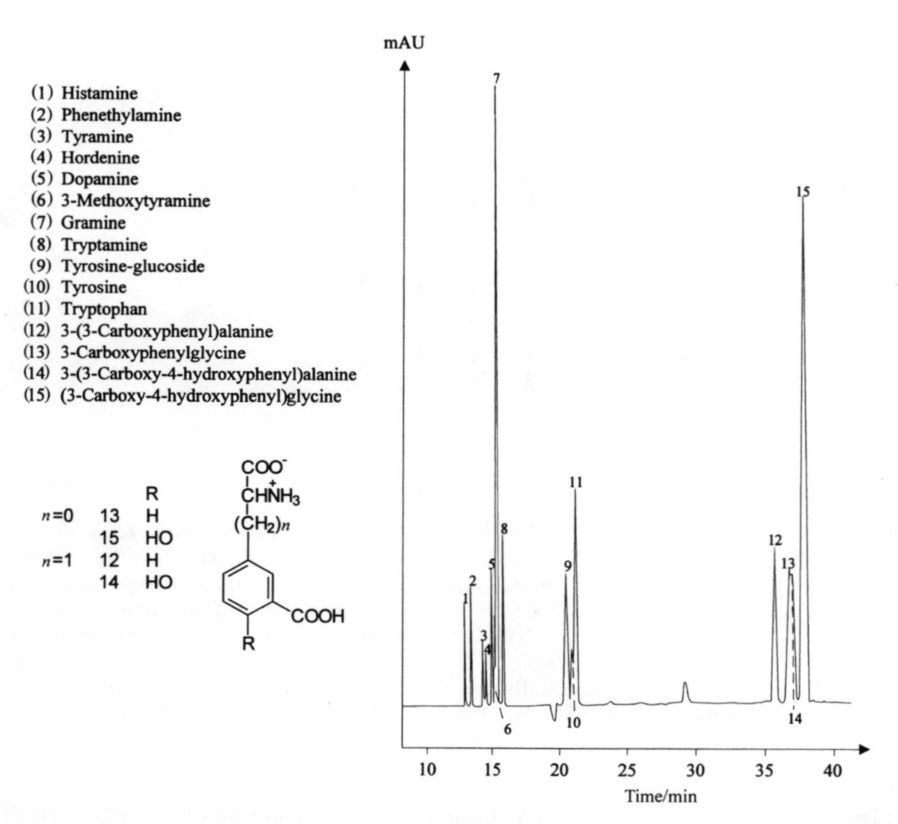

Figure 11.8 *MECC separation of biogenic amines (1–8), of neutral amino acids (9–11), and of acidic amino acids (12–15). Buffer: 100 mM Na_2HPO_4, 500 mM taurine, 35 mM sodium cholate, 2% propan-2-ol, pH 7.3; temperature: 25 °C; voltage: 10 kV; injection: hydrodynamic (vacuum) (+ to −); detection: UV, 205 nm; capillary: 50 µm i.d., 76 cm total length, 53 cm effective length*

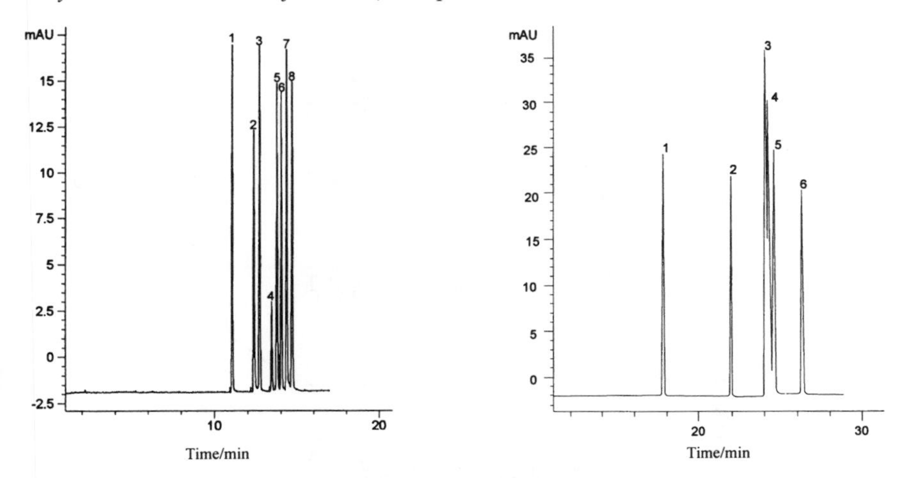

Figure 11.9 *MECC separation of biogenic amines, catecholamines, adrenalin, and noradrenalin. Peaks in left electropherogram: 1 = trigonellinamide, 2 = histamine, 3 = phenethylamine, 4 = tyramine, 5 = 2-amino-2-phenylpropanol, 6 = 3-methoxytyramine, 7 = 3,4-dimethoxyphenylethylamine, 8 = tryptamine. Peaks in right electropherogram: 1 = trigonellinamide, 2 = tyramine, 3 = noradrenalin, 4 = adrenalin, 5 = serotonine, 6 = 5-methoxytryptamine. Buffer: 100 mM Na_2HPO_4, 500 mM taurine, 35 mM sodium cholate, 2% propan-2-ol, pH 7.3; temperature: 40 °C (left electropherogram), 15 °C (right electropherogram); voltage: 7 kV; injection: hydrodynamic (vacuum) (+ to −); detection: UV, 215 nm; capillary: 75 μm i.d., 64.5 cm total length, 56 cm effective length*

appreciable effects, as illustrated in Figure 11.9. In both electropherograms, trigonellinamide is used as internal standard.

Anthocyanins

Anthocyanins are glycosides of the aglycones anthocyanidins (Figure 5.11), which in acidic solutions are cations owing to the flavylium cation structure. In addition, many native anthocyanins have acidic groups as benzoic and/or cinnamic acid derivatives (Figure 5.10) attached as ester groups to the carbohydrate part of the glycosides. With the intact flavylium cation structure, these compounds have red to blue colours (page 104) reflecting their chromophore system, and these aglycones in anthocyanins are responsible for the specific colours exhibited by various plants, *e.g.* flowers, grapes, red wine, red cabbage, and strawberry. Trivial names of aglycones in widely distributed anthocyanins are given in Figure 11.10.

Owing to the equilibria shown in Figure 11.10, anthocyanins are only stable in acidic solution, and several of them are easily oxidized (antioxidants). This means that extraction and purification require work in acidic solutions,[10] and separations based on A-column eluates (Section 7.6) followed by purifications

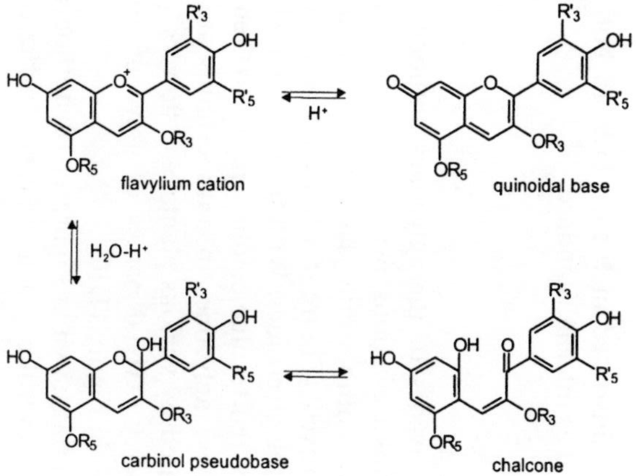

Substitution pattern in anthocyanidines

	R_3	R_5	R_6	R_7	$R_{3'}$	$R_{5'}$
Pelargonidin	**OH**	**OH**	**H**	**OH**	**H**	**H**
Aurantinidin	OH	OH	OH	OH	H	H
Margicassidin	OH	OH	*	OH	H	H
Cyanidin	**OH**	**OH**	**H**	**OH**	**OH**	**H**
Peonidin	**OH**	**OH**	**H**	**OH**	**OMe**	**H**
Rosinidin	OH	OH	H	OMe	OMe	H
Delphinidin	**OH**	**OH**	**H**	**OH**	**OH**	**OH**
Petunidin	**OH**	**OH**	**H**	**OH**	**OMe**	**OH**
Pulchellidin	OH	OMe	H	OH?	OH?	OH
Europinidin	OH	OMe	H	OH?	OMe	OH
Malvidin	**OH**	**OH**	**H**	**OH**	**OMe**	**OMe**
Hirsutidin	OH	OH	H	OMe	OMe	OMe
Capensinidin	OH	OMe	H	OH?	OMe	OMe

* alkenyl may be in pos. 8

Figure 11.10 *Anthocyanin equilibrium structures, and trivial names of the aglycones. The R_3 group is a carbohydrate, R_5 can be a carbohydrate or H, R'_3 and R'_5 can be H, OH, OMe, or a glycoside group. Bold indicates widely distributed anthocyanidins*

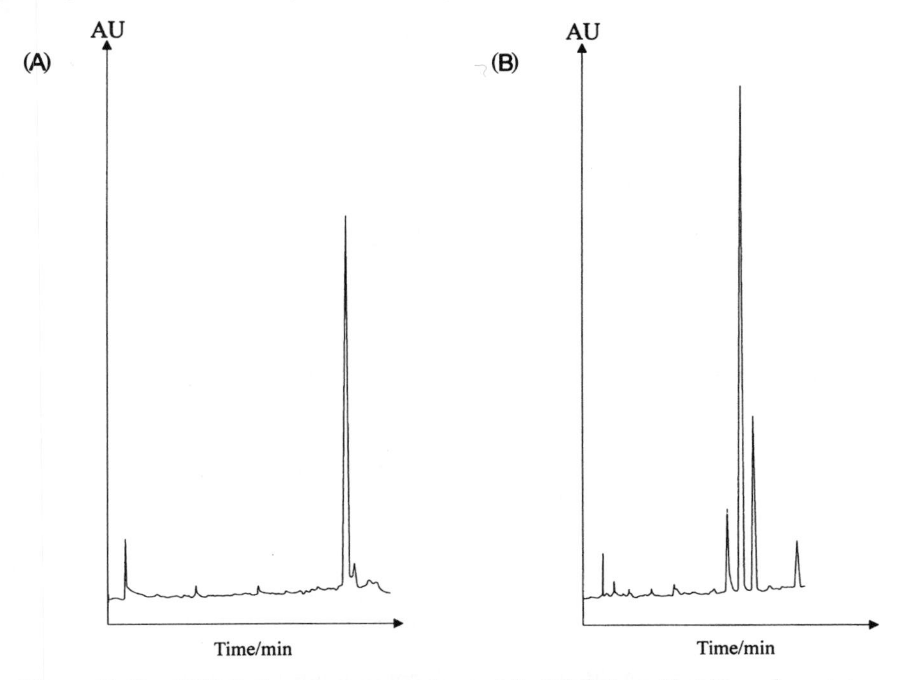

Figure 11.11 *HPLC chromatograms of petunidin-3,5-diglucoside* (A) *and a mixture of delphinidin-3-glycosides* (B) *isolated from aubergine. Nucleosil 5 C-18 column (250 × 4.6 mm) and linear gradient elution with solvent A:B (25:75) to solvent A:B (65:35) for 35 min. Solvent A: 90% MeOH in H_2O; solvent B: 5% HCO_2H in H_2O. Flow rate 1 ml min^{-1}. Detection: UV, 540 nm*

on a Bondapak C-18 column give samples which then can be analysed in HPCE or HPLC systems (Figure 11.11).

4 Ampholytes, Amino Acids, Peptides, and Heteroaromatics

Amino acids, peptides, heteroaromatic compounds, and other ampholytes (pages 103, 108 and 284–288) are present in appreciable quantities in extracts from all living tissues and at pHs below pI they will be retained on cation exchangers and released therefrom at a pH higher than pI.[11] The group of naturally occurring amino acids and peptides comprise at present about a thousand different compounds, and many more await identification. Glutathione (GSH) and its oxidized form (GSSG) are examples of ubiquitous γ-glutamyl peptides, which belong to a group of natural products containing more than a hundred different compounds.[12] In the majority of other peptides, amino acids are generally linked by peptide bonds between α-carboxy- and α-amino groups from Pro and the other 19 protein amino acids, with sequences in polypeptides and small proteins determined by the genetic code. However, several other types of bioactive peptides are produced *in vivo* by specific enzymes, and these peptides may contain other compounds than protein amino acids. Furthermore, owing to the processing conditions often used in feed and food production, and as a result of enzymatic *in vivo* processes, various changes of the amino acid side chains may occur in peptides and proteins. The result is many structurally different peptides and amino acids, and, among the naturally occurring non-protein amino acids, appreciable structural variation is found (Ref. 13 and refs cited therein).

There is thus a need for efficient methods of analyses for all of the above-mentioned compounds, where techniques for studies of proteins and other high-M_r compounds are considered in the Chapters 12–14. With peptides and small low-M_r proteins, no clear borderline exists, and traditional methods of protein analysis (Chapter 12) are often insufficient, whereas HPCE techniques (Chapter 10) seems to give efficient tools (*vide infra*) in this important, but otherwise often grey area with respect to analytical techniques.

Neutral Amino Acids

Group separations (page 161) will give the possibilities for isolation of neutral and acidic ampholytes including amino acids in the B-eluates (Figure 7.4) when crude extracts containing different low-M_r compounds are prepared by appropriate techniques (pages 69–74). Basic amino acids, peptides and hetero-aromatics with positive net charges at pH 3–5 will be found in the A-eluates. This separation in A- and B-eluates can be tested by methods of analyses, as shown in Figures 11.8, 11.12, and Figure 9.4. Analytical determination of aromatic amino acids can also be performed by HPLC (Figure 11.5) or MECC, as shown in Figures 11.6, 11.7, and 11.12. Qualitative amino acid analyses can be performed by PC (Figure 6.11 and Box 12.7). For quantitative analyses of protein amino acids, the methods of choice are often use of commercially available amino acid analyser instruments (Section 7.7 and page 332) and HPLC methods based on amino acid derivatization with, for example, *o*-phthalaldehyde (Box 12.8), dansyl chloride (Box 12.9), or other appropriate reagents.[9,13] For various reasons MECC is, however, an attractive alternative to HPLC (pages 235 and 284), and as a great number of non-protein amino acids (Figure 6.11; Ref. 13) and peptides are unstable under most of the derivatization procedures, this gives a need for direct analyses of underivatized compounds (Figures 11.6, 11.7, 11.8), *vide infra*.

Acidic Amino Acids

Acidic amino acids and peptides are initially isolated together with neutral amino acids in the B-eluates (*vide supra*) and analysis of this diverse mixture of compounds is actually possible in only one step using the MECC technique, as exemplified for a mixture of aromatic amino acids in Figure 11.12. The results from analysis directly on a crude extract of *Reseda odorato* seeds are also

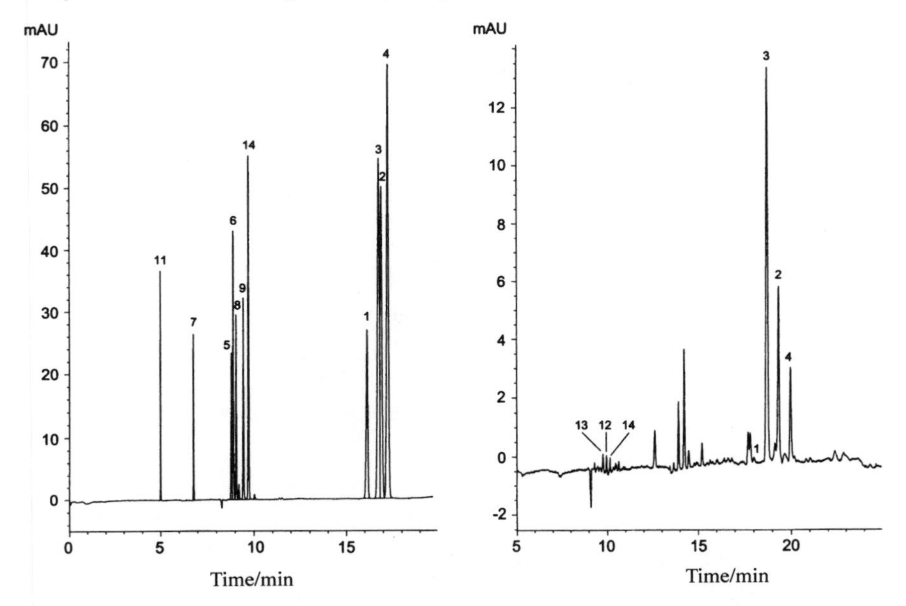

Figure 11.12 *MECC separation of acidic and neutral aromatic amino acids. Peaks 1–4 and 5,6,8,9 + 12–14 represent the acidic and neutral amino acids, respectively, whereas 7 is a basic amino acid, and the internal standard, trigonellinamide, is denoted 11. Peaks: 1 = 3-(3-carboxyphenyl)alanine, 2 = 3-carboxyphenylglycine, 3 =* m-*carboxytyrosine, 4 = 3-carboxy-4-hydroxyphenylglycine, 5 =* p-*aminophenylalanine, 6 =* m-*aminophenylalanine, 7 =* p-*aminomethylphenylalanine, 8 = 3-hydroxymethylphenylalanine, 9 =* m-*cyanophenylalanine, 12 = tyrosine, 13 = phenylalanine, and 14 = tryptophan. The electropherogram on the left is an artificial mixture of reference compounds. The right hand electropherogram is the neutral and acidic aromatic amino acids in a crude extract of* Reseda odorata. *Buffer: 75 mM Na_2HPO_4, 250 mM taurine, 75 mM sodium cholate, 4% propan-1-ol, pH unadjusted; temperature: 30 °C; voltage: 18 kV; injection: hydrodynamic (vacuum) (+ to −); detection: UV, 214 nm; capillary: 50 μm i.d., 64.5 cm total length, 56 cm effective length*

presented and show the acidic aromatic amino acids well separated from the neutral amino acids with the intact glucosinolates appearing in the area 12–15 min (Figure 11.12). In some cases, separations of the neutral and acidic compounds are, however, required prior to quantitative HPLC or HPCE in order to have reliable methods of analyses for complicated mixtures. This separation is easily obtained using the C-column technique (Figure 7.4).

Direct HPCE on crude extracts can also be possible for aliphatic compounds as shown for *Lathyrus sativus* amino acids (Figure 11.13), whenever the extracts or the samples have sufficiently high concentration of the amino acids of interest. The amino acids α-ODAP (α-oxalylamino-β-aminopropionic acid) and β-ODAP (β-oxalylamino-α-aminopropionic acid) have been devoted special attention owing to the effects they may have in relation to the disease lathyrism.

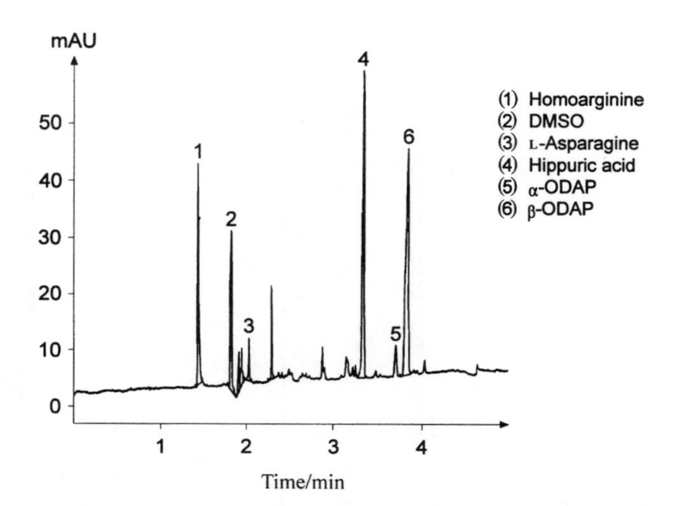

Figure 11.13 *Electropherogram of a crude ethanol extract of* L. sativus *obtained by FZCE. Buffer: 20 mM Na_2HPO_4, pH 7.8; temperature: 40 °C; voltage: 25 kV; injection: hydrodynamic (− to +); detection: UV, 195 nm; capillary: 50 μm i.d., 48.5 cm total length, 40 cm effective length*
(Reprinted with permission from *J. Agric. Food Chem.*, 1995, **43**, 942. Copyright (1995) American Chemical Society)

Peptides

Introduction of amide/peptide groups and additional chromophore groups in amino acid side chains, especially aromatic and heteroaromatic groups, increase the sensitivity of direct UV-detection at wavelengths around 200 nm for peptides (Tables 5.4 and 5.5). Thereby, it is possible to avoid derivatizations, which is otherwise often needed in connection with HPLC and HPCE methods of amino acid analyses (*vide supra*). Methods of analyses without derivatization procedures are also required for reliable determination of the great number of various labile non-protein amino acids, *e.g.* 3,4-substituted amino acids (Figure 6.11), γ-glutamyl derivatives, thiol-, *o*-dihydroxy-, and isoxazoline-derivatives (Figures 5.14 and 11.6; Refs. 9, 13). Figure 11.13 shows the possibilities of FZCE separation for the oxalyl amides of 2,3-diaminopropionic acid and Figure 11.14 shows the ability of peptide separations in MECC based on the cholate system. Simple peptides with a positive net charge will in this system appear in front of the solvent peak. Peptides with a zero net charge have higher migration times (*MT*) than the solvent peak and with *MT* mainly determined by the interaction of the analytes with the micelles. The negatively charged acidic, γ-glutamyl peptides (GSH and GSSG) have relatively high *MT*.

MECC with SDS as surfactant is also valuable for separation of simple peptides (Figure 11.15), but relative *MT* values obtained with this micellar system do not follow the same trend as seen for the separations in Figure 11.14.

Increasing the number of amino acid residues in the peptides to ten or more gives an appreciable increase in *MT* compared with *MT* for simple peptides with

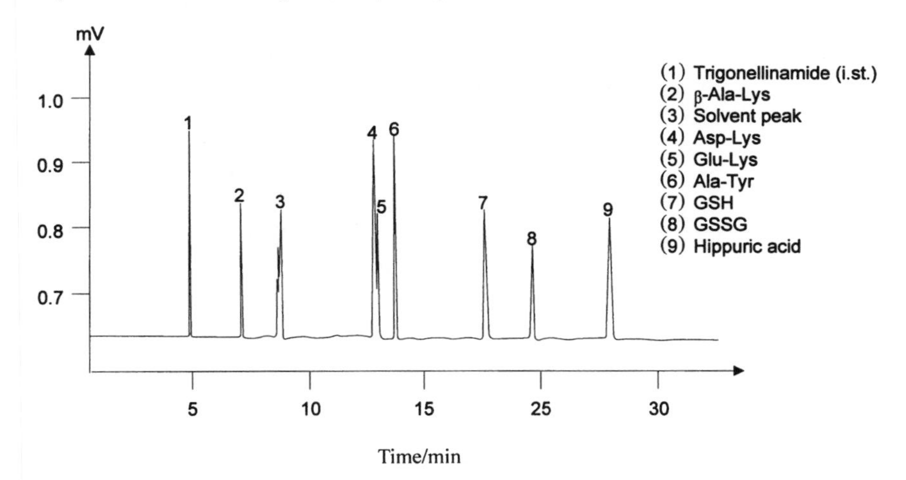

Figure 11.14 *Electropherogram of selected dipeptides, an acidic tripeptide γ-glutamyl-cysteinylglycine (GSH) and the corresponding hexapeptide (GSSG), separated by MECC (NaCh). Trigonellinamide as well as hippuric acid were used as internal standards. Buffer: 75 mM Na_2HPO_4, 50 mM taurine, 35 mM sodium cholate, 2% propan-2-ol, pH 8.0–8.1; temperature: 30 °C; voltage, 16 kV; injection: hydrodynamic (vacuum) (+ to −); detection: UV, 200 nm; capillary: 50 μm i.d., 61.4 cm total length, 53 cm effective length*

up to six amino acid residues (GSSG), when the SDS micellar system is used (Figure 11.16). This indicates that peptides of this size have three-dimensional structures, and that the ability of SDS binding, as is well known for proteins, results in highly negatively charged molecules with high *MT*.

In MECC with sodium cholate as surfactant, the decapeptides behave like the simple peptides, with well-separated angiotensins, and bradykinin with a positive net charge having mobility slightly smaller than the solvent peak (Figure 11.17).

Peptides with higher M_r than the decapeptides and proteins of relatively low-M_r, *e.g.* proteinase inhibitors, are also well suited for analytical studies in HPCE systems. The MECC system (Figures 11.14 and 11.17) has thus been found to be well suited for quantitative methods of peptide/protein analyses and for binding studies.[14] Figure 11.18 shows the results of MECC (NaCh) analyses of purified rapeseed protein-type proteinase inhibitors (RPPI) and Kunitz soybean trypsin inhibitor (KSTI). RPPI belong to the napin-type proteins and RPPI has thus a high pHi and M_r of *ca.* 6 kD. KSTI is a mixture of isoinhibitors with a M_r of *ca.* 21 kD and pHi values in the weakly acidic area, which are much lower pHi than found for RPPI. Peptides or proteins of this size have the possibilities of several types of interactions with the micelles. Therefore, their *MT* will not only follow the same simple rules concerning net charges of the molecule as found for the small peptides (Figure 11.17) even though the positively charged RPPI has a lower *MT* than the KSTI isoinhibitors.

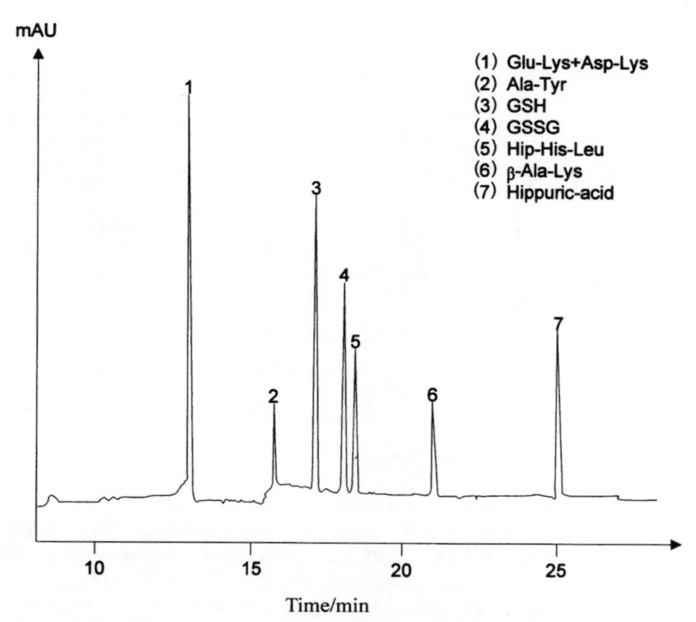

Figure 11.15 *Electropherogram of peptides as used in Figure 11.14, separated by MECC (SDS). Buffer: 100 mM boric acid, 150 mM SDS, pH 8.3; temperature: 27 °C; voltage: 15 kV; injection: hydrodynamic (vacuum) (+ to −); detection: UV, 216 nm; capillary: 50 μm i.d., 72 cm total length, 52 cm effective length*

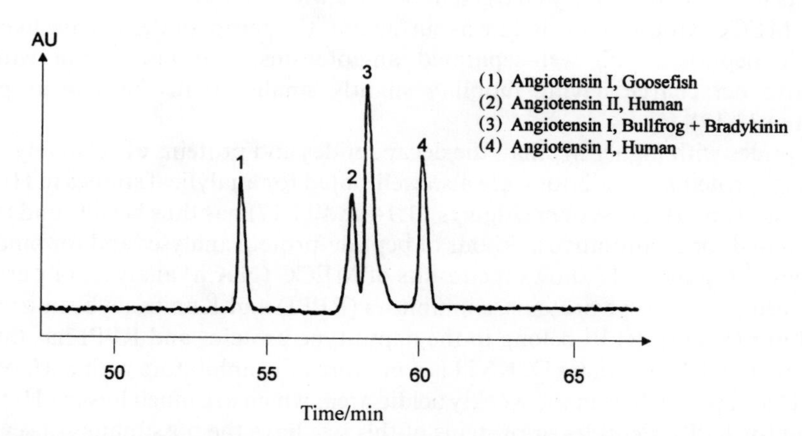

Figure 11.16 *MECC (SDS) electropherogram of decapeptides, different angiotensins and bradykinin, as used in Figure 11.17. Buffer: 100 mM boric acid, 150 mM SDS, pH 8.3; temperature: 27 °C; voltage: 15 kV; injection: hydrodynamic (vacuum) (+ to −); detection: UV, 216 nm; capillary: 50 μm i.d., 72 cm total length, 52 cm effective length*

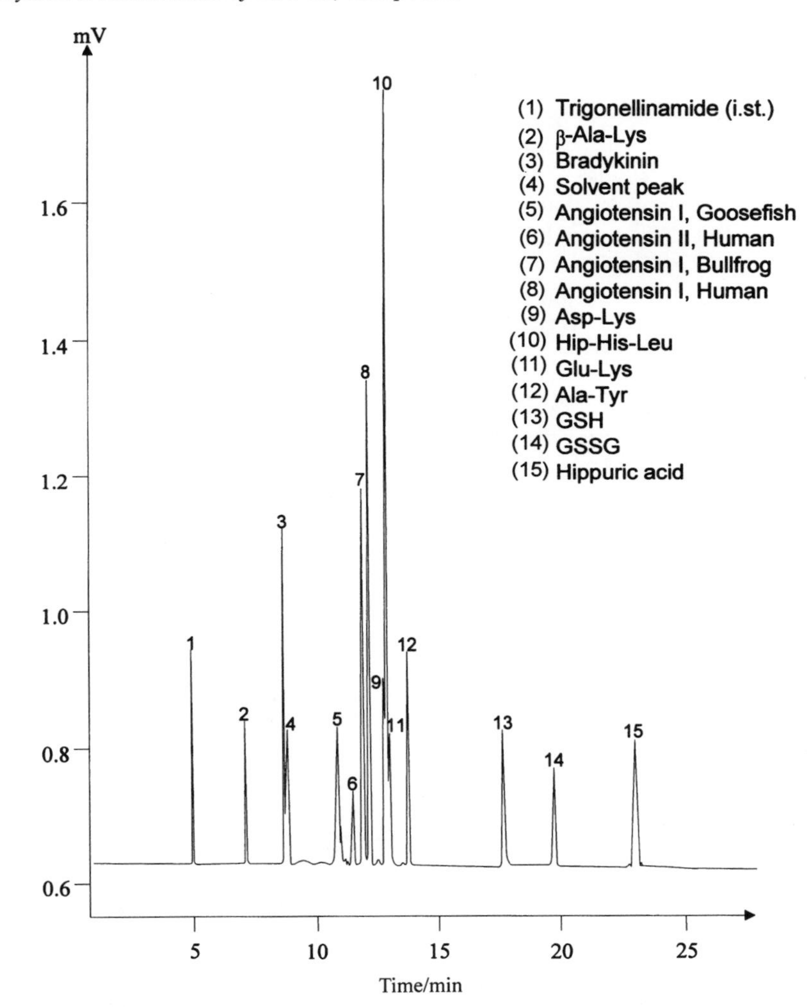

Figure 11.17 *Electropherogram of peptides separated by MECC (NaCh). Buffer: 100 mM Na_2HPO_4, 50 mM taurine, 35 mM sodium cholate, 2% propan-1-ol, pH 7.3; temperature: 30 °C; voltage, 20 kV; injection: hydrodynamic (vacuum) (+ to −); detection: UV, 214 nm; capillary: 50 μm i.d., 100 cm total length, 76 cm effective length*

Heteroaromatics

Heteroaromatic compounds generally have efficient chromophore systems (pages 108–111) which result in possibilities of sensitive UV detection. Owing to various protolytically active groups in the heteroaromatic compounds they can occur in different eluates from the group separation systems (Figure 7.4). However, it is often possible to use crude extracts for direct HPLC or HPCE methods of analyses owing to the efficient chromophore system in hetero-

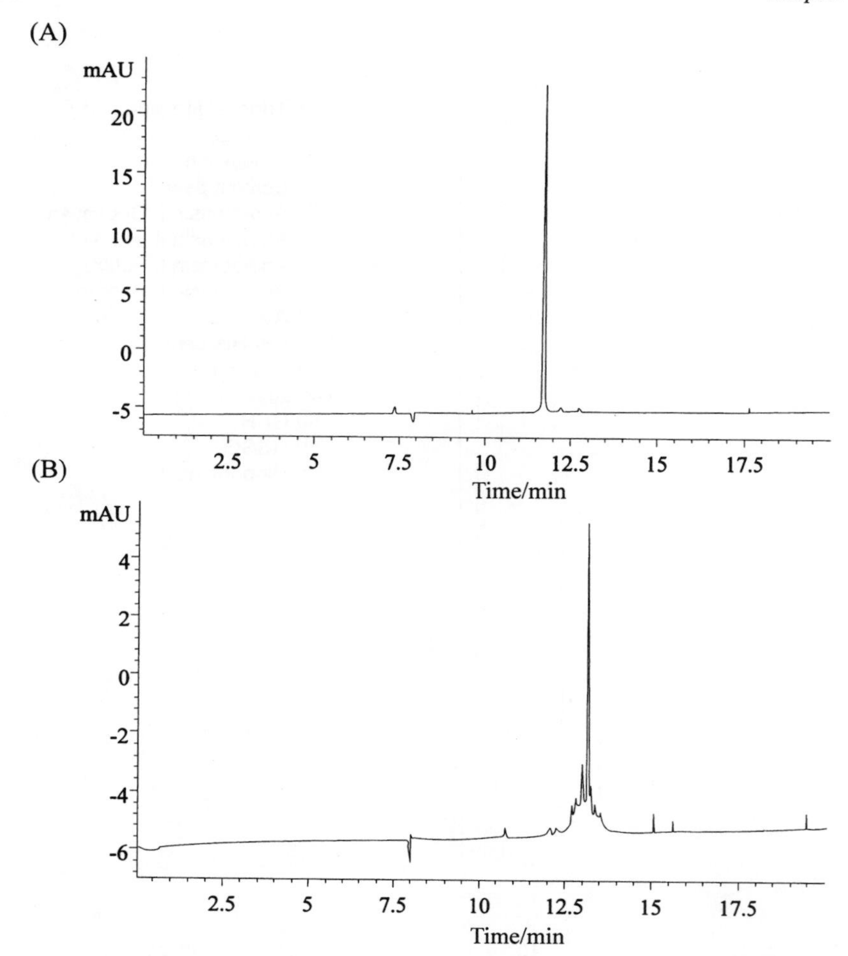

Figure 11.18 *Electropherograms of protein-type proteinase inhibitors [RPPI* (A) *and KSTI* (B)*] separated by MECC (NaCh). Buffer: 100 mM Na_2HPO_4, 50 mM taurine, 35 mM sodium cholate, 2% propan-2-ol, pH 7.3; temperature: 30 °C; voltage: 20 kV; injection: hydrodynamic (vacuum) (+ to −); detection: UV, 200 nm; capillary: 50 µm i.d., 61.4 cm total length, 53 cm effective length*

aromatics. Examples of LC methods of analyses for these compounds are shown in the Figures 11.5, 11.6, and 11.7.

5 Carboxylates, Phosphates, Sulfonates, and Sulfates

The low-M_r compounds considered in this section are those, which according to their protolytically active group will be bound to C-columns in the group separation system (Figure 7.4). Selective elutions of the anions are then performed in accordance with pK'_a-values for the compounds. When the anions are isolated as C-eluates, various LC methods of analyses can be used, *e.g.* HVE for a qualitative test (Figure 9.4) and, especially, HPLC and HPCE techniques for quantitative methods of analysis.

Aliphatic Carboxylates, Saponins

Isolation of carboxylates from crude extracts can be achieved by use of the C-column technique (Figure 7.4) and elution with acetic acid for isolation of dicarboxylates. These compounds can then be separated and determined quantitatively by MECC using cetyltrimethylammonium chloride (CTAC) as surfactant (Figure 11.19). Use of CTAC (Cl^- counterion) instead of CTAB (Br^- counterion) gives possibilities for UV detection at 205 mm as required for those aliphatic carboxylates, which only have weak chromophores (Table 5.4).

Aliphatic carboxylates and other anions with weak chromophores can also be analysed by FZCE using indirect UV detection (Figure 10.12). Saponins such as soyasaponin I which contain a carboxyl group (uronic acid) are possible

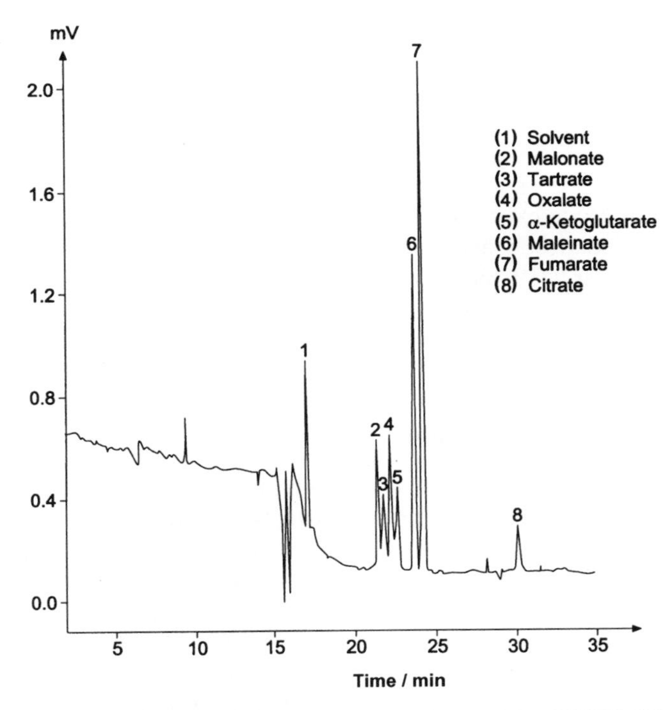

Figure 11.19 *Electropherogram of dicarboxylates separated in MECC (CTAC) using a Z-cell. Buffer: 50 mM Na_2HPO_4, 50 mM CTAC, 10% propan-1-ol, pH 7.0; temperature: 35 °C; voltage: 20 kV; injection: hydrodynamic (vacuum) (− to +); detection: UV, 205 nm; capillary: 75 μm i.d. (path length for Z-cell of 3 mm), 100 cm total length, 78 cm effective length*

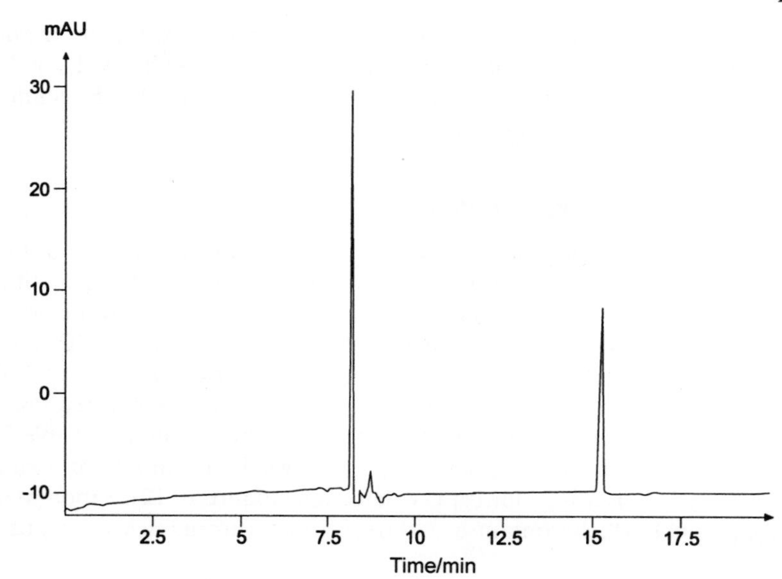

Figure 11.20 *Electropherogram of soyasaponin I using MECC (NaCh). Buffer: 75 mM Na_2HPO_4, 50 mM taurine, 35 mM sodium cholate, 2% propan-2-ol, pH 8.0–8.1; temperature: 30 °C; voltage: 16 kV; injection: hydrodynamic (vacuum) (+ to −); detection: UV, 200 nm; capillary: 50 μm i.d., 61.4 cm total length, 53 cm effective length*

to isolate from crude extracts by use of Dowex 1 × 8, 200–400 mesh (AcO^-) as C-column and acetic acid + methanol as eluent. The saponin dissolved in methanol–water (1:9) containing 5 mM sodium cholate (1.5 mg saponin per ml) can then be analysed in MECC with cholate as surfactant (Figure 11.20).

Aromatic Carboxylates, Phenolics

Phenolics containing a carboxylic acid group or aromatic carboxylates isolated as eluates from C-columns (Figure 7.4) can be separated and quantitatively determined by MECC as shown in Figure 10.19. Cinnamic acid and benzoic acid derivatives occur, however, most often bound to other compounds, *e.g.* dietary fibres (Chapter 14), and they need thus to be released from such compounds before they can be analysed by MECC (Figure 14.10). In other plant parts, such as cruciferous seedlings, which are often used as food, appreciable concentrations of cinnamoyl derivatives bound as malate esters may occur. HPLC analysis of malate esters accumulated in seedlings of *Raphanus sativus* has been described in Ref. 15. However, these compounds can now with advantages be determined by MECC at conditions that also allow efficient separation between the groups of mono- and dicarboxylates (Figure

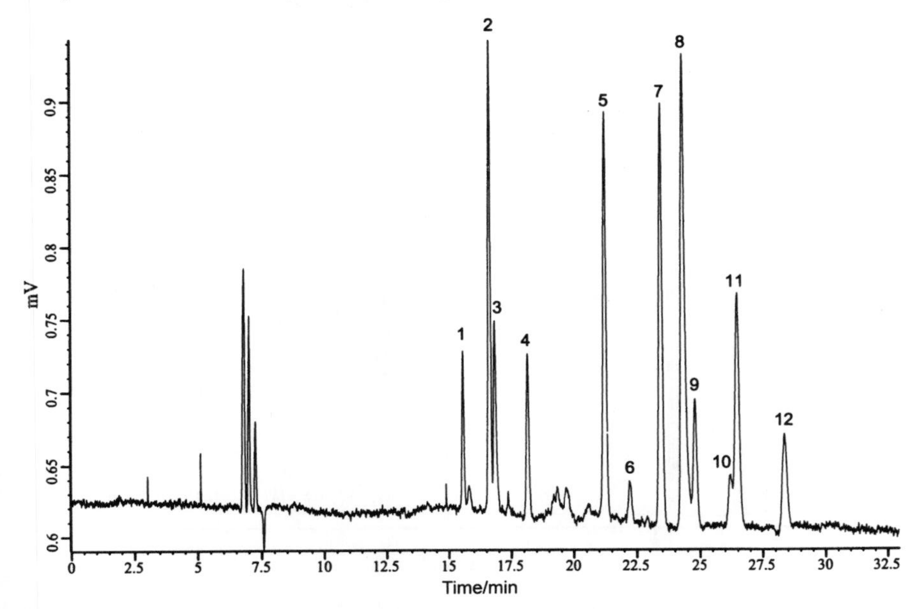

Figure 11.21 *Analyses of carboxylates containing aromatic groups by MECC (NaCh). Peaks: 1 = sinapic acid, 2 = ferulic acid, 3 = caffeic acid, 4 =* p-*coumaric acid, 5 =* trans-*2-*O-*sinapoyl-*L-*malate, 6 =* cis-*2-*O-*sinapoyl-*L-*malate, 7 =* trans-*2-*O-*feruloyl-*L-*malate, 8 =* trans-*2-*O-*cafferoyl-*L-*malate, 9 =* cis-*2-*O-*feruloyl-*L-*malate, 10 =* cis-*2-*O-*caffeoyl-*L-*malate, 11 =* trans-*2-*O-*(*p-*coumaroyl)-*L-*malate, 12 =* cis-*2-*O-*(*p-*coumaroyl)-*L-*malate. Buffer: 15 mM Na_2HPO_4, 400 mM taurine, 140 mM sodium cholate, 2% propan-1-ol, unadjusted pH; temperature: 30 °C; voltage, 20 kV; injection: hydrodynamic (vacuum) (+ to −); detection: UV, 325 nm; capillary: 50 µm i.d., 76 cm total length, 53 cm effective length*

11.21). Other groups of aromatic carboxylates, *e.g.* the compounds represented by various carboxylates produced from aromatic non-protein amino acids, can be determined as shown in Figure 11.22. The aromatic structure present in the compounds analysed facilitates UV detection compared to analysis of aliphatic dicarboxylates (Figure 11.19).

The numbered peaks in Figure 11.22 represents aromatic compounds having two carboxylate groups and a negative net charge −2, whereas the peak group denoted C contain the corresponding aromatic non-protein amino acids with only one negative charge (Nos. 1–4; Figure 11.12). The neutral non-protein amino acids (Figure 11.12) are found in peak group B whereas peak A represents the non-protein amino acids (No. 7, Figure 11.12) carrying one positive charge. Figure 11.22 illustrates well how the MECC technique allows analysis of a wide spectrum of compounds in a one-step procedure. Changed MECC conditions may provide better separation of peak groups B and C [neutral and acidic amino acids (−1)] as previously discussed (Figure 11.12).

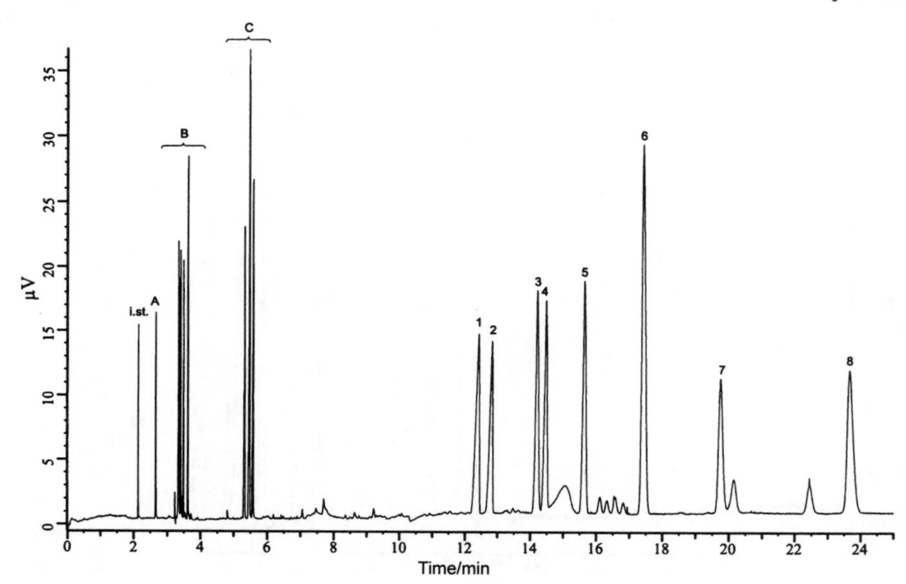

Figure 11.22 *Aromatic carboxylic acids analysed by MECC (NaCh). Peaks: 1 = 3-carboxyphenyllactic acid, 2 = 3-carboxycinnamic acid, 3 =* m-*carboxy-4-hydroxycinnamic acid, 4 = 3-carboxymandelic acid, 5 = 3-carboxyphenyl-acetic acid, 6 =* m-*carboxy-4-hydroxy-φ-acetic acid, 7 = 3-carboxy-glyoxylic acid, 8 =* m-*carboxy-4-hydroxy-φ-glyoxylic acid. The internal standard is trigonellinamide. Buffer: 25 mM Na_2HPO_4, 250 mM taurine, 75 mM sodium cholate, pH unadjusted; temperature: 30 °C; voltage: 30 kV; injection: hydrodynamic (vacuum) (+ to −); detection: UV, 214 nm; capillary: 50 μm i.d., 64.5 cm total length, 56 cm effective length*

Phosphates, Nucleotides, Phospholipids

Phosphates, mono- and diesters of orthophosphate are acidic compounds which can be isolated as eluates from the C-columns (Figure 7.4), if positively charged groups in the molecules do not give the compounds zero or positive net charges. Phosphate esters are ubiquitous, occurring in various types of low-M_r compounds as simple phosphate esters, as inositol phosphates (phytate), as nucleotides, and in various types of water-soluble vitamins, cofactors, and phospholipids (*vide infra*). This gives a need for selective methods of analyses that can give information on the individual compounds, but initially it may be of value with a total determination of phosphorus (page 280), or phosphate, which can be determined by HPCE (Figure 10.12). Phospholipids can be separated by MECC (Ref. 16; Figure 10.20), which will also be the method of choice for separating individual nucleotides,[17,18] although HPLC can also be used for some low-M_r phosphate esters (Figure 11.23).

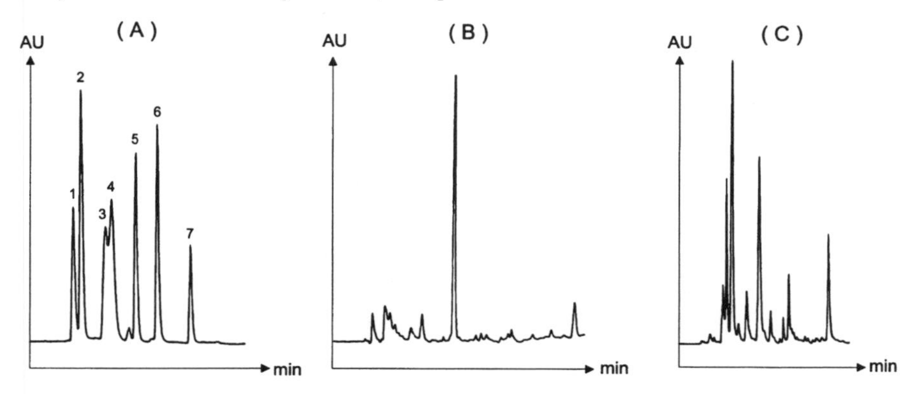

Figure 11.23 *HPLC of low-*M_r *phosphate esters:* (A) *a reference mixture of 1 = UDP-glucose, 2 = CMP, 3 = ADP-glucose, 4 = UMP, 5 = AMP, 6 = GMP, 7 = TMP.* (B) *and* (C) *are eluates from C-columns of extracts from pea* (Pisum sativum) *seed* (B) *and pea seed germinated for 9 days* (C). *Support: Nucleosil 5 C-18, 250 × 4.6 mm. Mobile phase, a linear gradient of solvent A-solvent B (1:99) to (99:1) for 10 min maintaining this final conditions for additional 5 min, flow rate 1 ml* min^{-1}. *Solvent A: 12.5 mM phosphate buffer (pH 2.0) modified with 25% MeOH. Solvent B: 10 mM phosphate buffer (pH 2.0). Recorder speed: 5 mm* min^{-1}. *Detection: UV, 280 nm*

Sulfonates, Sulfates, Glucosinolates

Sulfur is an essential part of all living cells. Not only as part of the protein amino acids Cys and Met, but the S-containing non-protein amino acids as garlic amino acids, as well as sulfate esters of various natural products as glucosinolates[19] are also important biomolecules affecting the quality of food and feed. Intact glucosinolates can be isolated and analysed as intact glucosinolates by MECC (Ref. 20; Figures 10.18, 10.34, and 10.35). Furthermore, on-column desulfatation followed by HPLC or MECC of desulfoglucosinolates is a valuable alternative technique as shown in Figure 11.24.

Quantitative determinations of glucosinolates can thus be performed with techniques based on HPLC as well as MECC and with the use of either intact glucosinolates or desulfoglucosinolates. In both cases, use of internal standards and relative response factors should be considered, as discussed in the above-mentioned references. The method of choice depends on the type of glucosinolates and the matrix these compounds have to be isolated from. However, it is possible to obtain acceptable results with both types of techniques when the compounds of interest are well separated and when they do not create problems in the desulfatation (Figure 11.25).

Other types of sulfate esters, *e.g.* glycosaminoglycans, can also be efficiently separated by MECC (Figure 11.26).

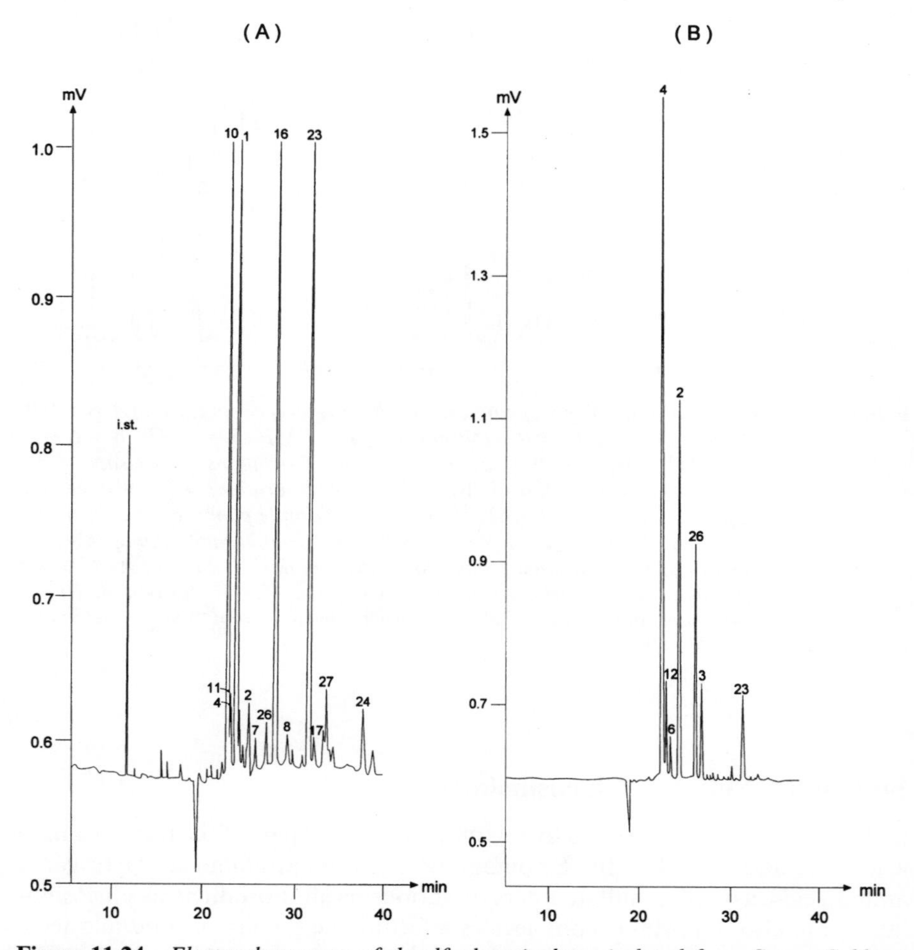

Figure 11.24 *Electropherogram of desulfoglucosinolates isolated from Savoy Cabbage* (A) *and from seed of oilseed rape* (B). *Peaks: 1 = sinigrin, 7 = glucoiberverin, 8 = glucoerucin, 10 = glucoiberin, 16 = glucotropaeolin. Other peak numbers as used for intact glucosinolates (Figure 10.18). i.s. = trigonellinamide. Buffer: 200 mM boric acid, 250 mM sodium cholate (NaCh), pH 8.5; temperature: 60 °C; voltage: 15 kV; injection: hydrodynamic (vacuum) (+ to −); detection: UV, 230 nm; capillary: 50 µm i.d., 100 cm total length, 76 cm effective length*

(Reprinted from C. Bjergegaard, S. Michaelsen, P. Møller, and H. Sørensen, *J. Chromatogr. A*, **717**, 325 (1995), Elsevier Science, Amsterdam)

6 Low-M_r Compounds without Protolytically Active Groups. Carbohydrates, Glycosides, Esters

Low-M_r compounds without protolytically active groups will appear in the effluent from the series of columns used in the group separation techniques

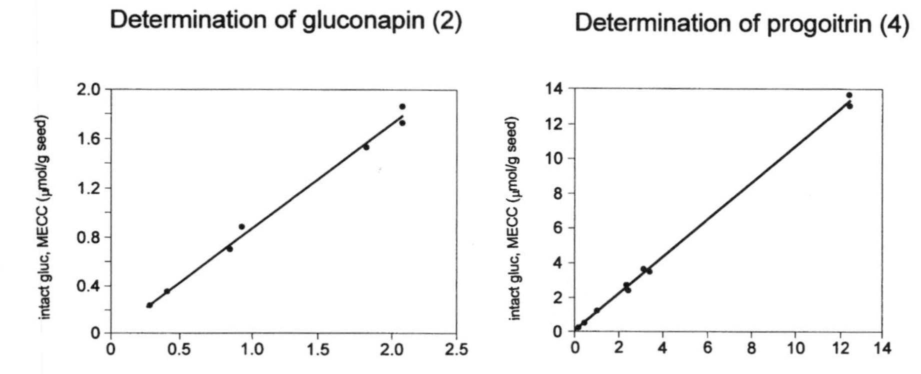

Figure 11.25 *Comparison of quantitative glucosinolate determinations based on HPLC and MECC as discussed in the text. Numbers as used for intact glucosinolates (Figure 10.18)*

(Figure 7.4). Among these are various carbohydrates, glycosides, esters, and natural products with other functional groups; all of these compounds may be of interest in relation to feed and food quality. Appreciable interest has therefore been devoted to methods of analyses which allow relatively simple, fast, cheap, and reliable determination of the individual compounds in each of the groups. Initial evaluations based on qualitative methods such as PC and TLC (Sections 6.5 and 6.1) are recommendable for the selection of appropriate methods of quantitative analyses. HPCE and, especially, MECC have been found promising for such types of analytical problems and a great number of applications have been described (Chapter 10;[5,21–23]). In addition to what is considered in this book, much more needs to be done and will without doubt appear in the future as new HPCE techniques.

Carbohydrates, Glycosides

In addition to monomer units in polysaccharides as starch and dietary fibres (Chapter 14), carbohydrates are as well important parts of glycoproteins, glycolipids, and various low-M_r glycosides. Depending on the functional groups, these glycosides are most often included in groups defined by the aglucone parts, *e.g.* as shown in Figures 5.11, 11.20, and 11.27, and discussed in subsections on pages 108, 303, 308, and 394. Carbohydrates without functional groups containing chromophore systems represent a particular challenge to HPCE techniques, as discussed in the subsection on page 406. Results obtainable for some monosaccharides and oligosaccharides by use of FZCE of carbohydrate–borate complexes (Figures 10.9, 10.10, and 11.27) are comparable to corresponding results obtainable by HPLC (Figure 11.27).[24]

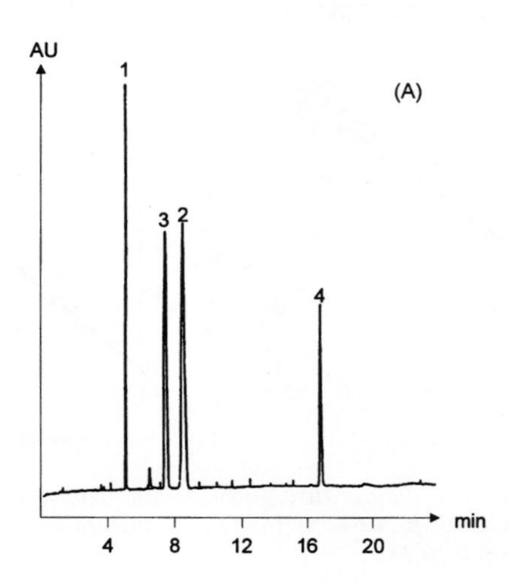
AU
1
(A)
3
2
4
min
4
8
12
16
20

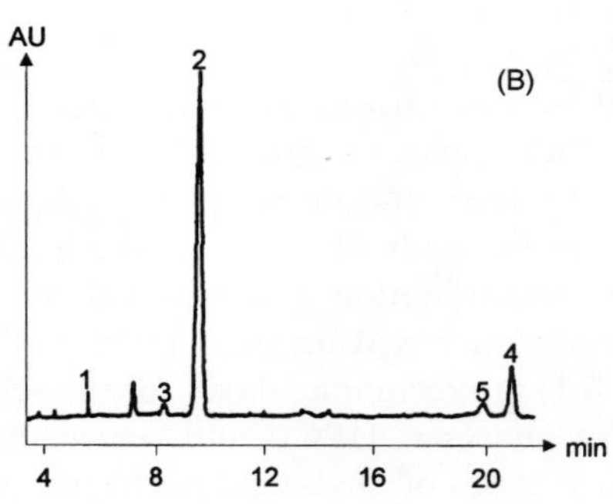
AU
2
(B)
1
3
5
4
min
4
8
12
16
20

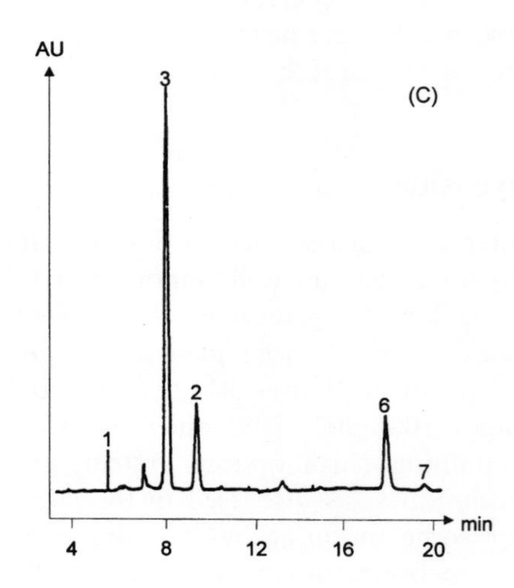
AU
3
(C)
2
6
1
7
min
4
8
12
16
20

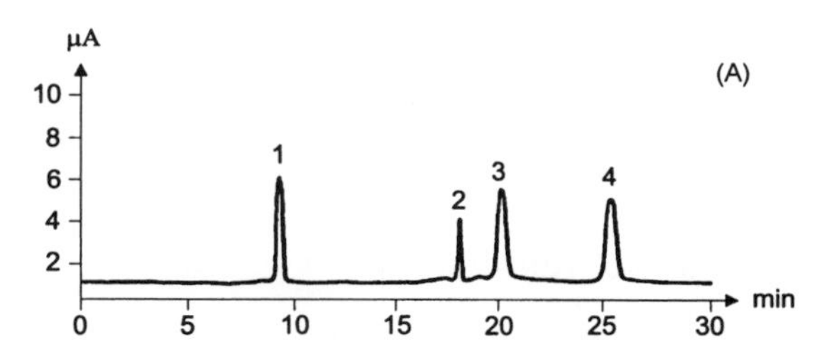

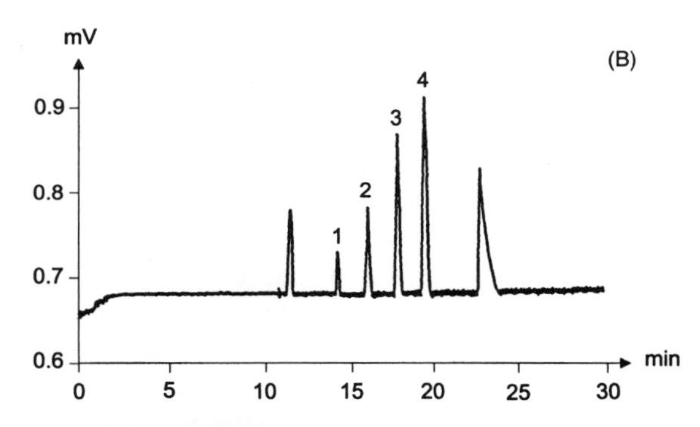

Figure 11.27 *HPLC* (A) *and FZCE* (B) *carbohydrate analyses of a RR RbRb pea seed sample. Peaks: 1 = sucrose, 2 = raffinose, 3 = stachyose, 4 = verbascose.* (A) *HPLC chromatogram (HPAC-PAD): CarboPak PA 100 (250 × 4.0 mm i.d.); eluent 145 mM NaOH.* (B) *Electropherogram (FZCE): buffer: 100 mM sodium tetraborate, pH 9.9; temperature: 50 °C; voltage: 10 kV; injection: hydrodynamic (vacuum) (− to +); detection: UV, 200 nm; capillary: 50 μm i.d., 50 cm total length, 43.2 cm effective length*
(Reprinted from J. Frias, K.R. Price, G.R. Fenwick, C. Hedley, H. Sørensen, and C. Vidal-Valverde, *J. Chromatogr. A*, **719**, 213 (1996), Elsevier Science, Amsterdam)

Figure 11.26 *Electropherogram of glycosaminoglycan (GAG) disaccharides:* (A) *GAG reference mixture;* (B) *GAGs from chondroitinase treated chondroitin sulfate B;* (C) *GAGs from chondroitinase treated chondroitin sulfate C. Peaks: 1 = Δ^4GlcUA-GalNAc, 2 = Δ^4GlcUA-4-O-sulfo-GalNAc, 3 = Δ^4GlcUA-6-O-sulfo-GalNAc, 4 = $\Delta^4$2-O-sulfo-GlcUA-sulfo-GalNAc, 5 = $\Delta^4$2-O-sulfo-GlcUA-GalNAc, 6 = Δ^4-2-O-sulfo-GlcUA-6-O-sulfo-GalNAc, 7 = Δ^4-GlcUA-4,6-bis-O-sulfo-GalNAc. Buffer: 18 mM disodium tetraborate, 30 mM Na_2HPO_4, 50 mM CTAB, pH 7.0; temperature: 30 °C; voltage: 20 kV; injection: hydrodynamic (vacuum) (− to +); detection: UV, 232 nm; capillary: 50 μm i.d., 75 cm total length, 52.2 cm effective length*
(Reprinted from S. Michaelsen, M.-B. Schrøder, and H. Sørensen, *J. Chromatogr. A*, **652**, 503 (1993), Elsevier Science, Amsterdam)

With respect to MECC methods of carbohydrate analyses, much more needs to be done, and the most promising approach is found within areas, where the methods are based on derivatives with an appropriate chromophore (pages 231 and 406).

Flavonoids, Carboxylic Acid Esters

Flavonoids including anthocyanins (pages 104 and 289 and Figures 5.11, 11.10, and 11.11) are widely distributed plant constituents, with more than 4000 known compounds included in this group, many with phenolics attached as ester groups on the carbohydrate part. Phenolic carboxylic acids, benzoic and cinnamic acid derivatives (Figure 5.10) occur in addition as esters of simple carbohydrates, as esters of choline (Figure 11.1), as esters of hydroxycarboxylic acids (Figure 11.22), and as esters of dietary fibres (Section 14.5). The phenolic carboxylic acids are, thus, rarely free (Figure 14.10), but released by hydrolysis of their derivatives. Flavonoids and other phenolics also call for attention as they are often considered as part of the group known as 'tannin', which includes lignin (page 397). Flavonoids and other phenolics are also of interest in relation to their contribution to taste and antioxidative properties. Flavonoids can be analysed by MECC based both on cation and anion surfactants (Figure 11.28).

7 Special Groups of Bioactive Compounds

In addition to the group of compounds considered in the preceding sections, other special groups of bioactive compounds also need attention in connection with methods of analyses required within the area of natural product chemistry, feed, and food analyses. It will, however, not be possible to consider the immense number of techniques described for all possible bioactive compounds, and the following sections are only a brief review of selected examples in this area. References to additional examples and practical applications for HPCE techniques can be found in the literature.[5,21–23,25,26]

Indolyls

The indolyl group from tryptophan is metabolically transformed into a great number of bioactive compounds such as alkaloids, biogenic amines (Section

Figure 11.28 (A) *Flavonoids analysed by MECC (CTAB). Buffer: 12 mM disodium tetraborate, 20 mM disodium hydrogenphosphate, 40 mM cetyltrimethylammonium bromide (CTAB), 4% propan-1-ol, pH 7.0; temperature: 50 °C; voltage: 26 kV; detection: UV, 350 nm.* (B) *Flavonoids analysed by MECC (NaCh). Buffer: 100 mM Na_2HPO_4, 500 mM taurine, 35 mM sodium cholate (NaCh), 6% propan-1-ol, unadjusted pH; temperature: 40 °C; voltage: 25 kV; detection: UV, 250 nm.* (A) *and* (B) *injection: hydrodynamic (vacuum) (CTAB: + to −; NaCh: − to +); capillary: 50 μm i.d., 72 cm total length, 50 cm effective length*
(Reprinted from C. Bjergegaard, S. Michaelsen, K. Mortensen, and H. Sørensen, *J. Chromatogr. A*, **652**, 477 (1993), Elsevier Science, Amsterdam)

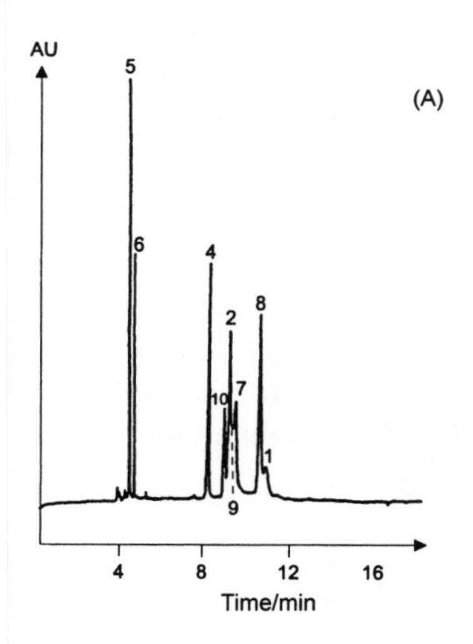

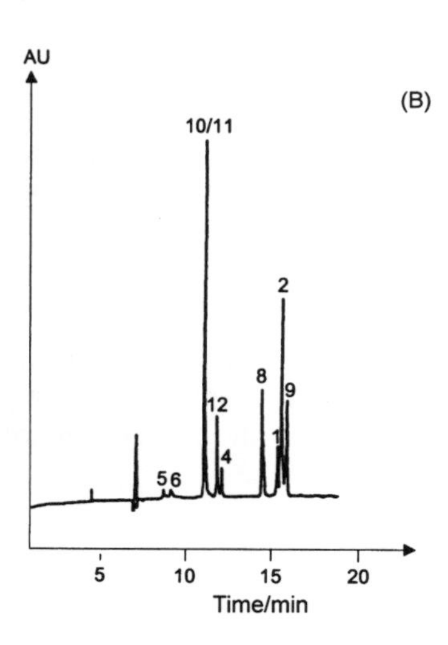

No.	Name	$R^{3'}$	R^3	R^7
1	Kaempferol	H	OH	OH
2	Kaempferol-3-glucoside	H	Glucoside	OH
3	Kaempferol-3,7-diglucoside	H	Glucoside	Glucoside
4	Rustoside	H	Xylopyranosyl-(1,2)-galacto-pyranosyl	OH
5	Kaempferol-3-sophoroside-7-glucoside	H	Sophoroside	Glucoside
6	Kaempferol-3-sinapoylsophoroside-7-glucoside	H	Sinapoyl-sophoroside	Glucoside
7	Kaempferol-3-(6″-carboxyglucoside)	H	6″-carboxy-glucoside	OH
8	Quercetin	OH	OH	OH
9	Quercetin-3-glucoside	OH	Glucoside	OH
10	Rutin	OH	Rutinoside	OH
11	Quercetin-3-(6″-carboxyglucoside)	OH	6″-carboxy-glucoside	OH
12	Isorhamnetin-3-(6″-carboxyglucoside)	OMe	6″-carboxy-glucoside	OH

Names, structures and numbering of flavonoids used.

11.3), and hormones (page 311). In food containing cruciferous plant products, various types of indolyl compounds are often produced from indolylglucosinolates (page 303). These compounds are considered to be involved in many both positive and negative physiological effects including smell and taste of food. Analyses for monomeric indolyls can be performed by MECC based on alkyltrimethylammonium surfactants (Figure 10.17) and for oligomeric indolyls by MECC based on sodium cholate surfactants.[27–29]

Vitamins

Vitamins are, structurally, a heterogeneous group of essential constituents of food or nutrients that are required in small amounts for normal growth, maintenance, and functioning of animal tissues. Many vitamins are precursors

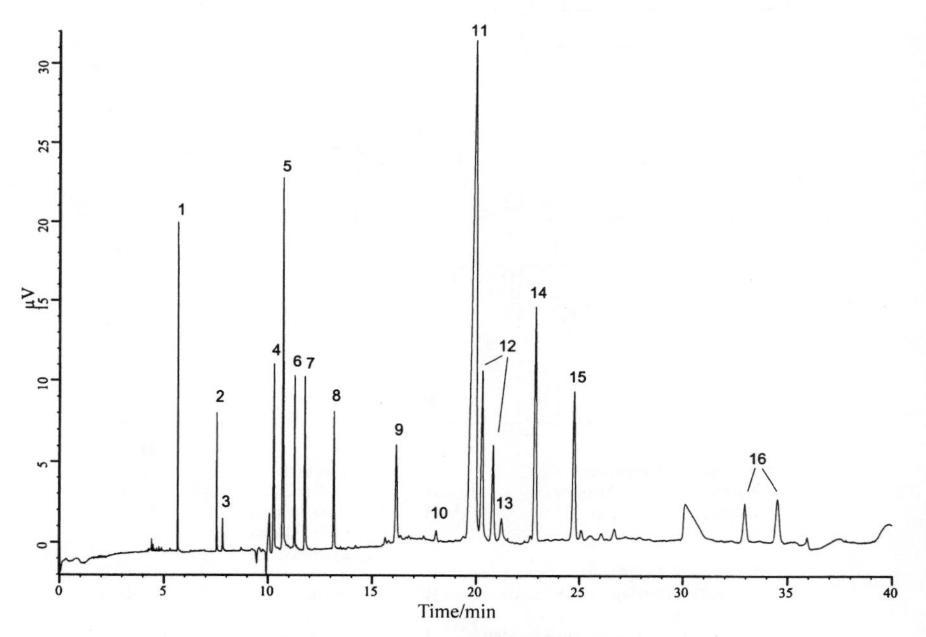

Figure 11.29 *Water-soluble vitamins analysed by MECC (NaCh). Peaks: 1 = trigonellinamide, 2 = thiamine, 3 = pyridoxamine, 4 = nicotinamide, 5 = pyridoxine, 6 = cyanocobalamin, 7 = riboflavin, 8 = NAD$^+$, 9 = pyridoxamine phosphate, 10 = biotine, 11 = riboflavin-5′-phosphate, 12 = riboflavin-5′-phosphate impurities, 13 = ascorbic acid, 14 = NADP$^+$, 15 = folic acid, 16 = riboflavin-5′-phosphate impurities. Buffer: 100 mM Na_2HPO_4, 75 mM sodium cholate (NaCh), 500 mM taurine, 2% propan-1-ol, unadjusted pH; temperature: 30 °C; voltage: 17 kV; detection: UV, 214 nm; injection: hydrodynamic (pressure) (+ to −); capillary: 50 µm i.d., 64.5 cm total length, 56 cm effective length*

(Reprinted from S. Buskov, P. Møller, H. Sørensen, J.C. Sørensen, and S. Sørensen, *J. Chromatogr. A*, **802**, 233 (1998), Elsevier Science, Amsterdam)

for enzyme cofactors and they are often relatively unstable. Therefore, vitamin losses occur easily during food preparation, storage, and processing, which gives rise to a need for efficient methods of analyses. This area of LC is still far from sufficiently covered and needs extensive development even though several HPLC and HPCE methods have been proposed. Reliable determination of vitamins, either by HPLC or HPCE, thus needs to be based on gentle extractions, group separation, and concentration procedures, and for these purposes vitamins need, initially, to be separated into their two main groups: water- and lipid-soluble vitamins. It is possible to extract and purify the latter by SFE (page 74 and Appendix), and as the majority of water-soluble vitamins are anionic compounds they can be separated by the group separation technique (Figure 7.4). For separation and quantification of the individual vitamins, FZCE and, especially, MECC seem to be the most promising techniques.[5,30–33] MECC based on the NaCh system thus give an efficient separation of water-soluble vitamins (Figure 11.29).

Hormones

Hormones are chemical messengers produced by ductless glands and transported to target cells or molecules, which they bind to in connection with regulation of metabolic processes. Several types of hormones occur in animals and they can, according to their structure, be divided into amines, amino acids, polypeptides, proteins, or steroids. The pineal gland is the site of melatonin (*N*-acetyl-5-methoxytryptamine) biosynthesis, and this hormone is important for diurnal rhythms as well as other functions. For analytical determinations, the MECC systems developed for monomeric indolyls (Figure 10.17) also allow the determination of melatonin. Steroidal hormones can be separated in systems corresponding to the MECC system used for saponins (Figure 11.20), and polypeptides or small proteins can be determined in the MECC systems described in the subsection on page 294. The principal hormones secreted by the thyroid gland are the iodine-containing amino acids L-thyroxine (T_4 = L-3,5,3′5′-tetraiodothyronine) and L-triiodothyronine (T_3 = L-3,5,3′-triiodothyronine). These hormones were efficiently separated in MECC based on NaCh as surfactant (Figure 11.30).

8 Conclusion

Liquid chromatographic methods of analyses, with high resolution systems such as HPCE and, especially, MECC, have in recent decades undergone great developments. These developments have resulted in new opportunities for relatively simple, fast, efficient, and cheap methods of analyses for the individual compounds in the various groups of low-M_r compounds that occur in feed, food, and as natural products. This trend seems set to continue, and it has obviously increasing importance in relation to the evaluation and control of the quality of feed and food, in studies of the bioavailability of nutrients, of

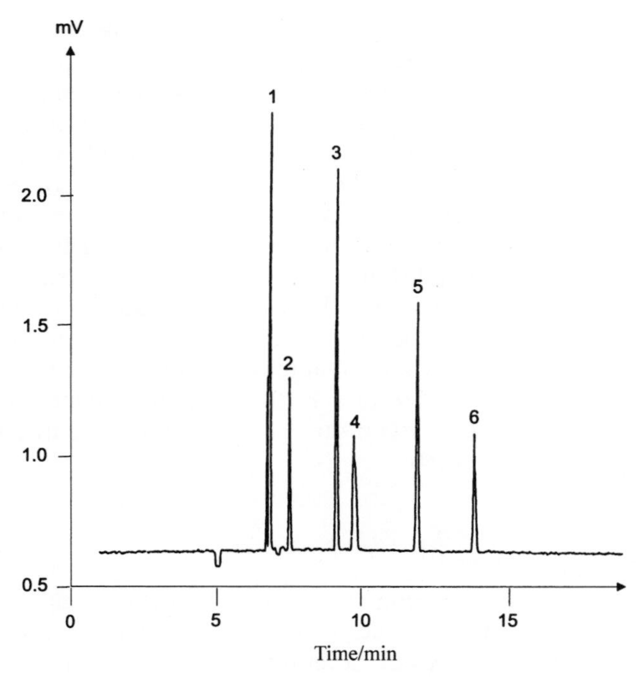

Figure 11.30 *Electropherogram of L-Trp (1), L-Tyr (2), 3-iodo-L-Tyr (3), 3,4-diiodo-L-Tyr (5), L-T_3 (4), and L-T_4 (6). Buffer: 100 mM Na_2HPO_4, 15 mM sodium cholate, 12% propan-1-ol, pH 7.0; temperature: 50 °C; voltage: 18 kV; injection: hydrodynamic (vacuum) (+ to −); detection: UV, 220 nm; capillary: 50 μm i.d., 76 cm total length, 53 cm effective length*

antinutrients, xenobiotica, and in studies of authenticity and adulteration of feed and food.

Success in this area of research and in analytical–biochemical work, as well as opportunities for success in use of the techniques in less advanced laboratories depends very much on the ability of analysts to utilize the required and appropriate group separation techniques prior to quantitative analytical determinations. Furthermore, some basic knowledge of chemotaxonomy is a great advantage. Chemotaxonomy will thus be of the upmost importance in relation to the design of an optimal strategy for the analytical procedures, and for evaluating which groups of compounds present may be relevant to consider. The majority of low-M_r natural products are of plant origin, and many of these compounds are specific for the plant or plant part producing them. For example, glucosinolates and aromatic choline esters occur only in plants of the order Capparales and in a few other plants, but they do not occur in cereals or legumes. Chemotaxonomy based on low-M_r compounds can also be an important tool in revealing authenticity and adulteration of food and feed. The trend toward more efficient methods of analyses for the determination of the great number of individual compounds in the various groups of natural products, food and feed additives is thus needed for many reasons. It is, therefore, our hope that the methods and techniques now described will be of value and will help to solve such analytical problems.

9 Selected General and Specific Literature

1. C. Bjergegaard and H. Sørensen, in 'Proceedings of 4th International Feed Production Conference', ed. G. Piva, Piacenza, Italy, 1996.
2. H. Sørensen and S. Sørensen, 'Proceedings of Bioavailability '97', Wageningen, Netherlands, 25–28 May 1997.
3. S. Michaelsen and H. Sørensen, *Pol. J. Food Nutr. Sci.*, 1994, **3**, 5.
4. P.F. Cancalon, *J. AOAC Int.*, 1995, **78**, 12.
5. J. Lindeberg, *Food Chem.*, 1995, **55A**, 73.
6. AOAC, Association of Official Analytical Chemists. Official Methods of Analyses (942.05), Arlington, VA, 1990, 70.
7. AOCS, American Oil Chemists' Society, Official Methods (Ca 12-55), Champaign, IL, 1992.
8. AOCS, American Oil Chemists' Society, Official Methods (Ca 126-92), Champaign, IL, 1994.
9. B.O. Eggum, N.E. Hansen, and H. Sørensen, in 'Absorption and Utilization of Amino Acids', ed. M. Friedman, CRC Press, Boca Raton, FL, 1989, vol. 3, ch. 5, p. 67.
10. H. Sørensen and P.M.B. Wonsbek, in 'Advances in the Production and Utilization of Cruciferous Crops', ed. H. Sørensen, Martinus Nijhoff Dr. Junk Publishers, Boston, 1985, ch. 11, p. 127.
11. B. Bjerg, O. Olsen, K.W. Rasmussen, and H. Sørensen, *J. Liq. Chromatogr.*, 1984, **7**, 691.
12. T. Kasai and P.O. Larsen, in 'Progress in the Chemistry of Organic Natural Products', ed. W. Herz, H. Grisebach, and G.W. Kirby, Springer Verlag, New York, 1980, vol. 39, p.173.
13. B.O. Eggum and H. Sørensen, in 'Absorption and Utilization of Amino Acids', ed. M. Friedman, CRC Press, Boca Raton, FL, 1989, vol. 3, ch. 17, p. 265.
14. H. Frøkiær, K. Mortensen, H. Sørensen, and S. Sørensen, *J. Liq. Chromatogr. Relat. Technol.*, 1996, **19**, 57.
15. J.K. Nielsen, O. Olsen, L.H. Pedersen, and H. Sørensen, *Phytochemistry*, 1984, **23**, 1741.
16. L. Ingvardsen, S. Michaelsen, and H. Sørensen, *J. Am. Oil. Chem. Soc.*, 1994, **71**, 183.
17. X. Huang, J.B. Shear, and R.N. Zare, *Anal. Chem.*, 1990, **62**, 2049.
18. A.-F. Lecoq, C. Leuratfi, E. Morafante, and S. Dibiese. *J. High Resolut. Chromatogr.*, 1991, **14**, 667.
19. H. Sørensen, in 'Canola and Rapeseed: Production, Chemistry, Nutrition, and Processing Technology', ed. F. Shahidi, Van Norstrand Reinhold, New York, 1990, ch. 9, p. 149.
20. S. Michaelsen, P. Møller, and H. Sørensen, *J. Chromatogr.*, 1992, **608**, 163.
21. S.F.Y. Li, 'Capillary Electrophoresis. Principles, Practice, and Applications', Journal of Chromatography Library, Elsevier, Amsterdam, 1992, vol. 52, 578 pp.
22. J. Lindeberg, *Food Chem.*, 1995, **55B**, 95.
23. K. Otsuka and S. Terrabe, in 'Capillary Electrophoresis Guidebook. Principles, Operation, and Applications', ed. K.D. Altria, Methods in Molecular Biology, Humana Press, Totowa, NJ, 1996, vol. 52, p. 125.
24. J. Frias, K.R. Price, G.R. Fenwick, C. Hedley, H. Sørensen, and C. Vidal-Valverde, *J. Chromatogr. A*, 1996, **719**, 213.
25. R. Kuhn and S. Hoffstetter-Kuhn, in 'Capillary Electrophoresis: Principles and Practice', Springer-Verlag, Berlin, 1993, 375 pp.
26. R. Weinberger, in 'Practical Capillary Electrophoresis', Academic Press, New York, 1993, 312 pp.

[27] C. Feldl, P. Møller, J. Otte, and H. Sørensen, *Anal. Biochem.*, 1994, **217**, 62.
[28] N. Agerbirk, C. Bjergegaard, C.E. Olsen, and H. Sørensen, *J. Chromatogr.*, 1996, **745**, 239.
[29] N. Agerbirk, C.E. Olsen, and H. Sørensen, *J. Agric. Food Chem.*, 1998, **46**, 1563.
[30] S. Fujiwara, S. Iwase, and S. Honda, *J. Chromatogr.*, 1988, **447**, 133.
[31] H. Nishi, N. Tsumagari, T. Kakimoto, and S. Terabe, *J. Chromatogr.*, 1989, **465**, 331.
[32] C.P. Ong, C.L. Ng, H.K. Lee, and S.F.Y. Li., *J. Chromatogr.*, 1991, **547**, 419.
[33] S. Boonkerd, M.R. Detaevernier, and Y. Michotis, *J. Chromatogr.*, 1994, **670**, 209.
[34] H. Sørensen and S. Sørensen, *European J. Clin. Nutr.*, 1998, in press.

CHAPTER 12

Protein Purification and Analysis

1 Introduction

Chromatography of proteins can be employed for various purposes, both of a qualitative and a quantitative nature. Thus, chromatography is applicable for analytical purposes with small sample volumes and for preparative purposes with large initial sample volumes, and where good recoveries from individual purification steps are needed. In most cases, food matrices of animal or plant origin contain only small amounts of specific proteins, often less than 0.1% of dry matter (DM). In general, it is therefore important to have an optimal strategy for the sequence of purification steps. In this connection, it is advisable to perform an appropriate extraction and initial group separation of the analytes and co-occurring compounds in the matrices to reduce interference from irrelevant components during chromatography and subsequent analyses.

Depending on the purpose of the study, the proteins can be divided into groups according to different criteria, *e.g.* their structure, function, solubility, or shape (pages 79–80). The divisions most often used in connection with protein purification are solubility and form (Figure 6.5, pages 126–130, and Section 12.2).

2 Proteins in Food Matrices

The first prerequisite for analysis of proteins is their solubilization from the matrix. Proteins are often associated with other biomolecules, for example in membranes (Figure 4.2), where they may be more or less integrated in the membrane structure. In particular, the proteins present in cellular and intracellular membranes interact with lipids, notably phospholipids. The interaction with the lipids is dependent on the length of the aliphatic chains and the hydrophobicity of the protein. The interaction with surfactants required for extraction and solubilization of such integral membrane proteins may modify the protein structure, and hence the solubilization of membrane-associated proteins may cause inactivation of the biological activity of the proteins. Peripheral proteins and storage proteins usually are more readily extractable. In the following, some general information on the composition of different food matrices is given.

Proteins of Plant Origin

Plants are one of the most important sources of proteins for human consumption. Edible plants have varying protein contents, but crops such as cereals, pulses, tubers, and bulbs may contain considerable amounts of proteins. In Table 12.1, a range of different plants are listed to illustrate the approximate levels of their protein, lipid, and carbohydrate contents.

Cereals, especially wheat, account for a large part of cultivated plants used as food and feed. The cereal proteins have been divided into groups according to their different solubilities (page 79), and this principle has also been applied to the grain legumes. In Table 12.2, the distributions in the different groups are listed together with specific names of the protein groups. For cereals, the albumins and globulins are primarily found in cytoplasm and other subcellular fractions, and these groups of proteins encompass a great part of the known enzymes found in cereals. The storage proteins in cereals are found in the prolamine and the glutelin fractions. Cereal proteins of particular interest include the wheat proteins (with regard to the baking qualities). For cereals in general and especially for barley used in breweries, different enzymes have for various reasons been investigated, *e.g.* amylase, lipase, phytase (in wheat), lipoxygenase, peroxidase, catalase, and glutathione dehydrogenase.

For grain legumes, three protein fractions are obtained: the albumin, globulin, and glutelin fractions (Table 12.2). For all grain legumes, the dominant globulin fraction can be divided into two main groups, the vicilins (*ca.* 7 S) and the legumins (*ca.* 11 S). The legumins in soybeans are called glycinin. In plant tissues some proteins act as storage proteins and they are often associated in oligomers as the case is for legumins and vicilins in grain legumes. Special attention is often paid to specific proteins in grain legumes represented by lipoxygenases, which influence the aroma, protein-type proteinase inhibitors, with trypsin and chymotrypsin inhibitor activity, and lectins, which are glycoproteins.

Vegetables contain in general 1–3% N-containing compounds, of which 35–80% is protein and the rest are amino acids, peptides, and other compounds. The proteins present in vegetables comprise a large number of enzymes, with representation of all the main groups of enzymes. In addition, protein-type enzyme inhibitors may be found in, for example, potato, cereals, crucifers, and legumes.

With respect to fruits, the composition varies to a large extent and is dependent on the species and ripeness. Most fruits have a DM content of 10–20% with only limited amounts of protein and lipid. Most of the protein fraction contains soluble proteins and enzymes, and, in general, the protein and enzyme composition to be analysed, *e.g.* by electrophoresis, is species specific. A special group is made up of shell fruits such as different types of nuts, which have water contents below 10%, N-containing compounds in the range of 20%, and lipids contributing approximately 50%.

Table 12.1 *Composition of selected cereals, leguminous seeds, cruciferous plants, and vegetables given as percent of DM content**†

	Latin name	*DM (%)§*	*Protein (N×6.25) (%DM)*	*Carbohydrate‡ (starch) (%DM)*	*Lipid (%DM)*	*Ash (%DM)*
Wheat	*Triticum sativum*	88	11–15	79.8 (68.2)	2.5	1.7
Rye	*Secale cereale*	88	11–15	80.0 (60.7)	2.0	2.2
Barley	*Hordeum vulgare*	88	10–14	81.3 (59.1)	2.4	2.6
Oats	*Avena sativa*	88	13–16	72.3 (46.1)	6.6	3.3
Maize	*Zea mays*	88	9–11	81.1 (71.5)	4.3	1.5
Rapeseed	*Brassica napus / B. campestris*	92	22–28	5–9 (1–3)	40–46	4–6
Pea	*Pisum sativum*	92	23–31	60–65 (20–50)	2–3	3.0
Chick pea	*Cicer arietinum*	92	20–25	66.3	5.0	3.0
Faba bean	*Vicia faba*	92	26–34	55–60 (40–50)	2–4	4
Soybean	*Glycine* max.	92	35–45	15–25 (0)	19.6	5.5
Peanut	*Arachis hypogeae*	92	22–28	24.6	47.9	3.0
Potato	*Solanum tuberosum*	22	7–11	85.1 (63.5)	0.9	5.0
Leek	*Allium porrum*	15	13–17	67.8	2.1	6.2
Curly kale	*Brassica oleracea* convar. *Acephala* var. *Sabellica*	17	31–39	43.4	5.2	8.7
Cabbage	*Brassica oleracea* convar *capitata* var. *Capitata* f. alba	8	15–19	60.5	2.6	9.2
Broccoli	*Brassica oleracea* convar. *Botrytis* var. *Italica*	11	30–36	40.4	–	10.1
Green bean	*Phaseolus vulgaris*	10	17–21	61.6	–	7.1
Green pea	*Pisum sativum* spp. *Sativum*	22	26–31	56.4	–	4.1
Tomato	*Lycopersicon lycopersicum*	7	15–19	64.6	3.1	7.7

Data modified after *H.-D. Belitz and W. Grosch, 'Lehrbuch der Lebensmittelchemie', Springer-Verlag, Berlin, 1992 and †C. Bjergegaard and H. Sørensen, in 'Proceedings of 4th International Feed Production Conference', Piacenza, Italy, 1996)

‡An appreciable part of the non-starch carbohydrates is dietary fibres (DF) and oligosaccharides (β-galactosides of sucrose). The actual figures for DF depend on the applied analytical procedure (Chapter 14). Total DF (%DM) for legume seeds are approximately 15–25; for rapeseed 18–25; for cereals 10–20

§Approx. mean values

Proteins of Animal Origin

Another quantitatively and qualitatively important proportion of food proteins are of animal origin. In Table 12.3, selected products are listed. In cow's milk, the dominant proteins belong to the caseins or the whey proteins (primarily lactalbumins and lactoglobulins). The hen egg proteins comprise the proteins found in egg white (ovalbumin, ovotransferrin, ovomucoid, and lysozyme as the main proteins) and the egg yolk, which contains lipoproteins [high and low

Table 12.2 *Protein distribution* given as percentage of the protein obtained by extraction according to page 79*

Crop	*Albumins*	*Globulins*	*Prolamines*	*Glutelins*
Wheat	14.7 (Leukosin)	7.0 (Edestin)	32.6 (Gliadin)	45.7 (Glutenin)
Rye	44.4	10.2	20.9 (Secalin)	24.5 (Secalinin)
Barley	12.1	8.4	25.0 (Hordein)	54.4 (Hordenin)
Oats	20.2	11.9 (Avenalin)	14.0 (Gliadin)	53.9 (Avenin)
Maize	4.0	2.8	47.9 (Zein)	45.3 (Zeanin)
Soybean	10	90		0
Peanut	15	70		10
Pea	21	66		12

*Modified after H.-D. Belitz and W. Grosch, 'Lehrbuch der Lebensmittelchemie', Springer-Verlag, Berlin, 1992

Table 12.3 *Composition of products of animal origin given as percent of DM content**

	DM (%)	*Protein (%DM)*	*Carbohydrate (%DM)*	*Lipid (%DM)*	*Ash (%DM)*
Milk, cow	13.0	27.7	38.5	28.5	5.4
Egg, hen, egg white	12.1	87.6	7.4	0.3	5.0
Egg, hen, egg yolk	51.3	32.4	2.0	63.6	2.1
Swine, ham	24.7	81.8	–	14.6	2.4
Cow, beef	23.6	92.4	–	3.0	5.1
Hen, breast	25.6	91.0	–	4.7	4.3
Salmon†	34.0	58.8	–	41.2	2.9
Mackerel†	37.0	51.4	–	32.4	3.5
Cod†	18.0	94.4	–	1.7	5.6

*Modified after H.-D. Belitz and W. Grosch, 'Lehrbuch der Lebensmittelchemie', Springer-Verlag, Berlin, 1992
†Given as percentages of edible part

density lipoproteins (HDL and LDL, respectively)], glycophosphoproteins, and water-soluble, globular proteins.

Muscle proteins can be divided into three major groups according to their solubility. The proteins of the contractile apparatus (myosin, actin, tropomyosin, and troponin) are extractable in concentrated salt solution. Myoglobin and enzymes are soluble in water and diluted salt solutions and the insoluble proteins are represented by the membrane proteins (mainly lipoproteins) and the connective tissue proteins (collagen).

The fish muscle tissue proteins can be largely divided into two groups: the sarcoplasmaproteins mainly containing a large number of enzymes (16–22% of the total protein), and the proteins of the contractile apparatus which account for *ca.* 75% of the total protein.

3 Handling of Protein Fractions

Working with proteins requires consideration of their handling. Storage of material containing proteins may be either as dry material or in solution. Dry seeds and flour can usually be kept at room temperature for years, whereas freeze-dried, partially purified proteins should be kept at 4 or −20 °C. With respect to solubilized proteins, efforts should be made to avoid fungi contamination and subsequent protein degradation. Often, freezing may be the choice of storage of solubilized proteins. Some proteins are, however, sensitive to freezing and may precipitate and/or denature, *e.g.* oligomers are usually sensitive to freezing and should be kept in the refrigerator. For short periods of storage (up to *ca.* two days) refrigeration of the protein solution without additives will usually be sufficient whereas longer storage periods at 4 °C may require addition of fungi inhibitors as *e.g.* NaN_3 (to a concentration of 0.02% w/v) or organic solvents such as toluene or benzene in closed bottles (ppm amounts). Care should be taken not to use NaN_3 in the storage of haemoproteins as it will act as a strong-field ligand for the haem iron and thereby it will inactivate many haem proteins (page 107; Figure 5.13). Nucleophilic compounds may also inactivate cofactors such as nicotinamide coenzymes (pages 108–111) and other heteroaromatics with corresponding structures. With use of the organic solvents toluene and benzene, the influence on UV absorbance of these compounds should be considered, if UV absorbance is used for protein quantifications. Before use after storage, the protein concentration in solution may be checked by measuring $A_{280\,nm}$ (page 99) or, if possible, the preservation of biological activity of the protein should be examined.

The use of protein kept as a powder may cause difficulties upon resolubilization. The solubility of proteins in water is a function of numerous parameters, mainly pH, ionic strength, solvent type (page 79), and temperature. Increasing the latter from 0 °C to 40–50 °C will increase the solubility of most proteins but higher temperatures may cause denaturation and aggregation. The solubility dependency of pH usually shows that proteins are least soluble at pHs near their pI (Figure 4.4), *e.g.* the solubility of seed proteins plotted as a function of the pH usually gives a V or U shaped curve, with minimum close to pI. Ions of neutral salts may also increase the solubility of proteins at salt molarities of 0.5–1 M (Figure 4.3). With salt concentrations above 1 M, the proteins may be 'salted out' as the salt ions and the protein molecules compete for the water molecules. Different salts also show different suitability for protein solubilization, according to Hofmeister's series (page 81). Non-aqueous solvents such as ethanol or acetone (pages 75, 84 and 324) favour solubilization of hydrophobic proteins and may cause aggregation and precipitation of less hydrophobic proteins.

Enzyme Handling

Special precautions should be taken to avoid denaturation and inactivation of enzymes. Usually, enzymes should be handled in the cold (0–4 °C) and checked for activity periodically. Freezing and thawing, especially of oligomeric enzymes in the presence of high salt concentration, can cause denaturation and loss of activity. Therefore, purified enzyme preparations should be stored in small aliquots to prevent too many thaw–freeze cycles. Detergents and preservatives may also affect enzyme activity (Sections 2.7 and 2.9). Sodium azide, for example, is an inhibitor of many enzymes with haeme groups (page 107) (*e.g.* peroxidase). Detergent concentrations above the critical micellar concentration (c.m.c.; Section 2.9) can result in formation of micelles which may entrap and denature the enzyme. Note that the c.m.c. may change as a function of temperature.

4 Strategy for Protein Purification

Defining a strategy for purification of a protein or a group of proteins will usually encompass a choice of method for detection of the particular protein(s) during the purification procedure, chosen on the basis of known characteristics of the protein(s). This may be information on, for example, biological activities of the protein(s), M_r, or pI. For measuring the efficiency of the purification scheme, it is valuable to be able to determine both the amount of total protein and the quantity and/or activity of the specific protein.

The first step in the purification procedure is the choice of a method suitable for extraction of the protein from the source (Sections 4.1 and 4.3). For each new protein subjected to purification efforts, the purification strategy should be considered in order to choose methods that take advantage of the specific characteristics of the protein considered. Furthermore, the order of different techniques should be considered to avoid, for example, unnecessary desalting and concentration steps between the different methods employed.

In most cases, satisfactory protein purification can not be accomplished in one step after the extraction, as this requires very specific purification methods. An exception is the use of immunoaffinity chromatography, which will be discussed in Chapter 13. In other cases, it will often be worthwhile to make a first crude separation that will not result in yield losses that are too high. For this purpose, group separations are well suited, and a technique should be chosen that will allow use of the resulting sample matrix for the next separation technique/step. Table 12.4 summarizes the characteristics of the different separation techniques.

The individual steps employed for purification of a protein should be evaluated for each purification step as illustrated in Table 12.5. The volume is determined before and after the purification step and the total protein content is determined for both the applied and the resulting protein fractions. In addition, measurement of the amount of the specific protein by, for example, biological activity (page 334) should be carried out if possible to determine the increase of specific activity given as the protein activity per weight unit of protein. The efficiencies of the different purification steps may then be evaluated from the increase in specific activity obtained by the actual purification step and from recoveries for the individual purification steps.

Table 12.4 *Essential sample parameters and compositional changes of resulting fractions after separation*

	Group separation		
		Resulting sample	
Method	*Requirements to sample*	*Ionic strength (I)*	*Volume*
Precipitation by $(NH_4)_2SO_4$	None	High	Small volume Concentration
Gel filtration	Moderate volume	Variable; buffer change (*e.g.* desalting)	Dilution
Flash Chromatography/ Ion exchange	Low *I*; correct pH Stable at various pH	High (salt gradient) or Low (pH gradient)	Concentration
Ultrafiltration	None	Unchanged/moderate	Concentration
Dialysis	None	Low	Autosolution dependent
Affinity chromatography	Specific binding conditions	Moderate I and/or high ligand concentration	Concentration
	High-resolution chromatography		
		Resulting sample	
Method	*Requirements to sample*	*Ionic strength (I)*	*Volume*
Affinity chromatography	Specific binding conditions	Moderate and/or high ligand concentration	Concentration
HPLC/FPLC	Small volumes	Depend on column; see below	
Hydrophobic interaction chromatography (HIC)	Stable at various *I*; high *I*	High	Concentration
Ion exchange	Low *I*; correct pH; stable at various pH	High (salt gradient) or Low (pH gradient)	Moderate concentration
Gel filtration	Moderate volume	Variable; buffer change (*e.g.* desalting)	Dilution
Chromatofocusing	Low *I*	Moderate; pH = pH_i + polybuffer	Concentration
HPCE	Low *I*; high concentration; small volumes	Moderate	Often only analytical technique

5 Sample Preparation

For application of sequences of purification procedures, it may be necessary to change, for example, the sample buffer or concentration between the different steps.

Table 12.5 *Scheme used in connection with purification of proteins (enzymes)*

	Purification step			
	1	2	3	4 .... *n*
Total volume (ml)				
Protein				
Sample volume (ml)				
Dilution factor				
Absorbance				
Protein (μg, mg, or mg ml^{-1})				
Total protein (mg)				
Activity				
Sample volume (ml)				
Dilution factor				
Absorbance				
Activity (U ml^{-1})				
Total activity (U)				
Specific activity (U mg^{-1})				
Recovery (%)				
Increment of specific activity				

Change of Sample Buffer

The sample buffer may be changed by gel filtration (Sections 6.6 and pages 121, 142–146, 175 and 325) although this procedure results in an increased sample volume. A gel should be chosen where the proteins are eluted at the void volume ($K_{av} = 0$) and the low-M_r substances are eluted near V_t ($K_{av} = 1$). For desalting purposes, Sephadex G-25 medium (Pharmacia) is often a good choice for desalting of molecules down to about $M_r = 5 \times 10^3$. For complete recovery of desalted sample after gel filtration, sample volumes of between 15–20% of the bed volume are recommended,[1] otherwise sample volumes up to 30% of the bed volume may be applied. To allow high flow rates, the gel should be packed in a short, wide column.

Another possibility is dialysis (Section 4.4). By dialysis, small ions migrate through a semipermeable membrane owing to a concentration gradient generated between the sample solution in the membrane sack and the solvent on the outside of the membrane sack. Dialysis can, therefore, be used for desalting or buffer change. Commercial dialysis tubes are available with a range of different pore sizes, and a retention limit of typically $M_r = 12–14 \times 10^3$. Chemically modified dialysis tubes for retention of molecules smaller than $M_r = 10 \times 10^3$ are also available.

Desalting is also possible by ultrafiltration using filters with a cut-off value smaller than the protein to be concentrated (Section 4.5).

Concentration of Sample

Concentration of sample is obtainable by several procedures, as revealed from Table 12.4, and, when possible, such steps should follow steps which result in dilution.

Dialysis may also be employed for concentration of a protein sample by using a 6–10% aqueous solution of polymer solution as outer solution (Section 4.4). However, depending on the protein, some of the protein may adsorb to the surface of the tube.

Ultrafiltration is another method for concentration (Section 4.5) in which proteins larger than the pore size of the ultrafiltration filter are retained during the filtration. Eppendorf vials with filters are available for filtration of small sample volumes.

Especially with small volumes, it is also possible to concentrate the protein without simultaneous salt concentration by adding dry Sephadex, which will then swell under uptake of water and salts. The chosen Sephadex should have a pore size small enough to prevent penetration of the proteins in the gel. However, this method will give variable loss of protein, often at least approximately 10–20%.

The sample may also be concentrated by lyophilization. In this case, the salt concentration after the concentration step should be considered in order to avoid protein denaturation or be too high for the following analyses and purification steps. Most often, a desalting step may therefore be advantageous prior to the concentration step.

Sample Filtration

For samples to be applied to liquid chromatography columns, it is usually necessary to make the sample free of particles. Centrifugation may cause sedimentation of the particles present in the samples but, otherwise, filtration may be a possibility. A large number of filters are available, both paper filters for Buchner funnels and commercially available filters for volumes from few microlitres to litres.

6 Group Separation

Homogenization and Extraction/Solubilization

Non-liquid food sources usually require homogenization prior to extraction to allow efficient recovery of the proteins from the material. For this purpose, milling, blending, homogenizing, or use of ultrasound are common methods. The homogenization step may be combined with extraction of the proteins, using suitable extraction buffers, which may give a selective extraction of specific groups of proteins; see Box 4.5 for strategies for protein extraction and solubilization.

The use of internal standards for calculation of recoveries should be considered at this point of sample preparation (page 73).

Precipitation

Precipitation is a method of high capacity and may be chosen if the protein of interest constitutes less than 1% of the total sample protein.

Precipitation at Isoelectric pH

Precipitation of proteins in aqueous solution is most easily obtained at a pH equal to their pI (page 82).

Ammonium Sulfate Precipitation

A very common technique is salt fractionation using ammonium sulfate (page 81, Box 4.6). This salt is highly soluble in water and has a stabilizing effect on some proteins but not for all, and it is therefore important to ensure that the proteins retain or renaturate to their native conformation. Each protein precipitates within a fairly limited concentration range of ammonium sulfate, but the precipitation is affected both by other constituents in the solution and the protein concentration. Proteins with a large surface-to-volume ratio, *e.g.* rodlike molecules, usually precipitate at low saturation conditions. This is also the case for membrane-bound proteins and lipoproteins, whereas small proteins and especially those with many charged groups on the surface and only few hydrophobic areas will stay in solution even at 100% saturation. Therefore, ammonium sulfate can be used either to precipitate the protein itself or the contaminating proteins present in the extract. Note that the presence of organic solutions in the extract will change the solubility conditions for both the extract and the salt.

Protamine Sulfate Precipitation

The presence of nucleic acids may interfere with the purification of proteins, as nucleic acids can bind significant amounts of proteins in large aggregates. Protamine sulfate is a polycation which binds to negatively charged compounds and consequently it is very efficient for the precipitation of nucleic acids. However, many anionic proteins will also be precipitated. Alternatively, the protein solution may be incubated for 30–60 min at 4 °C with commercially available DNAase to degrade the DNA into small pieces.

Organic Solution Precipitation

Especially for purification of some membrane-bound proteins, it may be useful to use precipitation by ethanol or acetone (page 84, Box 4.7). The membrane-bound proteins may by soluble in organic solutions whereas water-soluble proteins will precipitate. The proteins easily denature during this precipitation technique, and this is minimized by employing low temperatures during the precipitation (−20 °C to −10 °C). Redissolution must also be performed under cold conditions.

Gel filtration

Low pressure gel filtration (Section 6.6 and page 142) may be used as a group separation technique for separation of proteins from low-M_r and high-M_r components. If a gel is selected where the smaller molecules have a K_{av} approaching 1 and the larger molecules have a K_{av} equal to 0, sample volumes up to 0.3 × bed volume can be applied to the column, and at the same time, the sample can be transferred to an appropriate buffer for further purification. For such group separations, the gel should be packed in short, wide columns to allow high flow rates. If the purpose is to separate molecules with limited variation in M_r, it will require an application of a small volume or concentrated sample in a narrow band, typically with a sample volume of 1–5% of the bed volume. Filtration of crude extracts prior to application (pages 88 and 323) may protect the column in order to decrease the frequency of re-packing it.

Flash Chromatography

Flash chromatography is ion-exchange chromatography using a solid high capacity matrix suitable for high flow rates (Section 7.9). The commercially available columns are provided as prepared disks, and different ion-exchange materials are available with weak and strong cation- and anion-exchange columns. With these columns it is possible to apply large sample volumes in a short time and thereby group-separate on the basis of differences in charges. As the sample binds to the column, an appreciable concentration of the sample is possible with this technique. To avoid particles in crude extracts, the extract should be filtered (*e.g.* paper filter) or centrifuged prior to application (page 89). The sample should be buffered and pH of the sample checked to obtain optimum binding to the flash chromatography columns.

Dialysis

In dialysis, the protein solution is placed in a semipermeable sack suspended in a large volume of buffer (Section 4.4, Box 4.8). Small molecules (with M_r < cut-off value of the dialysis membrane) are separated from the sample matrix with a speed depending on the parameters described in Section 4.4.

Ultrafiltration

Ultrafiltration (Section 4.5) may be used both to separate small sample components from the proteins or small proteins from high-M_r components as for dialysis and, at the same time, it is a concentration step. However, the sample should not contain particles or polymers with high binding affinities to the filters as these will clog the ultrafiltration unit. To avoid this, different types of low binding filters are commercially available.

7 High Resolution Techniques

The theory and principles for these techniques have already been dealt with in the Chapters 6–11.

Affinity Chromatography

Affinity chromatography exploits specific biological interactions to achieve separation and purification. Purification can take place from complex mixtures and gives highly purified protein in a single step. The prerequisite for affinity chromatography is the reversible binding between a ligand coupled to an insoluble matrix and the compound to be purified. Preliminary knowledge is therefore required on the compound to be purified.

Examples of affinity chromatography ligands and their specificity are given in Table 6.7. Some matrices are commercially available, *e.g.* protein A conjugated matrices for antibody purification. Otherwise, preactivated support materials ready for ligand conjugation may be purchased and conjugated to relevant ligands, *e.g.* as described in Chapter 13. Affinity chromatography materials are available for use both on FPLC and for simple chromatographic procedures.

HPLC and FPLC

The sample volumes usually applied to HPCL and FPLC columns are in the 10 μl–2 ml range and, consequently, these separation techniques are not suitable for the large sample volumes often present in the first part of the purification scheme. However, preparative HPLC systems which allow washing with relatively large volumes are commercially available. In addition, the HPLC/FPLC columns are generally sensitive to impurities that may adsorb or bind to the column material. The commercially available columns are often relatively expensive, and therefore it may be advantageous to use other techniques for the initial group separation of the sample material.

For HPLC, column materials with small particles and high capacity are usually employed. Therefore, an increased or moderately high pressure is required to obtain flow through the columns. For separation and purification of proteins it is, however, often more advantageous to obtain a fast separation to preserve the native conformation of the proteins. For this purpose, FPLC and corresponding column materials have been developed for work with biopolymers and proteins (Chapter 8). A range of different column materials and prepacked columns are commercially available, and the column properties encompass all the different characteristics known from ordinary low-pressure chromatography.

HIC is also available as a FPLC method in which the molecules are separated according to differences in their hydrophobicity (pages 148 and 175).

For FPLC and HPLC, the samples should be filtered using a 0.2 or 0.45 μm filter to protect the columns from particles, and relatively low viscosity (page 126) salt concentrations are most often required.

HPCE

HPCE is a technique complementary to HPLC and FPLC as it involves both chromatographic and electrophoretic principles (Chapter 10). Additional information may therefore be obtained by using HPCE. The high resolving power of HPCE often allows better separation of closely related proteins and HPCE therefore offers an efficient tool for purity controls of peptides as well as proteins. The sample volumes injected in HPCE are often in the range of a few nanolitres and therefore HPCE is not a technique suited for prepative purification of proteins. However, some commercially available HPCE instruments now have devices for sample collection, either in buffer or on solid supports. With repeated runs and collections of sample it may be possible to collect enough sample to perform enzymatic and immunochemical tests as well as MS or sequence determination of the proteins. Applications are also available that describe the on-line analysis of HPCE separated samples by MALDI-TOF (Section 5.9).

The separation of proteins by HPCE is highly dependent on the characteristics of the proteins, especially the isoelectric points of the proteins, but also information on the differences between the proteins to be separated are important. In the following, a short summary of the most frequently used principles for proteins are given. For specific information on method optimization and suggestions for buffer components and specific reagents, the reader is referred to Chapter 10.

An important point to consider in relation to the choice of buffer systems used in analysing proteins is how to prevent or control the adsorption of proteins to the capillary wall of glass or fused silica. For buffers with a pH above 2.5, the silanol groups of the silica capillary surface become negatively charged. Positively charged sites on proteins will then interact with the capillary surface (Section 9.1), resulting in unreproducible results. The ionization of the silanol groups can be suppressed by working with background electrolytes with low pH-values. Alternatively, a buffer pH above the pI of the sample proteins may be employed, thus resulting in repulsion between the negatively charged capillary surface and the proteins. The interaction with the capillary wall may also be eliminated by coating of the capillary by derivatization with various detergents to make the surface uncharged at all buffer pHs. Coated capillaries may, however, not have stable coatings, and the degree of coating may not be easily reproducible. It is also possible to introduce a competition in binding to the wall by addition of alkali metal ions. This application raises demands on the proteins with respect to solubility in high salt concentrations and low binding to metal ions (pages 44 and 80). Alternatively, a dynamic coating of the capillary surface can be obtained by the use of zwitterions or detergents.

Preparative Electrophoresis

For separation of samples with techniques complementary to liquid chromatography methods such as FPLC and low pressure liquid chromatography, it is possible to run preparative gel electrophoresis with sample application in one well covering the entire gel. Separation is performed as slab gel electrophoresis, either as native gel electrophoresis as described previously (Box 9.9 and 9.10) or, for example, as SDS-PAGE as described in Ref. 2. The gel area containing the sample may be cut out and the sample eluted from the gel, although this may encompass difficulties for acrylamide gels as the sample may be detained in the gel structure. Otherwise, the sample may be transferred from the gel to solid supports (blotting), *e.g.* nitrocellulose or poly(vinylidene difluoride) membranes (Section 9.6; Figure 9.13), which can be used for amino acid analysis or sequence determination. For these purposes, care should be taken to avoid contaminating compounds during gel preparation and blotting procedures. Only limited sample volumes can be applied for gel electrophoresis (typically $< 100\,\mu l$) and the protein concentration should therefore be fairly high (0.1–2 mg/ml^{-1}). To diminish too high a voltage drop across the gel, the ionic strength of the sample should not be too high (preferably buffer concentrations < 0.2 M). Alternatively, separation based on isoelectric points may be employed, using IsoPrime™ (page 204).

8 Methods of Protein Analyses

For protein purification procedures, it is essential to have methods of analysis to determine the protein contents obtained after the individual purification steps. If specific determination of the purified protein is also possible, for example, by assay (*vide supra*) or by the RZ ratio (page 108) the increase in purification degree can be determined.

Total Protein Determination

For most of the methods for protein determination, a reference protein is included for quantification. BSA is a widely used reference protein (page 99), but for a true quantitative determination the pure protein itself should be used as reference. The best approximation for a true protein determination is probably obtained by amino acid analysis after protein hydrolysis. However, this is a time demanding and expensive procedure, resulting in a need for alternative procedures, especially if labile amino acids such as Trp, Cys, CysSSCys, Glu, Asn, Ser, and Thr are to be included correctly. When possible, the protein content should be determined by different techniques to diminish or reveal the interference from unwanted components.

UV-determination

Direct spectrophotometric measurement at 280 nm is based on the tyrosine and tryptophan content of the protein. An average value of absorption at 280 nm of a 1% solution (w/v) of protein gives $E^{1\%}_{1\,cm} = 10$, which may be employed for mixtures of proteins and when the extinction coefficient of a protein is unknown (page 99). Other components with UV absorption at 280 nm will interfere with this method, as is the case for nucleic acids (page 100), but as no protein determination methods give exact values for the protein content, corrections may represent unnecessary work. This method is typically not applicable for measurements in the first purification step, as crude extracts usually contain many non-protein components that absorb at 280 nm.

Lowry Determination

In the Lowry method[3] (Box 12.1), peptide bonds in the proteins form blue complexes with Cu^{2+} and some amino acid side chains are oxidized under reduction of the colour reagent to a strong blue colour. The colour intensity is measured and is proportional with the protein concentration. Concentrations of down to 5 μg protein ml^{-1} can be measured.

Various aromatic compounds, especially indolyls, ammonia, amines, and amino acids, interfere with the method by forming blue complexes with Cu^{2+}, and the presence of EDTA, sucrose, some detergents, and more than 0.15% ammonium sulfate will also influence the results. Phenols and SH-containing compounds will reduce the reagent and cause development of a blue colour and should thus be avoided.

With the high concentrations of phenolic compounds present in plant materials, this method is therefore not suitable for protein analyses of samples of plant origin until the proteins have been separated from the interfering compounds. The method may, however, be used for protein determination of extracts of animal and microbial origin.

Box 12.1 *Protein determination by the Lowry method*

Reagent list

Stock solution A: Na_2CO_3, 10% in 0.5 M NaOH
Stock solution B: $CuSO_4{\cdot}5H_2O$, 0.5% in 1% NaK tartrate
Stock solution C: phenol reagent/Folin reagent/Folin–Ciocalteu reagent; commercially available, dilution instructions are given on the bottle
Immediately prior to use, the following reagents are prepared:
Solution D: 10 ml A + 1 ml B
Solution E: Dilution in deionized water of C giving pH = 1

Procedure

A standard curve is prepared using BSA (0.5–0.005 mg protein/ml). The standard protein solution and the sample are diluted to 1.00 ml with deionized water and 1.00 ml solution D is added. After 10 min at ambient temperature, 4.00 ml solution E is added with immediate shaking. After 30 min at ambient temperature, or 10 min at 50 °C followed by cooling to ambient temperature, the absorbance is read at 540 and 750 nm

Box 12.2 *Total protein determination by Bicinchoninin acid method*

Reagent list

Stock solution A: BCA (10 g), $Na_2CO_3{\cdot}H_2O$ (20 g), sodium tartrate ($Na_2C_4H_4O_6{\cdot}2H_2O$; 1.6 g), NaOH (4 g), and $NaHCO_3$ (9.5 g) are dissolved in deionized water (1 litre) and the pH adjusted to 11.25
Stock solution B: $CuSO_4{\cdot}5H_2O$ (2 g) is dissolved in deionized water (50 ml)
Stock solutions A and B are stable for at least 12 months at room temperature
Standard solution: Stock solutions A and B are mixed 50 : 1. The solution is stable for a week

Procedure

Sample (150 μl) and standard solution (3 ml) are mixed and allowed to react for 2 h at room temp. or 30 min at 37 °C. Absorbance is measured at 562 nm

Biuret Method

The Biuret method uses the same principles as the Lowry method, but is less time consuming. However, the sensitivity is approximately 100 times poorer than for the Lowry method and so it will, therefore, not always be suitable.

Bicinchoninin Acid (BCA) Method

The BCA method (Box 12.2) also utilizes the reduction of Cu^{2+}, but it is simpler and with fewer interfering compounds compared with the Lowry method (Box 12.1). Salt levels up to 1 M, up to 40% non-reducing carbohydrate, up to 3 M urea, and 1% detergents do not interfere with the reagent. With this method, protein concentrations of a few μg protein/ml can be measured.

Coomassie Brilliant Blue Protein Determination

In this method (Box 12.3), the binding of the dye Coomassie Brilliant Blue G to proteins results in a shift in absorption maximum of the dye from 465 to 590 nm. The absorption at 590 nm is proportional to the protein concentration. Different proteins result in different extinction coefficients for the dye–protein complex; glyco- and lipoproteins in particular have lower extinction coefficients. The method is only influenced by a few components, such as strongly basic buffers and large amounts of detergents such as SDS and Triton-X-100. A popular variant of this type of protein determination is the Bradford technique in microtitre plates (Box 12.4).

Kjeldahl Analysis and Dumas Technique for Determination of Sample Nitrogen Content

With Kjeldahl analysis, the nitrogen content of a sample is determined. Usually, automated apparatus is employed, and the analysis involves boiling of the

Box 12.3 *Coomassie Brilliant Blue protein determination*

Coomassie Brilliant Blue structure:

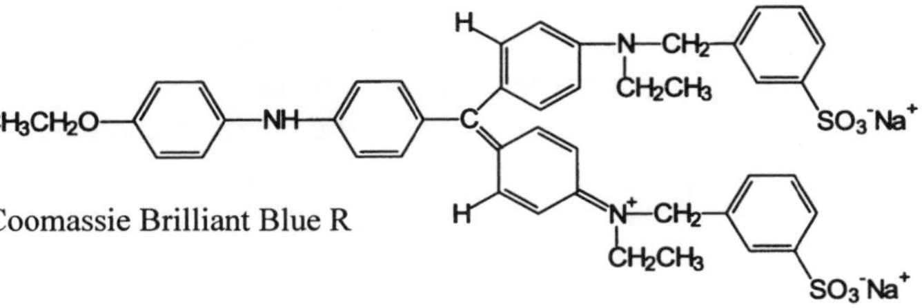

Coomassie Brilliant Blue R

Reagent list

Stock solution (commercially available):

Coomassie Brilliant Blue G (Sigma B-1131)	50 mg
EtOH, 96%	25 ml
Phosphoric acid [85% (w/v)]	50 ml
Deionized water	to 500 ml

Coomassie Brilliant Blue G is dissolved in EtOH with stirring. Phosphoric acid is added while stirring for additional 15 min. The solution is diluted to 500 ml with deionized water and filtered. The solution is stored away from light for max. of 1 month.

Diluted solution

Immediately before use, the stock solution is diluted using 1 ml stock solution to 4 ml deionized water

Procedure

A sample (100 μl) is pipetted into reagent glasses and 15 ml diluted reagent solution is added and mixed with the sample. After 2 min, the absorbances of the sample and a blank are read at 590 nm. Note that the glass used should be washed in EtOH and brushed with diluted HCl immediately after use to avoid colouring of the glass.

Box 12.4 *Bradford determination in microtitre assay*

The assay is performed in microtitre trays with 96 wells, which allows simultaneous determination of the protein content in samples, to produce a standard curve, and a range of protein samples using a microplate reader.

1. Two standard rows are prepared by pipetting 10 μl of appropriate standard protein solutions, 0.04–4.0 mg ml^{-1} to each of a number of wells. Blank samples are included as 10 μl buffer (same buffer as used for protein dissolution)
2. Samples (10 μl $well^{-1}$) are applied to the wells in different concentrations, *e.g.* prepared as dilution rows
3. To each well, apply 200 μl Bradford reagent (Bio-Rad protein assay; diluted 1 : 4 with deionized water and filtered on paper) and place the microtitre tray on a shaking table for 5 min
4. Read the absorbance at 600–630 nm using a microplate reader (depending on the options of the reader)

sample with concentrated sulfuric acid in the presence of digestion catalysts. The organic nitrogen is converted into ammonia which is trapped as ammonium sulfate, released by sodium hydroxide, collected, and titrated for quantification. The determination of protein content is usually found from the general assumption that the nitrogen content of proteins is 16% by weight. However, contamination with other compounds containing nitrogen such as, for example, DNA, influences the determination and, in addition, the nitrogen content may show variations for particular proteins.

The content of N may also be determined in automated apparatus according to the Dumas principle, in which N-oxides, produced by combustion at high temperatures (about 1000 °C) in an oxygen atmosphere, are reduced to N_2 by the passage through a copper-containing column; N_2 is detected in a thermal conductivity cell.

Amino Acid Analysis

With amino acid analysis, it is possible to obtain a relatively good determination of the total protein in a sample. This analysis is often performed as ion-exchange chromatography on HPLC after hydrolysis of the protein (Box 12.5) and derivatization of the resulting amino acids (Box 12.8). Not all of the amino acids can be determined in the same run due to different sensitivity to hydrolysis (Box 12.5). As the analysis is performed by HPLC, special requirements of sample purity, as described previously (pages 161 and 327), have to be fulfilled (Box 12.6).

Amino acids devoid of aromatic or heterocyclic groups detectable above 210 nm are dependent upon derivatization with a chromophore for either UV detection (page 103) or fluorimetric detection (Section 5.5). Labelling with ninhydrin is a simple and popular method for post-column derivatization of amino acids after separation on traditional commercially available amino acid analysers and for labelling of otherwise separated amino acids, peptides, and compounds with free amino groups (Box 12.7). However, various types of other derivatization procedures are now used prior to different HPLC and HPCE methods of analyses, *e.g.* derivatization with *o*-phthalaldehyde (OPA) (Box 12.8) and dansyl chloride (Dns-Cl) (Box 12.9).

Box 12.5 *Protein hydrolysis for amino acid analysis*

For proteins (part of a mg to few mg), hydrolysis is performed in 6 M HCl (few ml) with boiling (120 °C) under reflux for 20 h. During acid hydrolysis, certain amino acids undergo irreversible changes, where, for example, tryptophan is destroyed and cysteine, cystine, serine, threonine and tyrosine are partly destroyed. Analysis for these amino acids requires special techniques, which have been thoroughly investigated by M. Rudemo, V.C. Mason, and S. Bech-Andersen, *Z. Tierphysiol. Tierernähr. Futtermittelkd.*, 1980, **43**, 146 and related papers cited in B.O. Eggum and H. Sørensen, in 'Absorption and Utilization of Amino Acids', ed. M. Friedman, CRC Press, Boca Raton, FL, 1989, vol. 3, ch. 17, p. 265.

Box 12.6 *Purification after hydrolysis of crude protein extracts*

Amino acid analysis of crude protein extracts may require a purification step, *e.g.* by using the column system described in Section 7.6. The A-eluate contains the basic amino acids whereas the B-eluate contains the acidic and neutral amino acids. For analysis in one experiment, the eluates may be pooled.

1. The hydrolysed sample is evaporated followed by 2 rounds of evaporation after addition of 2 ml deionized water
2. The sample is redissolved in 2 ml deionized water and internal standards are added (10 μmol Norvalin/sample; 5 μmol 3,4-dimethoxyphenylammonium chloride/sample). The sample is stored at −20 °C
3. Sample (0.5 ml) is applied to a regenerated 2-column system [A-column; Sephadex CM, C-25 (H^+-form) and B-column; Dowex 50W × 8, 200–400 mesh (H^+-form), each 0.75 ml, see Section 7.4]
4. The columns are washed with 10 ml deionized water
5. The A-column is eluted with 20 ml 2 M HOAc–MeOH (1 : 1) and the eluate evaporated to dryness and resolubilized in 500 μl deionized water
6. The B-column is eluted with 10 ml 1 M pyridine and the eluate is evaporated to dryness and resolubilized in 500 μl deionized water

Box 12.7 *Derivatization of amino acids with ninhydrin after paper chromatography/thin layer chromatography*

Amines, amino acids, and peptides (primary and partly secondary amino groups) are detectable using ninhydrin (0.2% ninhydrin in acetone). The derivatization may be performed after paper chromatography (PC, Box 6.1), thin-layer chromatography (TLC, Box 6.1) or for detection of samples dot blotted on chromatography paper. The method has a detection range of 1 μg amino acids per dot in dot blots

Procedure

The chromatography paper is immersed in the ninhydrin solution or the paper or TLC plate is sprayed with the solution. The spots develop more slowly without the use of heating, but it is advisable to avoid heating above 50 °C. The developed spots may be preserved by immersing the paper in a solution of 10 ml saturated $Cu(NO_3)_2$, 2 ml 10% HNO_3, 50 ml deionized water, and 938 ml MeOH

Dns-Cl will also react with nitrogen in imidazole groups, thiol or sulfanyl groups, and phenol groups. Consequently, His, Tyr, and Cys will form didansyl deriatives. In Figure 10.15, an electropherogram of dansyl derivatized samples is shown.

Specific Protein Characterization

Determination of the amount of specific protein present is determined using knowledge of the specific protein, such as its biological activity, M_r, pI, or RZ ratio (page 108), or through the availability of specific detection methods, *e.g.* as antibodies with specificity for the protein. Biological activity may offer a specific

Box 12.8 *Amino acid derivatization with* o-*phthalaldehyde (OPA) for HPLC analysis*

o-Phthalaldehyde reacts with primary amino groups in the presence of a thiol:

$$\text{OPA} + HS{-}CH_2{-}CH_2{-}X + H_2N{-}R \longrightarrow \text{(1-}S{-}CH_2{-}CH_2{-}X\text{-substituted isoindole)}N{-}R + 2H_2O$$

OPA

OPA stock solution: 50.0 mg OPA, 4.0 ml MeOH, 0.5 ml potassium borate (1 M from boric acid in deionized water, pH adjusted to 10.4 with KOH) and 50.0 μl sulfanylpropionic acid
OPA working solution: OPA stock solution (200 μl) + 1.0 ml potassium borate (1 M, pH 10.4)
Derivatization: the derivatization is performed using HPLC apparatus, with use of 25 μl sample, 30 μl OPA working solution, and a reaction time of 2 min before injection

HPLC analysis of OPA-derivatives:

Buffer A: 0.025 M $NaH_2PO_4{\cdot}H_2O$, 0.025 M $Na_2HPO_4{\cdot}2H_2O$, and 0.02% NaN_3 to 1 L with Milli Q H_2O. Buffer B: CH_3CN (50%) in A-buffer

HPLC may be performed using a 3 μm C-18 column (*e.g.* Sperisorb ODS2, Pharmacia) and a buffer gradient as follows: 0 min, 0% buffer B; 0–20 min, 20% buffer B; 20–40 min, 35% buffer B; 40–50 min, 60% buffer B, 50–55 min, 0% buffer B; and 55–65 min, 0% buffer B

tool of analysis for identifying a compound present in a mixture with other proteins. Examples of such biological activity are enzymatic activity measurable with an appropriate substrate, inhibitor, or haemolysis (some glyco-alkaloids), or agglutinating activity (lectins).

Enzymatic Assays

Determination of enzyme activity requires a knowledge of enzyme kinetics (pages 53 and 67), the reaction catalysed by the enzyme, and the optimal reaction conditions for the enzyme such as pH optimum, buffer type and concentration, demands for cofactors, and temperature optimum.

For enzyme measurements, the enzyme activity is expressed in enzyme units (U), with 1 enzyme unit often defined as the activity that catalyses the transformation of 1 μmol of substrate into product(s) per time unit under defined conditions, *i.e.* $1\,U = 1\,\mu mol\,min^{-1}$. The specific enzyme activity is defined as the number of enzyme units per mg protein, *i.e.* $U\,mg^{-1}$. Using the SI system (Section 1.1), the denomination for U becomes $mol\,s^{-1}$, called katal (kat). However, $1\,kat = 6 \times 10^7\,U$ and therefore $1\,U = 16.67\,nkat$ may be used instead.

Box 12.9 *Amino acid derivatization with dansyl chloride (Dns-Cl) for HPCE analysis*

Dansyl chloride (1-dimethylaminonaphthalene-5-sulfonyl chloride) reacts with amino acids under weak alkaline conditions. A method has been developed for HPCE analysis of dansyl chloride derivatized amino acids (S. Michaelsen, P. Møller, and H. Sørensen, *J. Chromatogr.*, 1994, **680**, 299) where acidic and neutral amino acids are separated in a MECC system with SDS, and basic amino acids are separated in an MECC system with cholate

$N(CH_3)_2$ COO^- HO^- H_2O $N(CH_3)_2$

H_2N-C-H + Cl^-

R

SO_2-Cl SO_2 COO^-

$H-N-C-H$

Dansyl chloride R

Derivatization of amino acids

Acetonitrile–water (1:2) containing lithium carbonate (40 mM; pH 9.5), Dns-Cl (3.3 mg ml^{-1}) and amino acids (0.5 μM of each) is allowed to react at room temperature for 2 h. The reaction is stopped by adding 4% ethylamine, and the derivatized sample is dried with air and redissolved in 20% aqueous MeOH

HPCE for amino acid analysis

Fused-silica capillary, 720 mm × 50 μm i.d.; detection 520 nm from injection end; on-column UV detection at 216 nm

HPCE conditions for separation of acidic and neutral amino acids

Buffer: 100 mM boric acid, 150 mM SDS; pH 8.3; 22 °C; 15 kV

HPCE conditions for separation of basic amino acids

Buffer: 50 mM Na_2HPO_4, 50 mM sodium cholate; pH 8.0; 30 °C; 20 kV

Determination of enzyme activity implies measurement of the velocities of enzyme catalysed processes:

$$S + E \underset{k_2}{\overset{k_1}{\rightleftharpoons}} ES \underset{k_4}{\overset{k_3}{\rightleftharpoons}} E + P$$

where S = substrate, E = enzyme, and P = product.

By using initial velocities (v_0), there is no significant reaction between enzyme and product corresponding to a reaction from right to left, therefore a possible k_4 can be neglected in determinations based on v_0:

$$v_0 = \left(\frac{\mathrm{d}s}{\mathrm{d}t}\right)_{t=0} = \left(\frac{\mathrm{d}p}{\mathrm{d}t}\right)_{t=0} = \frac{\text{change in concentration } (\Delta c)}{\text{change in time } (\Delta t)} \tag{12.1}$$

Using Lambert–Beers law (page 98), the enzyme activity in an enzyme solution can be calculated from (enzyme activity)/(enzyme volume):

$$\frac{v_0}{\text{enzyme volume}} = \frac{\Delta c}{\Delta t} \cdot \frac{1}{\text{enzyme volume}} = \frac{a}{\varepsilon_{\lambda_{\max}} l b 10^{-3}} \cdot \frac{\Delta A_{\lambda_{\max}}}{\Delta t} \tag{12.2}$$

where a = the total assay volume (ml), b = the enzyme volume used for the assay (ml), ε = the molar extinction coefficient ($\text{M}^{-1}\,\text{cm}^{-1}$), and l = the path length of the cuvette (cm). Δt is given in min and the factor 10^{-3} represents conversions of denominations in order to have enzyme activity/enzyme volume in U ml^{-1}.

The use of a spectrophotometer offers a fast and simple method for following the transformation of substrate into product. The change (ΔA/min) may either be followed directly during the assay by continuously measuring the absorption of the product/substrate or ΔA can be measured after a certain time of reaction (end point determination). In both cases, coupled assays are often used, in which the product is transformed quickly and quantitatively in a new enzymatic process into a compound with an appropriate absorption. The change in absorbance (ΔA/min) should be determined in the range with a constant change, *i.e.* at the first, rectilinear part of the absorbance time curve. Therefore, development of an assay always requires determination of the curve of absorbance plotted against time (Figure 12.1).

The λ_{max} and corresponding ε_{max} for the compound in question may either be determined experimentally by recording a spectrum with varying wavelengths or by looking up the values in a handbook. It is advisable to record the spectrum of both substrate and products in the assay mixture to find the best value for the measurements.

An appropriate enzyme concentration or amount of catalytic activity is determined from preliminary experiments where a suitable velocity (ΔA/min) for a spectrophotometric determination could be in the range of 0.06–0.5 absorbance units min^{-1}, depending on the desired assay time. In general, absorbance values of 0.1–0.8 and assay times of 1–5 min are preferable. The linear dependency between ΔA/min and amount of enzyme should always be confirmed.

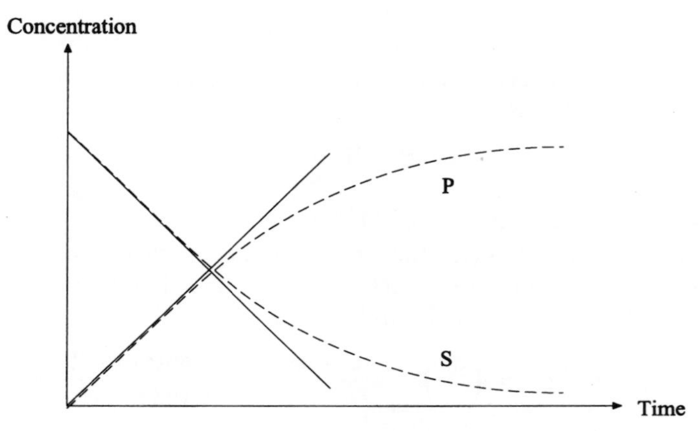

Figure 12.1 *Concentrations of substrate (S) and product (P) as function of the reaction time (without appreciable concentration of ES)*

The amount of substrate present during the reaction is of decisive importance, as a large amount of substrate will result in the presence of all enzyme (E_T) as ES under which conditions the process will proceed with maximal velocity $V_{max} = k_3 \cdot [E_T]$. This condition is a prerequisite for direct proportionality between velocity and enzyme concentration (Figure 3.14). Knowledge of the Michaelis–Menten constant (K_m) of the enzyme for the substrate in question is valuable because the substrate concentration should be 10–100 times larger than K_m. As a guideline, K_m for mono- and pseudomonosubstrate reactions are 10^{-2}–10^{-5} M in the order of magnitude. Working with high substrate concentrations may result in substrate inhibition or parallel, non-enzymatic transformation of the substrate, resulting in high blind values.

The choices of pH, buffer type and concentration (Chapter 2) are made from a series of experiments as the pH-optimum for enzymes often lies within a narrow pH-range which is again dependent on the type of buffer, ionic strength, temperature and substrate concentration. The buffer should be chosen with pH $\pm$ 1 from the pK_a-value of the buffer. Especially for phosphate and citrate buffers, care should be taken to examine whether necessary divalent metal-activators are complexed (Section 2.7 and pages 44–47). It should also be considered whether the buffer may act as substrate or product in the reaction and thus may result in substrate or product inhibition.

The choice of temperature is often free, with plant enzymes usually determined at 20 °C and animal enzymes determined at 20 or 37 °C. At temperatures above 40 °C, denaturation may occur. Enzyme-catalysed reactions are as all other chemical reactions sensitive to temperature changes, and a change of 1 °C may result in a 10% rise in activity.

The assay may be built up in different ways, *e.g.* with the substrate solution constituting approximately 10% of the total volume, the volumes of enzyme solution and, for example, cofactors being as small as possible, and adjustment of the final volume made with buffer. A reference solution should be included to take into account possible interfering compounds in the solution. With biological solutions, the interfering compounds may be present in the substrate, the enzyme, and the additional reagents. Therefore, it may be necessary to investigate several different reference solutions. Blind values that are too high should be avoided as the sample contribution to the absorbance will then be determined as a small difference between two large absorbance readings.

From the plots of absorbance as a function of time, different curves may be obtained as illustrated in Figure 12.2.

Curve a in Figure 12.2 represents too much enzyme; the enzyme concentration should be lowered. For curve b, there is also too much enzyme, and the product is unstable; the enzyme concentration and possibly also the substrate concentration should be lowered, and the assay time has to be short. For curve c, the bending of the curve could be a result of substrate deficiency, excess of enzyme, product inhibition, or measurement of unstable product; if the bending is unavoidable, the slope of the tangent at $t = 0$ should be used. Curve d seems sensible; the bending is commented on for curve c. Check whether double/half amount of enzyme results in double/half absorbance. The absorbance plotted

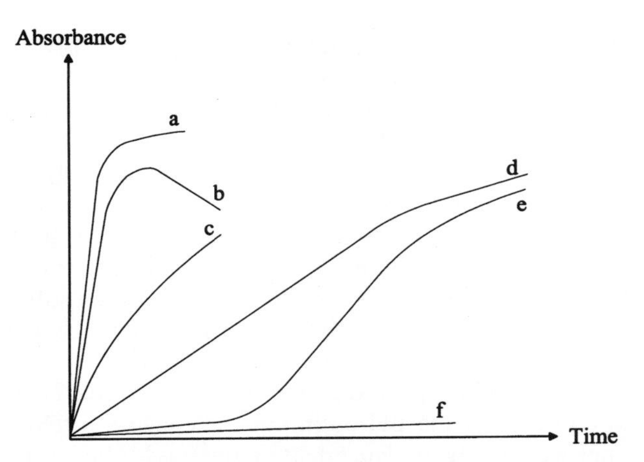

Figure 12.2 *Curves obtainable from enzyme-catalysed reactions when absorbance is plotted as a function of time*

against the amount of enzyme at constant substrate concentration and time should give a straight line. For curve e, the initial velocity is lowered, indicating the presence of an inhibitor (page 61), presumably the substrate. Another possibility is the presence of an allosteric enzyme (pages 49–53) which will result in a lag phase. The substrate concentration should be lowered and/or the enzyme concentration increased. Curve f represents the presence of too small an amount of enzyme; prolonged assay time may result in parallel chemical reactions and high blind values. The enzyme concentration should be larger and the temperature increased if possible.

Optimization of all the variable parameters in an assay results in a large number of experiments, corresponding to the solution of a matrix. To simplify matters, the parameters are optimized one at a time, often in the order enzyme concentration, substrate concentration 1, substrate concentration 2, pH, buffer type, buffer concentration, and other variables. It may be necessary to make experiments with simultaneous variation of 2–3 substrate/enzyme concentration ratios and perhaps further investigate the connection between pH optimum and buffer concentration.

Commercially Available Enzyme Systems for Food Analysis

An enzymatic assay can also be used for quantitative determination of a compound (substrate) in a mixture, *e.g.* a component in a foodstuff, and many commercially available systems have been carefully described.[4,5] The substrate concentration is determined from an assay under conditions where the velocity v is proportional to the substrate concentration (1. order area; pages 53–56) using substrate concentrations $\leq K_m$ and relatively large amounts of enzyme to obtain an acceptable velocity v or a complete transformation within an acceptable time.

In cases where quantitative assays are not necessary, a fast and simple qualitative determination of the activity may be performed by mixing substrate and buffer with a small amount of the solution to be tested for enzyme activity. This is especially applicable for fast examination of fractions from, for example, chromatographic analysis of enzymes.

Immunological Methods

An increasing number of antibodies against food components are becoming commercially available. As described in the following chapter on immunochemical techniques, antibodies can provide sensitive and selective assays even when the protein or hapten in question is present at low concentration in a mixture with many other compounds. These assays may both comprise assays in solutions, on solid supports, and in gels. For development of quantitative assays, appropriate protein standards are required, as discussed in Chapter 13.

M_r *Determination*

Slab gel electrophoresis may be employed for determination of M_r of proteins (pages 199–203 and 243–246; Figure 9.11). Identification of specific bands can be supported by M_r determination by gel filtration on FPLC (pages 142, 175–177) where the individual fractions can be tested by other methods. Some discrepancies may be seen as the M_r determination by FPLC from a standard mixture of globular proteins implies a globular structure of the proteins in question and, in several cases, adsorption problems can also give incorrect M_r determinations. Similarly, for analysis in slab gel electrophoresis, some proteins do not bind the expected amount of SDS, *e.g.* owing to glycosylation of the proteins, thus resulting in incorrect determination of subunit M_r in SDS–PAGE. The M_r determined by SDS–PAGE gives only the subunit M_r, whereas M_r based on, for example, FPLC gel filtration may give that M_r for the native structures of the compounds.

Methods for the determination of M_r of proteins are also possible using applications developed for HPCE techniques (pages 243–246).

MALDI-TOF (Section 5.9) is another new opportunity for M_r determination of proteins. For this technique, it is essential that the sample protein is highly purified to be able to interpret the resulting data.

pI Determination

Isoelectric focusing is another method for the characterization of proteins (page 190). In agarose slab gels, the proteins may be identified by immunological methods by blotting the proteins to solid supports followed by immunostaining (Chapter 13). Chromatofocusing (pages 166 and 174) also gives information on pI. Methods for determining the pI of proteins are also possible using HPCE techniques (pages 245–250).

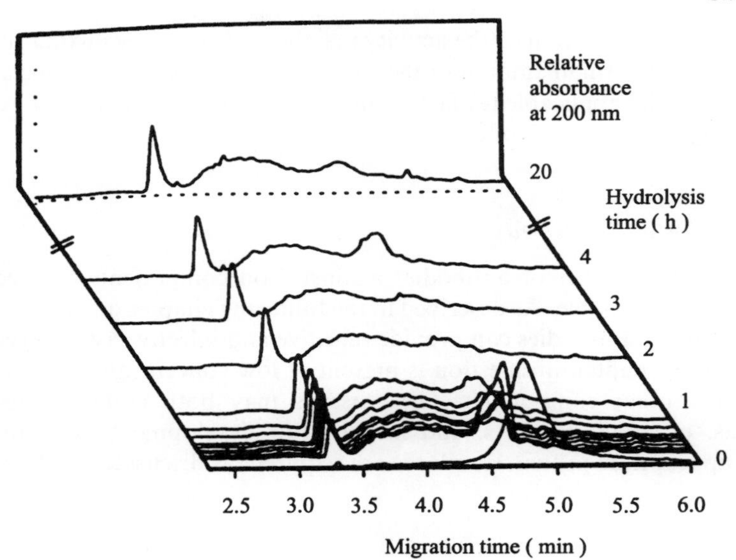

Figure 12.3 *Electropherograms of isolated soy protein hydrolysates treated with a fungal protease at pH 8.5. The peak at 4.0 min was unique for the protease. Electrophoreses were performed in a fused-silica tube, 70 cm, 75 μm i.d., with run buffer 20 mM borate, pH 8.5, at 30 °C, with a voltage of 25 kV and on column detection at 200 nm*
(Reprinted from T.M. Wong, C.M. Carey, and S.H.C. Lin, *J. Chromatogr. A*, **680**, 413 (1994), Elsevier Science, Amsterdam)

Protein Composition

Protein-containing fractions may in addition be characterized by their amino acid composition, determined as described previously (page 332). For proteins with known amino acid composition, such analyses give information on the purification degree obtained (Table 12.5).

To determine the amino acid sequence of a protein, the protein must be highly purified and in addition degradable into peptides that are analysable by the sequencer.

Another possibility for protein characterization is the enzymatic degradation of the protein, which will result in a specific peptide pattern ('fingerprint') for that specific protein. Owing to the high separation efficiency in HPCE (Chapter 10), this method may be efficient for the evaluation of such enzymatic degradation (Figure 12.3).

9 Selected Examples of Protein Purification Procedures

In the following sections, selected examples are given of relevant applications of the different separation techniques for protein purification. Table 12.6 summarizes the methods used for the individual protein purification procedures.

Table 12.6 *Guide to link between methods and examples of practical usage of the methods given as Box numbers and figure numbers (in parenthesis)*

Method	*Myrosinase*	*α-Galactosidase*	*Peroxidase*	*Alkaline phosphatase*	*Acid phosphatase*	*Trypsin*	*Protein-type proteinase inhibitor*	*Phenylalanine ammonia-lyase*	*Ovomucoid*
				Group separation					
Precipitation with $(NH_4)_2SO_4$			12.17	12.19	12.21			12.28	
Precipitation with organic solvents									12.29
Gelfiltration	12.11 (12.4)	12.14	12.17 (12.8)	12.19	12.21	12.23	12.26	12.28	12.29
Flash Chromatogr.		12.14	12.17	12.19	12.21	12.23	12.36		12.29
Ion exchange			12.17	12.19					
Dialysis			12.17	12.19	12.21	12.23	12.23		
				High-resolution chromatography					
Affinity chromatogr.	12.11								
HPLC/FPLC									
HIC								12.28	
Ion exchange		12.14 (12.7)				12.23			
Gel filtration		12.14 (12.7)					12.26		12.29
Chrom.focus.	12.11 (12.5)								
HPCE	(12.6)						(12.9)		
				Methods of analysis					
Enzyme assay	12.10	12.12 12.13	12.15 12.16	12.18	12.20	12.22	12.24 12.25	12.27	12.24
Electrophoretic specific detection	12.11	12.14	12.17						

Selected procedures for a range of different enzymatic assays are also given. For further general descriptions and suggestions for assays for various enzymes, Refs. 6 and 7 may be consulted.

Myrosinase

Myrosinase (EC 3.2.3.1) is an enzyme that catalyses the hydrolysis of the β-thioglucoside binding in glucosinolates. Myrosinase can be isolated from various sources within the plant order Capparales and from a few other plants. It is a glycoprotein and exists as different numbers of isoenzymes depending on plant species, the particular plant part, and developmental stage. Both intact glucosinolates and their degradation products are strongly correlated to the quality of both rapeseed (*Brassica napus* L. and *B. campestris* L.) and other glucosinolate-containing plants used for food and feed.[8] In human nutrition, the organoleptic properties of vegetables containing glucosinolates depend strongly on the myrosinase-catalysed cleavage products of the glucosinolates.

Myrosinases are β-thioglucoside glucohydrolases that cleave the thioglucoside binding in glucosinolates in the presence of water. The resulting aglycone is unstable in water, and therefore the myrosinase assay (modification of Ref. 9) measures the decrease in substrate (here; sinigrin) concentration (Box 12.10). To have a high concentration of sinigrin and at the same time avoid too high an absorbance of the reaction solution, cuvettes with path lengths of 0.5 cm are recommended. The ascorbic acid used for the assay also contributes to the absorption and as the ascorbic acid is gradually degraded, the assay has to be corrected for the value contributed by the ascorbic acid.

Box 12.10 *Myrosinase assay with sinigrin as substrate*

Reaction process

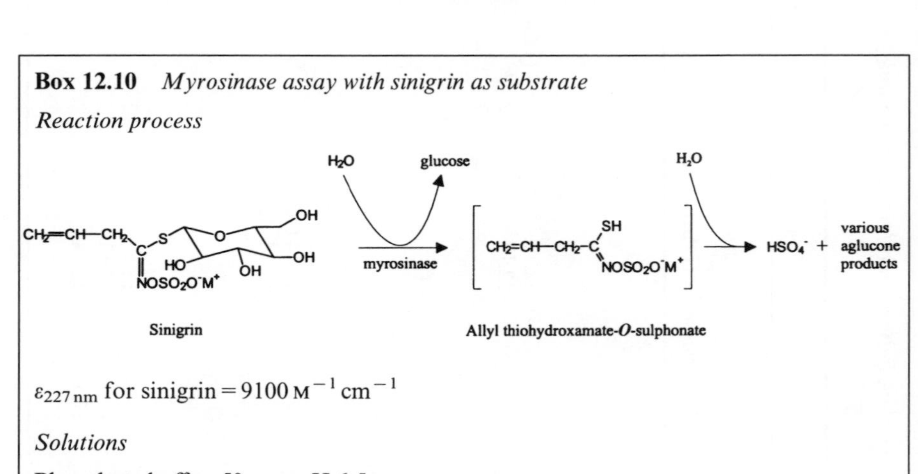

$\varepsilon_{227\,nm}$ for sinigrin = 9100 $M^{-1}\,cm^{-1}$

Solutions

Phosphate buffer; 50 mM, pH 6.5
Ascorbic acid solution; 10 mM in deionized water, prepared daily
Sinigrin; 1.2 mM in deionized water (*ca.* 0.48 mg ml^{-1})
Myrosinase solution; the myrosinase is dissolved in deionized water or assay buffer, with a concentration that gives ΔA/min = 0.02–0.08

(*continued opposite*)

Assay

Myrosinase solution $x\,\mu l$
Ascorbic acid solution $75\,\mu l$
Phosphate buffer $z\,\mu l$
$x + 75\,\mu l + z = 1000\,\mu l$, incubation at 30 °C for 5 min followed by addition of
Sinigrin $500\,\mu l$
Assay volume 1.5 ml

Measurement

Absorbance at 227 nm is measured continuously using a cuvette with 0.5 cm path length.
The myrosinase activity is calculated according to Equation 12.2, with $\Delta A/\Delta t$ as $\Delta A_{227\,nm}/\Delta t$ corrected for the change in absorbance caused by ascorbic acid, which is measured in an assay without myrosinase

Assay in microtitre plates

For the microtitre assay the myrosinase activity is measured by determination of the liberated amount of glucose using a commercially available kit for measuring glucose concentration (Sigma)

Reaction process

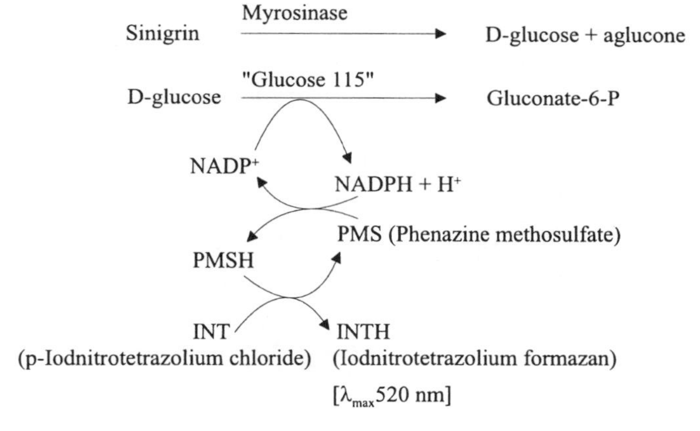

Sample with myrosinase ($50\,\mu l$) and sinigrin ($50\,\mu l$, 1.2 mM as above) are mixed in microtitre plates and left at ambient temperature for 1 h. Glucose reagent ($20\,\mu l$; Glucose 115 reagent, Sigma) is added and the red colour development is read at 520 nm after 10 min

Gel filtration of a Con-A eluate of *B. napus* cv. Doral resulted in isolation of three different myrosinase isoforms (Figure 12.4).[10] Fractions containing the isolated isoforms were ultrafiltrated to obtain a concentration applicable for FPLC chomatofocusing (Figure 12.5). Chromatofocusing of the sample showed that there were more than three enzyme forms with different pI-values. Recordings of the absorbance at 280 nm showed that the samples did not only contain active myrosinases, which may partly be owing to inactivation of part of the myrosinases during ultrafiltration.

Box 12.11 *Isolation and characterization of myrosinase (β-thioglucoside glucohydrolase)*

Extraction

1. Select the starting material for isolation of myrosinase, *e.g.* rapeseed (*Brassica napus* L.)
2. Weigh 10 g rapeseed to a centrifuge tube and add deionized water (*ca.* 67 ml)
3. Place the centrifuge tube in a waterbath with ice and extract for 20 min by homogenization with Ultra-Turrax. The homogenization should be interrupted every second minute to avoid heating of the extract
4. Add *ca.* 30% chloroform (v/v) and extract for additional 2 min to defat the extract
5. Centrifuge for 30 min (4 °C, 4200*g*). The extract will now separate into three distinct phases: the water phase at the top, the lipid phase in the middle, and the chloroform phase and sediment at the bottom of the centrifuge tube
6. Decant the water phase into another centrifuge tube and repeat centrifugation for 20 min (4 °C, 15000 rpm) to remove impurities
7. Determine protein content (page 328 and Box 12.3) and enzyme activity (Box 12.10) in the clear water phase (crude extract)

Affinity chromatography

Glycoproteins can be retained on a column containing Sepharose with Concanavalin A attached (Con-A-Sepharose; Pharmacia). Concanavalin A binds α-D-glucopyranosyl and α-D-mannopyranosyl with free hydroxy groups at the C-3, C-4, and C-5 positions, including the glycoprotein myrosinase. Compounds not bound are removed from the column by simply washing with buffer, whereas retained glycoproteins are liberated by adding methyl-α-D-mannopyranoside, thus shifting the equilibrium.

8. Prepare the column; *e.g.* use a Pasteur pipette as column and add a bit of glass wool to the bottom
9. Add *ca.* 2 ml Con-A-Sepharose to the column and allow to settle
10. Equilibrate the column with *ca.* 2 ml buffer A (0.02 M TRIS-HCl + 0.5 M NaCl, pH 7.4). Repeat 3–4 times. After the last addition of buffer, leave the column overnight
11. Next day, degas the crude extract and then pass it through the equilibrated column (max. 30 ml h^{-1}). Wash with buffer A until the $A_{280\,nm}$ of the eluate is close to zero. To obtain this, about 15 to 25 ml buffer is needed
12. Add 1 ml 0.25 M methyl-α-D-mannopyranoside solubilized in buffer A. Leave the column overnight
13. Next day, elute the column with 1 ml portions of the solution used in 12. Stop the flow for 30 min between each fraction. Measure enzyme activity (Box 12.10) of every fraction and stop the elution when the activity approaches zero
14. Combine the fractions with enzyme activity and determine protein content (page 328) and total enzyme activity

Gel filtration

Gel filtration forms the basis for separation of myrosinase isoenzymes according to size

15. Use a column with dimensions 5 × 85 cm
16. Add swelled and degassed Sephadex G-200 material to the column and equilibrate with buffer B (0.05 M phosphate + 0.15 M NaCl, pH 6.8). Use four times the column volume

(*continued overleaf*)

17. Add the combined eluate obtained by affinity chromatography to the column. Use buffer B for the elution (flow rate: 25–30 ml h^{-1}, fraction size: 15 ml, detection: 280 nm)
18. Test the obtained fractions for enzyme activity (Box 12.10) and combine the fractions according to the results. Determine protein content (page 328) and total enzyme activity

Concentration and desalting

Concentration/desalting is needed prior to further purification, for example, by FPLC and electrophoresis. It should be noted that concentration/desalting procedures such as dialysis and ultrafiltration imply a risk for considerable loss of enzyme activity. This loss may be reduced by using 10 mM imidazole–HCl buffer, pH 6.0 for the concentration. Alternatively, the sample may be concentrated by affinity chromatography (8.–14.)

FPLC chromatofocusing

19. Samples with high ionic stremgth (*I*) should be buffer exchanged prior to FPLC. Apply 2.5 ml sample to a prepacked PD-10 column (Pharmacia) containing 9.1 ml Sephadex G-25 material, and elute the myrosinase with 3.5 ml buffer C (0.025 M Bis-TRIS, pH 7.1)
20. Add buffer exchanged samples to a prepacked Mono P HR 5/20 column equilibrated with buffer C
21. Elute the column with 3 ml buffer C, 46 ml buffer D [10% Polybuffer 74 (Pharmacia), pH 4.0] and 3 ml buffer C (flow rate: 1 ml min^{-1}, fraction size: 1 ml, detection: 280 nm)
22. Determine protein content (page 328) and enzyme activity (Box 12.10) and pH in the relevant fractions

Electrophoresis

Many electrophoretic techniques are suitable for further characterization of the purified myrosinase isoenzyme fractions, *e.g.* isoelectric focusing (Boxes 9.3 or 9.5) or SDS-PAGE (Box 9.8)

Specific myrosinase detection in electrophoresis gels

Gels from isoelectric focusing are placed in a solution of 1 mg sinigrin, 5 mg barium chloride, 0.5 mg ascorbic acid, and 0.1 ml acetic acid per ml of solution (D.B. MacGibbon and R.M. Allison, *Phytochemistry*, 1970, **9**, 541). Myrosinase bands are seen as white precipitates of Ba_2SO_4 after 8–36 h

Enzyme kinetics

Myrosinase is a hydrolase exhibiting pseudomonosubstrate kinetic (pages 53–59 and 61)

The purification of myrosinases may also be followed by HPCE as illustrated in Figure 12.6.

α-Galactosidase

α-Galactosidases (EC 3.2.1.22) are hydrolases that catalyse the hydrolytic release of galactose from α-D-galactosides,[11,12] including oligosaccharides from

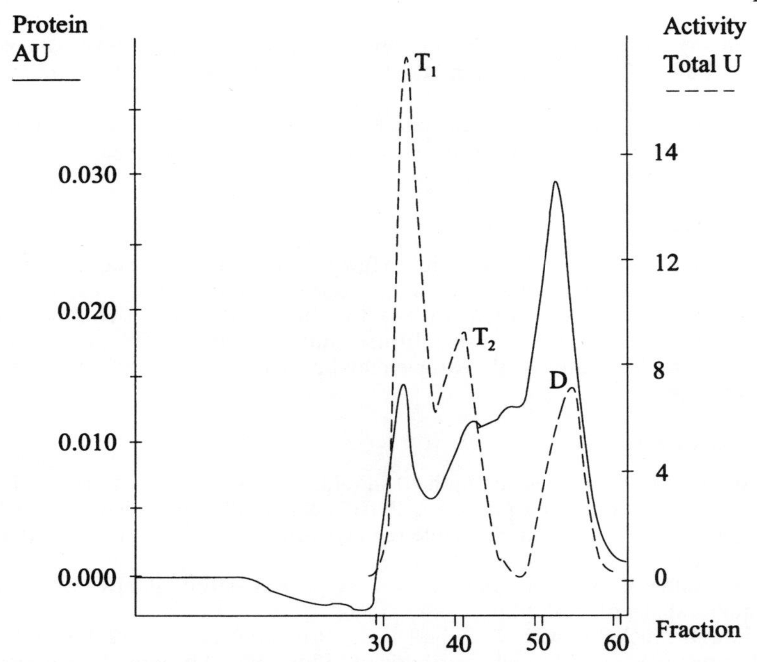

Figure 12.4 *Gel filtration of Con-A purified myrosinase from* Brassica napus *cv. Doral using a Sephadex G-200 column*
(Reproduced by permission from S. Michaelsen, K. Mortensen, and H. Sørensen, 'Myrosinases in *Brassica*: Characterization and Properties', GCIRC-Congress, Saskatoon, Canada, 1991, vol. 3, p. 905)

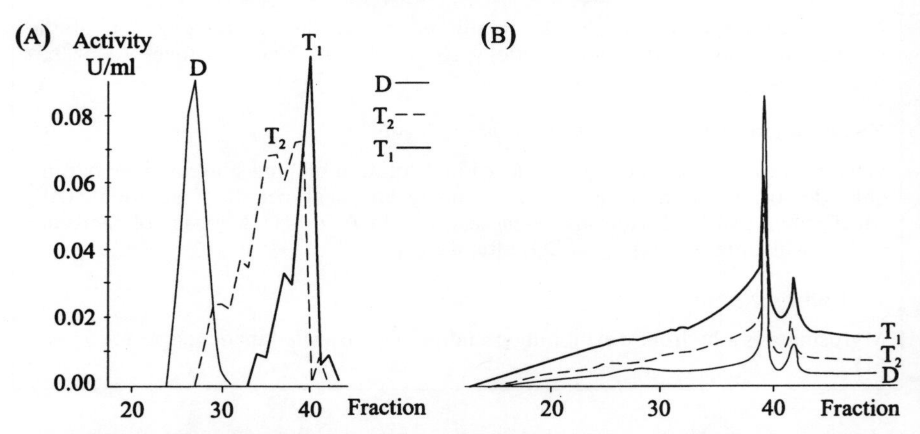

Figure 12.5 *FPLC chromatofocusing of the enzyme forms T1, T2 and D obtained after gel filtration of Con-A purified myrosinase from* Brassica napus *cv. Doral.* (A) *Activity.* (B) *Absorbance at 280 nm*
(Reproduced by permission from L. Buchwaldt, L.M. Larsen, A. Plöger, and H. Sørensen, *J. Chromatogr.*, 1986, **363**, 71; S. Michaelsen, K. Mortensen, and H. Sørensen, 'Myrosinases in *Brassica*: Characterization and Properties'. GCIRC-Congress, Saskatoon, Canada, 1991, vol. 3, p. 905)

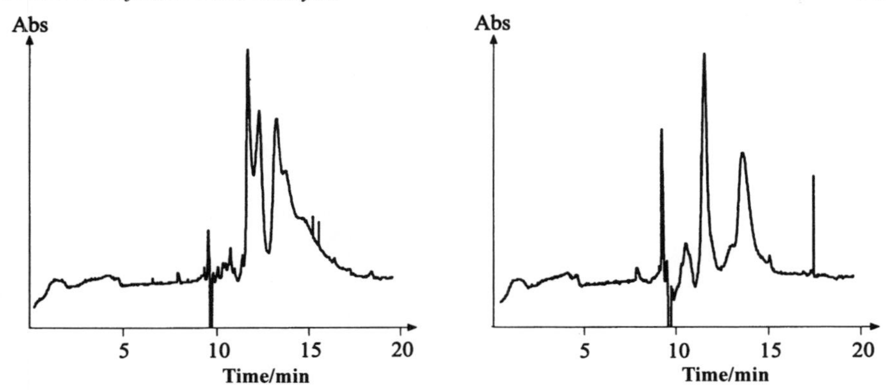

Figure 12.6 *HPCE of Con-A purified myrosinase from* Sinapis alba. *HPCE conditions: Buffer: 35 mM sodium cholate, 50 mM taurine, 100 mM Na_2HPO_4 and 2% propan-1-ol; temperature: 30 °C; voltage: 15 kV; detection: 214 nm; capillary: 50 μm i.d., 76 cm total length, 53 cm effective length*

the raffinose family. Oligosaccharides are undesirable constituents of food and feed because they are not hydrolysed by animal enzymes and therefore escape absorption in the small intestine. In the distal part of the digestive tract, the oligosaccharides are degraded in the presence of α-galactosidases of microbial origin and intestinal gases are produced thus contributing to the development of flatulence.[13] α-Galactosidase, which is a glycoprotein, is typically isolated from leguminous seeds, *e.g.* pea (Figure 12.7), soybean, and faba beans (Boxes 12.12–12.14).

Peroxidases

Peroxidases (EC 1.11.1.7) comprise a heterogeneous group of enzymes found widespread in the nature and it is thus possible to isolate peroxidases from various sources including, among others, milk (Box 12.17), liver, higher plants such as horseradishes (Box 12.17), potatoes, legumes, vegetables, and fruits.[6] The best studied peroxidase is present in horseradish.[14] Peroxidases are haem proteins and may, in addition to the methods described below, be characterized by the RZ-value $= A_{403\,\mathrm{nm}}/A_{280\,\mathrm{nm}}$. Peroxidases oxidize a substrate while reducing H_2O_2. For assays of peroxidases isolated from different sources, the conditions for the peroxidase assay should be optimized individually. In Boxes 12.15 and 12.16, two assays with different substrates are described.

Alkaline Phosphatase

Alkaline phosphatases (EC 3.1.3.1.) comprise a heterogeneous group of enzymes that are nonspecific and cleave many different phosphate esters. This group of enzymes are found in higher animals, bacteria, and fungi. In animals,

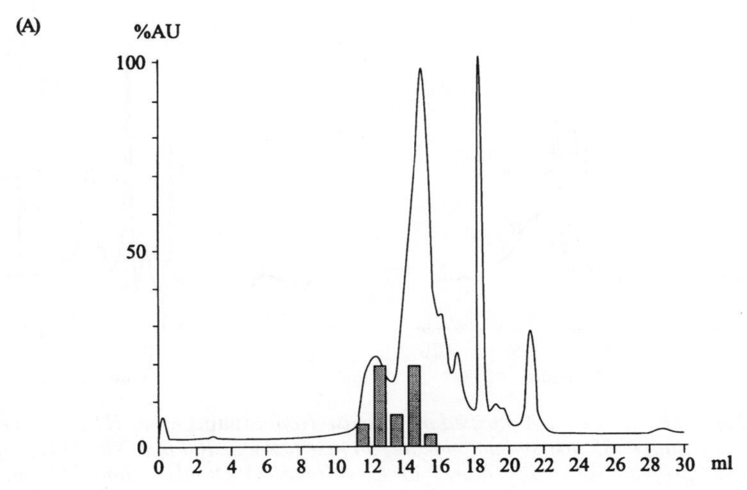

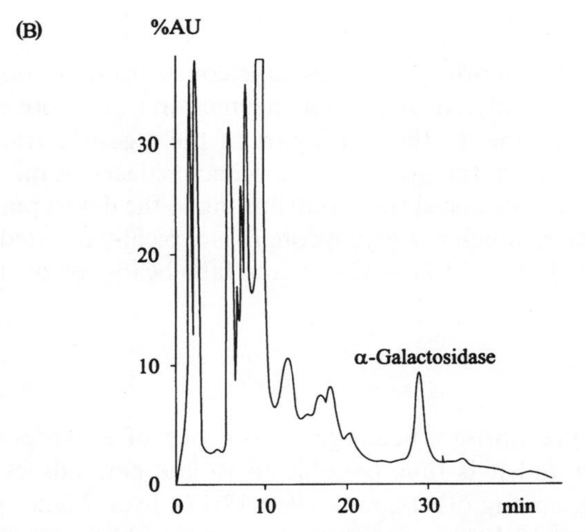

Figure 12.7 (A) *FPLC gel filtration (Box 12.14) of α-galactosidase purified from pea (*Pisum sativum *cv. Bodil) by flash chromatography (ZetaPrep SP, Box 12.14) and gel filtration (Sephadex G-200, Box 12.14). Bars indicate α-galactosidase activity.* (B) *FPLC cation exchange (Mono S, Box 12.14) of the last eluting α-galactosidase peak from* (A)

the enzymes are typically found in intestinal mucosa but may also be isolated from liver, placenta, kidney, and milk (Box 12.19).

Alkaline phosphatases are phosphoric monoester phosphohydrolases. Alkaline phosphatases have pH optimums of 7–12. The pH-optimums are generally very narrow, and it may therefore be necessary to use 1 M assay buffers when the enzyme is kept in buffers with pH-values that diverge from the optimal assay pH. Phosphatases require, generally, either Zn^{2+}, Mg^{2+}, or Mn^{2+} as activators

Box 12.12 *α-Galactosidase assay with α-galactopyranosyl-*p*-nitrophenol as substrate*

Reaction process

NO_2 + H_2O $\xrightarrow{\alpha\text{-galactosidase}}$ D-galactose + HO NO_2

HO NO_2 + HO^- ⟶ H_2O + ^-O NO_2

$\varepsilon_{405\,nm}$ for *p*-nitrophenolate = 18500 $M^{-1}\,cm^{-1}$
The temperature optimum for the process is 30 °C

Solutions

Na_2CO_3; 5% (w/v) in deionized water
Citrate; 0.2 M, pH 4.5
McIlvaine buffer; 0.1 M, pH 4.5 (Handbook of Biochemistry, ed. H.A. Sober, CRC Press, Cleveland, OH, 1973)
α-Galactopyranosyl-*p*-nitrophenol (Serva); 12 mM in McIlvaine buffer
α-Galactosidase solution; a concentration is chosen that results in ΔA/min = 0.02–0.08. Changes in enzyme volume for the assay are corrected for by adjusting the citrate volume to keep the assay volume constant

Assay

α-Galactosidase solution	100 μl
Citrate buffer	200 μl
α-Galactopyranosyl-*p*-nitrophenol	100 μl

Reaction for 10 min followed by addition of

Na_2CO_3	2.3 ml
Assay volume	2.7 ml

Meaurement

Absorbance is measured at 405 nm. The α-galactosidase activity is calculated according to Equation 12.2

Assay in microtitre plates

α-Galactosidase solution (10–30 μl), citrate buffer (20 μl), and substrate (10 μl) are mixed in wells of microtitre trays and left covered at ambient temperature for several min or hours. The reaction is stopped by addition of 60 μl Na_2CO_3

(for assay, see Box 12.18). Excess Zn^{2+} is strongly inhibiting as is the product of the reaction. Chelating buffers (pages 26–28 and 44–46) are therefore usually not suitable for phosphatase assays. If enzyme solutions obtained from a series of purifications have lost their enzyme activity, addition of Mg^{2+} may reactivate the enzymes.

Box 12.13 *α-Galactosidase assay with natural substrates*

The oligosaccharides of the raffinose family (raffinose, stachyose, verbascose, and ajugose) are α-(1→6)-galactosides linked to the glucose moiety of sucrose, containing one to four α-(1→6)-galactoside moieties. In the assay using natural substrates, the liberated D-galactose is measured using a method modified after Boehringer Mannheim Biochemica, 'Methods of Enzymatic Bioanalysis and Food Analysis', 1995 and the Boehringer Mannheim kit (Raffinose UV method)

Reaction process

$$\text{Raffinose} + H_2O \xrightarrow{\alpha-\text{galactosidase}} \text{D-galactose} + \text{sucrose}$$

$$\text{D-galactose} + NAD^+ \xrightarrow{\text{galactose dehydrogenase}} \text{D-galactonic acid} + NADH + H^+$$

$\varepsilon_{340\,nm}$ for NADH = 6.3 mM^{-1} cm^{-1}

Solutions

Citrate; 10 mM, pH 4.0
TRIS-HCl, 0.75 M, pH 8.6
NAD^+; 14 mM (Boehringer Mannheim kit)
α-Galactosidase; minimum 0.6 U ml^{-1}

Substrate; raffinose, stachyose, verbascose, or ajugose in deionized water with concentrations that will not exceed 3 μmol galactose per ml

β-Galactose dehydrogenase (Boehringer Mannheim)

Assay

Citrate	200 μl
Substrate	200 μl
α-Galactosidase	30 μl

Incubation at 30 °C for a defined period followed by addition of

TRIS-HCl	1000 μl
NAD^+	100 μl
Deionized water	1500 μl

Measurement of reference absorbance at 340 nm (A_1)

β-Galactose dehydrogenase	20 μl

Incubation for 1 h at 30 °C. Measurement of absorbance at 340 nm (A_2).

Assay volume	3.05 ml

Calculation of galactose concentration (mM):

$$c = \frac{V(A_2 - A_1)}{\varepsilon_{340\,nm} l v}$$

where c = concentration of galactose (mM)
V = total volume of assay mixture (ml)
v = sample volume (ml)
l = light path-length (cm)
$\varepsilon_{340\,nm}$ = absorption coefficient of NADH at 340 nm (M^{-1} cm^{-1})
A_1 = reference absorbance
A_2 = final absorbance

Box 12.14 *Isolation and characterization of α-galactosidase*

Extraction

1. Select the starting material for isolation of α-galactosidase, *e.g.* seeds of pea (*Pisum sativum* L.)
2. Germinate seeds of pea for 3 days in Petri dishes on moist filter paper under plastic (laboratory conditions)
3. Lyophilize the cotyledons (testas removed) and grind into meal
4. Extract 40 g of meal in 150 ml deionized water using Ultra-Turrax for 2 min intervals of 30 s to avoid heating
5. Centrifuge for 20 min (3000*g*), keep the supernatant and re-extract the sediment as described in 4. using 100 ml deionized water
6. Centrifuge for 20 min (3000*g*), keep the supernatant and combine it with the supernatant from 5. Adjust pH to 4.0 by HCl and leave overnight at 4 °C
7. Next day, adjust pH to 5.3 by NaOH and centrifuge for 20 min (13 000*g*). Keep the supernatant (crude extract)
8. Determine enzyme activity (Box 12.12 or 12.13) and protein content (page 331) in the crude extract

Flash chromatography

9. Filtrate the crude extract through a ZetaPrep disposable filter or through a filter for water-based buffers (0.22 μm). Buffer exchange the filtered crude extract to buffer A (0.01 M CH_3CO_2Na; pH 5.3)
10. Prepare a strongly acidic cation-exchanger (ZetaPrep SP, KE) by washing (flow rate 4 ml min^{-1}) with 100 ml buffer B (0.01 M CH_3CO_2Na + 0.35 M NaCl, pH 5.3). Continue with wash with buffer A
11. Pump the buffer-exchanged crude extract through the column at a flow rate of 4 ml min^{-1}
12. Elute the column with buffer A (flow rate: 4 ml min^{-1}, fraction size: 10 ml, detection: 280 nm) until the UV-signal stabilizes at zero. Then start salt gradient elution with 250 ml of each of buffers A and B in the gradient mixer. The salt concentration of the elution buffer is to be continuously increased from 0 (pure buffer A) to 0.35 M (pure buffer B). Continue the elution with buffer B until the UV-signal stabilizes at zero
13. Test the obtained fractions for enzyme activity (Box 12.12) and combine the relevant fractions. Determine protein content (pages 328 and 331) and total enzyme activity

Concentration and affinity chromatography

Concentration of the sample is advantageous prior to further purification; however, α-galactosidase is not very stable. Concentration therefore implies a risk for considerable loss of enzyme activity and should be performed under gentle conditions. Ultrafiltration is acceptable for this purpose only with some membranes, but affinity chromatography by Con-A-Sepharose, as described for myrosinase (Box 12.11), has also proved suitable. The affinity chromatography procedure for α-galactosidase is typically performed after gel filtration to obtain an appropriate enzyme concentration for further studies

Gel filtration

14. Use a column with dimensions 2.5 × 100 cm

(*continued overleaf*)

15. Add swelled and degassed Sephadex G-200 material to the column and allow to settle
16. Equilibrate the column with buffer C (0.05 M NaH_2PO_4 + 0.15 M NaCl, pH 5.3) at a flow rate of 8 ml h^{-1}
17. Add 30–40 ml of the concentrated supernatant to the column. Elute with buffer C (flow rate: 8 ml h^{-1}, fraction size: 10 ml, detection: 280 nm)
18. Test the obtained fractions for enzyme activity (Box 12.12 or 12.13) and combine the relevant fractions, adjust pH to 4.0 and store at 4 °C. Determine protein content (page 328) and total enzyme activity

FPLC

FPLC gel filtration and FPLC cation exchange have been tested for α-galactosidase purification with good results.

FPLC gel filtration

19. Use a Superose 12 gel-filtration column (Pharmacia) equilibrated with buffer D (0.05 M NaH_2PO_4 + 0.15 M NaCl pH 6.0)
20. Inject 200 μl filtrated sample (0.22 μm filter) onto the column. Elute the sample with buffer D (flow rate 0.2 ml min^{-1}, detection 280 nm) and perform test as in 24

FPLC cation exchange

21. Samples should be buffer exchanged with the equilibration buffer prior to analysis
22. Use a Mono S HR 5/5 column (Pharmacia) equilibrated with buffer E (0.05 M CH_3CO_2Na + 0.02 M NaCl, pH 5.5)
23. Inject 200 μl of filtrated sample (0.22 μm filter) onto the column. Use salt gradient elution according to the following programme: 0–3 min: 0% buffer F; 3–6 min: linear increase to 40% buffer F; 6–24 min: linear increase to 55% buffer F; 24–32 min: linear increase to 75% buffer F; 33–35 min: 100% buffer F; 35 min: 0% buffer F (flow rate: 0.8 ml min^{-1}, fraction size: 1 ml, detection: 280 nm). Buffer F: 0.05 M CH_3CO_2Na + 0.35 M NaCl, pH 5.5)
24. Test the obtained fractions for enzyme activity (Box 12.12) and protein (page 328)

Electrophoresis

Isoelectric focusing has been the method of choice for characterization of α-galactosidase, but other techniques can also be used. Electroblotting of focused protein bands (pages 204–207) followed by specific α-galactosidase detection can, for example, be combined with isoelectric focusing (Box 9.3 or 9.5) in the PhastSystem (Pharmacia)

Specific detection of α-galactosidase after electrophoresis

Focused protein bands can be blotted onto an immobilizing PVDF membrane (PhastTransfer, Pharmacia). The membrane is washed with water for 2–5 min and covered with 4 layers of lens cleaning tissue soaked in sun substrate solution [25 mg α-galactopyranosyl-*p*-nitrophenol in 8 ml McIlvaine citric acid–phosphate, pH 4.5 (Handbook of Biochemistry, ed. H.A. Sober, CRC Press, Cleveland, OH, 1973)]. The membrane is incubated in a closed chamber for 2–3 h followed by development of *p*-nitrophenol in an NH_3 atmosphere generated by placing a small container with few drops of concentrated ammonium solution in the chamber

Enzyme kinetics

α-Galactosidase is a hydrolase exhibiting pseudomonosubstrate kinetics (pages 53–59 and 61)

Box 12.15 *Peroxidase assay with guajacol as substrate*

Reaction process (uncertain stoichiometry)

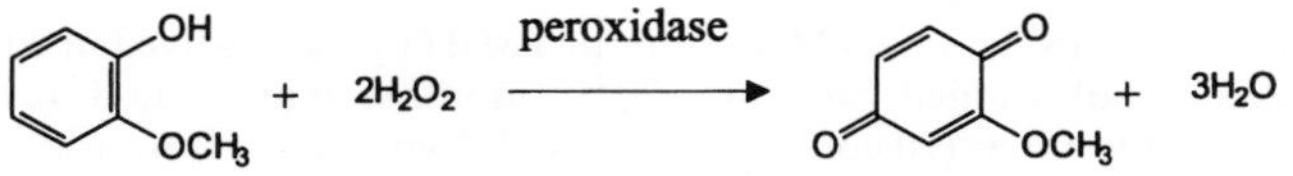

$$\text{guaiacol} + 2H_2O_2 \xrightarrow{\text{peroxidase}} \text{product} + 3H_2O$$

$\varepsilon_{436\,nm} = 25500\ \text{M}^{-1}\,\text{cm}^{-1}$

Solutions

Phosphate buffer; 0.1 M, pH 7.0
Guajacol solution; 0.05 M in 10% EtOH (commercially available)
H_2O_2 solution; 0.03% = 8.8 mM
Enzyme; if the specification on commercially available enzyme does not allow calculation of an expected optimal concentration, an initial concentration of 0.04 mg enzyme ml^{-1} may be used. Preferably dilute in deionized water instead of buffer

Assay

Buffer	2.25 ml
Guajacol solution	250 μl
H_2O_2 solution	250 μl
Enzyme solution	250 μl
Assay volume	3.00 ml (adjusted with buffer, if one of the volumes are changed)

Measurement

Absorbance measurement at 436 nm, either continuously or by reading at intervals of 20 s for 2 min. The peroxidase activity is calculated according to Equation 12.2

Box 12.16 *Peroxidase assay in microtitre plates with 3,3′,5,5′-tetramethylbenzidine (TMB)*

The assay is optimized with respect to analysis of peroxidase from horseradish (H. Gallati and I. Pracht, *J. Clin. Chem. Clin. Biochem.*, 1985, **23**, 453). Analysis of peroxidases from other sources may require an optimization of the reaction conditions (pH, temperature, concentrations). The TMB solution is light sensitive, and the reaction mixtures should therefore be protected from light.

Reaction process

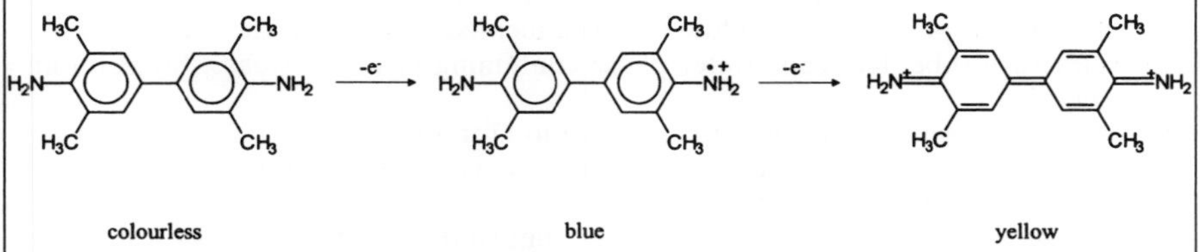

$$\text{TMB} \xrightarrow{-e^-} \text{TMB}^{\bullet +} \xrightarrow{-e^-} \text{TMB}^{2+}$$

colourless — blue — yellow

$\varepsilon_{370\,nm}$ is determined experimentally under the actual assay conditions

(*continued overleaf*)

Solutions

H_2O_2 buffer: Potassium citrate (0.21 M; pH 3.95), 3.15 mM H_2O_2. Store at 15–25 °C (a few weeks)
3,3′,5,5′-Tetramethylbenzidine/TMB solution: TMB (1 g) is dissolved in 20 ml acetone and 180 ml methanol is added. Store at −20 °C (months) or at 15–25 °C (a few weeks).
Assay buffer: H_2O_2 buffer (20 vol.) are mixed with 1 vol. of TMB solution (use within 2–3 h)
Phosphoric acid: 2 M

Assay

Peroxidase solution	10 μl
Assay buffer	100 μl

Reaction for 10 min followed by addition of

Phosphoric acid	100 μl

Measurement

Absorbance is measured at 450 nm. Calculation of activity demands a knowledge of the light path length in the microtitre plates, which is dependent on the actual assay volume

Alternatively

The change in absorbance is measured continually at 370 nm for a mixture of peroxidase solution and assay buffer, *e.g.* in a cuvette

Box 12.17 *Isolation and characterization of peroxidase*

Extraction

The exact extraction procedure depends on the type of starting material chosen

Extraction from fat-containing material, e.g. *milk*

1a. Select the starting material for isolation of peroxidase, *e.g.* milk
2a. Add 240 ml ice-cold milk to a beaker placed in icebath and apply drop by drop 80 ml ice-cold chloroform under magnetic stirring
3a. Homogenize the mixture with Ultra-Turrax for 5 × 30 s and cool in icebath inbetween
4a. Centrifuge for 30 min (4 °C, 12000*g*). Be sure that the centrifuge tubes used are resistant to chloroform. Three phases will appear: the water phase at the top, the chloroform phase in the middle, and sedimented material in the bottom of the centrifuge tube. Keep the water phase containing lactoperoxidase. Store on an icebath
5a. Remove possible impurities by centrifugation for 30 min (4 °C, 20000*g*). Keep the supernatant (crude extract) and determine enzyme activity (Box 12.15 or 12.16) and protein content (pages 331 and 347)
6a. Be aware that precipitation with ammonium sulfate should be performed at the same day as the extraction and that all remaining organic phase should be removed to maintain enzyme activity

(*continued opposite*)

Extraction from plant material

1b. Select the starting material for isolation of peroxidase, *e.g.* horseradish
2b. Cut 50 g horseradish into pieces and add 1 to 5 vol. of deionized water
3b. Homogenize the mixture with Ultra-Turrax for 3–5 × 30 s
4b. Filter the homogenized mixture to remove coarse material
5b. Centrifuge the filtrate for 15 min (4 °C, 5000*g*)
6b. Decant the supernatant (crude extract) through a filter and determine enzyme activity (Box 12.15 or 12.16) and protein content (pages 331 and 347). Store at 4 °C

Precipitation with ammonium sulfate (solid)

7. Slowly add an amount of ammonium sulfate corresponding to 313 $g l^{-1}$ to the crude extract kept on ice and with magnetic stirring
8. Allow to equilibrate for 30 min followed by centrifugation for 15–30 min (4 °C, 12000*g*). Keep both the supernatant and the precipitate. Filtrate the supernatant to remove impurities prior to the following precipitation step
9. Slowly add an amount of ammonium sulfate corresponding to 302 $g l^{-1}$ to the supernatant from 8., kept on ice and with magnetic stirring
10. Allow to equilibrate for 30 min followed by centrifugation for 15–30 min (4 °C, 12000*g*). Keep both the supernatant and the precipitate. Filtrate the supernatant to remove impurities
11. Redissolve the precipitates from 8. and 10. in deionized water (as little as possible)
12. Determine protein content (pages 328 and 347) and enzyme activity (Box 12.15 or 12.16) in the redissolved precipitates and in the supernatant from 10.
13. Continue with the solutions having appreciable enzyme activity compared with the content of protein

Concentration and desalting

Ultrafiltration or dialysis are commonly used techniques. If the enzyme activity is found in a small vol., use dialysis or ultrafiltration against water overnight. If the enzyme activity is found in a large vol., use dialysis or ultrafiltration against polymer overnight.

Gel filtration

14. Use a column with the dimensions 2.5 × 100 cm
15. Add swelled and degassed Sephadex G-75 material to the column and allow to settle
16. Equilibrate the column with deionized water at a flow rate of 80 $ml h^{-1}$
17. Add max. 10 ml of the concentrated sample to the column. Elute with water to obtain sample conditions ready for flash chromatography (flow rate: 80 $ml h^{-1}$, fraction size: 8 ml, detection: 280 nm). Collect 60 fractions in total
18. Test the obtained fractions for enzyme activity (Box 12.15 or 12.16) and combine the relevant fractions. Determine protein content (pages 328 and 347) and total enzyme activity

Flash chromatography

Flash chromatography of peroxidase is performed on a strongly acidic cation-exchanger

19. Buffer exchange the sample with buffer A (0.01 M CH_3CO_2Na, pH 3.0)
20. Equlibrate a strongly acidic cation-exchanger (ZetaPrep SP, KE) by washing (flow rate 4 $ml min^{-1}$) with 100 ml buffer A

(continued overleaf)

21. Pump the buffer exchanged sample through the column at a flow rate of 4 ml min^{-1}
22. Elute the column with buffer A (flow rate: 4 ml min^{-1}, fraction size: 10 ml, detection: 280 nm) until the UV-signal is decreased significantly. Then start gradient elution with buffer A and buffer B (0.01 M Na_2HPO_4, pH 8.3) in the gradient mixer. Finally, use isocratic elution with buffer C (0.01 M Na_2HPO_4 + 0.35 M NaCl, pH 8.3)
23. Test the obtained fractions for enzyme activity (Box 12.15 or 12.16) and combine the relevant fractions. Determine protein content (pages 328 and 347) and total enzyme activity

Ion-exchange, normal column chromatography

24. Use a column with dimensions 2.5 × 50 cm
25. Add swelled and degassed CM-Sephadex C-50 material to the column and equilibriate with buffer B (0.005 M phosphate, pH 6.2)
26. Add the combined fractions obtained from gel filtration. Use the lowest possible volume
27. Elute with a gradient of 250 ml buffer B and 250 ml buffer C (0.1 M phosphate, pH 6.2) (flow rate: 20 ml h^{-1}, fraction size: 8 ml, detection: 280 nm)
28. Test the obtained fractions for enzyme activity (Box 12.15 or 12.16) and combine the relevant fractions. Determine protein content (pages 328 and 347) and total enzyme activity

FPLC

As for most of the other enzymes described, FPLC of peroxidase can be recommended for further purification and characterization

Electrophoresis

In general, electrophoresis of enzyme fractions serve as an efficient tool for checking the purification degree obtained in connection with each step in the isolation procedure. Many electrophoretic techniques are suitable for identification and further characterization of the purified peroxidase fractions, *e.g.* isoelectric focusing (Box 9.3 or 9.5) or SDS-PAGE (Box 9.8)

Specific detection of peroxidase activity after electrophoresis

The gels obtained after non-denaturing electrophoresis may be developed for detection of enzyme activity by incubating the gels in a solution of 2 ml 0.10 M guajacol, 0.5 ml 0.3% H_2O_2, and 27.5 ml deionized water until the peroxidase-containing bands turn brown

Enzyme kinetics

Peroxidase is an oxidoreductase exhibiting Theorell–Change bisubstrate kinetic (page 61)

Acid Phosphatase

Acid phosphatase (EC 3.1.3.2) is widespread in nature, and can be isolated from both animal and plant material, *e.g.* wheat germ, potato, and milk.

Acid phosphatases are orthophosphoric monoester hydrolases and some may be very specific concerning the actual substrate. Acid phosphatases have most

Box 12.18 *Alkaline phosphatase assay with* p-*nitrophenylphosphate as substrate*

Reaction process

$$p\text{-}O_2N\text{-}C_6H_4\text{-}O\text{-}PO_3^{2-} + H_2O \xrightarrow[OH^-]{\text{alkaline phosphatase}} p\text{-}O_2N\text{-}C_6H_4\text{-}O^- + HPO_4^{2-}$$

p-Nitrophenylphosphate *p*-Nitrophenolate

$\varepsilon_{405\,nm} = 18\,500\,\text{M}^{-1}\,\text{cm}^{-1}$

Solutions

TRIS-HCl buffer; 0.1 M, pH 8.5, with 1 mM $MgCl_2$
p-Nitrophenylphosphate solution; 3.8 mM in assay buffer, prepared daily
Enzyme solution; a concentration is chosen that gives ΔA/min = 0.15 when *ca.* 100 μl enzyme solution is used

Assay

TRIS-HCl with Mg^{2+}	1.90 ml
p-Nitrophenylphosphate solution	1.00 ml
Enzyme solution	100 μl
Assay volume	3.00 ml

Measurement

Absorbance measurement at 405 nm, either continuously or by reading at intervals of 1 min for 3–4 min. The alkaline phosphatase activity is calculated according to Equation 12.2. For optimization studies, Mg^{2+} should be added to the enzyme, buffer, and substrate solutions to a final concentration of 1 mM.

Box 12.19 *Isolation and characterization of alkaline phosphatase*

Extraction

The exact extraction procedure depends on the type of starting material chosen, and may be the water extraction procedure described for peroxidase (Box 12.17) or with organic solvents, as described below

Extraction from e.g. *milk*

1. Select the starting material for isolation of alkaline phosphatase, *e.g.* milk
2. Add 200 ml ice-cold milk to a beaker placed in icebath and apply, drop by drop, 100 ml ice-cold n-butanol under magnetic stirring
3. Homogenize the mixture with Ultra-Turrax for 5 × 30 s cooling in an icebath between runs
4. Centrifuge for 30 min (4 °C, 12 000*g*). Be sure that the centrifuge tubes used are resistant to n-butanol. Three phases will appear: the n-butanol phase at the top, the fat containing phase in the middle, and the water phase in the bottom of the centrifuge tube. Carefully decant away the n-butanol phase. The water phase, in which alkaline phosphatase is present, is obtained by use of a water suction pump, as illustrated below

(*continued overleaf*)

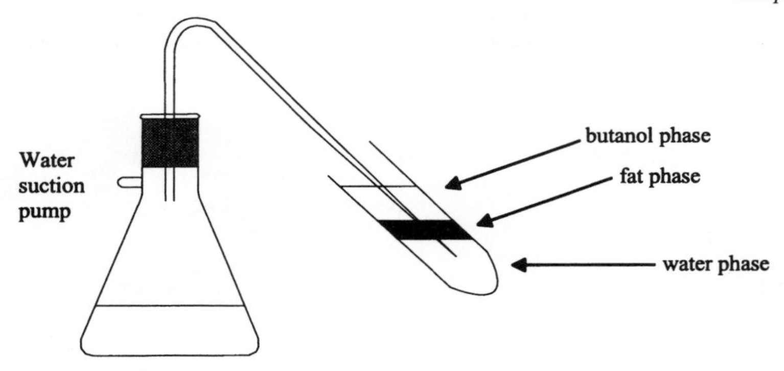

5. Remove possible impurities in the water phase by centrifugation for 30 min (4 °C, 20000*g*). Keep the supernatant (crude extract) and determine enzyme activity (Box 12.18) and protein content (pages 328–329) herein
6. Be aware that precipitation with ammonium sulfate should be performed on the same day as the extraction and that all remaining organic solvent is removed to maintain enzyme activity

Precipitation with ammonium sulfate (solid)

7. Slowly add an amount of ammonium sulfate corresponding to 176 g l^{-1} to the crude extract, which is kept on ice, with magnetic stirring
8. Allow to equilibrate for 30 min followed by centrifugation for 15–30 min (4 °C, 12000*g*). Keep both the supernatant and the precipitate. Filtrate the supernatant to remove impurities prior to the following precipitation step
9. Slowly add an amount of ammonium sulfate corresponding to 356 g l^{-1} to the supernatant from 8., which is kept on ice, with magnetic stirring
10. Allow to equilibrate for 30 min followed by centrifugation for 15–30 min (4 °C, 12000*g*). Keep both the supernatant and the precipitate, which is not always a definite precipitate but more of a creamy phase. Filtrate the supernatant to remove impurities
11. Redissolve the precipitates (creamy phase) from 8. and 10. in deionized water (as little as possible)
12. Determine protein content (pages 328–329) and enzyme activity (Box 12.18) in the redissolved precipitates and in the supernatant from 10.
13. Continue with the solutions having appreciable enzyme activity compared with the content of protein

Concentration and desalting

Ultrafiltration or dialysis are commonly used techniques. If the enzyme activity is found in a small vol., use ultrafiltration or dialysis against water overnight. If the enzyme activity is found in a large volume, use ultrafiltration or dialysis against polymer overnight

Gel filtration

Gel filtration is performed as described for peroxidase (Box 12.17)

(*continued opposite*)

Flash chromatography

Both anion- and cation-exchange flash chromatography (described below) are applicable to alkaline phosphatase. The anion exchange procedure is the same as described for trypsin (Box 12.23).

14. Centrifuge the sample for 30 min (4 °C, 10 000*g*), filter the supernatant, and adjust pH to 3.0 with HCl. Finally filter the pH-adjusted sample through a ZetaPrep disposable filter
15. Equlibrate a strongly acidic cation-exchanger (ZetaPrep SP, KE) by washing (flow rate 4 ml min^{-1}) with 100 ml buffer A (0.01 M CH_3CO_2Na, pH 3.0)
16. Pump the filtered sample through the column at a flow rate of 4 ml min^{-1}
17. Elute the column with buffer A (flow rate: 4 ml min^{-1}, fraction size: 10 ml, detection: 280 nm) until the UV-signal is decreased significantly. Then start salt gradient elution with 250 ml each of buffer A and buffer B (0.01 M CH_3CO_2Na + 0.35 M NaCl, pH 5.3) in a gradient mixer. The salt concentration of the elution buffer is to be continuously increased from pure buffer A to pure buffer B. Continue the elution with buffer B until the UV-signal stabilizes at zero
18. Test the obtained fractions for enzyme activity (Box 12.18) and combine the relevant fractions. Determine protein content (pages 328–329) and total enzyme activity

Ion exchange column chromatography

19. Use a column with the dimension 2.5 × 50 cm
20. Add swelled and degassed CM-Sephadex A-50 material to the column and equilibrate with buffer C (0.1 M TRIS, pH 7.2)
21. Add the combined fractions obtained from gel filtration. Use the lowest possible volume
22. Elute with a gradient of 250 ml buffer C and 250 ml buffer D (0.1 M TRIS + 0.25 M NaCl, pH 7.2) (flow rate: 40 ml h^{-1}, fraction size: 8 ml, detection: 280 nm)
23. Test the obtained fractions for enzyme activity (Box 12.18) and combine the relevant fractions. Determine protein content (pages 328–329) and total enzyme activity.

FPLC

As for most of the other enzymes described, FPLC of alkaline phosphatase can be recommended for further purification and characterization

Electrophoresis

In general, electrophoresis of alkaline phosphatase, as well as other enzyme fractions serves as an efficient tool for checking the purification degree obtained in connection with each step in the isolation procedure

Enzyme kinetics

Alkaline phosphatase is a hydrolase exhibiting pseudomonosubstrate kinetics (pages 53–59 and 61)

often pH-optima between 4 and 6. The pH-optima are generally very narrow and, as for alkaline phosphatase, it may therefore be necessary to use 1 M assay buffers when the enzyme is kept in buffers with pH-values diverging from the optimal assay pH (for assay, see Box 12.20). Phosphatases generally require either Zn^{2+}, Mg^{2+}, or Mn^{2+} as activators. Excess Zn^{2+} is strongly inhibiting as is the product of the reaction. Chelating buffers (pages 26–28 and 44–46) are therefore usually not suitable for phosphatase assays. If enzyme solutions obtained from a series of purifications (Box 12.21) have lost their enzyme activity, addition of Mg^{2+} may reactivate the enzymes.

Box 12.20 *Acid phosphatase assay with* p*-nitrophenylphosphate as substrate*

Reaction process

$$p\text{-Nitrophenylphosphate} + H_2O \xrightarrow{\text{Acid phosphatase}} p\text{-Nitrophenol} + H_2PO_4^-$$

$$p\text{-Nitrophenol} \xrightarrow{OH^-} p\text{-Nitrophenolate}$$

$\varepsilon_{405\,nm} = 18\,500\ \text{M}^{-1}\,\text{cm}^{-1}$

Solutions

Acetate buffer; 0.1 M, pH 5.7
$MgCl_2$ solution; 0.1 M
p-Nitrophenylphosphate solution; 6.2 mM in assay buffer, prepared daily
Enzyme solution; a concentration is chosen that gives $\Delta A/\text{min} = 0.06$ with use of *ca.* 200 μl enzyme solution

Assay

Acetate buffer	500 μl
$MgCl_2$ solution	500 μl
p-Nitrophenylphosphate solution	1.00 ml
Deionized water	3.00 ml
Enzyme solution	200 μl
Assay volume	5.20 ml

Measurement

Samples (800 μl) are taken with appropriate intervals between $t = 0$ to 25 min from the assay mixture and transferred to 2.00 ml NaOH (0.1 M) and the absorbance read at 405 nm. Centrifugation prior to measurements, to eliminate cloudiness in the solution, may be necessary. The acid phosphatase activity is calculated according to Equation 12.2. For optimization studies, Mg^{2+} should be added to the enzyme, buffer, and substrate solutions to a final concentration of 10 mM

Box 12.21 *Isolation and characterization of acid phosphatase*

Extraction

Extraction with water is as described for peroxidase (Box 12.17). Wheat germ may be a possible starting material

Precipitation with $MnCl_2$

1. Place a beaker with crude extract on icebath with weak magnetic stirring
2. Slowly add 1 M $MnCl_2$ (2 ml/100 ml crude extract) in portions of 0.5 ml
3. Centrifuge for 10 min (4 °C, 10 000*g*). Filtrate the supernatant and determine protein content (pages 328 and 331) and enzyme activity (Box 12.20)

Precipitation with ammonium sulfate (saturated solution)

4. Prepare a saturated ammonium sulfate solution by dissolving 150 g ammonium sulfate in 200 ml deionized water under heating. Transfer the solution (not too hot) to a bottle and store at 4 °C. Await formation of crystals at the bottom of the bottle. The saturated solution should be ice-cold when used
5. Place a beaker containing the supernatant from the precipitation with $MnCl_2$ on an icebath under magnetic stirring
6. Slowly add 54 ml ice-cold saturated ammonium sulfate solution/100 ml supernatant. This is best done by pipette, dripping the saturated solution into the supernatant over a time interval of 10 to 15 min. Continue stirring for further 10 to 15 min to avoid foaming
7. Centrifuge for 10 min (4 °C, 10 000*g*). Keep both the supernatant and the precipitate. Filtrate the supernatant to remove impurities and determine protein content (pages 328 and 331) and enzyme activity (Box 12.20)
8. Place a beaker containing the filtrated supernatant from the first precipitation with ammonium sulfate on an icebath under magnetic stirring
9. Slowly add 51 ml ice cold saturated ammonium sulfate solution/100 ml supernatant as described in 6.
10. Prepare a 70 °C waterbath and place on it the mixture from 9, with magnetic stirring and a thermometer
11. When the temperature in the mixture has reached 60 °C, keep this temperature for 2 min, and then quickly place the heated mixture in an icebath with magnetic stirring
12. When the temperature has decreased to 6–8 °C, centrifuge for 10 min (4 °C, 10000*g*). Keep both the supernatant and the precipitate. Filtrate the supernatant to remove impurities and determine protein content (pages 328 and 331) and enzyme activity (Box 12.20) herein
13. Redissolve the precipitate from 7. in 0.05 M CH_3CO_2Na (as little as possible). Redissolve the precipitate from 12. in deionized water (4 °C). Remove insoluble material by centrifugation and determine protein content (pages 328 and 331) and enzyme activity (Box 12.20) in the supernatants obtained
14. Continue with the solutions having the highest enzyme activity compared with the protein content. Store at 4 °C

Concentration and desalting

15. Dialyse or ultrafiltrate overnight against 0.005 M EDTA solution at 4 °C

Gel filtration

Gel filtration is performed as described for peroxidase (Box 12.17), with the exception that 15 ml concentrated/desalted sample should be added to the column

(*continued overleaf*)

Flash chromatography

Flash chromatography is performed as described for alkaline phosphatase (Box 12.19)

Ion-exchange column chromatography

Ion exchange is performed as described for alkaline phosphatase (Box 12.19), with the exception that the CM-Sephadex A-50 material should be replaced by DEAE-Sephadex A-50 material

FPLC

As for most of the other enzymes described, FPLC of acid phosphatase can be recommended for further purification and characterization

Electrophoresis

In general, electrophoresis of acid phosphatase, as well as of other enzyme fractions, serves as an efficient tool for checking the purification degree obtained in connection with each step in the isolation procedure

Enzyme kinetics

Acid phosphatase is a hydrolase exhibiting pseudomonosubstrate kinetics (pages 53–59 and 61)

Trypsin

Trypsin (EC 3.4.21.4) is a protease that catalyses the hydrolysis of proteins at peptide linkages containing adjacent arginine or lysine residues. Trypsin is produced as a proenzyme in pancreas from where it is isolated (Box 12.23). *In vivo*, activation of the proenzyme takes place in the small intestine, whereas activation *in vitro* has to be induced. For *in vitro* studies of protein degradation, it may be essential to isolate and use trypsin from relevant animal species, and for this purpose, methods for measuring (Box 12.22) and isolating trypsin are described.

Box 12.22 *Trypsin assay with* N_α*-benzoyl-L-arginine-4-nitroanilide (L-BAPA) as substrate*

Reaction process

N_α-benzoyl-L-arginine-4-nitroanilide-(L-BAPA) + H_2O $\xrightarrow{\text{trypsin}}$ N_α-benzoylargine + 4-nitroaniline + H^+

(*continued opposite*)

$\varepsilon_{410\,nm}$ for 4-nitroaniline = 8800 $M^{-1}\,cm^{-1}$

Solutions

TRIS-HCl; 0.05 M, pH 8.2
$MgCl_2$; 2 M
N_α-Benzoyl-L-arginine-4-nitroanilide (L-BAPA, Merck, Darmstadt); 4.3 mg L-BAPA is dissolved in 200 μl dimethyl sulfoxide (DMSO) and mixed with 19.6 ml TRIS-HCl buffer and 200 μl 2 M $CaCl_2$
Enzyme solution: diluted in 1 mM HCl, a concentration is chosen that results in ΔA/min = 0.05–1.0

Assay

TRIS-HCl	250 μl
Trypsin	50 μl

Incubation at 37 °C for 10 min followed by addition of

L-BAPA	2.00 ml
Assay volume	2.30 ml

Measurement

Absorption is read continuously at 410 nm for 1 min. The trypsin activity is calculated according to Equation 12.2

Assay in microtitre plates

An assay volume of 275 μl is used, with enzyme concentration chosen that results in ΔA/min = 0.05–1.0. Enzyme and buffer account for 125 μl and substrate for 150 μl. The absorbance is measured continuously at 405 nm for 100 s in a microplate reader

Box 12.23 *Isolation and characterization of trypsin*

Extraction of proenzyme

1. Weigh fresh or defrozen pancreas (15 g) into an appropriate glass
2. Add 40 ml 0.2 M TRIS-HCl buffer (pH 8.0) including 0.05 M $CaCl_2$ and homogenize with Ultra-Turrax for 5 × 20 s at room temperature
3. Centrifuge for 10 min in a table centrifuge (room temperature, 1000*g*). Keep the supernatant and the sediment
4. Reextract the sediment as described in 2. Keep the supernatant and combine it with the supernatant from 3.
5. Centrifuge the combined supernatants for 20 min (4 °C, 12 000*g*). Keep the supernatant

Activation of proenzyme

6. Leave the supernatant from 5. for up to 2 days at room temperature
7. Check the enzyme activity during the activation process
8. When the activity seems to have reached a maximum, continue the isolation and purification procedure

Flash chromatography

Flash Chromatography (FC) using cation as well as anion exchangers are applicable to the activated supernatant

(*continued overleaf*)

Cation exchange FC

9a. Centrifuge the sample for 30 min (4 °C, 10 000*g*), filtrate the supernatant, and adjust pH to 4.0 with HCl. Finally filter the pH-adjusted sample through a ZetaPrep disposable filter
10a. Equlibrate a strongly acidic cation exchanger (ZetaPrep SP, KE) by washing (flow rate 4 ml min^{-1}) with 100 ml buffer A (0.01 M CH_3CO_2Na, pH 4.0)
11a. Pump the filtered sample through the column at a flow rate of 4 ml min^{-1}
12a. Elute the column with buffer A (flow rate: 4 ml min^{-1}, fraction size: 16 ml, detection: 280 nm) until the UV-signal is decreased significantly. Then start salt gradient elution with 250 ml of each of buffer A and buffer B (0.01 M CH_3CO_2Na + 0.30 M LiCl, pH 4.3) in a gradient mixer
13a. Test the obtained fractions for enzyme activity (Box 12.22) and combine the relevant fractions. Determine protein content (pages 328–329) and total enzyme activity

Anion exchange FC

9b. Centrifuge the sample for 30 min (4 °C, 10 000*g*), filter the supernatant, and adjust pH to 8.8 with NaOH. Finally filter the pH-adjusted sample through a ZetaPrep disposable filter
10b. Equlibrate an anion exchanger (ZetaPrep DEAE) by washing (flow rate 4 ml min^{-1}) with 100 ml buffer C (0.02 M TRIS-HCl, pH 8.8)
11b. Pump the filtered sample through the column at a flow rate of 4 ml min^{-1}
12b. Elute the column with buffer C (flow rate: 4 ml min^{-1}, fraction size: 16 ml, detection: 280 nm) until the UV-signal is decreased significantly. Then start salt gradient elution with 250 ml of each of buffer C and buffer D (0.02 M TRIS-HCl + 1.0 M NaCl, pH 7.5) in a gradient mixer
13b. Test the obtained fractions for enzyme activity (Box 12.22) and combine the relevant fractions. Determine protein content (pages 328–329) and total enzyme activity

Concentration and desalting

Desalting should be performed prior to gelelectrophoresis and FPLC ion-exchange chromatography. Ultrafiltration and dialysis with traditional filters/dialysis bags are not recommended owing to the low M_r of trypsin. Use instead special small-pore sized material

Gel filtration

14. Use a column with dimensions 2.5 × 100 cm
15. Add swelled and degassed Sephadex G-75 material to the column and allow to settle
16. Equilibrate the column with deionized water at a flow rate of 100 ml h^{-1}
17. Add the concentrated/desalted sample to the column. Elute with deionized water (flow rate: 100 ml h^{-1}, fraction size: 15 ml, detection: 280 nm)
18. Test the obtained fractions for enzyme activity (Box 12.22) and combine the relevant fractions. Determine protein content (page 328) and total enzyme activity

FPLC

FPLC gel filtration, FPLC on cation exchanger, FPLC on anion exchanger, and FPLC chromatofocusing are all techniques tested for trypsin purification. An example of FPLC on a selected anion exchange column is shown below.

(*continued opposite*)

19. Use a Mono Q HR 5/5 column (Pharmacia) equilibrated with buffer C for 30 min at a flow rate of 0.5 ml min^{-1}
20. Inject 200 μl sample filtered through a disposable filter (0.2 μm) on to the column. Use gradient elution according to the following programme: 0–3 min: 0% buffer E; 3–20 min: linear increase to 80% buffer E; 20–22 min: linear increase to 100% buffer E; 22–24 min: 100% buffer E; 24–26 min: 0% buffer E (flow rate: 1 ml min^{-1}, fraction size: 1 ml, detection: 280 nm). Buffer E: 0.02 M TRIS-HCl + 1.0 M NaCl, pH 8.8)
21. Test the obtained fractions for enzyme activity (Box 12.22) and combine the relevant fractions. Determine protein content (page 328) and total enzyme activity

Electrophoresis

In general, electrophoresis of trypsin, as well as other enzyme fractions, serve as an efficient tool for checking the purification degree obtained in connection with each step in the isolation procedure. Good results have been obtained with HPCE, both as FZCE and MECC (pages 389–390)

Enzyme kinetics

Trypsin is a hydrolase exhibiting pseudomonosubstrate kinetics (pages 53–59 and 61)

Protein-type Proteinase Inhibitor

Protein-type proteinase inhibitors are present in leguminous and cruciferous seeds such as pea, soybean, potato and rapeseed.[15–18] It is proteins with low M_r, typically below 2×10^4, which are of special interest in connection with the utilization of proteins as these inhibitors may specifically inhibit the proteinases present in the digestive tract. The following gives an example of isolation and characterization (Box 12.26) of protein-type proteinase inhibitors from pea (PPI = pea proteinase inhibitor) with inhibitor activity for chymotrypsin (Box 12.25) and trypsin (Box 12.24).

Box 12.24 *Trypsin inhibitor assay with* N_α*-benzoyl-L-arginine-4-nitroanilide (L-BAPA) as substrate*

$\varepsilon_{410\,nm}$ for 4-nitroaniline = 8800 $M^{-1}\,cm^{-1}$
1 inhibitor unit (I U) = the amount of inhibitor required to reduce the enzyme activity by 1 U under the defined assay conditions with determination in the range of 30–70% inhibition

Solutions

TRIS-HCl; 0.05 M, pH 8.2
$MgCl_2$; 2 M
N_α-Benzoyl-L-arginine-4-nitroanilide (L-BAPA, Merck, Darmstadt); 4.3 mg L-BAPA is dissolved in 200 μl dimethyl sulfoxide (DMSO) and mixed with 19.6 ml TRIS-HCl buffer and 200 μl 2 M $CaCl_2$
Enzyme solution; diluted in 1 mM HCl, a concentration is chosen that results in ΔA/min = 0.05–1.0
Inhibitor solution

(continued overleaf)

Trypsin assay:

TRIS-HCl	$y\,\mu$l	
Trypsin	$x\,\mu$l	$(x+y=300\,\mu\text{l})$

Incubation at 37 °C for 10 min followed by addition of

L-BAPA	2.00 ml
Assay volume	2.30 ml

Measurement

Absorption is read continuously at 410 nm for 1 min. The enzyme activity in this reference assay (Akt_{ref}) is calculated as illustrated below. The enzyme volume is not included in the calculation as this is identical in the reference and inhibitor assays

$$Akt_{ref} = \frac{a}{\varepsilon_{\lambda_{max}} l \times 10^{-3}} \cdot \frac{\Delta A_{ref,\lambda_{max}}}{\Delta t}$$

where a = total assay volume (ml), $\varepsilon_{\lambda max}$ = the molar extinction coefficient for 4-nitroaniline ($\text{M}^{-1}\,\text{cm}^{-1}$), l = path length of the cuvette (cm), $\Delta A_{ref,\lambda max}$ = absorbance change of the enzyme solution, and Δt = assay time in minutes.

The established enzyme concentration and volume is then used in an assay where a volume of inhibitor is added together with the enzyme and incubated for 10 min. The inhibitor volume (v_i) is adjusted to obtain inhibition in the range of 30–70% of the enzyme activity

Inhibitor assay

TRIS-HCl	$y\,\mu$l	
Trypsin	$x\,\mu$l	
Inhibitor	$z\,\mu$l	$(x+y+z=300\,\mu\text{l})$

incubation at 37 °C for 10 min followed by addition of

L-BAPA	2.00 ml
Assay volume	2.30 ml

Measurement

Absorption is read continuously at 410 nm for 1 min. The inhibitor activity (IA) per inhibitor volume (IU ml^{-1}) is calculated according to

$$\frac{\text{inhibitor activity}}{\text{inhibitor volume}} = \frac{IA}{v_i} = Akt_{ref}\frac{l}{v_i}\left(1 - \frac{\Delta A_{ref},\lambda_{max}}{\Delta t} \Big/ \frac{\Delta A_{inh},\lambda_{max}}{\Delta t}\right)$$

where $\Delta A_{inh,\lambda_{max}}$ = absorbance change of the inhibitor assay. If several inhibitor assays are performed with different inhibitor volumes, the inhibitor volume corresponding to 50% inhibition may be calculated by extrapolation together with the absorbance change at 50% inhibition and these values may then be used for the calculation of inhibitor activity

The purified inhibitors may be characterized by HPCE as illustrated in Figures 13.5 and 13.6 by mixing different ratios of enzyme with inhibitor. The electrophoretic pattern of the complex formation gives information on the number of binding sites and the specificity of the inhibitors.

Box 12.25 *Chymotrypsin assay and chymotrypsin inhibitor assay with α-Glutaryl-L-phenylalanine-4-nitroanilide (GLUPHEPA) as substrate*

For the chymotrypsin inhibitor assay, the appropriate amount of chymotrypsin is determined in an assay without inhibitor.

Chymotrypsin assay

Reaction process

α–glutaryl-L-phenylalanine-4-nitroanilide-(GLUPHEPA) + H_2O $\xrightarrow[\text{trypsin}]{\text{chymo-}}$ α-glutaryl-L-phenylalanine + 4-nitroaniline

$\varepsilon_{410\,nm}$ for 4-nitroaniline = 8800 $M^{-1}\,cm^{-1}$

Solutions

TRIS-HCl; 1 M, pH 7.6
TRIS-HCl; 0.05 M, pH 7.6
$MgCl_2$; 2 M
α-Glutaryl-L-phenylalanine-4-nitroanilide (GLUPHEPA); 8 mg GLUPHEPA is dissolved in 400 μl warm methanol and mixed with 19.4 ml TRIS–HCl buffer (0.05 M) and 200 μl 2 M $CaCl_2$
Enzyme solution; diluted in 1 mM HCl, a concentration is chosen that results in $\Delta A/\text{min} = 0.06–0.08$

Assay

TRIS-HCl (1 M)	60 μl
Chymotrypsin	50 μl
Deionized water	390 μl

Incubation at 37 °C for 10 min (for use of other enzyme volumes, correction is made in the deionized water volume) followed by the addition of

GLUPHEPA	2.00 ml
Assay volume	2.50 ml

Measurement

Absorption is read continuously at 410 nm for 100 s. The chymotrypsin activity is calculated according to Equation 12.2

(*continued overleaf*)

Chymotrypsin inhibitor assay

A volume of inhibitor is added together with the enzyme, prior to the 10 min incubation. The inhibitor volume is adjusted to obtain inhibition in the range 30–70% of the enzyme activity without inhibitor. The inhibitor activity is calculated according to the formula given in Box 12.24

Assay in microtitre plates

An assay volume of 275 μl is used, with enzyme concentration chosen that results in ΔA/min = 0.06–0.08. Enzyme, inhibitor and buffer accounts for 125 μl and substrate for 150 μl. The absorbance is measured continuously at 405 nm for 100 s on a microplate reader

Box 12.26 *Isolation and characterization of protein-type proteinase inhibitors*

Extraction

1. Homogenize 40 g meal from pea in 150 ml deionized water by Ultra-Turrax for 2 × 6 min with 0.5 min intervals, resulting in an effective time of extraction of 2 × 3 min
2. Centrifuge for 20 min (3000*g*). Keep the supernatant and reextract the sediment as described in 1. using 100 ml deionized water
3. Centrifuge for 20 min (3000*g*). Keep the supernatant and combine it with the supernatant from 2. Adjust pH to 3.0 by HCl and leave overnight at 4 °C
4. Next day, readjust pH of the combined supernatants to 3.0 and centrifuge for 20 min (13 000*g*). Filter the supernatant through a ZetaPrep disposable filter (0.22 μm)
5. Determine protein content (pages 328 and 331) and enzyme inhibitor activity (Box 12.24 or 12.25) in the filtered supernatant (crude extract)

Gel filtration

6. Use a column with dimensions 2.5 × 100 cm
7. Add swelled and degassed Sephadex G-75 material to the column and allow to settle
8. Equilibrate the column with deionized water at a flow rate of 80 ml h^{-1}
9. Add the crude extract to the column. Elute with deionized water (flow rate: 80 ml h^{-1}, fraction size: 10 ml, detection: 280 nm)
10. Test the obtained fractions for enzyme inhibitor activity (Box 12.24 or 12.25) and combine the relevant fractions. Determine protein content (pages 328 and 331) and total enzyme inhibitor activity

Flash Chromatography

11. Buffer exchange the sample with buffer A (0.01 M CH_3CO_2Na, pH 3.0)
12. Equilibrate a strongly acidic cation-exchanger (ZetaPrep SP, KE) by wash (flow rate 4 ml min^{-1}) with 100 ml buffer A
13. Pump the buffer exchanged sample through the column at a flow rate of 4 ml min^{-1}

(*continued opposite*)

14. Elute the column with buffer A (flow rate: 4 ml min^{-1}, fraction size: 10 ml, detection: 280 nm) until the UV-signal is decreased significantly. Then start gradient elution with 250 ml of each of buffers A and B (0.01 M Na_2HPO_4, pH 8.3) in a gradient mixer. If the elution has not finished after 500 ml buffer, continue the gradient elution with 250 ml buffer B and 250 ml buffer C (0.01 M Na_2HPO_4 + 0.1 M NaCl, pH 8.3)
15. Test the obtained fractions for enzyme activity (Box 12.24 or 12.25) and combine the relevant fractions. Determine protein content (page 328) and total enzyme activity

Concentration and desalting

Concentration/desalting is as described for trypsin (Box 12.23)

FPLC

Further purification of PPI may include FPLC gel filtration as well as FPLC on cation and anion exchangers. An example of the procedure for FPLC gel filtration (preparative; 16 consecutive runs) is given below

16. Use a Superose 12 column (Pharmacia)
17. Add 200 μl sample and elute with 0.05 M phosphate + 0.15 M NaCl, pH 7.2 (flow rate: 0.75 ml min^{-1}, fraction size: 1 ml, detection: 280 nm)
18. Test the obtained fractions for enzyme activity (Box 12.24 or 12.25) and combine the relevant fractions. Determine protein content (page 328) and total enzyme activity

Immunochemical methods

mAbs have been raised in mice against pea proteinase inhibitors using a purified PPI preparation, and non-competitive and competitive ELISA have been tested with good results using the obtained antibodies (A.M. Arentoft, H. Frøkiær, H. Sørensen, and S. Sørensen, *Acta Agric. Scand., Sect. B, Soil Plant Sci.*, 1994, **44**, 236)

Electrophoresis

In general, electrophoresis of protein-type proteinase inhibitors, as well as of other enzyme fractions, serves as an efficient tool for checking the purification degree obtained in connection with each step in the isolation procedure. HPCE (MECC), using zwitterions as buffer additives, is recommendable as an analysis tool (Figure 12.9; H. Frøkiær, H. Sørensen, and S. Sørensen, *J. Liq. Chromatogr. Relat. Technol.*, 1996, **19**, 57)

Phenylalanine Ammonia-lyase

Phenylalanine ammonia-lyase (PAL) (E.C. 4.3.1.5) is an enzyme catalysing the transformation of phenylalanine into *trans*-cinnamic acid (for assay, see Box 12.27). PAL is found in all green plants and also in some bacteria and fungi. It is a large enzyme with a relative molecular mass in the range 240–338 × 10^3, and has a high proportion of hydrophobic amino acids. During extraction of the enzyme, phenyl methyl sulfonyl fluoride (PMSF) is added to inhibit proteases and thereby prevent degradation of the enzyme. Sulfanylethanol is added to protect the enzyme against oxidation of thiol groups which is a prerequisite for extraction of PAL (Box 12.28).

Box 12.27 *Phenylalanine ammonia-lyase (PAL) assay*

Reaction process

$$H_3N^+-CH(COO^-)-CH_2-C_6H_5 \longrightarrow C_6H_5-CH{=}CH-COO^- + NH_4^+$$

Phenylalanine → *trans*-Cinnamate

$\varepsilon_{290\,nm}$ for cinnamate = *ca.* 1120 $M^{-1}\,cm^{-1}$

Solutions

Borate buffer; 0.1 M $Na_2B_4O_7$, pH 9.0
Substrate; 0.1 M phenylalanine in 0.1 M borate buffer

Assay

Borate buffer	2.7 ml
Substrate	200 μl
Enzyme	200 μl
Assay volume	2.9 ml

The assay is carried out at 25 °C

Measurement

Absorbance is read at 290 nm for 5–10 min after a lag time of 30 s. Alternatively, a more specific determination of both substrate and product can be obtained by MECC as in Figures 11.12 and 11.22. The PAL peroxidase activity is calculated according to Equation 12.2

Box 12.28 *Isolation and characterization of PAL*

Extraction

1. Select the starting material for isolation of PAL, *e.g.* potato tubers (*Solanum tuberosum* L.)
2. Peel *ca.* 500 g potatoes and cut them into 1 mm slices. Wash the slices in 1% sodium hypochlorite and place them on a layer of wet paper in trays covered with transparent film. Place the trays under an UV lamp overnight (to induce the enzyme)
3. Next day, wash the sliced potatoes in deionized water, which is allowed to drip off. The sliced potatoes are then put into a plastic bag and stored on ice until blending
4. Blend in two batches, each with 500 ml buffer A (0.1 M $Na_2B_4O_7$ + 0.1 M NaCl, pH 8.7 with 0.001 M phenylmethylsulfonyl fluoride (PMSF) and sulfanylethanol (0.8 ml L^{-1})). Use a commercial blender
5. Filter the obtained suspension through 4 layers of gauze and adjust the pH of the filtrate to 5.5 with 1 M HCl
6. Keep the filtrate on ice and stir it for 10 min followed by centrifugation for 10 min (4 °C, 7000*g*). Keep the supernatant (crude extract), and determine protein content (pages 328 and 331) and enzyme activity (Box 12.27)

(*continued opposite*)

Precipitation with ammonium sulfate (solid)

7. Slowly add an amount of ammonium sulfate corresponding to 191 $g L^{-1}$ to the crude extract, which is still kept on ice and with magnetic stirring
8. Allow to equilibrate for 30 min followed by centrifugation for 10 min (4 °C, 7000*g*). Keep the supernatant
9. Add an amount of ammonium sulfate corresponding to 92 $g L^{-1}$ of the supernatant
10. Allow to equilibrate for 30 min followed by centrifugation for 10 min (4 °C, 7000*g*). Discard the supernatant
11. Dissolve the precipitate in buffer B (0.1 M $Na_2B_4O_7$ + 0.1 M NaCl, pH 8.7) (as little as possible, 15–25 ml)
12. Centrifuge for 10 min (4 °C, 27 000*g*) and keep the supernatant. Determine protein content (pages 328 and 331) and enzyme activity (Box 12.27)

Gel filtration

13. Use a column with dimensions 2.5 × 100 cm
14. As column material, use Sephadex G-200 regular. Dissolve 13 g in 450 ml deionized water containing 0.05% NaN_3
15. Wash the column material 4 times with 500 ml buffer B and leave to equilibrate overnight with an additional 500 ml
16. Next day, degas the suspension, transfer the material to the column and allow to settle. Further settling is performed by pumping buffer B through the system at a flow rate of 18 ml h^{-1} for a minimum of 2 h
17. Remove buffer above the Sephadex G-200 material and add the supernatant from the ammonium sulfate precipitation to the column
18. Apply pressure to the column by pumping air through the system at a flow rate of 15 ml h^{-1} until all of the supernatant is in the column material
19. Elute the column with buffer B (flow rate: 15 ml h^{-1}, fraction size: 9 ml, detection: 280 nm)
20. Test the obtained fractions for enzyme activity (Box 12.27). Determine the protein content (page 328) and total enzyme activity

FPLC

Different separation techniques exist within FPLC depending on the chosen column. Here the purification of PAL by FPLC HIC is described

21. Equilibrate the column (Phenyl Superose, Pharmacia) with 20 ml of a mixture consisting of 72.5% buffer C (0.05 M NaH_2PO_4 + 1.7 M $(NH_4)_2SO_4$, pH 7.0) and 27.5% buffer D (0.05 M NaH_2PO_4, pH 7.0) with a resulting ammonium sulfate concentration of 30% saturation
22. Prepare the sample by mixing 300 μl of the concentrated desalted fractions from gel filtration with 300 μl 60% ammonium sulfate. Add only one drop at a time, mix and continue. Centrifuge the mixture for 10 min (2000*g*). Keep the supernatant
23. Inject 500 μl of the supernatant on to the column and elute for 5 min with the buffer mixture described in 21. Increase the proportion of buffer D linearly from 27.5 to 100% over the next 25 min and finally flush the column with pure buffer D for 5 min [25 °C, flow rate: 0.5 ml min^{-1}, fraction size: 1 ml (in peaks, 0.5 ml), detection: 280 nm]
24. Test the obtained fractions for enzyme activity (Box 12.27) and protein (page 328)

(continued overleaf)

Electrophoresis

Among the electrophoretic techniques suitable for further characterization of the purified PAL fractions are isoelectric focusing (Box 9.3 or 9.5) and SDS–PAGE (Box 9.8).

Enzyme kinetics

PAL is a lyase exhibiting monosubstrate kinetics (pages 53–59)

Ovomucoid

Ovomucoid is a small glycoprotein ($M_r = 27 \times 10^3$) present in egg white. Ovomucoid contains 8 disulfide bridges and is stable upon heating. It exhibits trypsin inhibitor activity and may act as an allergen (anaphylactic reaction). The ovomucoid inhibitor activity is determined as described in Box 12.24.

Box 12.29 *Isolation and characterization of ovomucoid*

Extraction

1. Slowly add a 0.5 M TCA–acetone solution (1 : 2) to 45 ml fresh egg white under magnetic stirring. Measure pH continuously and stop the addition when the pH reaches 3.5. Then continue the stirring for 30 min. Leave overnight at 4 °C
2. Next day, centrifuge the mixture for 30 min (4 °C, 13 000*g*). Keep the supernatant
3. Heat the supernatant to 80 °C for 5 min and recentrifuge for 10 min (4 °C, 13 000*g*). Keep the supernatant (crude extract)

Precipitation with acetone

4. Cool the crude extract to 4 °C
5. Add 180 ml ice-cold acetone under magnetic stirring. Isolate the precipitate by filtration and dissolve it in deionized water (as little as possible, *ca.* 25 ml)
6. Use acetone in 2.5 times the vol. of the dissolved precipitate to reprecipitate ovomucoid
7. Centrifuge for 30 min (4 °C, 13 000*g*). Keep the sediment
8. Redissolve the sediment in 20 ml deionized water. Adjust pH to 3.5
9. Determine the protein content (page 328) and inhibitor activity (Box 12.24)

Flash Chromatography, gel filtration, FPLC, and electrophoresis

The above-mentioned techniques should be performed according to the procedures described for PPI (Box 12.26)

10 Selected General and Specific Literature

[1] Pharmacia, 'Gel Filtration Theory and Practice', Rahms, Lund, Sweden, 1984.
[2] H. Schägger and G. von Jagow, *Anal. Biochem.*, 1987, **166**, 368.
[3] O.H. Lowry, N.J. Rosebrough, A.L. Farr, and R.J. Randall, *J. Biol. Chem.*, 1951, **193**, 265.
[4] Boehringer Mannheim Biochemica, 'Methods of Enzymatic BioAnalysis and Food Analysis', 1995.
[5] Sigma Chemical Company, 'Biochemicals, Organic Compounds and Diagnostic Reagents', 1996.

[6] 'Methods of Enzymatic Analysis', ed. H.U. Bergmeyer, Verlag Chemie, Weinheim, 1983, vols 3–5.
[7] Methods in Enzymology, eds. in chief S.P. Colowick and N.O. Kaplan, Academic Press, New York, London.
[8] H. Sørensen, in 'Canola and Rapeseed: Production, Chemistry, Nutrition, and Processing Technology', ed. F. Shahidi, Van Nostrand Reinhold, New York, 1990, ch. 9, p. 149.
[9] S. Schwimmer, *Acta Chem. Scand.*, 1961, **45**, 535.
[10] S. Michaelsen, K. Mortensen, and H. Sørensen, 'Myrosinases in *Brassica*: Characterization and Properties'. GCIRC-Congress, Saskatoon, Canada, 1991, vol. 3, p. 905.
[11] P.M. Dey and J.B. Pridham, in 'Advances in Enzymology', ed. A. Meister, Interscience, New York, 1972, p. 91.
[12] J.B. Pridham and P.M. Dey, in 'Plant Carbohydrate Chemistry', ed. J.B. Pridham, Academic Press, New York, 1974, p. 83.
[13] A.M. Arentoft and H. Sørensen, '1st European Conference on Grain Legumes', Angers, France, 1–3 June, 1992, p. 457.
[14] D. Keilin and E.F. Hartree, *Nature (London)*, 1951, **49**, 88.
[15] I. Svendsen, D. Nicolova, I. Goshev, and N. Genov, *Carlsberg Res. Commun.*, 1989, **54**, 231.
[16] H.-D. Belitz and J.K.P. Weder, *Food Rev. Int.*, 1990, **6**, 151.
[17] M. Richardson, in 'Methods in Plant Biochemistry', ed. L.J. Rogers, Academic Press, London, 1991, vol. 5, p. 259.
[18] F. Ceciliano, F. Bortolotti, E. Menegatta, R. Severino, P. Ascen, and S. Palmieri, *FEBS Lett.*, 1994, **342**, 221.

CHAPTER 13

Immunochemical Techniques

1 Introduction

The use of immunochemical techniques offers the potential of highly specific methods of analyses because antibodies recognize and bind to corresponding antigens with high specificity. This enables detection and characterization systems which complement traditional methods of analyses. The use of antibodies for recognition and binding also offers the potential of highly selective methods which may be applied to mixtures containing only small amounts of the component of interest. Consequently, sample pretreatment, such as purification steps, may be omitted, and thus handling time is shortened and the risk of low recoveries in a series of purification procedures is reduced, giving immunochemical techniques advantages over alternative methods of analyses.

Antibodies usually have a very high affinity for the corresponding antigens and when used together with sensitive detection systems it is possible to run analyses in microtitre wells with total volumes of, for example, 300 μl. This enables analysis of large numbers of samples and may offer a low-cost method with respect to time and reagent consumption for routine analyses.

2 Antibodies

Antibodies are proteins belonging to the group of immunoglobulins (Ig) (Figure 13.1). An antibody belongs to one of five classes (IgG, IgM, IgA, IgD or IgE) which differ from each other with respect to the constant region of the heavy chains. In addition, some immunoglobulin classes can be subdivided into subclasses, each with its own constant region. For example, the mouse IgG class can be subdivided into four subclasses (IgG1, IgG2a, IgG2b, and IgG3). The variable parts of the antibody contain the antigen binding sites.

The antibodies employed for immunochemical techniques are produced by humoral responses in an animal after immunization by injection of foreign organisms or macromolecular compounds (antigens) into its tissues. After immunization, the antibodies produced can be isolated from serum by taking blood samples (antibody concentration *ca.* 10 mg ml^{-1}). This results in a population of different antibodies (polyclonal antibodies; pAbs) of which the

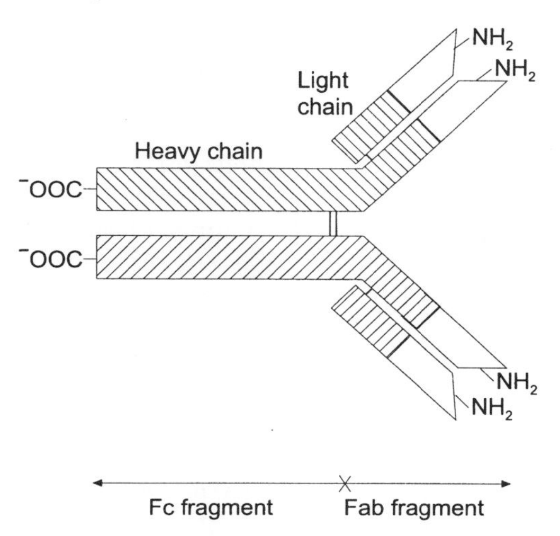

Figure 13.1 *Structure of an IgG molecule illustrating the two heavy chains linked together with two disulfide bridges and linked to two light chains. The two Fab fragments each contain an antigen binding site composed of the amino acid residue at the N-terminal regions of the light and heavy chains. The Fc fragment comprises constant regions with respect to amino acid sequence, which are post-translationally glycosylated. Hatched areas represent constant regions and white areas are variable regions*

major part has a binding specificity for the injected antigen. Alternatively, antibodies may be produced *in vitro* in culture after fusion of antibody-producing cells with appropriate fusion partner cells, which are able to grow in culture. After clonal selection of individual, antibody-producing cells, and subsequent growth in appropriate culture media, monoclonal antibodies (mAbs) can be isolated from the culture media (mAb concentration *ca.* 1–50 $\mu g\ ml^{-1}$).[1] The preparation of mAbs in cultures originating from a single cell means that the preparation contains mAbs with identical amino acid sequences, but with different post-translational glycosylations, giving rise to several protein bands in isoelectric focusing. The antibody-binding site on the antigen is called the *epitope* or *antigen determinant*.

Small antigens (*haptens*) may not be able to elicit an immune response upon immunization of animals. Such haptens require, for example, covalent coupling[2,3] to an immunogenic molecule (*carrier*), which can participate in the generation of an immune response.

Purification of Antibodies

Suitable methods for purification of antibodies depend on:

- the intended use of the antibodies;
- the species in which the antibodies were raised;

- the source of starting material;
- and for mAbs, the class/subclass of the antibodies.

In the following sections, selected methods for antibody purification are given. For further details on applicability and for information on additional methods the reader is referred to Ref. 4. As for purification of other proteins with biological activity (pages 333 and 382–387), the antigen binding activity of a purified antibody should be examined in relation to the antibody present in the solution.

Precipitation

In general, antibodies are precipitated in a 35–50% saturated solution of ammonium sulfate. Most antibodies are precipitated at 40% saturation and higher $(NH_4)_2SO_4$ concentrations increase contamination by other proteins, especially transferrin and albumin. The antibodies obtained by ammonium sulfate precipitation will, however, not be pure owing to contamination with other high-M_r proteins and proteins trapped in the flocculant precipitates. Consequently, ammonium sulfate precipitations are only useful for concentration and partial purification of antibody from various sources and species (Box 13.1).

Box 13.1 *Ammonium sulfate precipitation of antibodies from sera*

This procedure for precipitation of antibodies does not result in complete purification of the antibodies but the preparation is suitable for most purposes. Lipids and lipoproteins are removed by dialysis and precipitation against acetate buffer. The purification steps may be performed at 4 °C to avoid contamination

1. $(NH_4)_2SO_4$ (2.5 g 10 ml^{-1} rabbit serum) is slowly dissolved under magnetic stirring (ambient temperature) and the mixture is left overnight
2. The solution is centrifuged at 3000g for 30 min at ambient temperature
3. The sediment is washed 2 times with 1.75 M $(NH_4)_2SO_4$ (2.5 ml 10 ml^{-1} rabbit serum)
4. The sediment is dissolved in smallest possible vol. deionized water and transferred to a dialysis tube
5. Dialysis is performed for 2 × 12 h against deionized water
6. Dialysis is continued for 24 h against acetate buffer (4.1 g CH_3CO_2Na L^{-1}; 1.2 g CH_3CO_2H L^{-1})
7. Repeated dialysis as 5. and 6.
8. The content of the dialysis tube is centrifuged at 3000g for 40 min
9. The supernatant is concentrated 5 times under a stream of cold air
10. Dialysis for 24 h against 50 mM NaCl
11. Centrifugation as 8. Concentration and purity of the antibody preparation (supernatant) is determined as described on pages 378 and 382–387. The supernatant is kept at −20 °C

Low-pressure Chromatography

The pI of most antibodies is higher than that of the majority of other serum proteins and this may be utilized for purification of the antibodies by ion-exchange chromatography. One method employs DEAE-cellulose and a 10 mM TRIS buffer (pH 8.5) as sample solute and application buffer with subsequent elution using a NaCl gradient in 10 mM TRIS buffer (pH 8.5). This procedure used alone does not result in pure antibody and is usually used together with other techniques, often after ammonium sulfate precipitation.

IgM antibodies may be purified by thiophil adsorption on modified thiophilic gels (containing thioether and sulfone bonds), *e.g.* by using the column material Thiosorp M (Biotech IgG, England). Another possibility for the separation of IgM antibodies is the application of gel filtration using column material with an exclusion limit of 300–500 kDa. Owing to their large M_r, IgM molecules will be eluted in the interstitial volume and thereby separated from most other proteins. It is advisable to use this technique in combination with others, such as ion-exchange chromatography and ammonium sulfate precipitation, in, for example, sample concentration prior to gel filtration.

Affinity Chromatography

Antibodies (except IgM) from most species (*e.g.* mouse, rabbit, pig, cow, and human) may be purified using matrices coupled with protein A (Section 6.7), which is a cell wall protein from *Staphylococcus aureus* with high affinity for the Fc portion of the antibody molecule.[5] Some antibodies such as mouse IgG1 molecules bind with low affinity to protein A at low salt concentrations, whereas increasing the salt concentration also increases the purified amount substantially (Box 13.2).

Box 13.2 *Affinity purification of IgG on protein A matrix*

This procedure employs a high salt concentration in the binding buffer to increase the antibody yield of the purification. Use fractionation equipment fitted with a peristaltic pump and UV-detector, preferably mounted in refrigerator or cool laboratory

1. A column (*e.g.* C10/10 column, Pharmacia) is packed with a protein A matrix (*e.g.* Prosep A from Biotech IgG)
2. Culture supernatant is filtered on filterpaper and pH adjusted to 8.5–8.8
3. The column is washed with 5 vol. buffer A (Borate 0.1 M, 0.15 M NaCl, pH 8.5)
4. The column is washed with 5 vol. buffer B (NaCl, 193 g L^{-1}; glycine, 124 g L^{-1}; NaOH, 8 g L^{-1}; pH 8.83)
5. The culture supernatant is applied on to the column
6. The column is washed with 10 vol. buffer B
7. The column is washed with 10 vol. buffer A
8. Antibodies are eluted with buffer C (citric acid, H_2O, 0.1 M, pH 3.0) and collected as 1 ml fractions in glasses containing 33 μl 1 M TRIS to obtain neutral pH in the eluted fractions
9. The column is regenerated with 5 vol. 0.1 M HCl and 5 vol. buffer A
10. The column is kept in 70% ethanol

Protein A coupled FPLC columns are also commercially available, and are applicable for partly purified antibody fractions that have been centrifuged and filtered prior to application.

Concentration of Antibodies

Antibody preparations may be concentrated by ammonium sulfate precipitation (Box 13.1), ultrafiltration, dialysis against a polymer (Section 4.4), or by evaporation of the solvent under a stream of cold air or by vacuum evaporation.

Storage of Antibodies

In general, antibodies comprise a group of fairly stable proteins in up to 150 mM salt solutions and at neutral pH, although some mAbs may be sensitive to freezing. Problems with storage of antibodies at 4 °C are most often encountered in connection with contamination of the antibody solution by bacteria or fungi and this can be avoided by storage in 0.02% NaN_3. At 4 °C, antibodies are often stable for more than 6 months. However, NaN_3 may block biological assays where the cofactor may be sensitive to nucleophiles, *e.g.* haemoproteins (page 107) and heteroaromatics (page 110), and it may also interfere with coupling methods. Alternatively, the antibody solution can be filter sterilized and treated aseptically. Antibodies that are not sensitive to freezing may be stored at −20 °C and are usually stable for years provided too many thaw–freeze cycles are avoided. For purified antibodies, the protein concentration should preferably be above 1 mg ml^{-1}, obtained, for example, by addition of 1% bovine serum albumin (BSA) to the antibody solution.

Characterization of Antibody Preparations

Investigation of employed antibodies is necessary to evaluate the results obtained by use of those antibodies. For both polyclonal and monoclonal antibodies, it is important to include appropriate controls to evaluate the background binding, unspecific binding, and possible cross-reactions. For comparison of antigen binding affinities of different antibody preparations, it is important to know the concentration of antibody. For purified antibody preparations, this may be determined by an $A_{280\,nm}$ measurement. Average extinction coefficients[6] may be used, pAbs: $E^{1\%}_{1\,cm} = 14.3$; IgM: $E^{1\%}_{1\,cm} = 11.8$; IgG: $E^{1\%}_{1\,cm} = 14.3$.

Alternatively, an enzyme-linked immunosorbent assay (ELISA) may be used for quantification by including appropriate standards and this method is also applicable for antibody solutions that have not been purified.

Owing to the nature of mAbs, they may be specifically characterized. Knowledge of class and subclass can be obtained either by the Ouchterlony precipitin reaction (*vide infra*) or ELISA (less time consuming) by use of class/subclass specific secondary antibodies. This information is valuable for selecting appropriate methods of purification. The binding characteristics of mAbs may also be examined by investigating the simultaneous/non-simultaneous binding to an antigen by different antibodies produced against this antigen. Such knowledge is useful for the development of assays.

3 Immunochemical Methods of Analyses

Precipitin Reactions

The precipitin reactions take advantage of the ability of antibodies to bind to two epitopes and thereby contribute to the formation of a lattice that precipitates at an appropriate size. A prerequisite for the formation of a lattice is the presence of the correct ratio between antibody and antigen molecules. With an antibody excess, the addition of antigen leads to an increased amount of precipitate, whereas antigen excess will result in dissolution of the complexes formed. Consequently, the maximum precipitation is seen in the zone of equivalence.

Precipitin reactions in free solutions have for most purposes been superseded by precipitin reactions in gels (immunodiffusion), which are more easily evaluated. Note that a mAb can not form lattices except with antigens containing repeated epitopes.

Immunodiffusion

With the diffusion of molecules from a solution into a gel, a concentration gradient is formed in the gel. At an appropriate point in this concentration gradient, precipitation lines are formed if the antibody has specificity for the antigen.

Single radial immunodiffusion (Mancini)

With this technique, a gel containing antibodies is cast on a solid support and antigen solution is applied in punched holes in the gel. Precipitin lines are formed around the wells with diameters corresponding to the antigen concentrations in the individual wells.

Double radial immunodiffusion (Ouchterlony)

With the Ouchterlony technique, both antigen and antibody are applied in wells in an agarose gel, thus resulting in diffusion of both reagents (Box 13.3). This technique may be used to determine which sera, culture supernatants, or other fractions contain a particular antigen/antibody.

The appearance of precipitin lines using this technique may also give information about partial or complete identity of antigens. If the antigen solutions are placed in individual wells at an equal distance from an antibody-containing well, the precipitin lines formed for each of the antigen wells will fuse if the antigens are identical. If the antigens have different epitopes, the two

Box 13.3 *Ouchterlony for antibody class and subclass determination*

The Ouchterlony technique may be used for class and subclass determination of antibodies. For this purpose, antibodies with specificities for the individual classes and subclasses are used

1. An agarose gel [1% (w/v) in TRIS–veronal buffer (diethylbarbituric acid, 4.48 g L^{-1}; TRIS, 8.86 g L^{-1}; pH 8.6] is cast on glass plates and wells are punched in the gel, with sets that are made up of one well in the middle surrounded by six wells
2. Anti-subclass/anti-class antibodies are placed in the central wells and antibodies to be tested are placed in the surrounding wells. The gel is kept in a humid chamber for 1–3 days
3. Precipitation lines indicate the subclass of the antibodies

precipitin lines will cross each other, whereas partial identity will result in one of the precipitin lines extending from the point of fusion between the two precipitin lines.

Immunoelectrophoresis

An immunoelectrophoresis technique may be designed in various ways. In general, the electrophoresis is carried out in agarose gels, with electrophoretic separation of the antigen sample. The antigens may then be examined for their reaction with antibodies and hence formation of precipitin lines. For a qualitative immunoelectrophoresis, long troughs may be cut along the separation line of the antigens after the end of electrophoresis and filled with antibody solutions. Precipitin lines will then form between the separated antigens and the troughs from which corresponding antibodies diffuse into the gel.

Alternatively, the antibodies may be contained in the gel. When current is applied, the antigens will migrate into the antibody-containing gel and precipitin lines will form that have heights proportional to the concentration of the antigen in the wells (rocket immunoelectrophoresis).

Crossed immunoelectrophoresis is another application, which gives much information from a single electrophoresis analysis (Box 13.4). As a first step, the antigen components are separated by electrophoresis. Then the gel containing

Box 13.4 *Crossed immunoelectrophoresis*

Crossed immunoelectrophoresis is employed for identification of electrophoretically separated bands by subsequent electrophoresis into a gel containing immunoglobulins

First dimension electrophoresis

Apparatus; see Figure 9.9B

1. The agarose gel solution (1% HSA agarose (Litex, Denmark) in TRIS-veronal buffer (73.5 mM TRIS L^{-1}, 24 mM 5,5-diethylbarbituric acid L^{-1}, pH 8.6) is brought to the boil and placed on water bath at 56 °C

(*continued opposite*)

2. Agarose gel solution (15 ml) is poured on to a 10 × 10 cm glass plate. When the agarose gel is set, five holes with diameters of 2.5 mm are punched out as illustrated below
3. The glass plate is placed on the electrophoresis apparatus (10–15 °C). Samples and marker (bromophenyl blue/human serum albumin) are applied in the holes
4. Eight wicks wedded with TRIS-veronal buffer (1.) are placed on each side of the gel. The electrophoresis is run at 10 V cm^{-1} until the marker has reached the end line (see illustration)

Second dimension electrophoresis

5. The gel is cut according to the illustration and placed on a 5 × 7 cm glass plate
6. A spacer gel of 1 cm may be cast [0.7 ml agarose gel solution (1.)]
7. For the antibody-containing gel, agarose solution is warmed to 56 °C, mixed with antibody, and poured on to the plate, with a final volume of agarose solution of 3.4 ml (or 4.1 ml if no spacer gel is cast)
8. The plates are placed in the electrophoresis apparatus and 5 wicks wedded with TRIS-veronal buffer (1.) are placed on each side. A glass plate is placed on top to avoid condensation of water
9. The electrophoresis is run at 1–2 V cm^{-1} overnight

Wash and Coomassie staining of the gels

10. The application holes are filled with water and a thin wedded filter paper is placed on the gel followed by 10 layers of adsorbent tissue and a glass plate. The plates are pressed for 2 × 10 min with shifts of adsorbent tissue
11. The plates are washed for 30 min with 0.1 M NaCl and pressed for 10 min
12. The plates are washed for 30 min with 0.1 M NaCl and pressed for 2 × 10 min
13. The plates are dried under a stream of cold air
14. The plates are left for 15 min in Coomassie staining solution (5 g Coomassie Brilliant Blue R-250 (Sigma B-0630) L^{-1}, 45% (v/v) EtOH and 10% (v/v) acetic acid in deionized water)
15. The plates are destained, minimum of twice, for 15 min in destaining solution (45% EtOH (v/v) and 10% (v/v) acetic acid in deionized water) until the background is destained

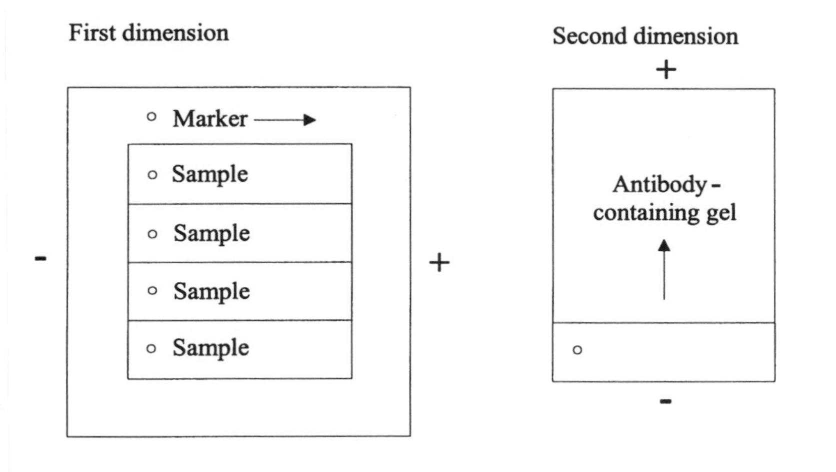

the separated antigen components is transferred to a new glass plate and a gel containing antibodies is cast next to the antigen-containing gel. An electric field is then applied to force the separated antigens to migrate into the antibody-containing gel where the antibodies will bind specifically to their corresponding antigens. At an appropriate ratio of antigen to antibody in the gel, large antigen–antibody complexes will be formed and precipitated in the gel. With the presence of several different antibodies and antigens, a range of different precipitin lines will be formed. These precipitation lines can then be visualized by protein staining of the gel after soluble proteins have been washed out. In 1% agarose gel at pH 8.6, antibodies in most antisera do not migrate because of a balance between the anodic electrophoretic migration of the antibodies and the backflow due to electro-endo-osmotic flow towards the cathode.

Solid Phase Immuno-detection

In immunoassays, the first layer of the immuno-complex is bound to a solid phase. The subsequent formation of non-covalent binding between antigens and antibodies in the immuno-complex may then be separated from unbound components by washing of the solid phase. The immunoassay may be performed using different sorts of solid supports whereto the first layer of reactants are either adsorbed or covalently coupled. A practical approach is the use of microtitre plates with wells made of polymers, *e.g.* polystyrene. Different sorts of plates are commercially available, with different immobilization characteristics. However, microtitre plates have a low binding capacity compared with nitrocellulose paper (NC).

Detection and perhaps quantification are performed by use of complex formation with a reagent (*e.g.* antibody) covalently bound to detector molecules, *e.g.* easily assayed enzymes. A range of different principles can be employed for setting up an assay. See Figure 13.2 for examples of the different configurations. The analyte-specific binding may involve a labelled reagent (direct assay) or an additional layer may be applied to the assay with binding of another labelled molecule, often commercially available (indirect assay). In addition, the analyte may be present in solution, resulting in a competition in binding of the analyte-specific molecule between, for example, soluble and immobilized analyte (competitive assay).

For enzyme-labelled techniques, the choice of staining procedure depends on the support media employed. For staining in solution (*e.g.* in microtitre wells) soluble chromogenic substrates are chosen whereas the use of membranes involves precipitating chromogenic substrates. Commonly used soluble substrates for horseradish peroxidase-labelled detection are 3,3′,5,5′-tetramethylbenzidine (acidified TMB product, λ_{max} = 450 nm), *o*-phenylenediamine (OPD product, λ_{max} = 492 nm) and 2,2′-azinodi(ethylbenzthiazoline)sulfonate (ABTS product, λ_{max} = 410/650 nm) and for alkaline phosphatase a substrate such as a *p*-nitrophenyl phosphate (*p*-NPP product, λ_{max} = 410 nm) may be employed. For insoluble products, diaminobenzidene (DAB, brown) and TMB (blue) may be used with horseradish peroxidase and, for example, bromochloroindolyl

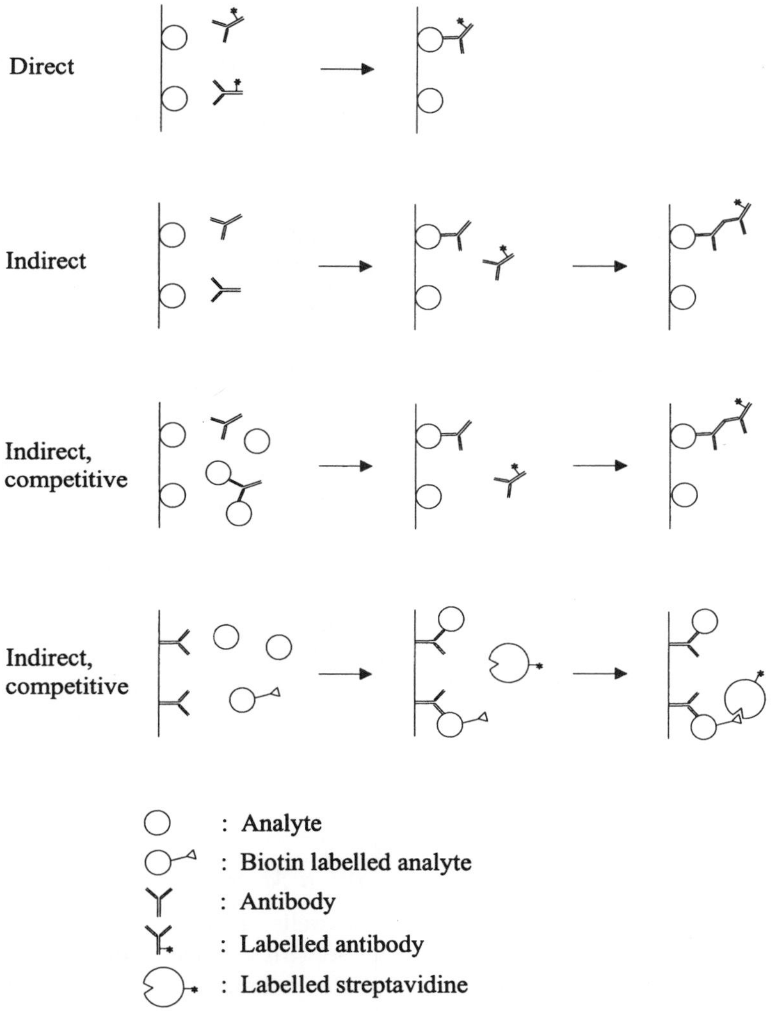

Direct

Indirect

Indirect, competitive

Indirect, competitive

: Analyte
: Biotin labelled analyte
: Antibody
: Labelled antibody
: Labelled streptavidine

Figure 13.2 *Selected configurations of enzyme-labelled immunoassays on solid supports. The first reagent/analyte is immobilized on the solid support and residual binding sites are blocked by appropriate washing procedures. After each of the subsequent incubation steps, the support is washed to remove unbound reagents/analytes before the next incubation step. Eventually, labelled reagent is detected by an appropriate substrate reaction*

phosphate/Nitro Blue Tetrazolium (BCIP/NBT, purple) may be used with alkaline phosphatase.

Enzyme labelling may be difficult and the resulting conjugate may not have a stable enzyme activity. Alternatively, additional layers with appropriate, commercially available, enzyme conjugates may be employed or reagents may be covalently conjugated to biotin (Box 13.5). Biotin has a very strong binding to streptavidine which is commercially available as enzyme-labelled streptavidine.

Box 13.5 *Biotinylation of proteins*

Biotinylation may be perfomed using a biotin succinimide ester for conjugation of biotin to primary amino groups, *e.g.* on a protein. For subsequent use of biotinylated compounds in immunochemical reactions it should be noted that biotinylation may occur at the binding sites if they contain amino groups. The biotinylation can be stopped by either dialysis to remove unreacted reagent or by addition of ethanolamine.

1. A protein solution (1 mg ml^{-1} in phosphate-buffered saline, pH 7.2) is mixed with biotin-*N*-hydroxysuccinimide (5 mM in dimethylformamide) in the ratio 1:0.2 ml. The reaction mixture is incubated with gentle agitation for 2–4 h
2. The reaction is stopped by addition of ethanolamine solution (1 M ethanolamine, pH 8.0 with HCl), with 10 μl ethanolamine solution per ml reaction mixture

Dot Blotting

Nitrocellulose paper (NC) may act as solid support for adsorption of antibodies/antigens in dots applied to the paper. Special equipment with sample application under vacuum is available in microtitre plate formats, where subsequent washing and incubation steps are performed in the wells of the filtration apparatus. Remaining binding sites are blocked with extraneous protein (*e.g.* BSA or casein) or a detergent (*e.g.* Tween-20) followed by incubation with relevant immunoreagents. The last incubation step involves a labelled immunoreagent that is detectable with a substrate forming insoluble products. The dot blotting may be performed essentially as described in Box 13.8. NC is a high capacity carrier and therefore better suited for impure or scarce coating materials than microtitre plates. However, NC is less easily washed and gives a high background binding.

Enzyme-linked Immunosorbent Assay in Microtitre Plates

For the ELISA, microtitre plates with 96 wells are used. Different types of microtitre plates are commercially available with a range of different binding characteristics but in general, plates with high binding capacities should be chosen for ELISA. However, the surface of wells in even high capacity plates will generally be saturated at solutions of 5 μg protein ml^{-1}. Consequently, impure antigen solutions should not be used for coating microtitre plates as only minute amounts (and not necessarily a ratio representative of the sample) will be immobilized on the plastic surface.[7] For impure antigen solutions, the use of dot blotting is more likely to be able to immobilize an antigen in amounts that are adequate for detection.

The use of a multichannel pipette is advantageous for simultaneous pipetting of solutions in several wells and for the simultaneous preparation of dilution rows directly in the wells of several rows or columns of the microtitre plate. In this way, dilution curves covering a range of concentrations are made, which is recommended for reliable interpretation of the data.

For quantitative purposes, it is advantageous to optimize the ELISA for the specific analyses, *e.g.* with respect to detection limit, working range, *etc.* In general, assays in which all the constituents other than the sample are in excess will be more precise and have a higher sensitivity. Optimization of the ELISA may be performed from the bottom of the immuno-complex, *i.e.*, the coating concentration is optimized first. The detection ranges of sample not immobilized on the microtitre plate are usually in the 1–1000 ng ml^{-1} range. Commercially available enzyme-labelled antibodies or, for example, streptavidin are usually diluted 1000–5000 times.

ELISA may be used for determination of a range of different compounds, and a number of assays are commercially available. This technique has also proven valuable for analysis of low-M_r compounds that are otherwise difficult to analyse for, *e.g.* saponins, which are compounds present in, for example,

Box 13.6 *Enzyme-linked immunosorbent assay in microtitre plates*

Between the different incubation steps, the plates are washed with a buffer containing Triton X-100, which will cause blocking of residual binding sites on the well surfaces. For dilute protein solutions, the dilution buffer may contain 0.3% extraneous protein (*e.g.* BSA or casein) for stabilization of the protein in solution

Solutions

Buffer A (coating): Na_2CO_3 (1.59 g L^{-1}); $NaHCO_3$ (2.93 g L^{-1}) in deionized water, adjusted to pH 9.6
Buffer B (dilution): NaCl (29.2 g L^{-1}); KCl (0.2 g L^{-1}); KH_2PO_4 (0.2 g L^{-1}); $Na_2HPO_4 \cdot 2H_2O$ (1.15 g L^{-1}); Triton X-100 (1 ml L^{-1}) in deionized water
Buffer C (wash buffer): dilution buffer diluted 10 times in deionized water
Buffer D (substrate solution): potassium citrate (0.2 M); H_2O_2 (3.15 mM) in deionized water, pH adjusted to 4.0
TMB solution: TMB (1 g) is dissolved in 20 ml acetone and diluted with 180 ml MeOH. Kept in aliquots at −20 °C
Buffer E (stop solution): 2 M H_3PO_4

Procedure

1. The wells in the microtitre plates are each coated with 100 μl of protein solutions of 0.1–10 μg ml^{-1} in buffer A and left for incubation overnight (4 °C) or for 1 h at ambient temperature
2. The plates are washed 4 times with buffer C (300 μl $well^{-1}$)
3. The plates are incubated with antigen, antibody or other molecules with high, specific binding affinity. Dilution is performed in buffer B, usually 5 ng ml^{-1}–5 μg ml^{-1}. The plates are left for incubation for 1 h at ambient temperature
4. As 2.
5. Both 3. and 4. are repeated with different layers of molecules until the appropriate assay configuration is obtained. The last layer employs molecules linked to horseradish peroxidase; a range of such conjugates are commercially available and most of these conjugates should be diluted 1000–5000 times in buffer B before application
6. As 2.
7. To each well, 100 μl substrate–TMB solution is added (buffer D:TMB solution; 35:1) and left for 5–20 min, until appropriate colour development
8. The colour development is stopped by addition of 100 μl buffer E to each well
9. Absorbance is read at 450 nm

legumes as triterpenoid glycosides. Saponins are suspected to influence the palatability of leguminous seeds but they are difficult to detect owing to their low extinction coefficient. Furthermore, isolation and concentration of the saponins often result in low recoveries owing to their ability to form micelles and hence change their solubility characteristics. Soyasaponin I is a saponin present in most if not all leguminous seeds, and with a monoclonal antibody against this saponin it has been possible to develop an ELISA method for measurement of the soyasaponin I content in extracts of leguminous seeds (Figure 13.3). Preparation of monoclonal antibodies against saponins require conjugation of saponin to an appropriate carrier molecule (*e.g.* BSA) to elicit an immune response. The saponin–BSA conjugate may also be employed for immobilization in ELISA as it is not possible to immobilize saponin to microtitre plates by hydrophobic interactions that allow subsequent binding of antibodies to the immobilized saponin (Figure 13.3). The assay is performed as a competitive ELISA in microtitre plates with immobilization of a saponin–BSA conjugate (2 μg ml^{-1}). After immobilization, flour extracts are diluted in the wells, resulting in 100 μl well^{-1}. The dilution step is followed by addition of a mAb, with 100 μl well^{-1} of a solution of 150 ng ml^{-1}. The plates are left for binding to occur between the mAbs and either the saponin in solution or the saponin present in the immobilized saponin–BSA conjugates. After washing the plate, incubation is initiated with a horseradish peroxidase conjugated rabbit-anti-mouse antibody. The amount of mAb is then determined by enzymatic

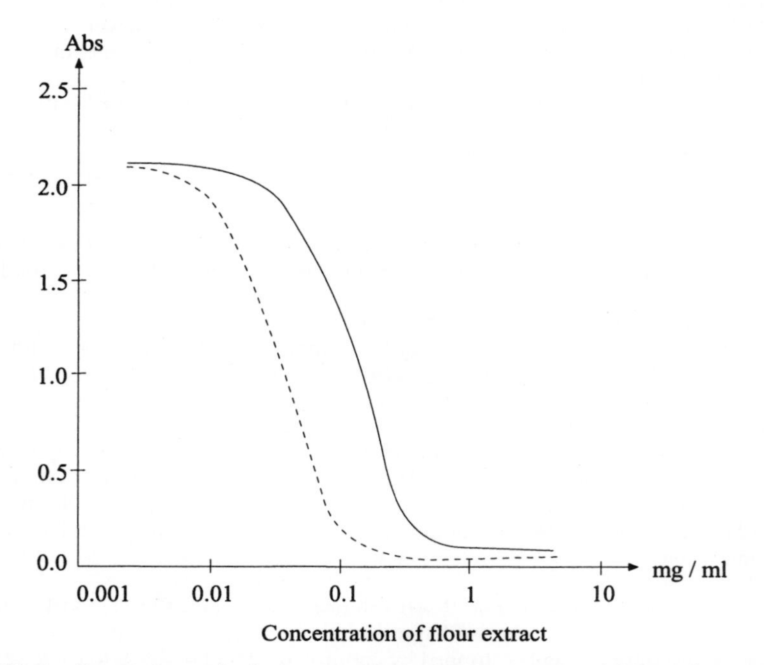

Figure 13.3 *Comparison of relative saponin content in extracts of flour from soybean (- - - - -) and pea (———).*

cleavage of a chromogenic substrate (TMB) measured spectrophotometrically at 450 nm. From dilution curves, it is seen that the investigated soybean flour contains approximately four times more soyasaponin I than the pea flour.

Immunostaining of Blots from Electrophoresis

Enzyme-linked immunodetection may also be used for identification of components separated electrophoretically. The separated components are transferred to membranes as, for example, NC or poly(vinylidene difluoride) (PVDF) membranes. The binding of proteins to the membranes may differ depending on the actual proteins and therefore it may be advantageous to examine the performance of other membranes in the case of inadequate binding. As these membranes have a high protein-binding capacity, care should be taken to use virgin membranes and gloves for handling.

Blotting may be performed either by diffusional blotting or by electrophoretic transfer.[8] With diffusional blotting, no special equipment is required (Box 13.7), and no restrictions are made in the use of buffers for the transfer. However, the diffusion time is long and some resolution may be lost. In contrast, electrophoretic transfer is fast, using a device in which the gel, blotting membrane, and appropriately wetted filter papers are placed in a sandwich. For this procedure, a setup must be chosen where the proteins migrate from the gel to the membrane upon application of an electric field. Electrophoretic transfer is often employed after gel electrophoresis in polyacrylamide gels, which have small pores that do not allow easy diffusion. The use of SDS-PAGE leaves the proteins negatively charged owing to the SDS on the proteins, and the proteins will migrate towards the anode. The presence of SDS may, however, also cause denaturation of the proteins, resulting in failure of antibody binding to the proteins.

After blotting of proteins onto membranes it is important to block the residual binding sites to prevent non-specific binding of reagents to the membrane. Blocking may be performed with serum, BSA, or detergents such as Tween-20, and these blocking steps may be involved in partial renaturation of denatured proteins.[8]

Box 13.7 *Blotting from agarose gels to nitrocellulose*

This procedure may be employed after gel electrophoresis, *e.g.* as described in Boxes 9.9 and 9.10.

1. A piece of nitrocellulose, 1 cm larger on each side than the gel, is wetted for 30 min in deionized water
2. The gel is washed in deionized water
3. The gel is placed on top of a glass plate and the nitrocellulose is placed on the gel, without air bubbles between the gel and the nitrocellulose. A piece of wedded chromatography paper is placed on the membrane, then *ca.* 0.5 cm adsorbent tissue layer is placed on top followed by a glass plate and a *ca.* 1 kg weight in an arrangement such as shown in Figure 9.13. The gel is pressed for *ca.* 5 min
4. The nitrocellulose membrane is carefully removed from the gel by use of pincers

Box 13.8 *Immunostaining of nitrocellulose blots with BCIP/NBT*

For this immunostaining procedure, a NC blot prepared as described in Box 13.7 may be used.

The primary antibody used in immunostaining should be diluted to 1 μg ml^{-1}–10 μg ml^{-1} depending on the actual antibody. The secondary antibody should be an alkaline-phosphatase-labelled antibody, diluted according to the manufacturer's instructions

1. Unoccupied sites on the NC membrane are blocked for 15 min with a blocking buffer (6.06 g Trizma-base (Sigma), 8.76 g NaCl, and 2.0 g Tween-20 (Merck) to 1000 ml deionized water, pH 10.3 with NaOH)
2. The membrane is air-dried
3. The membrane is incubated overnight with primary antibody diluted in incubation buffer (6.06 g Trizma-base (Sigma), 8.76 g NaCl, and 0.5 g Tween-20 (Merck) to 1000 ml deionized water, pH 10.3 with NaOH)
4. The membrane is washed 3 times in incubation buffer
5. The membrane is incubated for 1 h with secondary antibody (diluted in incubation buffer, *e.g.* 1500 times)
6. The membrane is washed 3 times in incubation buffer
7. The membrane is stained for 5–15 min with staining solution [1.0 ml NBT stock solution [1 g NBT L^{-1} in ethanolamine buffer (0.63% ethanolamine (v/v) pH 6.3)], 0.15 ml substrate stock solution [10 ml methanol, 5 ml acetone and 60 mg 5-bromo-4-chloroindol-3-yl phosphate (Sigma, B-8503)], and 0.04 ml MgCl solution (1 M)
8. The gel is washed in deionized water and air-dried

The immunostaining of blots of proteins depends on the ability of the employed antibodies to recognize the immunoblotted proteins. A high specificity of mAbs is desirable for identification of specific proteins, but this specificity also renders the mAbs sensitive to partially denatured proteins. In general, incubation times with antibodies should be optimized to obtain the best immunostaining in accordance with the different binding constants of the antibodies to their corresponding antigens.

For total protein detection, the protein bands may be visualized with, for example, gold staining (Box 13.9), Amido black, or ink staining.[8]

Box 13.9 *Gold staining of nitrocellulose blots*

Gold staining is a procedure for staining all protein bands. With gold staining of protein on nitrocellulose blots a sensitivity of *ca.* 50 pg/protein band is obtainable.

1. The membrane is blocked with blocking buffer (Box 13.8) for 2 min followed by washing in incubation buffer (Box 13.8) for 5 min
2. The membrane is incubated overnight in citrate buffer (50 mM, pH adjusted to 3.0 with NaOH)
3. The membrane is washed with deionized water
4. The membrane is incubated overnight with gold colour solution [75 ml gold colour stock solution (0.015% tetrachlorogold(III) in deionized water is brought to boil, 4.5% (v/v) 1% trisodium citrate is added and the solution is boiled under reflux for 30 min), 25 ml citrate buffer (50 mM, pH 3.0), and 0.1 ml Tween-20]
5. The membrane is washed with deionized water
6. The membrane is air-dried

High-performance Capillary Electrophoresis (HPCE)

HPCE may be a valuable complementary method for different immunochemical techniques. This method gives a special opportunity for following coupling reactions between protein carriers and haptens which may otherwise require several individual experiments and the use of larger amounts of reagents. An example of this application is shown in Figure 13.4, where the reaction time for conjugation of soyasaponin I to Kunitz soybean trypsin inhibitor (KSTI) is investigated. Briefly, soyasaponin I is a triterpenoid glycoside, and vicinal hydroxy groups of the carbohydrate part of the soyasaponin may be oxidized to aldehyde groups which can then react with amino groups of a protein (here KSTI) to give Schiff bases. In HPCE, the coupling of soyasaponin molecules to KSTI results in changes in the migration time of the KSTI conjugates compared with native KSTI.

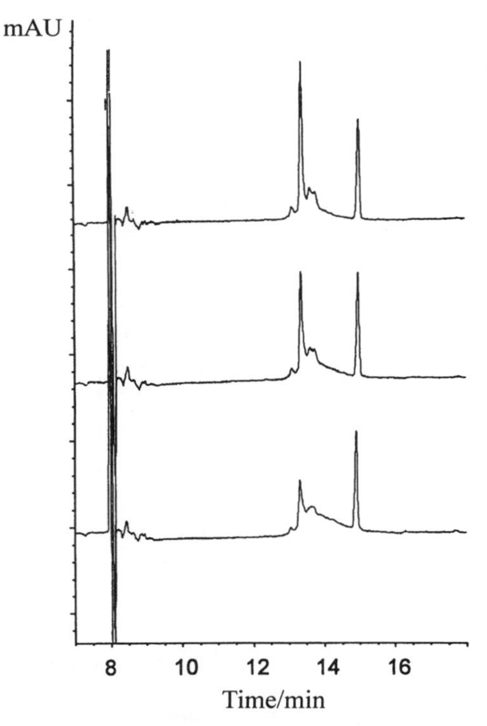

Figure 13.4 *Electropherograms of soyasaponin I conjugation to KSTI after 60 min of preincubation with 0.1* M *$NaIO_4$. Conjugation after 10 (top), 90 (middle), and 140 min (bottom). The migration times of the soyasaponin I and KSTI are 14.8 and 13.2 min, respectively, whereas the migration time of the complex is an intermediate value. Electrophoreses were performed in a fused-silica tube, 614 mm, 50 µm i.d., with a run buffer of 75* mM *Na_2HPO_4; 50* mM *taurine; 35* mM *cholic acid, pH 8.0; at 30 °C, with a voltage of 16 kV, and column detection at 200 nm*

(Reprinted from H. Frøkiær, H. Sørensen, J.C. Sørensen, and S. Sørensen, *J. Chromatogr. A*, **717**, 75 (1995), Elsevier Science, Amsterdam)

Immunochemical binding reactions may also be studied in HPCE when buffer conditions are used that allow the binding to occur between the relevant molecules. By HPCE, the binding takes place in solution, thus avoiding the possible denaturation or alteration in protein structure caused by the hydrophobic interactions involved in techniques with immobilization steps. HPCE has been employed to investigate the binding site on KSTI for a mAb produced against KSTI, compared with the inhibitor site on KSTI. This was performed by examination of whether is was possible to obtain simultaneous binding of KSTI to the mAb and to trypsin (Figure 13.5). The left-hand side of the figure shows that the KSTI–mAb complex has a different migration time to the individual components. On mixing the KSTI–mAb complex with trypsin (right-hand side of Figure 13.5) it is seen that a trypsin–KSTI–mAb complex is formed with yet

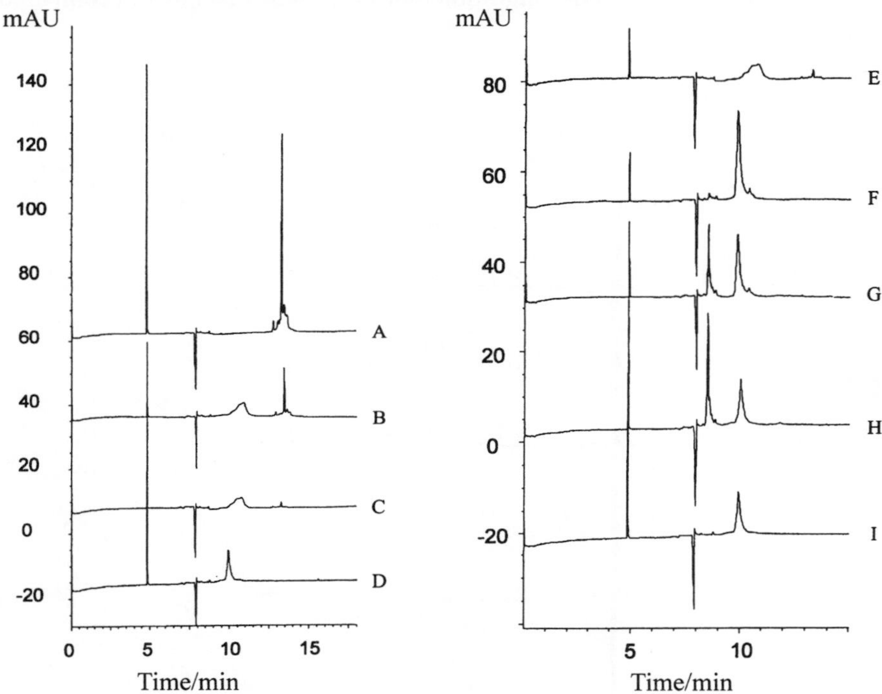

Figure 13.5 *Left: Complex formation of KSTI and a monoclonal antibody with specificity for KSTI. The peak at 4.8 min is an internal standard (trigonellinamide). A: KSTI 2 mg ml^{-1}; B: KSTI : mAb in a molar ratio of 3.6 : 1; C: KSTI : mAb in molar ratio 1.8 : 1; D: mAb 1.07 mg ml^{-1}. Right: E: mixture of mAb and KSTI (1 : 1.8); F: KSTI : mAb: trypsin in molar ratio 1 : 0.56 : 1.5; G: KSTI : mAb : trypsin in molar ratio 1 : 0.56 : 3; H: mAb 0.84 mg ml^{-1} and trypsin 0.5 mg ml^{-1}; I: mAb 1.07 mg ml^{-1}. Electrophoreses were performed in a fused-silica tube, 614 mm, 50 μm i.d., with a run buffer of 75 mM Na_2HPO_4; 50 mM taurine; 35 mM cholic acid, pH 8.0; at 30 °C, with a voltage of 16 kV and on column detection at 200 nm*
(Reprinted from H. Frøkiær, K. Mortensen, H. Sørensen and S. Sørensen, *J. Liq. Chromatogr. Relat. Technol.*, 1996, **19**, 57, by courtesy of Marcel Dekker Inc.)

another migration time. Consequently, the mAb employed binds to a site on KSTI that is different from the trypsin inhibitor site.

4 Immunoaffinity Chromatography

Immunoaffinity chromatography offers the possibility of specific purification of antigens from mixtures by binding of the antigens to columns with antibodies immobilized to the column material. In this way, a high degree of purification is possible in a single step.

A range of different procedures are available for preparation of immuno-affinity columns. The column materials used for conjugation to antibodies may be preactivated in different ways, depending on the actual column material and the desired subsequent conjugation procedure. Furthermore, a range of commercially available preactivated column materials are available for direct conjugation of antibodies.

In general, mAbs are most convenient to use for immunoaffinity chromatography but affinity purified polyclonal IgG can also be used. High-affinity antibodies are preferable to low-affinity antibodies owing to the significantly more efficient binding of antigen in solution to high-affinity antibodies.

The coupling efficiency of the procedure in Box 13.10 may be determined by examination of the antibody contents present in the washing buffers and other

Box 13.10 *Coupling of antibodies to activated agarose*

This procedure describes the conjugation of mAbs to commercially available preactivated agarose

1. Transfer an appropriate amount of activated agarose (Mini Leak Low, Kem-en-Tek, Denmark) to a Buchner funnel with filter and wash the agarose twice with 2–3 vol. of deionized water. Dry the agarose until cracks are formed
2. Transfer the agarose to a centrifuge tube (1 g L^{-1}–10 mg protein). Add the antibody solution (>1 mg ml^{-1}) to the agarose and measure the volume of the protein/agarose-solution
3. Add buffer A [poly(ethylene glycol) (PEG, M_r 20 000), 300 g L^{-1}; $NaHCO_3$ (0.6 M), 500 ml L^{-1}; pH approximately 8.6] to a final concentration of PEG of 5–7%
4. Leave the mixture on rocking table, or something similar, overnight at ambient temperature or 4 °C
5. Centrifuge the solution at 400*g* for 15 min and remove the supernatant
6. Stop the reaction by addition of 2 vol. of buffer B (ethanolamine, 0.2 M, pH 9.0 with HCl) and leave on rocking table for 3–5 h
7. As 5.
8. Wash the agarose with 3 vol. of buffer C (K_2HPO_4, 0.1 M, pH 11 with NaOH) by incubating the mixture for 5 min on rocking table. Centrifuge as in 5. and remove the supernatant
9. Wash the agarose with 3 vol. of buffer D (glycine, 0.1 M pH 3.0 with HCl) by incubating the mixture for 5 min on rocking table. Centrifuge as in 5. and remove the supernatant
10. The agarose is kept in 0.1 M NaCl with 15 mM NaN_3

steps during the conjugation procedure, compared with the amount of antibody initially employed in the coupling procedure.

The binding and elution conditions must be examined for each antibody used. This may be done in small-scale columns with examination of a range of different buffers. In general, as mild a set of conditions as possible should be

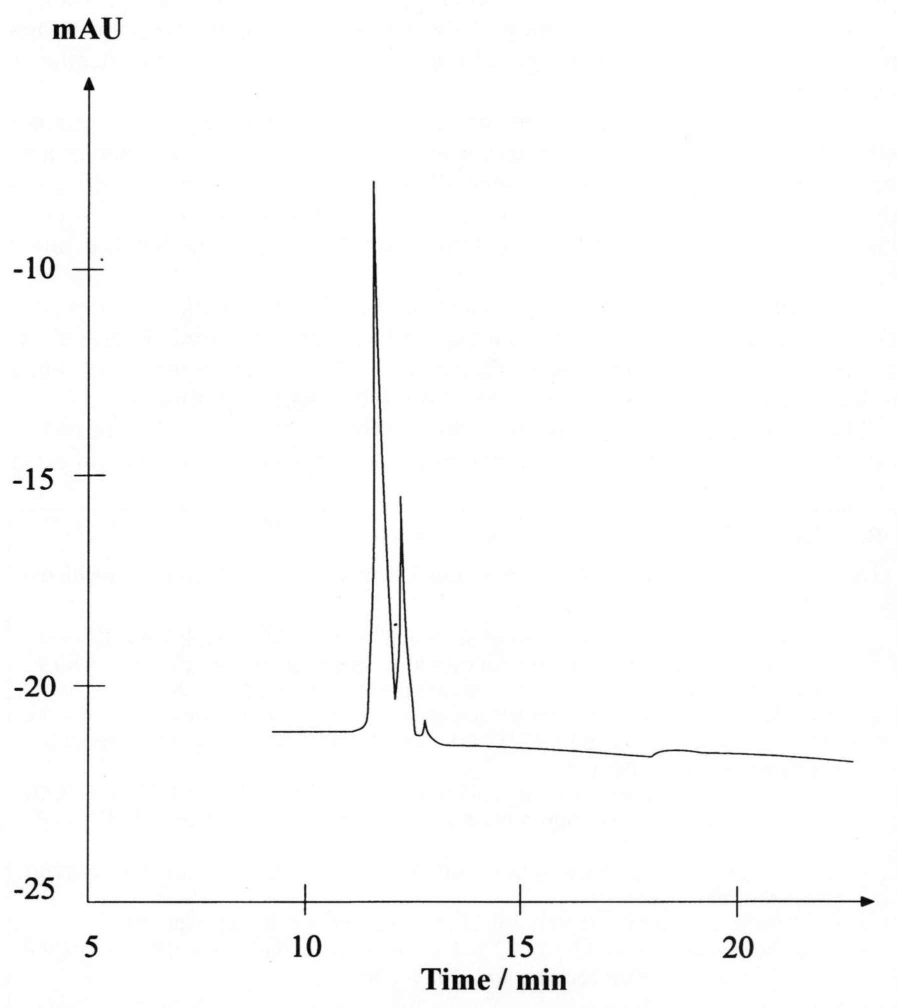

Figure 13.6 *Electropherogram of a protein-type inhibitor (PPI) isolated by immunoaffinity chromatography on a column with a mAb against the inhibitor. Isoelectric focusing of immunoaffinity purified PPI has shown that the mAb has binding affinity for several isoinhibitors, corresponding to the presence of at least two peaks in the electropherogram. Electrophoreses were performed in a fused-silica tube, 614 mm, 50 μm i.d., with a run buffer of 75 mM Na_2HPO_4; 50 mM taurine; 35 mM cholic acid, pH 8.0; at 30 °C, with a voltage of 16 kV and column detection at 200 nm.*

Box 13.11 *Immunoaffinity chromatography of antigens*

This procedure is a general procedure for elution of antigens bound on a Mini Leak affinity column

1. The column is washed with 5 vol. buffer A (boric acid, 0.1 M; NaCl, 0.15 M; pH 8.5 with 5 M NaOH)
2. The antigen solution is mixed with buffer A, filtered and applied to the column
3. The column is washed with buffer A until baseline is obtained
4. The column is washed with 2 vol. buffer B ($Na_2HPO_4 \cdot 12H_2O$, 3.0 g L^{-1}; KH_2PO_4, 0.2 g L^{-1}; KCl, 0.2 g L^{-1}; NaCl, 8.0 g L^{-1}; pH 7.5)
5. Elution is performed with buffer C (citric acid, deionized water, 0.1 M, pH 3.0)
6. The column is washed with 5 vol. of buffer B and kept in buffer B with 0.02% (w/v) NaN_3

used to avoid denaturation of the antigen and the antibody if the column is to be reused. A possible order in which to check different elution conditions is:[4]

> Acid (pH 3–1.5); base (pH 10–12.5); $MgCl_2$ (3.5 M); LiCl (5–10 M), water; ethylene glycol (25–50%); dioxane (5–20%); thiocyanate (1–5 M); guanidine (2.5 M); urea (2–8 M); and SDS (0.5–2%)

If none of these buffers are suitable for elution of the antigen, combinations of the buffers may be used, or alternatively another antibody should be used for preparation of the immunoaffinity chromatography column (Box 13.11).

The procedures described in Boxes 13.10 and 13.11 have been employed for purification of a protein-type inhibitor present in pea (PPI) that was obtained directly from an extract of pea flour. The eluted PPI fractions were tested for purity in HPCE as shown in Figure 13.6.

5 Selected General and Specific Literature

[1] J.W. Goding, 'Monoclonal Antibodies: Principles and Practice', 3rd Edition, Academic Press, London, 1996.
[2] 'Handbook of Experimental Immunology. Immunochemistry', ed. D.M. Weir, Blackwell Publications, Oxford, 1986.
[3] S.S.Wong, 'Chemistry of Protein Conjugation and Cross-linking', CRC Press, Boca Raton, FL, 1991.
[4] E. Harlow and D. Lane, 'Antibodies. A Laboratory Manual', Cold Spring Harbor Laboratory Press, Plainview, NY, 1988.
[5] P.L. Ey, S.J. Prowse, and C.R. Jenkin, *Immunochemistry*, 1978, **15**, 429.
[6] J.E. Coligan, A.M. Kruisbeek, D.H. Margulies, E.M. Shevach, and W. Strober, eds. 'Current Protocols in Immunology', Wiley-Interscience, New York, 1995.
[7] L.A. Cantarero, J.E. Butler, and J.W. Osborne, *Anal. Biochem.*, 1980, **105**, 375.
[8] H. Towbin and J. Gordon, *J. Immunol. Methods*, 1984, **72**, 313.
[9] D.M. Kemeny, *J. Immunol. Methods*, 1992, **150**, 57.

CHAPTER 14

Analysis of Dietary Fibre

1 Introduction

Dietary fibre (DF) comprises plant cell wall and epidermis constituents that are not digested by the alimentary enzymes in the digestive tract of monogastrics. It is a complex group of compounds, dominated by non-starch polysaccharides (NSP) and the polyphenolic lignins, but also contains proteins as well as other minor components, *e.g.* phenolics and other low-M_r compounds, some being found in close association with the lignin NSP constituents.

DF is known as an important constituent of a nutritionally balanced diet for humans, possibly involved in the prevention of a wide range of diseases, including colon cancer and coronary heart disease. DF may reduce the availability of nutrients in the diet, but is also considered to contribute, to a lesser extent, to the energy value of food owing to fermentation in the lower part of the digestive tract. The physiological effects of DF are determined by their physico-chemical properties, including binding capacity towards components in the diet and in the digestive tract (nutrients, vitamins, minerals, bile acids, xenobiotics, water, *etc.*). It should be emphasized that the composition or structure of DF and thereby also its properties will vary depending on the DF source (plant-type, -part, -development).

At present, the most widely used methods of DF analysis are based on gravimetric or chromatographic principles. The gravimetric methods determine DF as the remaining residue after *in vitro* enzymatic treatment, whereas the chromatographic analyses focus on the constituent monosaccharides in NSP, generally determined by GLC, HPLC, or HPCE. This chapter focuses mainly on methods of DF analyses within the chromatographic domain; however, a brief overview of gravimetric methods and the possibilities for further characterization of DF are included.

2 Composition of DF

The NSP fraction comprises a wide range of polysaccharides of high complexity and variability. The dominant monosaccharides are shown in Figure 14.1.

DF polysaccharides differ in the proportion of constituent monosaccharides

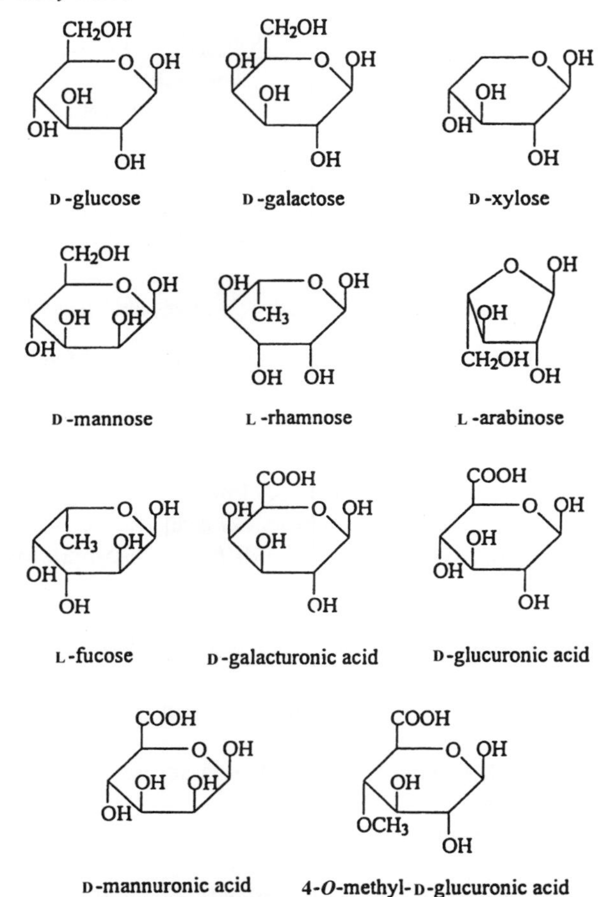

Figure 14.1 *The major monosaccharides of NSP*

as well as the type of *O*-glycosidic bond between monosaccharides and degree of branching of the polysaccharide chain. The classical fractionation scheme divides the polysaccharides into cellulose, hemicelluloses, pectic material, and other polysaccharides, a classification based on solubility properties and/or chemical composition, two criteria which may be somewhat contradictory.

Cellulose is a highly polymerized linear β-(1→4)-D-glucan. Cellulose has a fibrillar appearance, generated by a side-by-side alignment of cellulose chains, stabilized in a crystalline structure by inter- and intramolecular hydrogen bonds and with varying degrees of amorphous regions incorporated (Figure 14.2).

Hemicelluloses cover a wide spectrum of complex heteropolysaccharides containing a minimum two types of sugar residues. The dominant polysaccharides in the hemicellulose fraction comprise:

- xylans [β-(1→4)-D-xylose backbone];
- xyloglucans [β-(1→4)-D-glucose backbone];

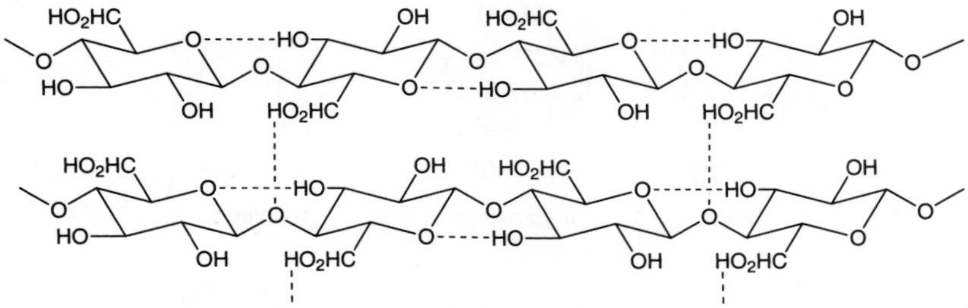

Figure 14.2 *Cellulose, part of a microfibril. Notice the inter- and intramolecular hydrogen bonds*
(Adapted from R.R. Selvendran, *The Chemistry of Plant Cell Walls*, in 'Dietary Fibre', eds. G.G. Birch and K.J. Parker, Applied Science Publishers, 1983, pp. 95–147)

- arabinoxylans [β-(1→4)-D-xylose backbone];
- glucuronoxylans [β-(1→4)-D-xylose backbone];
- mannans [rare, β-(1→4)-D-mannose backbone];
- glucomannans [β-(1→4)-D-glucose and β-(1→4)-D-mannose backbone (1:3)];
- galactomannans [β-(1→4)-D-mannose backbone];
- galactoglucomannans [β-(1→4)-D-glucose and β-(1→4)-D-mannose backbone (1:3)];
- glucuronomannans [α-(1→4)-D-mannose and β-(1→2)-D-glucuronic acid backbone];
- arabinogalactans II [β-(1→3)-β-(1→6)-D-galactose backbone];
- β-(1→3)-D-glucans;
- β-(1→3)-,β-(1→4)-D-glucans.

The β-D-glucans included in this group differ from cellulose in the type of β-linkages in their glucose backbone. These β-D-glucans are, however, not to be considered as hemicelluloses in the strict sense owing to different solubility properties compared with the traditional hemicellulose fraction (Section 14.5).

Pectic material is characterized by a high content of D-galacturonic acid, L-rhamnose, L-arabinose, and D-galactose. The dominant polysaccharides are:

- rhamnogalacturonans [α-(1→4)-D-galacturonic acid and α-(1→2)-L-rhamnose backbone];
- arabinans [α-(1→5)-L-arabinose backbone];
- galactans [β-(1→4)-D-galactose backbone];
- arabinogalactans I [β-(1→3)-β-(1→6)-D-galactose backbone].

The rhamnogalacturonans have relatively large side chains, which are structurally very similar to that of the arabinans, galactans, and arabinogalactans I.

Other polysaccharides include, among others, gums and mucilages, non-structural polysaccharides which will not be described further here.

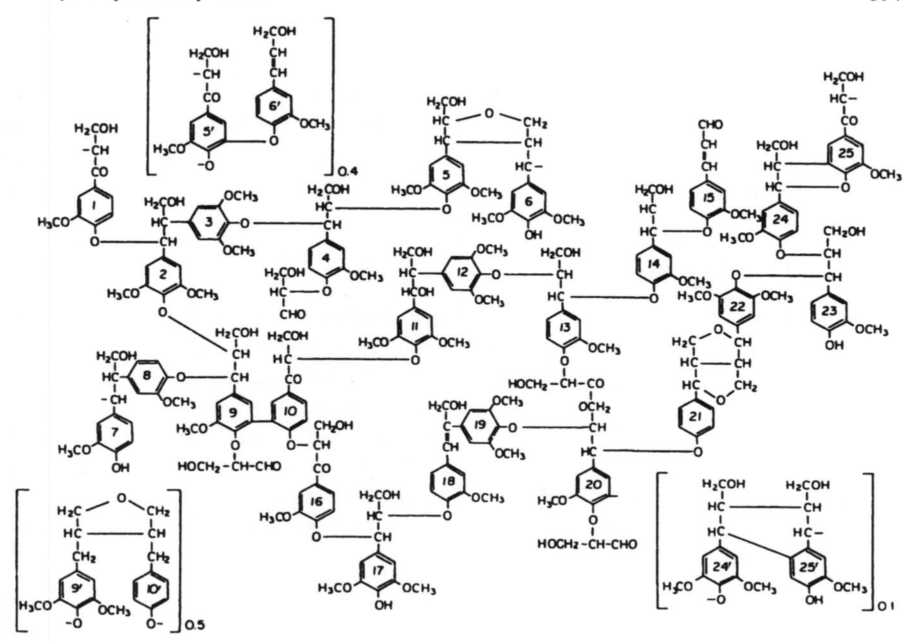

Figure 14.3 *Part of a lignin molecule; structural model of beech lignin* (Reproduced by permission from H.H. Nimz, *Angew. Chem., Int. Ed. Engl.*, 1974, **13**, 313)

Starch and resistant starch (RS), where part of the starch (amylose, amylopectin) can occur as RS for structural reasons, binding to DF, or owing to the presence of amylase inhibitors; RS or indigestible starch may therefore be considered as DF.

Lignins make up the traditional non-carbohydrate fraction of DF. Lignins are complex aromatic polymers, consisting mainly of repeating substituted phenylpropane residues (Figure 14.3). These residues are derived from three main monomers, the cinnamic acid derivatives *p*-coumaryl-, coniferyl-, and sinapyl-alcohol, which are held together with variable types of chemically resistant linkages, including ether and carbon–carbon bonds.

Non-carbohydrate components that differ from lignin comprise, for example, cell-wall proteins, oligosaccharides of the raffinose family, tannins, and low-M_r phenolics. These types of compounds will generally not dominate the DF fraction, although, for example, proteins and oligosaccharides can occur in appreciable amounts in DF.

3 Gravimetric Methods of Analyses

Crude Fibre and Detergent Methods

The crude fibre method implies extraction by dile acid and alkali with subsequent isolation of the insoluble residue by filtration, whereas the detergent methods are based on boiling with detergent in acidic (ADF) or neutral (NDF) solution, possibly supplemented with an amylase treatment to prevent filtration problems.[1,2] Soluble DF components are lost with these methods; however, the detergent method results in more well-defined residues than the crude fibre method, and should be preferred. The detergent methods are primarily used for analyses of forage.

Enzymatic Methods

The analytical approach of more recent gravimetric methods, using enzymes in combination with chemicals for extraction, comprises recovery of both insoluble (IDF) and soluble dietary fibres (SDF). These methods have, however, been criticized for overestimating the DF content by including components in the residue that differ from the traditional DF components (Section 14.2). The most widely used methods at present are the AOAC procedures (Association of Official Analytical Chemists[3]) developed and improved by Prosky and co-workers, *e.g.* Ref. 4. The primary steps involved in these procedures are illustrated in Box 14.1.

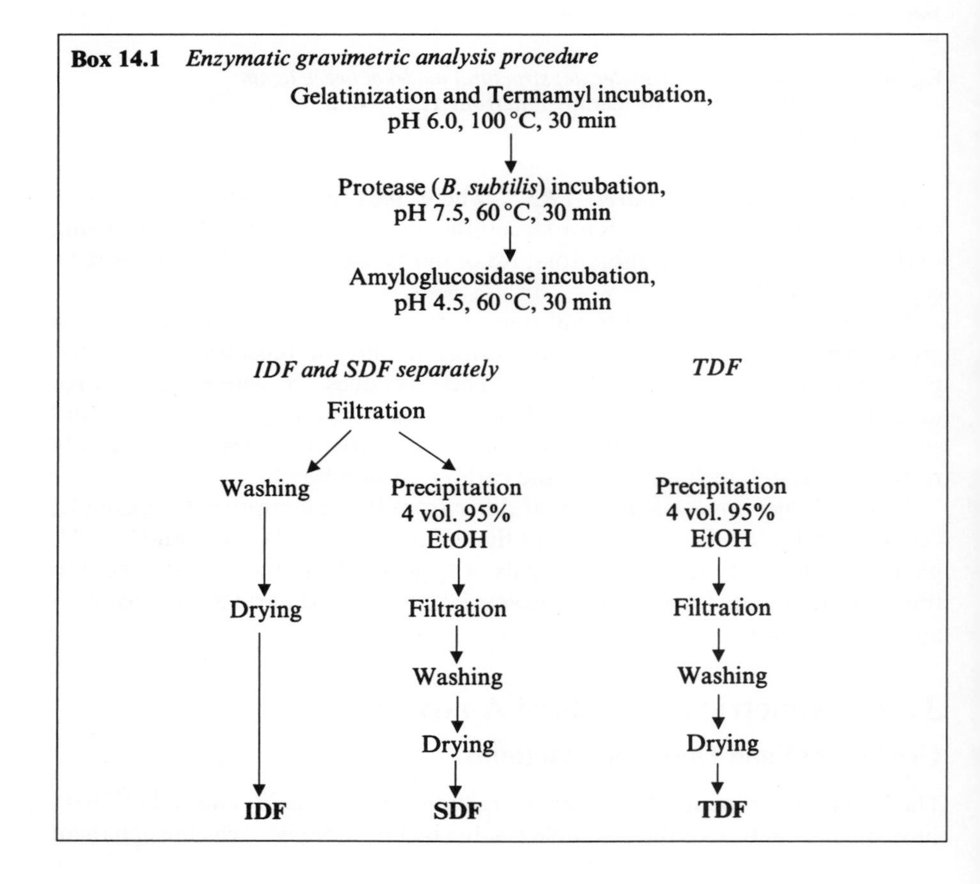

All analyses include a correction for ash and protein in the isolated residues. Determination of TDF (total dietary fibre), IDF, and SDF constitute a good starting point for further investigation of the DF fraction.

Lignin Analysis

Few methods exist for lignin determination, the most well-known being the isolation of lignin as the residue that is insoluble in 72% sulfuric acid (Klason lignin). Other methods are the permanganate oxidation method ($KMnO_4$), in which lignin is determined as the weight difference between oxidized and untreated material,[5] and the acetyl bromide method[6] in which lignin, after prior removal of interfering phenolic material, is measured by dissolving the sample in 25% acetyl bromide in acetic acid, followed by determination of the absorbance at 280 nm. All methods are limited by severe interference problems. A new and more promising approach of lignin analysis has recently been introduced (Ref. 7), which uses the HPCE separation of phenolic aldehydes, ketones, and acids produced from lignin degradation (Chapter 10).

4 Chromatographic Methods of Analyses

The chromatographic approach to DF analyses comprises an initial enzymatic/chemical removal of starch followed by measurement of NSP as the sum of its constituent sugars released by acid hydrolysis. Lignin is either determined separately, usually as Klason lignin, or omitted from the analysis, but HPCE determination of phenolics produced in lignin degradation may be an attractive supplementary technique.

Analysis of monosaccharides includes a wide spectrum of methods ranging from simple techniques such as paper (PC) and thin-layer chromatography (TLC) to colorimetry to more complicated methods involving the use of advanced apparatus such as in GLC, HPLC, and HPCE. Both PC and TLC (Chapter 6) can be used for qualitative purposes but are otherwise of limited interest in DF analysis.

Hydrolysis of Polysaccharides

The glycosidic bonds involved in the formation of polysaccharides differ widely in their susceptibility to acid hydrolysis, and the resulting liberated monomers may be unstable to various extents, depending on the actual reaction conditions. Acid-catalysed hydrolysis of polysaccharides is thus a compromise between complete liberation and destruction of monosaccharides.

A wide range of methods of hydrolysis exists, differing in type and concentration of acid as well as temperature and time used for hydrolysis. Sulfuric acid is widely used, but trifluoroacetic acid and hydrochloric acid have also been applied to the hydrolysis of polysaccharides. An advantage of trifluoroacetic acid over sulfuric acid is that the former is readily evaporated, thus avoiding the need for neutralization of the hydrolysate. However, incomplete recovery of glucose from the relatively acid-resistant cellulose fraction may be a problem. Hydrolysis of polysaccharides by sulfuric acid is generally performed by an initial treatment of the sample with a strong acid (sulfation of monosaccharides) followed by hydrolysis in dilute acid (*vide infra*, Box 14.3). In a comprehensive study using HPLC and GLC for sugar determination,[8] it was found that the fastest analysis and best yields were obtained with 12.0 M sulfuric acid for 1 h at 35 °C, followed by treatment with 2.0 M sulfuric acid for 1 h at 100 °C. The overall recovery of neutral sugars under these conditions was determined to be 52% (rhamnose), 89% (xylose), 92% (mannose), 94% (galactose, glucose), 95% (arabinose), and 96% (fucose). It is possible to achieve high yields of particular sugars by adjustment of the experimental conditions. For quantitative determination of sugars, variations in recovery according to differences in structure, method of hydrolysis, *etc.* have to be taken into account.

Polysaccharides rich in uronic acids constitute a special problem concerning hydrolysis. Typically, uronic acids are liberated as their aldobiuronic acids – disaccharides with an uronic acid in combination with another monosaccharide, *e.g.* xylose, which is very resistant to acid hydrolysis. However, the high stability of the adjacent uronic acid glycosidic bond partly protects uronic acids from degradation, as free uronic acids easily decarboxylate in free form, a property also utilized for their analyses (page 408). Attempts to optimize conditions of hydrolysis for polysaccharides rich in uronic acids have been described in Ref. 9.

Colorimetry

A wide range of colorimetric methods exist for qualitative or semiquantitative determination of monosaccharides; the results are expressed either as total sugar content or by division into groups of neutral pentoses, neutral hexoses, and uronic acids, *e.g.* Refs 10 and 11. It should, however, be noted that different groups of monosaccharides which interfere with each other may be a serious problem with mixtures of unknown composition. Interference by non-sugar components may also provide a problem. Colorimetric methods are not much used in analyses of DF, except for the determination of uronic acids (pages 111–112, 408) and for determination of starch/resistant starch (Section 14.5).

GLC

GLC, based upon partitioning of compounds between a stationary liquid and a mobile gaseous phase, is a widely used method for the quantitative determination of neutral monosaccharides. The GLC methods most often referred to in connection with DF determination are those of Theander, *e.g.* Ref. 12, and Englyst, *e.g.* Ref. 8. Differences between these methods are found mainly in the sample preparation (starch degradation), the Englyst group including a dimethyl sulfoxide (DMSO) step in addition to the enzymatic treatment in order to dissolve RS.

The polar and non-volatile nature of monosaccharides imply some preparatory steps prior to the actual GLC analysis (Figure 14.4).

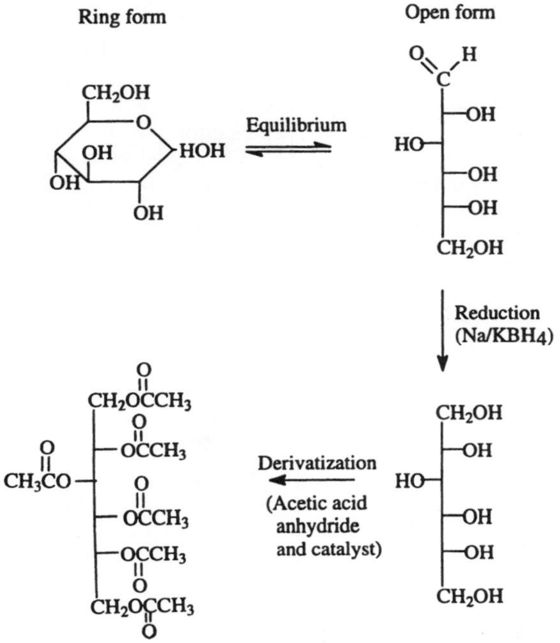

Figure 14.4 *Sample preparation prior to determination of monosaccharides by GLC. For experimental guidance see AOAC, Association of Official Analytical Chemists, Official Methods of Analysis, Washington D.C., 1995, vol. 16*

The first step after the acid hydrolysis of polysaccharides is a reduction (Na/KBH_4), which is performed to avoid multiple peaks resulting from derivatization of both of the different forms of anomeric rings (furanose, pyranose, α/β) and the open form of a given monosaccharide. The subsequent derivatization of the reduced monosaccharides are most commonly performed by transformation of the alditols into their corresponding alditol acetates, but trimethylsilyl derivatives and aldononitrile acetate derivatives are also used. *N*-Methylimidazole is used as catalyst for the acetylation of alditols by acetic anhydride, and the concentration in the final mixture is important with respect to completion of acetylation.

The actual GLC operating conditions, including type of column (length, diameter), column/injection temperature, carrier gas and carrier gas flow rate, type of stationary phase, and detection system differ among methods. The AOAC official method for determining TDF as neutral sugar residues (alditol residues), uronic acid residues, and Klason lignin proposes that shown in Box 14.2.

Figure 14.5. shows the results of a GLC analysis of hydrolysed SDF isolated from partly yellow seeded rapeseeds. The analysis was performed with aldononitrile acetate derivatives analysed on a DB-Wax wide-bore capillary column (30 m × 0.54 mm i.d.) (J & W Scientific, CA, USA) with a flame ionization

Box 14.2 *GLC operation conditions, AOAC proposal*

A gas chromatograph equipped with a flame-ionization detector, computing integrator, and capillary column, in split mode (ratio 1:50). The capillary column should be 50% cyanopropylphenylmethylpolysiloxane, 15 m × 0.25 mm i.d., 0.15 μm film thickness, DB 225 or equivalent. An autosampler is recommended. Operating conditions are specified as follows; injector: 200 °C, detector: 250 °C, helium carrier gas (linear flow velocity 0.8 m s^{-1}). Compressed air, nitrogen, and hydrogen are used as detector gases. A suitable temperature programme could be 160 °C (6 min initial) to 220 °C (4 min final) at a rate of 4 °C min^{-1} (ethyl acetate as solvent). Calibrate gas chromatograph with calibration standard solution prior to use.

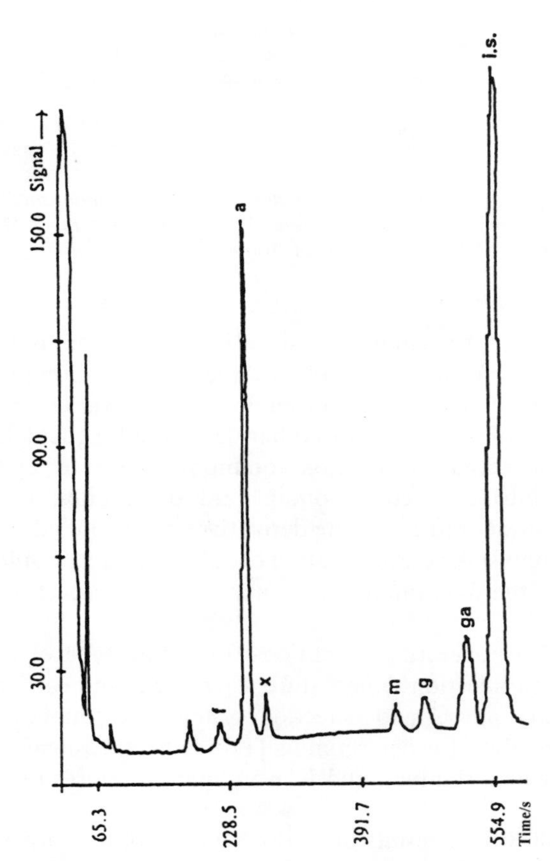

Figure 14.5 *GLC chromatogram of monosaccharides in SDF from partly yellow seeded rapeseeds; f = fucose, m = mannose, ga = galactose, a = arabinose, x = xylose, g = glucose, and i.s. = internal standard*

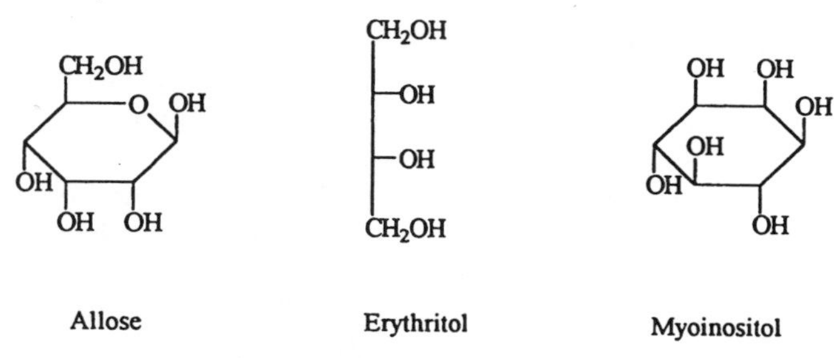

Figure 14.6 *Possible internal standards in GLC*

detector. The column was maintained at 195 °C for the first 2 min, then increased at 4 °C min^{-1} to 220 °C, the final temperature, which was held for 4 min. Helium (25 psi) was used as carrier gas.

Quantification of the neutral monosaccharide level should be based on averaging results from analyses of duplicate samples. Correction for losses of monosaccharides during acid hydrolysis and derivatization is necessary, and response factors for the different sugars in the actual GLC method should be included in the calculation. Theoretically, interference from uronic acids may occur, with galacturonic and, especially, glucuronic acid being partly reduced to galactitol and glucitol, respectively. Possible internal standards are allose, erythritol, and myoinositol (Figure 14.6).

The cyclohexitol structure of myoinositol used in the AOAC method does not permit any estimation of the reduction efficiency. Allose, however, will participate in all steps, and should be preferred. Equation 14.1 shows how calculation of the level of neutral monosaccharides (NM) is performed:

$$\mathrm{NM}(\%) = \frac{RF_{\mathrm{m}} PA_{\mathrm{m}} W_{\mathrm{is}}}{PA_{\mathrm{is}} S} \times 100 \qquad (14.1)$$

where RF_{m} = relative response factor for monosaccharide, PA_{m} = peak area for the monosaccharide in the sample solution, PA_{is} = peak area for the internal standard in the sample solution, W_{is} = weight (mg) of internal standard in the sample solution, and S = weight (mg, dry matter) of the original sample. A final conversion of the results from a monosaccharide into a polysaccharide basis is usually performed by multiplication by a factor of 0.88 (pentoses/deoxyhexoses) or 0.90 (other hexoses).[3]

HPLC

HPLC for monosaccharide analysis is attractive compared with GLC as no need for derivatization exists. Various methods have evolved over the last twenty years, including the use of normal phase systems with amino-bonded or amino-modified silica/vinyl alcohol copolymer, *e.g.* Ref. 13, reverse phase systems, generally with C_{18} material, *e.g.* Ref. 14, and various forms of cation, *e.g.* Ref. 15 and anion exchangers, *e.g.* Ref. 16. Separation parameters such as temperature, flow rate, and type of mobile phase, elution principle, *etc.* depend on the actual system. This also applies to the possibilities for detection of monosaccharides.

Separation in the normal phase system is based on the formation of hydrogen bonds between amines that are either covalently bonded to the column or are part of the mobile phase (dynamic coating) and the hydroxy groups of the monosaccharides. Acetonitrile and water, normally in a 3:1 ratio, constitute the mobile phase. The silica-based columns suffer from a relatively short lifetime owing to dissociation into the mobile phase, and vinyl alcohol copolymer material has proven to be superior to silica in this respect. A limitation, especially of primary amine columns, exists owing to the potential for Schiff base formation with reducing sugars.

The high stability C_{18} columns used in reverse phase systems for monosaccharide separations is thought to result in a longer lifetime than for the silica-based columns used in normal phase system. Dynamic column modification by use of n-alkyl amine (C_{12}, C_{14}) in the mobile phase as well as precolumn derivatizations that result in hydrophobic complexes have been used to improve the selectivity of the system.

The separation principle in the cation exchange of carbohydrates is based on interaction between the counterion (*e.g.* Na^+ or Ca^{2+}) and the hydroxy groups of the analytes. The size of sample molecules (monosaccharides, oligosaccharides) may affect their penetration into the resin and thereby also the separation. Mobile phases generally have water as the single component, but acetonitrile may also constitute part of the eluent.

Effective separation of monosaccharides by anion exchange implies introduction of a negative net charge on the sample molecules. This may be done by inclusion of borate in the eluent system, as described for HPCE (*vide supra*), or by utilizing the slightly acidic nature of carbohydrates, resulting in deprotonization of hydroxy groups in strong basic solutions. The inductive effect of the ring oxygen enables the hydrogen atom of the anomeric carbon hydroxy group to be the first to dissociate. Stabilization of the anionic oxygen will occur by intramolecular hydrogen bonds with adjacent OH groups.

Recently, a method based on the anion exchange principle for separating neutral sugars released upon acid hydrolysis of NSP has been presented.[9] Promising results were obtained, with data comparable to GLC results of the same standard mixture (Figure 14.7). The use of a guard column (AG5) prior to the analytical column (Dionex Carbopac PA1, 10 μm; 250 $\times$ 4 mm i.d.) prevents sulfate ions from the hydrolysate interfering with the separation of analytes, and protects the HPLC column from impurities.

Box 14.3 shows more details of the possible experimental procedure, supplemented with results from the analysis of monosaccharides in pea IDF hydrolysates.

The anion exchange principle gives the opportunity to replace the generally used refractive index (RI) detection with pulsed amperometric detection, which gives a considerably higher sensitivity. Detection by UV (190 nm) is possible but

Box 14.3 *HPLC (anion exchange) of monosaccharides in IDF from peas*

Hydrolysis

1. Weigh 100 mg IDF into 25 ml flasks
2. Add 1 ml 12 M sulfuric acid
3. Keep at 35 °C for 1 h (agitation after 30 min)
4. Add 11 ml deionized water
5. Boil under reflux for 2 h
6. Cool and remove insoluble material by filtration
7. Dry at 100 °C overnight before weighing
8. Neutralize the filtrate with saturated barium hydroxide solution
9. Remove precipitated barium sulfate by centrifugation
10. Add water to a known volume

HPLC operating conditions (proposal)

A Waters LC Module 1 HPLC gradient system (Millipore Corporation), including autoinjector and Waters 464 electrochemical detector with gold electrode (Millipore Corporation) was used. Detector settings (pulsed electrochemical detection): E_1 = 145–165 mV, E_2 = 500 mV, and E_3 = −800 mV. Separations were carried out on a Dionex CarboPac PA1 column using gradient elution with NaOH solution (10 mM, 0.0–3.8 min) and plain water (3.8–38 min) and a flow at 1 ml min^{-1}. A pre-column is generally recommended to guard the analytical column from impurities. Flushing with 300 mM NaOH (0.6 ml min^{-1}) should be performed after each sample followed by an equilibration period (15 min) to obtain constant retention times.

HPLC chromatogram of monosaccharides in IDF isolated from Pisum sativum *(cv.* Solara*)*

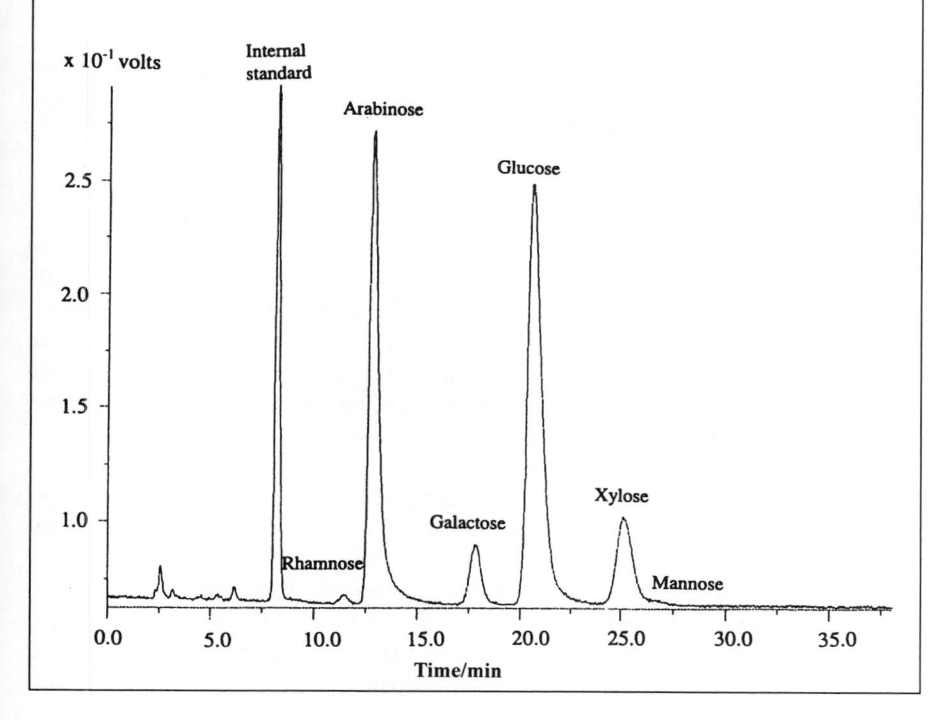

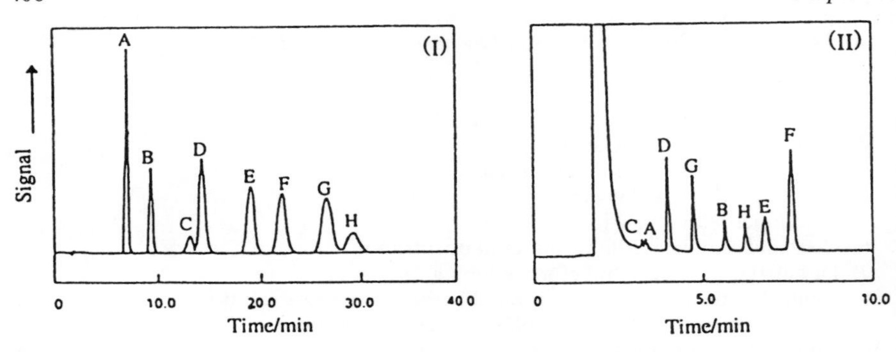

Figure 14.7 *Chromatograms of a standard mixture of neutral monosaccharides.*
(I) *HPLC: A = fucose, B = deoxygalactose (internal standard), C = rhamnose, D = arabinose, E = galactose, F = glucose, G = xylose, and H = mannose*
(Reproduced by permission from M.E. Quigley and H.N. Englyst, *Analyst (Cambridge)*, 1994, **119**, 1511)
(II) *GLC: A = fucose, B = allose (internal standard), C = rhamnose, D = arabinose, E = galactose, F = glucose, G = xylose, and H = mannose.*
(Reproduced by permission from H.N. Englyst, M.E. Quigley, G.J. Hudson, and J.H. Cummings, *Analyst (Cambridge)*, 1992, **117**, 1707)

is sensitive to impurities, and detection at higher wavelengths after pre- or postcolumn derivatization of monosaccharides, to improve sensitivity, reduces the advantage of HPLC. Quantification is, as in GLC, based on peak areas of the sample molecules and internal standards together with response factors for the individual monosaccharides.[9]

HPCE

Analysis of monosaccharides by HPCE (Chapter 10) is not generally used in DF connections; however, the method exists as an alternative to GLC and HPLC. Complexation of monosaccharides and borate is often used as a basis for separation in HPCE, giving rise to differences in charge density of the complexes and thereby different migration times in an electric field (Figure 14.8). A suitable proximity of hydroxy groups in the molecule favours complexation.

It is probably the tetrahydroxyborate ion, rather than boric acid itself, which is complexed with the polyhydroxy compounds. The complex may undergo further reactions, and the actual structure will be closely related to the stability of the complex in aqueous solution.

Separation in HPCE is very much dependent on the experimental conditions, *e.g.* applied voltage, temperature, buffer composition, pH, length of capillary, *etc.* A method suitable for the analysis of mannose, galactose, glucose, and xylose (base line separation) has been developed[17] (Figure 10.29). Figure 14.9

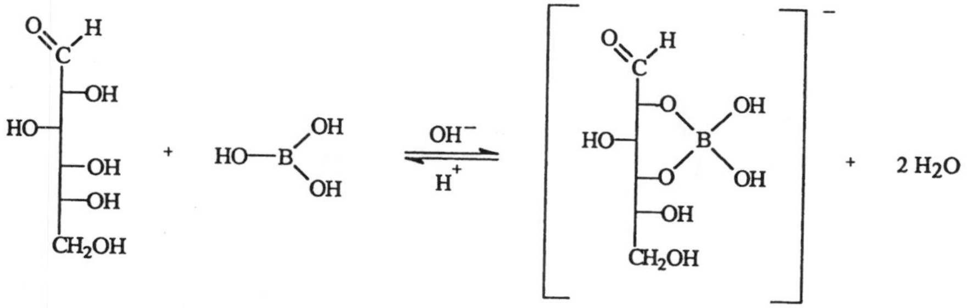

Figure 14.8 *Example of borate complexation with the open-chain form of glucose* (Reproduced with modifications by permission from S. Honda, S. Iwase, A. Makino, and S. Fujiwara, *Anal. Biochem.*, 1989b, **176**, 72)

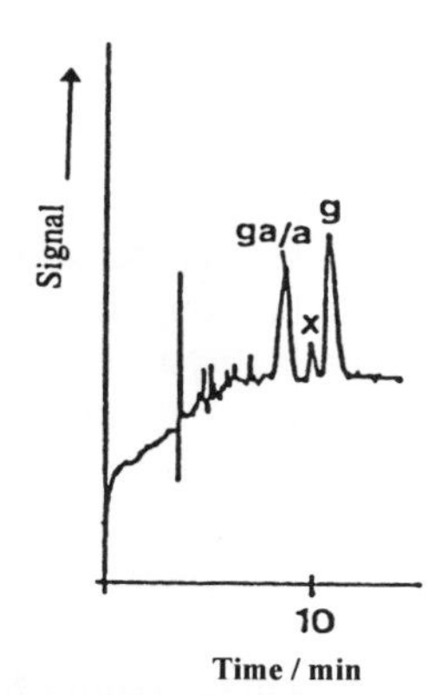

Figure 14.9 *Monosaccharides in IDF from yellow seeded rapeseeds analysed by a HPCE method. Peaks: ga = galactose, a = arabinose, x = xylose, and g = glucose; buffer: 50 mM sodium tetraborate, pH 9.9; temperature: 60 °C; voltage: 18 kV; injection: hydrodynamic (vacuum) (+ to −); detection: UV, 195 nm; capillary: 50 μm i.d., 63 cm total length, 40.5 cm effective length* (Modified from S. Hoffstetter-Kuhn, A. Paulus, E. Gassmann and H.M. Widmer, *Anal. Chem.*, 1991, **63**, 1541)

shows an electropherogram of monosaccharides in a hydrolysate of IDF isolated from yellow seeded rapeseeds.

Monosaccharides complexed with borate have a 2–20 times higher absorption than the uncomplexed monosaccharides. Other improvements in detection limit include derivatization by use of chromophoric compounds such as 3-(4-carboxybenzoyl)-2-quinolinecarboxaldehyde (CBQCA), 4-aminobenzoic acid, 4-aminobenzonitrile, 2-aminopyridine, and 9-aminopyrene-1,4,6-trisulfonate. The best sensitivity (attomol) was obtained with CBQCA by use of laser-induced fluorescence (argon-ion laser).[18] UV monitoring generally allows detection at the pico- to femtomol level. Indirect detection methods, comprising fluorescence as well as UV, have also been demonstrated, *e.g.*

Ref. 19. Separation of monosaccharides with indirect detection methods based on charge displacement has been carried out according to differences in pK_a'-values of analytes. Only a minor part of the neutral monosaccharides has been tested in these systems, and it seems that the selectivity of the methods is below that of the borate system used for derivatized monosaccharides. Recently, investigations with borate buffers containing β-cyclodextrins have also shown potential for separating fluorescent derivatives of sugar enantiomers,[20] which may also be separated by HPCE after derivatization with S-(−)-1-phenylethylamine.[21]

Quantification of sugars by HPCE may be complicated by the dynamic equilibria of borate complexation as the position of the equilibrium determines the magnitude of the absorption. However, this seems to be a minor problem compared with the relatively low sensitivity of the method. Use of a sugar as internal standard and relative normalized areas for calculation purposes should be standard, although it does not solve the above-mentioned problem.

Uronic Acids

Quantification of uronic acids constitutes a special problem in DF determination. Colorimetric methods include, among others, the carbazole and the dimethylphenol reaction.[22,11] Another analytical concept is the decarboxylation method, in which uronic acids are decarboxylated by boiling with hydroiodic acid in nitrogen, followed by quantification of the liberated CO_2.[23] An advantage of the decarboxylation method is that aldobiuronic acids give the same response as the free uronic acids. Interference, however, constitutes a problem in both types of analysis.

Chromatographic methods are the only possible way to distinguish different types of uronic acids. A HPLC method has been introduced that shows good separation of galacturonic acid, glucuronic acid, and mannuronic acid.[24] Disaccharide subunits, consisting of a uronic acid linked to a hexosamine and with various numbers of sulfate groups attached, can be liberated enzymatically from glycosaminoglycans, and separated successfully by HPCE[25] (Chapter 10). The problems with insufficient hydrolysis of polysaccharides and low stability of the free uronic acids is discussed in Ref. 24; however, work still has to be done in this area. At present, various correction factors have to be used for hydrolytic losses, the actual size depending on the state of the uronic acids (free form, oligomers, polymers). The use of acid anhydrous methanol or ethanol for polysaccharide hydrolysis, yielding the relatively stable (m)ethyl glycosides, could be a way of avoiding degradation problems.

5 Further Characterization of DF

Information about the monomeric composition of DF polysaccharides is widely used as a first step in the investigation of the DF fractions. However, the information may be difficult to interpret without preliminary knowledge of the main type of polysaccharides present. Fractionation of DF, based on solubility properties,[26] can be performed, *e.g.* on the TDF residue obtained from the enzymatic gravimetric DF analyses methods. Chromatographic methods may then provide additional information on the extracted compounds.

Extraction of pectic polysaccharides can be performed in a range of extraction media, *e.g.* water, ammonium oxalate, and different chelating agents such as cyclohexanediaminetetraacetic acid (CDTA), ethylenediaminetetraacetic acid (EDTA), and hexametaphosphate. The actual procedures differ in choice of molarity, pH, temperature, and duration of extraction. Dilute alkali (0.05 M NaOH) has also been used, either alone (20 °C) or in combination with EDTA. After dialysis and possible concentration, the solubilized components can be further studied by a wide range of different methods, including gel filtration and ion-exchange chromatography (Chapter 7).

Hemicelluloses are extracted with a range of alkali types and strengths, but 5 and 24% (w/v) KOH (0.9 and 4.3 M) or 4 and 10% (w/v) NaOH (1.0 and 2.5 M) are common. The extraction of hemicelluloses should preferably be carried out in the absence of oxygen, and a strong reducing agent as $NaBH_4$ may be added. Different subfractions of hemicelluloses may be obtained by sequential extraction with alkali, *e.g.* using 1.0 M KOH at different temperatures (1 °C and room temperature). Other extractants include DMSO and aqueous chaotropic agents (page 81; perchlorate, urea, and guanidinum thiocyanate). In strongly lignified tissue, extraction of hemicelluloses should preferably be performed after delignification. Further studies of the extracted components can be carried out by a wide range of techniques, as described for pectic polysaccharides. Fractionation of branched and unbranched xylans on a cellulose column by affinity chromatography (hydrogen bonds to cellulose) is described in Ref. 27.

The classical procedure for extracting cellulose involves the use of 72% sulfuric acid (w/w). Other less common extractants are cadoxen (1,2-aminoethane + cadmium oxide) and dry DMSO + paraformaldehyde. Further investigations of extracted cellulose are most likely limited to determination of the degree of polymerization and studies of the non-glucans present.

The Klason lignin residue, obtained after extraction of cellulose, is the classic end product of the fractionation scheme. Further studies of the intact lignin polymers is complicated by their very complex structure. Evaluation of the constituent phenolics may, however, be possible (pages 308 and 399; Ref. 7).

Non-traditional DF components such as glycoproteins may be extracted from the cell wall by use of cell-wall degrading enzymes or by the procedures described in Chapter 4. Further studies of the protein part can with advantage be initiated by group separation based on affinity chromatography followed by flash chromatography and fast polymer liquid chromatography (FPLC) (Chapter 8), isoelectric focusing (IEF) (Chapter 9), amino acid analysis or other protein-based techniques (Chapter 13). Low-M_r phenolics, such as benzoic and cinnamic acid derivatives, (monomers, dimers) are typically studied by use of either HPLC or HPCE (Figure 14.10), and ester and ether-linked phenolics can be determined separately by sequential alkaline- and acid-catalysed hydrolysis.

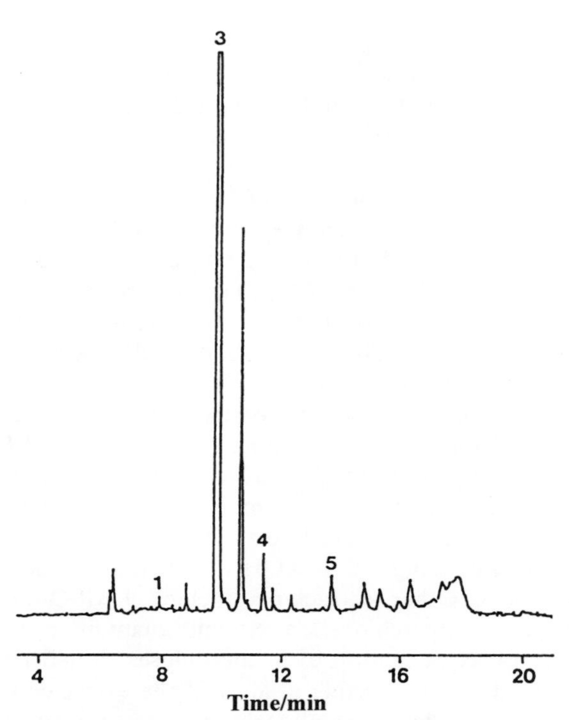

Figure 14.10 *Electropherogram of ester-linked cinnamic and benzoic acid derivatives in IDF from rapeseed. Peaks: 1 = 4-OH-benzoic acid, 3 = sinapic acid, 4 = ferulic acid, 5 = coumaric acid. The phenolic acids were obtained by alkaline hydrolysis of IDF (2 M NH_3, 1 h, 70 °C) followed by cooling and evaporation to dryness as a first step. The next step entails redissolution in 1 M HCl and evaporation to dryness followed by extraction three times in ethyl acetate with centrifugations and mixing of the resulting supernatants. The final steps involve evaporation to dryness of the mixed supernatant, redissolving in 1 M NH_3, evaporation to dryness and redissolution in water*
(Reprinted from C. Bjergegaard, S. Michaelsen, and H. Sørensen, *J. Chromatogr. A*, **608**, 403 (1992), Elsevier Science, Amsterdam)

Oligosaccharides present in the DF fraction will usually be found adsorbed to polysaccharides and may be extracted with boiling methanol–water. Analysis of individual oligosaccharides can then be performed by HPLC and HPCE techniques (Figure 10.9; Ref. 28).

Determination of starch or resistent starch by use of an ELISA reader is a newly developed technique that comprises four steps: extraction of plant flour, boiling of extract to dissolve starch, addition of iodine (blue complex), and reading of absorbance at 630 nm. The analysis is performed by use of microtitre plates and a computer-aided ELISA reader, allowing determination of eight dilutions for each of up to eleven samples at the same time, with as little as 100 μl sample material for each determination (Figure 14.11).

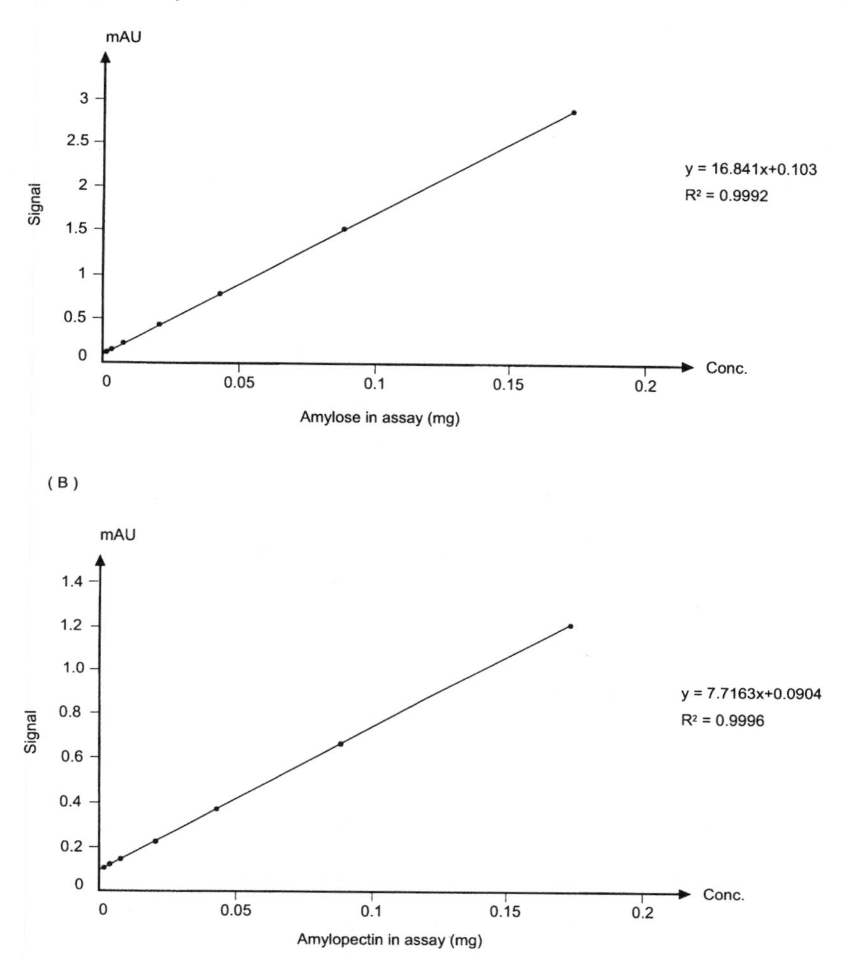

Figure 14.11 *Linear regression curves and resulting regression coefficients for* (A) *amylose and* (B) *amylopectin (seven dilutions), determined by use of a computer-aided ELISA reader*

6 Conclusion

Chromatographic techniques are well suited for analysis of DF, including the determination of the NSP monosaccharide composition as a first step, but also in respect to oligo- and polysaccharides as well as non-traditional DF components such as low-M_r phenolics and oligosaccharides. To date, GLC and HPLC have been the techniques preferred for monosaccharide analysis; however, HPCE provides a good alternative. The enzymatic gravimetric methods for determining IDF, SDF, and TDF are a good starting point for chromatographic investigations and further fractionation studies, and are also valuable in themselves.

7 Selected General and Specific Literature

[1] P.J. Van Soest, *J. Assoc. Off. Anal. Chem.*, 1963, **46**, 829.
[2] P.J. Van Soest and R.H. Wine, *J. Assoc. Off. Anal. Chem.*, 1967, **50**, 50.
[3] AOAC, Association of Official Analytical Chemists, in 'Official Methods of Analysis'. Washington D.C., 1995, vol. 16, (985.29, 991.42, 993.19, 994.13).
[4] L. Prosky, N.G. Asp, T.F. Schweizer, J.W. DeVries, and I. Furda, *J. AOAC Int.*, 1992, **75**, 360.
[5] H.K. Goering and P.J. Van Soest, U.S. Department of Agriculture, in 'Agricultural Handbook No. 379', U.S. Government Printing Office, Washington D.C., 1970.
[6] I.M. Morrison, *J. Sci. Food Agric.*, 1972, **23**, 455.
[7] O. Maman, F. Marseille, B. Guillet, J.-R. Disnar, and P. Morin, *J. Chromatogr. A*, 1996, **755**, 89.
[8] H.N. Englyst, M.E. Quigley, G.J. Hudson, and J.H. Cummings, *Analyst*, 1992, **117**, 1707.
[9] M.E. Quigley and H.N. Englyst, *Analyst*, 1992, **117**, 1715.
[10] M. Dubois, K.A. Gilles, J.K. Hamilton, P.A. Robers, and F. Smith, *Anal. Chem.*, 1956, **28**, 350.
[11] R.W. Scott, *Anal. Chem.*, 1979, **51**, 936.
[12] O. Theander, P. Åman, E. Westerlund, and H. Graham, in 'New Developments in Dietary Fiber – Physiological, Physicochemical, and Analytical Aspects', eds. I. Furda and C.J. Brine, Plenum Press, New York, 1990, p. 273.
[13] T. Akiyama, *J. Chromatogr.*, 1991, **588**, 53.
[14] S. Honda, E. Akao, S. Suzuki, M. Okada, K. Kakehi, and J. Nakamura, *Anal. Biochem.*, 1989, **180A**, 351.
[15] T. Kawamoto and E. Okada, *J. Chromatogr.*, 1983, **258**, 284.
[16] T.J. Paskach, H.-P. Lieker, P.J. Reilly, and K. Thielecke, *Carbohydr. Res.*, 1991, **215**, 1.
[17] S. Hoffstetter-Kuhn, A. Paulus, E. Gassmann, and H.M. Widmer, *Anal. Chem.*, 1991, **63**, 1541.
[18] J. Liu, O. Shirota, D. Wiester, and M. Novotny, *Proc. Natl. Acad. Sci. U.S.A.*, 1991, **88**, 2302.
[19] T.W. Garner and E.S. Yeung, *J. Chromatogr.*, 1990, **515**, 639.
[20] M. Stefansson and M. Novotny, Abstract in 'Proceedings of the Fifth International Symposium on High Performance Capillary Electrophoresis', 25–28 January, 1993, Orlando, FL.
[21] C.R. Noe and J. Freissmuth, *J. Chromatogr. A*, 1995, **704**, 503.
[22] T. Bitter and H.M. Muir, *Anal. Biochem.*, 1962, **4**, 330.
[23] O. Theander and P. Åman, *Swedish J. Agric. Res.*, 1979, **9**, 97.
[24] M.E. Quigley and H.N. Englyst, *Analyst*, 1994, **119**, 1511.
[25] S. Michaelsen, M.-B. Schrøder, and H. Sørensen, *J. Chromatogr. A*, 1993, **652**, 503.
[26] R.R. Selvendran and P. Ryden, in 'Methods in Plant Biochemistry', series eds P.M. Dey and J.B. Harborne, Academic Press, 1990, vol. 2, p. 549.
[27] S.C. Fry, in 'The Growing Plant Cell Wall: Chemical and Metabolic Analysis', ed. M. Wilkins, Longman Scientific & Technical, London, 1988, p. 102.
[28] J. Frias, K.R. Price, G.R. Fenwick, C.L. Hedley, H. Sørensen, and C. Vidal-Valverde, *J. Chromatogr.*, 1996, **719**, 213.
[29] 'New Developments in Dietary Fiber – Physiological, Physicochemical, and Analytical Aspects', eds. I. Furda and C.J. Brine, Plenum Press, New York, 1990, 325 pp.
[30] D.A.T. Southgate, 'Dietary Fibre Analysis', The Royal Society of Chemistry, Cambridge, UK, 1995, 174 pp.
[31] L. Prosky and J.W. DeVries, 'Controlling Dietary Fiber in Food Products', Van Nostrand Reinhold, New York, 1992, ch. 3, p. 91.

APPENDIX

Supercritical Fluid Extraction (SFE) and Supercritical Fluid Chromatography (SFC)

1 Introduction

Supercritical fluid techniques (SFT = SFE + SFC) call for special attention in relation to analytical and preparative methods of interest for disciplines as biochemistry, natural product chemistry, feed- and food-analyses.[1–3] SFT seems thus to have the potential to fill in the 'analytical grey area' between the areas, covered by GC and LC (Figure A1).

The analytical area between GC and LC comprises various types of lipids and amphiphilic compounds, which often create problems with respect to determination of many of these individual and intact compounds. This is due to the special properties of lipids and amphiphilic compounds, which are often bound in membranes (pages 74–79), in dietary fibres (Chapter 14), and more generally in matrix systems. This may create problems both with respect to extraction and analysis of the native compounds. However, SFT seem to have the potential as efficient tools for covering at least part of these important areas.

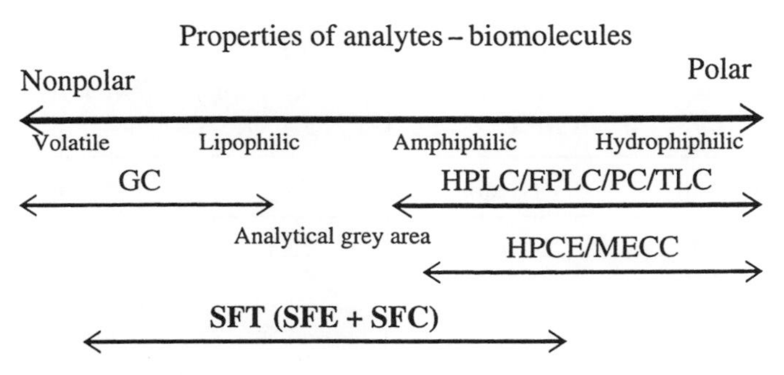

Figure A1 *Illustration of the areas where different types of analytical techniques are suitable as methods of analysis for compounds with a certain degree of polarity*

2 Theory

As basis for use of SFT, a brief description of the theory and behaviour of an ideal gas and supercritical fluids is considered to be of value and necessary for optimal use of SFT. The behaviour of an ideal gas can be described by the equation of state (EOS):

$$P\bar{V} = RT \quad \text{or} \quad PV = nRT \tag{A1}$$

The ratio $P\bar{V}(RT)^{-1} = 1$ holds good only for ideal gases and it is known as the compressibility and deviates from unity for non-ideal gases. The product of $P\bar{V}$ will thus only be 22.414 L atm mol^{-1} at all pressures for ideal gases at 0 °C.

Carbon dioxide is a non-ideal gas, which for various reasons is of special interest in connection with SFT, and important relations derived for CO_2 by use of the EOS (equation A1) are shown in Table A1.

It is seen that the compressibility factor can be below unity, which indicates that CO_2 is more compressible than an ideal gas, but at high pressures CO_2 is less compressible than an ideal gas. The critical state of a pure substance is the state of temperature and pressure where the gas and liquid phases become nearly alike, where they cannot exist as separate phases. The critical temperature (T_c) is the highest temperature where the gas and liquid can exist as separate phases, and the critical pressure (P_c) is the pressure at T_c. The critical volume $\bar{V}_c$ is the molar volume under these conditions. For CO_2, the critical constants are:

$$P_c = 72.8 \text{ atm or } 7.38 \text{ MPa}$$
$$T_c = 304.2 \text{ K or } 31\ ^\circ\text{C}$$
$$\bar{V}_c = 0.0942 \text{ L mol}^{-1}$$

Mixtures will give a more complicated behaviour as liquid and vapour phases will then be of different composition. An illustration of the relations between solid, liquid, gas, and supercritical fluid as a function of pressure and temperature is shown for pure CO_2 in Figure A2.

Points along the P–T curves represent conditions with equilibrium between the phases. By approaching the CP along the vaporization or boiling line, the

Table A1 *Pressure–molar volume and compressibility relations for CO_2 at 0 °C (273.15 K)*

P (atm)	$\bar{V}$ (L mol^{-1})	$P\bar{V}$	$P\bar{V}$ (RT)$^{-1}$
0.1	224.100	22.41	1.00
1.0	22.262	22.26	0.99
50	0.047	2.34	0.10
100	0.045	4.50	0.20
300	0.042	12.46	0.46
1000	0.037	36.87	1.65

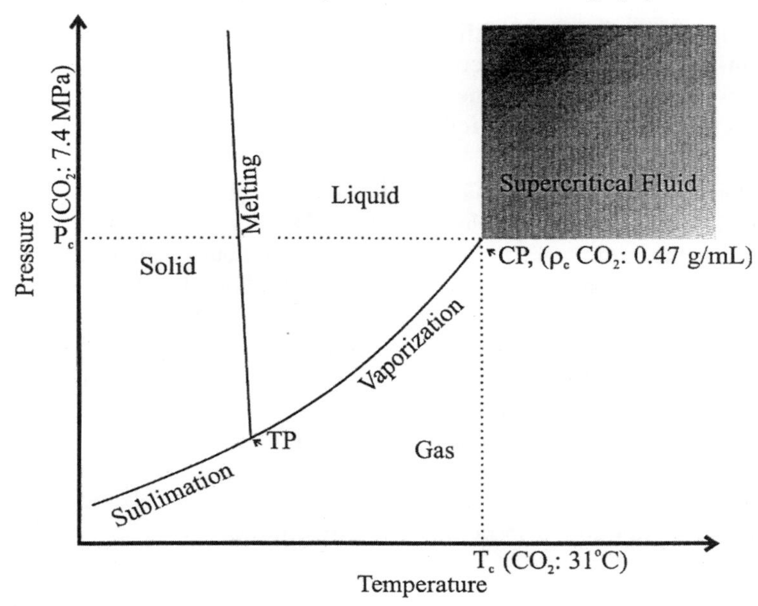

Figure A2 *Three-phase diagram for CO_2 extended to the supercritical region (the dark region) where* $T > T_c$ *and* $P > P_c$. *Also indicated are the triple point = TP, the critical point = CP (*$T = T_c$ *and* $P = P_c$*), critical temperature =* T_c *and critical pressure =* P_c

liquid- and vapour-phases become more and more alike, and beyond the CP only one phase exists – the supercritical phase. This phase is compressible like a gas, has solvating properties like a liquid and has a diffusion rate between those of the corresponding liquid- and gas-phases. However, although the supercritical fluid region is defined as shown in Figure A2, the boundaries to the adjacent liquid and vapour areas do not represent phase boundaries. Consequently, the properties of the SF are continuously changed from the supercritical phase to the liquid or vapour phase properties, respectively, when the pressure or temperature is changed. The only limitations for using conditions different from the supercritical fluid region is that the liquid–vapour phase boundary is not crossed.[4]

The supercritical fluid has several advantages compared with traditional liquids used as organic solvents, thus making SF attractive for many types of extraction (SFE) and chromatographic (SFC) purposes. These advantages are mainly due to the ease of change of density, and thereby the solvating power, as well as selectivity of the SF. In addition, the supercritical CO_2 fluid has many other beneficial properties both with respect to SFC and for extractions needed prior to LC (HPLC and HPCE). These properties comprise:

- high solvating power, high density (ρ);
- low viscosity (η);

- high diffusivity – large diffusion coefficient;
- non- or low chemical reactivity;
- non-toxic to the environment;
- easy to remove after extraction.

The density of the fluid turns liquid-like above C (Figure A2) due to the high compressibility near CP (Table A1). The viscosity (η; pages 126–127) and diffusion coefficient (D) of SFs are, despite the high density (ρ), still appreciably lower and higher, respectively, compared with liquids (Table A2). The low η and high D of supercritical CO_2 fluid compared with liquids (Table A2) gives a better penetrability and mass transport compared with liquid extraction.

Pure supercritical CO_2 has a low polarity owing to lack of dipole moment (Table A3), which makes this SF ideal for extraction of lipophilic compounds such as triglycerides (oils/fat) and fat-soluble vitamins without co-extraction of lipophilic/amphiphilic compounds such as phospholipids, *etc.*

Carbon dioxide is a nearly non-polar solvent, but the solvating power for amphiphilic compounds can be increased by addition of polar modifiers. The

Table A2 *Physical properties of gases, liquids and supercritical CO_2; ρ = density; η = viscosity; D = diffusion coefficient*

	ρ (g mL^{-1})	η (g cm^{-1} s^{-1})	D (cm^2 s^{-1})
Gases (ex. He)	$(0.2–2) \times 10^{-3}$	$(1–3) \times 10^{-4}$	0.1–0.4
SF (ex. CO_2); $T > T_c$; $P > P_c$	0.2–1.0	$(1–9) \times 10^{-4}$	$(200–700) \times 10^{-6}$
Liquids (ex. H_2O)	0.6–1.6	$(0.2–0.3) \times 10^{-2}$	$(2–20) \times 10^{-6}$

Table A3 *Critical constants and dipole moments for CO_2 and frequently used modifiers**

Modifier	T_c (K)	P_c (bar)	$\bar{V}_c$ (cm^3 mol^{-1})	*Dipole moment* (Debye)
CO_2	304.2	73.8	98.2	0.0
Methanol	513.1	80.9	118.0	1.7
Ethanol	516.2	61.4	167.1	1.7
1-Propanol	536.8	51.7	218.0	1.7
2-Propanol	508.8	47.6	220.0	1.7
2-Methoxy ethanol	597.6	52.8	263.0	2.8
Tetrahydrofuran	540.1	51.9	224.0	1.6
Acetonitrile	548.0	48.3	173.0	3.5
Dichloromethane	510.0	63.0	178.0	1.6
Chloroform	536.4	53.7	238.9	1.1
Water	647.4	221.2	57.1	0.998
Acetone	508.1	47.0	209.0	2.9

* L.T. Tailor, 'Introduction to Supercritical Fluid Extraction', R & D Magazine, 1996

increased interaction of the solute molecules is reflected in the size of the dipole moment (Table A3) of the modifier employed. Another aspect of the co-solvent or modifier is the Lewis acid–base properties,[3] where, *e.g.* methanol is both a strong hydrogen donor and acceptor, acetone is an acceptor and chloroform is a hydrogen donor. The modifier should thus be chosen in accordance with the solute molecules to be dissolved. The solubility of a solute in the modifier is often a measure of the efficiency of the modifier for solute extraction.

For extraction of amphiphilic compounds the solvating power can be further increased by adding small amounts (0–10%) of a polar modifier like methanol or ethanol (Table A3). Note should be taken that not all polar modifiers are miscible with SFs and, for example, water is only slightly soluble in SF CO_2 (less than 3–4% soluble). At higher modifier concentrations (10–60%), the modifier effects on the solvent strength or solvating power are more dominating, which is necessary for extractions (SFE) of amphiphilic compounds like phospholipids. The increase in solvating power by incorporation of the modifier into the SF may affect the mass transfer and result in a reduced diffusion coefficient. Increasing the modifier concentration may also change the selectivity or retention order for individual compounds (SFC). Moreover, the modifier affects both the stationary phase, as in SPE (pages 149–150), and the binding properties of the analytes to the matrix system.

In a mixture of, for example, CO_2 and a modifier, a binary system is obtained with a two-phase region which should be avoided in extraction and chromatography.[4] For each composition of the binary system, the SF region will represent an area with parameters defined in the range between the limits of the region of the two individual, pure components of the binary system.[4] The nature of mixed fluid solvent systems is thus complicated by the increased complexity of the phase behaviour comprising changes in T_c and P_c by use of a modifier. However, acceptable approximations for these mixed fluid systems can be calculated by, for example, the Bender virial EOS (*vide infra*).

3 SFE

For extraction of lipophilic (and amphiphilic) compounds from solid samples, Soxhlet extraction is at present the most widely used technique. Consequently, comparison of SFE with traditional techniques involves comparison with Soxhlet extraction. In general, some of the most pronounced advantages of using SFE compared with other separation techniques for analytical purposes are higher efficiency and selectivity, decreased extraction time and increased safety.[3]

3.1 Equipment

Various supercritical extraction units for analytical purposes are commercially available, and the instrument components are shown in Figure A3.

An oven regulates the temperature in the extraction chamber and a restrictor regulates the pressure. A restrictor is typically a narrow capillary made of fused

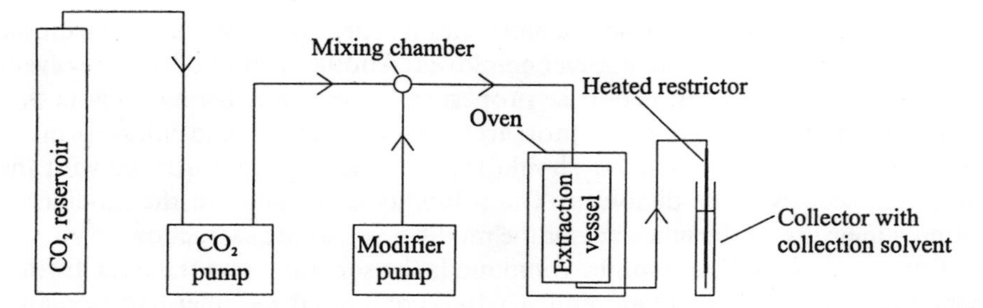

Figure A3 *Analytical SFE instrument showing CO_2 reservoir, CO_2 pump and modifier pump. The extraction vessel is thermostated by the oven and extractable material is collected from the outlet restrictor in a collection vial*

silica or stainless steel which regulates the flow from the extraction cell as well as the back pressure and transports the extractable material to the collection vial. The restrictor is usually heated to 60–80 °C to avoid plugging from extracted material or water, which may freeze at the low temperatures obtained at the restrictor tip when the SF rapidly expands. However, this heating is not expected to influence, for example, heat sensitive/labile extracted compounds, as the retention time for the extractables in the restrictor is short.

3.2 Extraction Strategy

Sample requirements

Samples subjected to SFE should usually consist of a solid matrix. For liquid samples, pretreatment prior to SFE could therefore be lyophilization or use of adsorbents for generation of a solid matrix.

Moisture in solid samples may change the properties of analytes or solutes corresponding to the use of modifier, and consequently reproducibility poses demands for control of the sample moisture.

Extraction

The use of sequential SF extractions may separate families of compounds by sequentially changing the parameters for the extraction. The SF solvating power can thus be increased, *e.g.* in pressure steps or by increasing the modifier addition. The CO_2 purity needed generally depends on the required sensitivity of the analysis and high purity CO_2 ($>99.7\%$) is usually used.

Collection

The most common method for separation of the extracted solutes from the SF is by lowering the pressure which reduces the solubility. This is achieved by the depressurization of the SF through the restrictor. The solutes may be collected by passing the depressurized SF through an appropriate volume of an appropriate solvent. Depending on the solutes, considerations should be made with respect to the recovery during collection. In this connection, the temperature of the collection solvent is important because lowering of the solvent temperature results in an increased surface area to volume ratio of the SF bubbles since an increase in the solvent viscosity leads to smaller bubbles. In addition, a lower temperature reduces the solvent volatility and thereby the risk of aerosol formation. Similarly, the flow of the SF is usually kept to a maximum of 1–2 mL min^{-1} as a flow of 1.0 mL CO_2 SF min^{-1} translates into 500 mL gas min^{-1}, and therefore too large a flow increases the risk of aerosol formation. Other collection modes used particularly for volatile analytes comprise collection on a sorbent material, *e.g.* as a bed or column packing of solid material, with subsequent elution of the analytes.

Quantification

SFE is an extraction technique similar to liquid solvent extraction, and as such, SFE is often the first step in a row of separation and characterization steps. For determination of the amount of extracted material, gravimetric determination is often valuable, with determination of both the amount of analyte extracted (determined as weight gain of collection vial after evaporation of solvent) and the sample weight difference of the sample residual in the extraction chamber.

3.3 Examples of Applications

When extracting oil from different kinds of matrix systems with great variations in oil content[5–6] it was seen that SFE generally gives slightly lower values than Soxhlet extraction. This difference has been found to be caused by amphiphilic compounds, especially phospholipids, extracted together with fat in the Soxhlet procedure but not in SFE[6] (Figures A4 and A5). The reproducibility obtained by SFE was found to be at the same level as for Soxhlet extraction.

The amount of tocopherols, carotenoids, chlorophylls and compounds with corresponding lipophilic–amphiphilic properties occur in nearly equal amounts in SFE and Soxhlet extracted oils. The extraction time including sample preparation is about 40 min for SFE compared with up to 20 h for Soxhlet extraction. Furthermore, no hazardous solvents, only CO_2 and maybe ethanol, are used in the SFE procedures.

4 SFC

SFC is an efficient technique for both analytical and preparative work with lipophilic–amphiphilic compounds, which often create problems when studied by LC, HPLC and HPCE.

4.1 Theory

The lower viscosity and higher diffusivity properties of SFs (Table A2) compared with liquids generally used as the mobile phase in HPLC and

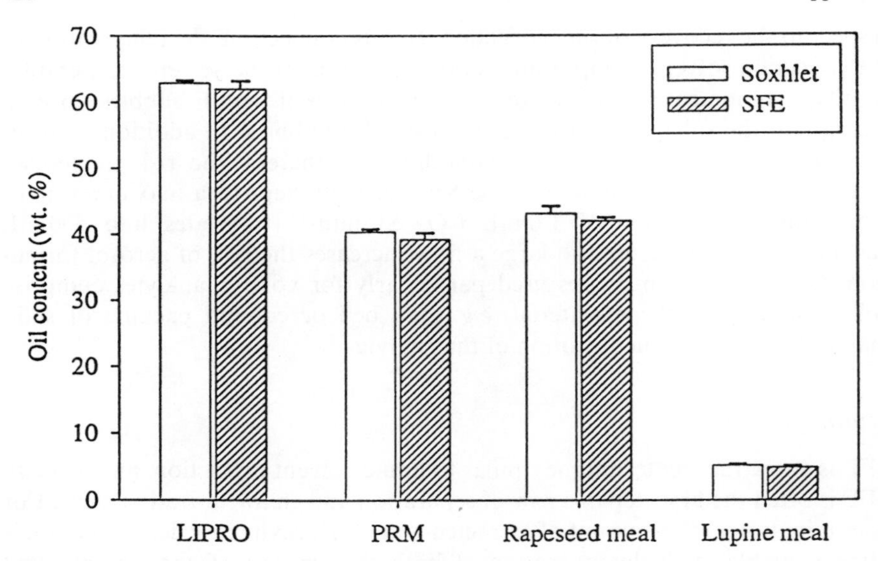

Figure A4 *Comparison of results obtained by determination of lipids in matrix systems with great variation in their lipid content using SFE and Soxhlet extraction. LIPRO = lipid binding protein; PRM = protein rich meal from dehulled rapeseeds. The oil extractions were performed with either a Spe-ed SFE instrument with 10 mL extraction vessels (Applied Separations, Allentown, PA, USA), or an Isco SFE unit consisting of Isco 260D syringe pumps, Isco series D controller, Isco SFX™ 2–10 SF extractor and Isco restrictor temperature controller (Lincoln, Nebraska, USA). See Box A1 for detailed extraction procedure*

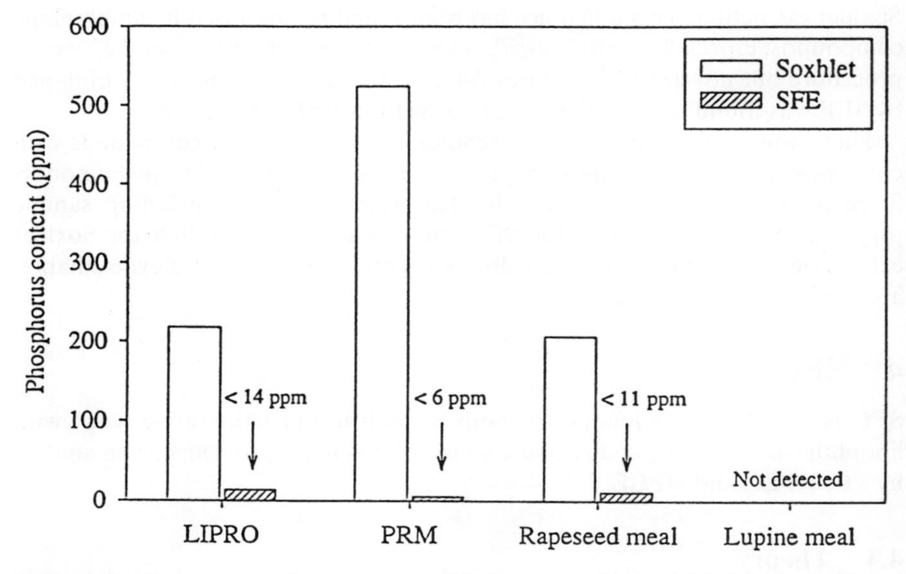

Figure A5 *Comparison of phospholipid content determined as phosphor content in oil obtained by SFE and Soxhlet extraction of the samples mentioned in Figure A4. Extraction was performed as described in Box A1*

Box A1

Extraction of lipids

1. A weighed amount (1–5 g) is mounted in the extraction chamber
2. Extraction is performed with a flow of 1–2 ml SF CO_2 min^{-1} for 30 min, at 50 MPa and 75 °C. Extractables are collected in pre-weighed glass collection vials containing 5–10 ml ethanol or like as solvent for the extractables
3. The ethanol solvent is evaporated, *e.g.* by passing an air stream over the solvent surface. The dry weight of the extractables is determined
4. The weight of the residual sample in the extraction chamber is determined
5. From the gravimetric measurements performed in 3 and 4, the lipid content of the sample is calculated. Further analyses may comprise TLC (Box 6.1) and SFC (*vide infra*)

Extraction of amphiphilic compounds

For sequential extraction of different compound groups, 1 and 2 are performed and followed by extraction with a polar modifier. Extraction of phospholipids:

6. Extraction for an additional 30 min with 20% ethanol as modifier is performed with sample collection in another pre-weighed collection vial
7. Analyses may comprise TLC (Box 6.1), phosphorous determination (page 280), HPCE (Figure 10.20), and SFC

HPCE–MECC give better penetrability and mass transport. Therefore, the pressure drop along the chromatographic columns used in HPLC (Chapter 9) and SFC is thus found to be much lower in SFC compared with HPLC. This results in the possibility of a much higher linear velocity of the mobile phase in SFC compared with HPLC for generating the same height equivalent to a theoretical plate[7] (HETP; page 135) (Figure A6).

The flat van Deemter curve in SFC (Figure A6) shows that relatively high flow rates (u) can be used and still give low HETP. The time needed for equilibration after parameter changes is thus much shorter in SFC than in HPLC, resulting in the prospect of fast changes in parameters such as pressure and mobile phase composition, which need to be tested and evaluated for obtaining optimal SFC procedures. Evaluation of SFC procedures comprise use of the following parameters as for HPLC or LC in general (pages 132–135):

(1) the retention factor [$k'_A = (t_A - t_0)/t_0$]; t_A and t_0 are the retention times for analyte and unretained compounds, respectively;
(2) the separation factor $\alpha = (k'_B/k'_A)$ for two compounds A and B;
(3) the resolution [$R_s = 1.177\,(t_A - t_B)/(w_{A½} + w_{B½})$] for two compounds A and B with retention times (t_A and t_B) and peak widths $w_{A½}$ and $w_{B½}$ at half peak height (all in time units);
(4) the number of theoretical plates [$N = 5.54\,(t_A/w_{A½})^2$] for each analyte (A);
(5) the height equivalent to a theoretical plate for a column of length (L) where HPLC and packed column SFC give a simplified van Deemter equation for HETP, with D_A = diffusion coefficient for the analyte, t_d = desorption time for the analyte and linear velocity of the mobile phase (u). The

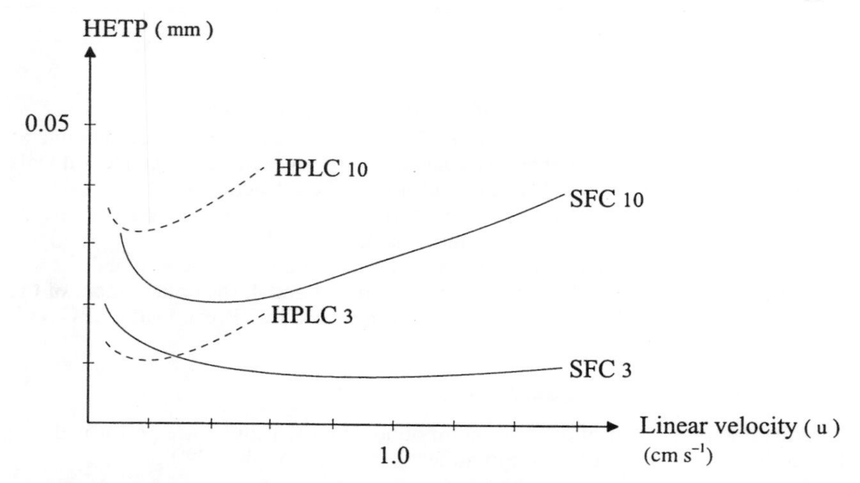

Figure A6 *Comparison of HPLC and SFC with respect to HETP as function of linear velocity for the mobile phase using equal types of columns packed with* 10 μm *and* 3 μm *particle size materials, respectively*

variable F is a function of u, particle diameter (d_p) and column diameter (d_c) with $F = f(d_p^2, d_c^2, u)$.[8]

$$\mathrm{HETP} = L\,N^{-1} = (2t_d\,k'\,u)(1+k')^{-2} + F\,u\,D_A^{-1} \tag{A2}$$

The minimal HETP can thus be obtained when:

- u is within the flat part of the van Deemter plot (Figure A6);
- D_A in the mobile phase is high;
- the particle size of the column material is small (Figure A6).

From the van Deemter equation and experiences from LC (HPCE) the size of HETP depends upon the column diameter (d_c). However, d_c has been found not to be of importance for the efficiency of packed column SFC due to the high diffusivity and efficient mass transfer properties of SF.[9] Pressure and temperature effects are thus more important factors for SFC.

Pressure (P) of the fluid phase has been found to be one of the most important parameters in SFC as t_A shows strong dependence on P, with decreasing t_A at increasing P. This pressure dependence is also reflected in the density (ρ), which affects t_A in SFC because the solvent strength or solubility parameter (δ) is related to the solubility of the analyte in the fluid and δ comprises interactions caused by non-polar, dipole and hydrogen bonding types.[10] Temperature changes have, however, been found to have less influence on δ (*vide infra*), whereas δ varies linearly with the density (ρ) as shown by Giddings *et al.* (1968).[11] With ρ_r = reduced density (ρ/ρ_c) of supercritical fluid; $\rho_{r,\mathrm{liquid}}$ =

reduced density of the liquid state, which at the boiling point is 2.66, $\rho_{c(CO_2)} = 0.469$ g mL^{-1}, and with P_c in atmospheres, the resulting equation is:

$$\delta = 1.25\, P_c^{1/2} \rho_r (\rho_{r,\text{liquid}})^{-1} \tag{A3}$$

$$\delta_{CO_2} = 1.25 \times 72.8^{1/2}\, \rho_{CO_2} (1.248)^{-1} = 8.61\, \rho_{CO_2} \tag{A4}$$

The units of δ are given in (cal cm^{-3})$^{1/2}$. The use of density in the description of the retention factor k' at constant temperature (subscript T) has been shown to be more important than the use of P since an approximately linear relation between log k' and ρ could be found over a limited range, where the slope = α and the intercept = β:

$$\log k'_T = \beta - \alpha \rho_T \tag{A5}$$

$$\log \alpha_T = (\beta_B - \beta_A) - (\alpha_A - \alpha_B)\rho_T \tag{A6}$$

The separation factor or selectivity between compound (A) and (B) will thus not change with ρ when the slopes α_A and α_B are equal. Generally, maximum R_s is achieved, especially for homologous series, at minimum values for ρ of the mobile phase, where the selectivity is highest.

The effects of temperature (T) at constant P on retention times for solutes are rather complex due to influence of T on both the fluid phase ρ and solute vapour pressure. As D increases with increasing T at constant ρ it should be expected that HETP would decrease and R_s increase with increasing T. However, for different types of analytes, this is not always what is observed due to differences in physical properties like vapour pressure.

Calculation of $P\bar{V}T$ relations of CO_2 in the supercritical state cannot be neglected due to the importance of T and fluid ρ both in relation to retention mechanism in SFC and to solubility, as well as extraction rate, in SFE. Calculation of $P\bar{V}T$ relations can be based on modified van der Waals EOS as the Bender virial EOS,[12] corresponding to the use of virial coefficients for other types of non-ideal solution (Box 4.9) or just interpolating $P\bar{V}T$ data from available tables.[13,14] Use of the Bender virial EOS with a model using virial coefficients fitted to available data such as those from IUPAC[13] gives an equation as,

$$P\bar{V}(RT)^{-1} = 1 + B'\rho + C'\rho^2 + D'\rho^3 + \ldots \tag{A7}$$

where B', C' and D' are the virial coefficients which are simply functions of T. This equation is selected as a type of EOS which gives good accuracy in calculation of $P\bar{V}T$ relations, with the pressure in MPa and density in g mL^{-1}.

$$P = T\rho[R + B\rho + C\rho^2 + D\rho^3 + E\rho^4 + F\rho^5 + (G + H\rho^2)\,\rho\, e^{-n_{20}\rho^2}] \tag{A8}$$

where

$$B = n_1 + \frac{n_2}{T} + \frac{n_3}{T^2} + \frac{n_4}{T^3} + \frac{n_5}{T^4}$$
$$C = n_6 + \frac{n_7}{T} + \frac{n_8}{T^2}$$
$$D = n_9 + \frac{n_{10}}{T}$$
$$E = n_{11} + \frac{n_{12}}{T}$$
$$F = \frac{n_{13}}{T}$$
$$G = \frac{n_{14}}{T^3} + \frac{n_{15}}{T^4} + \frac{n_{16}}{T^5}$$
$$H = \frac{n_{17}}{T^3} + \frac{n_{18}}{T^4} + \frac{n_{19}}{T^5}$$

As seen from Figure A7, the Bender virial EOS (Table A4) gives a good accuracy for estimation of $P\bar{V}T$ data in a wide P and T range, when calculated $(P - \rho)$ isotherms are compared with available data.

4.2 Equipment and Selected Applications

A schematized supercritical fluid chromatograph is shown in Figure A8.

SFC in the selected applications has been performed by use of Gilson SF3 Supercritical Fluid Chromatography System (Gilson, Middleton, WI, USA)

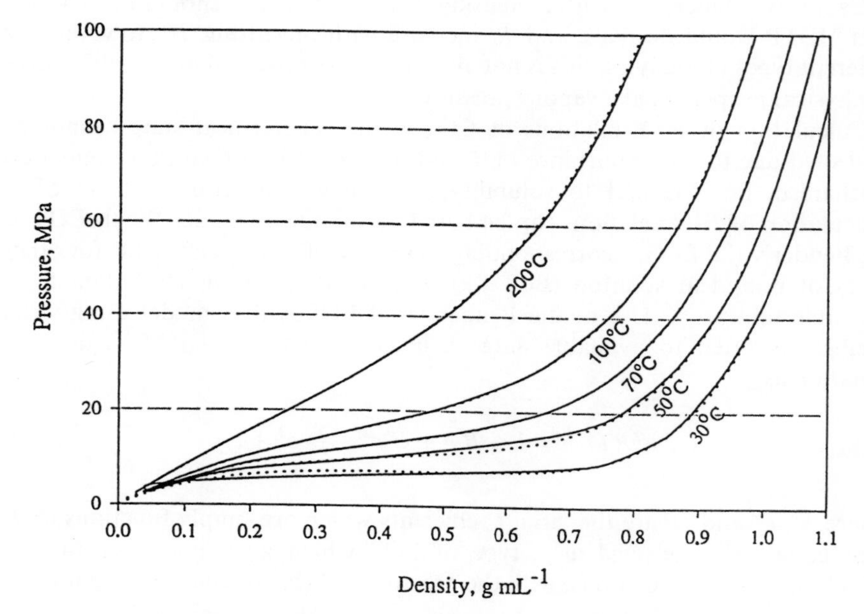

Figure A7 *Comparison of measured* (P − ρ) *data from ref. 14 with data calculated by the Bender virial EOS;*[12] *(−) measured and (.....) calculated data*

Table A4 *Virial coefficients used in the Bender EOS*[9] *to calculate data for CO_2*

$R = 0.188\,918$	
$n_1 = 0.224\,885\,58$	$n_{11} = 0.121\,152\,86$
$n_2 = -0.137\,179\,65 \times 10^3$	$n_{12} = 0.107\,833\,86 \times 10^3$
$n_3 = -0.144\,302\,14 \times 10^5$	$n_{13} = 0.439\,633\,6 \times 10^2$
$n_4 = -0.296\,304\,91 \times 10^7$	$n_{14} = -0.365\,055\,45 \times 10^8$
$n_5 = -0.206\,060\,39 \times 10^9$	$n_{15} = 0.194\,905\,11 \times 10^{11}$
$n_6 = 0.455\,543\,93 \times 10^{-1}$	$n_{16} = -0.291\,867\,18 \times 10^{13}$
$n_7 = 0.770\,428\,40 \times 10^2$	$n_{17} = 0.243\,586\,27 \times 10^8$
$n_8 = 0.406\,023\,71 \times 10^5$	$n_{18} = -0.375\,465\,30 \times 10^{11}$
$n_9 = 0.400\,295\,09$	$n_{19} = 0.118\,981\,41 \times 10^{14}$
$n_{10} = -0.394\,360\,77 \times 10^3$	$n_{20} = 0.500\,000\,00 \times 10^1$

equipped with a Gilson 119 UV–Vis detector, a Gilson 306 pump with a cooled pump head, a Gilson 306 pump for modifier, and a Gilson 811C dynamic mixer with a 1.5 mL chamber. Injections were done by use of 20 μl loop and a Gilson 233 XL auto injector (5 μl of sample injected, with air gap at both sides). The separation columns used were a 150 × 4.6 mm i.d. Spherisorb S3 ODS2, particle size 3 μm with pore diameter of 80 or 150 Å. The columns were thermostated using a Gilson 831 temperature controller. Column back-pressure was maintained by a Gilson 821 pressure regulator (nozzle regulation). Instrument programming and data processing were carried out using Unipoint system software, version 1.71.

Figure A9 shows SFC results obtained with oil produced by SFE. As seen from these results, it is possible to obtain a relatively fast separation of triacylglycerol derivatives according to the type of fatty acid in these neutral lipids. The triglyceride oils with the more saturated and long chain fatty acids have t_A values in the range 20–30 min as seen for the high erucic acid rapeseed

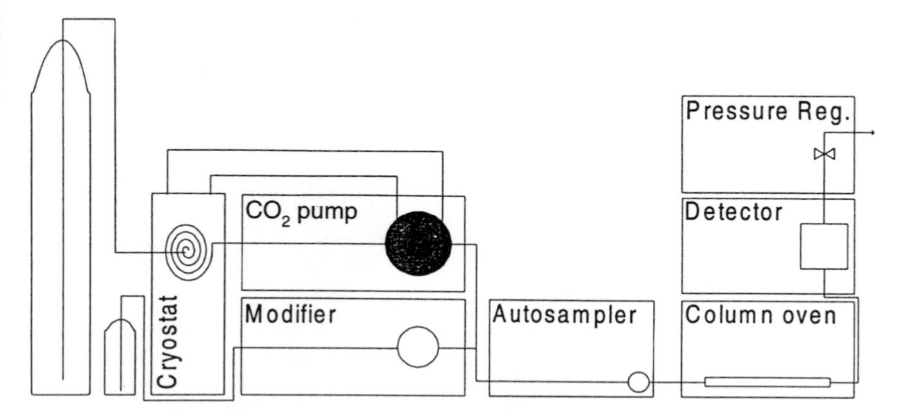

Figure A8 *Analytical SFC instrument*

oil, whereas soybean oil and oil from double low oilseed rape are quantitatively dominated by triglycerides containing C-18 fatty acids. These triglycerides have t_A values in the range 10–15 min.

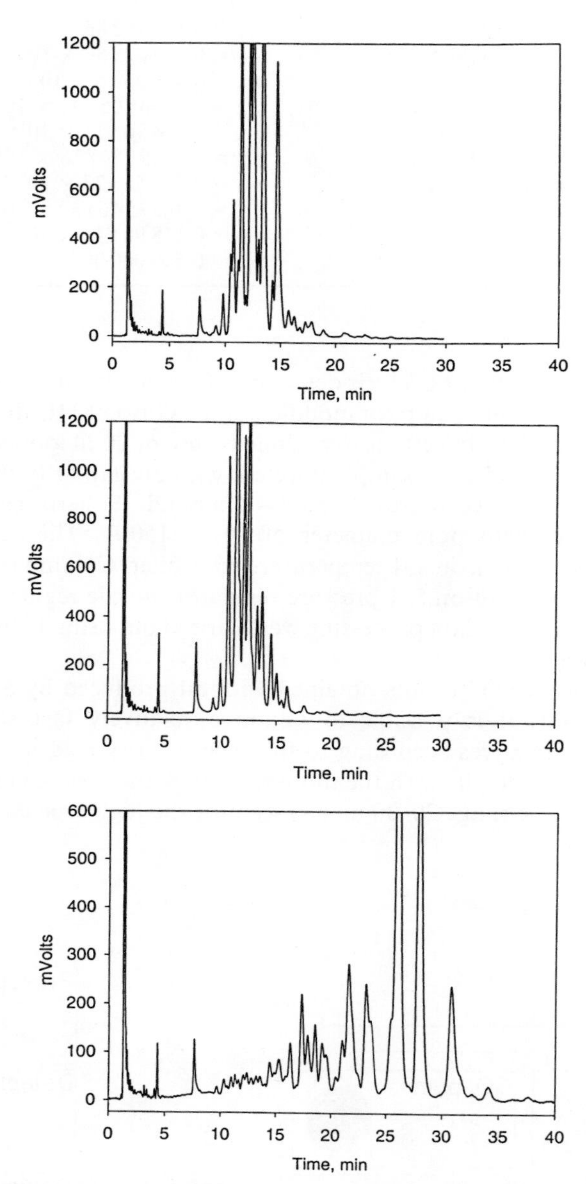

Figure A9 *SFC-separated triglycerides isolated by SFE from double low oilseed rape* (top), *soybean oil* (middle), *and high erucic acid rapeseed* (bottom). *SFC conditions: pressure* 30 MPa; *fluid phase composition* 96% CO_2, 4.0% CH_3CN; *temperature* 40 °C; *flow rate* 1.5 mL min^{-1}; *UV-detection at* 200 nm; *column* 150 × 4.6 mm *Spherisorb S3 ODS2 (Waters)*

Determination of individual tocopherols in the unsaponifiable oils extracted from various types of seeds by SFE has been performed by SFC (Figure A10).

In addition to natural antioxidants as phenolics and tocopherols, the content of chlorophyll and especially their degradation products have attracted atten-

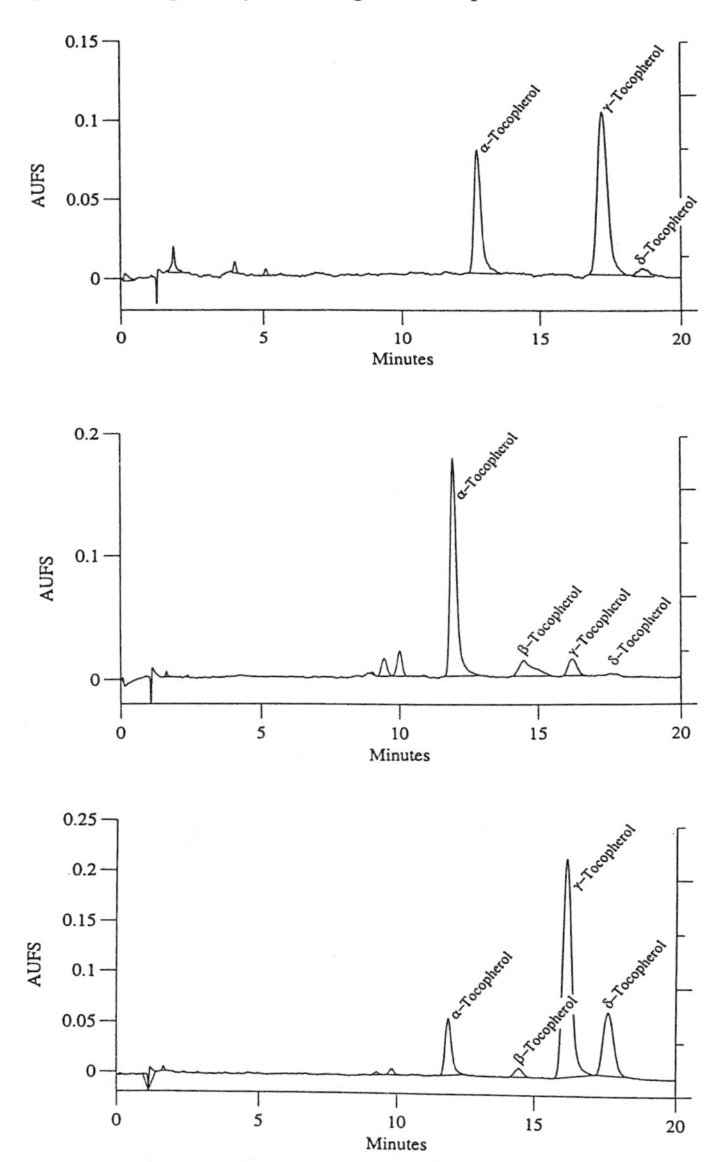

Figure A10 *SFC used for determination of individual tocopherols occurring in vegetable oils. Rapeseed oil* (top), *sunflower oil* (middle), *and soybean oil* (bottom). *SFC conditions: pressure* 15 MPa; *fluid phase composition* 99.4% CO_2, 0.6% CH_3OH; *temperature* 40 °C; *flow rate* 1.5 mL min^{-1}; *UV-detection at* 280 nm; *column* 150 × 4.6 mm *Hichrom ExcelRange S3 ODS2*

tion in relation to the quality of vegetable oils. Determination of such types of compound has been performed by use of SFC as shown in Figure A11.

In addition to the above mentioned examples, SFC has been found to be an efficient method of analysis for various types of lipids and amphiphilic compounds, *e.g.* oligomers of indolyls. Owing to its wide application range, SFC has for these and various other reasons gained increasing interest, as has

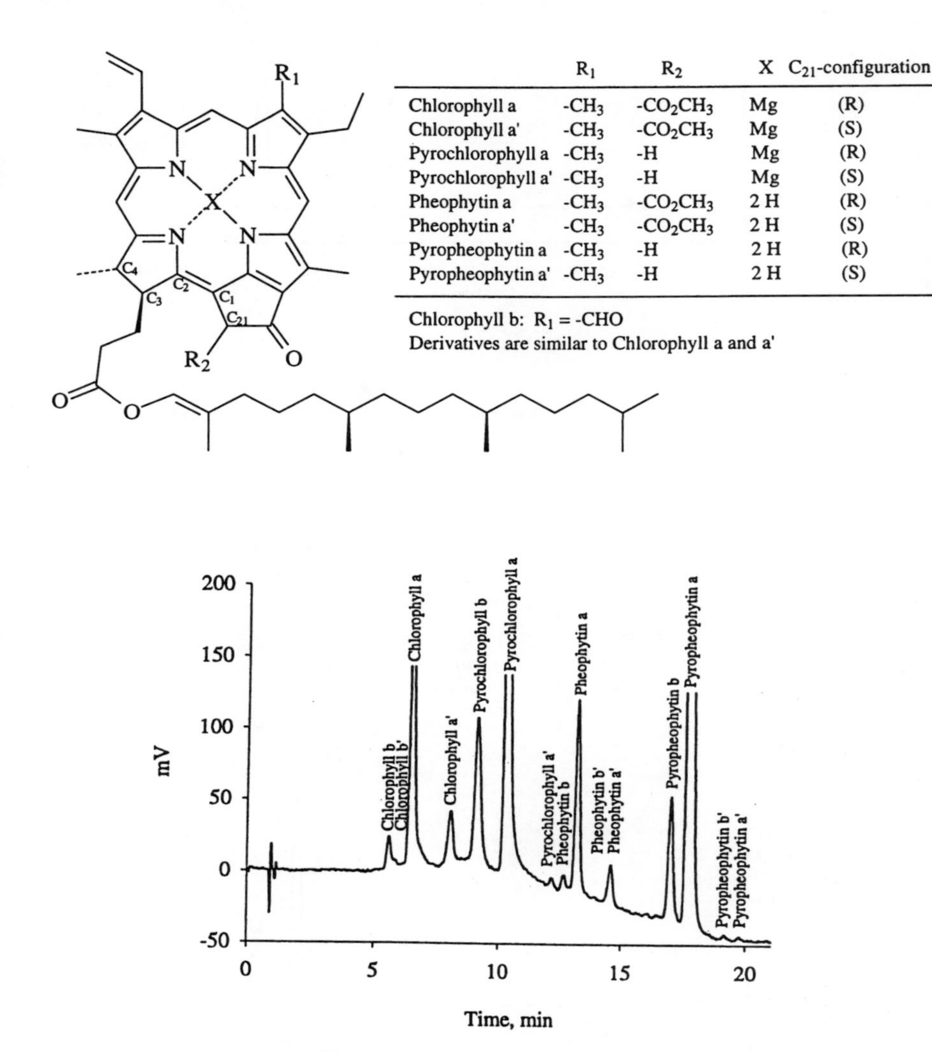

	R_1	R_2	X	C_{21}-configuration
Chlorophyll a	$-CH_3$	$-CO_2CH_3$	Mg	(R)
Chlorophyll a'	$-CH_3$	$-CO_2CH_3$	Mg	(S)
Pyrochlorophyll a	$-CH_3$	-H	Mg	(R)
Pyrochlorophyll a'	$-CH_3$	-H	Mg	(S)
Pheophytin a	$-CH_3$	$-CO_2CH_3$	2 H	(R)
Pheophytin a'	$-CH_3$	$-CO_2CH_3$	2 H	(S)
Pyropheophytin a	$-CH_3$	-H	2 H	(R)
Pyropheophytin a'	$-CH_3$	-H	2 H	(S)

Chlorophyll b: R_1 = -CHO
Derivatives are similar to Chlorophyll a and a'

Figure A11 *Structures and names of chlorophylls and their degradation products. SFC has been developed as a method, which can be used for determination of these compounds. SFC conditions: pressure* 30 MPa; *fluid phase composition was a gradient consisting of* 6% CH_3OH *linear increasing to* 7% *in* 8 min, *then up to* 25% *in* 9 min *followed by* 3 min *at isocratic conditions; temperature* 40 °C; *flow rate* 1.5 mL min^{-1}; *UV-detection at* 655 nm *after* 5 µl *of sample was injected into the mobile phase*

SFE. A particular application area for SFC is that comprising amphiphilic–lipophilic compounds, where SFC has some advantages compared to GC and LC (HPLC and HPCE). High flow rates are possible in SFC owing to the high diffusivity of supercritical fluids compared with liquids used in LC, which also give increased possibilities of using longer columns, not only capillary columns used in cSFC but also with packed columns used in pSFC. SFC also has some advantages compared with HPLC and CEC (page 239) when changing fluid phase composition, as the equilibration time required in SFC to achieve constant chromatographic conditions after changing solvent composition is much faster for SFC than for HPLC and CEC. SFC will, in addition to filling the gap between GC and LC (Figure A1), also compete with both GC, HPLC, and HPCE in a wide area of applications.

5 Selected General and Specific Literature

[1] J.W. King and G.R. List, 'Supercritical Fluid Technology in Oil and Lipid Chemistry', AOCS Press, Champaign, IL, 1996.

[2] B. Wenclawiak, 'Analysis with Supercritical Fluids: Extraction and Chromatography', Springer Verlag, Berlin, 1992.

[3] M.D. Luque de Castro, M. Valcáecel and M.T. Tena, 'Analytical Supercritical Fluid Extraction', Springer Verlag, Berlin, 1994.

[4] T.L. Chester, *Anal. Chem.*, 1997, **69**, 165A.

[5] AOCS, Oil in oilseeds: Supercritical fluid extraction method. AOCS Official Method Am 3–96, 1996, in 'Official Methods and Recommended Practices of the American Oil Chemist Society', 4th edn., AOCS Press, USA.

[6] S. Buskov, H. Sørensen, J.C. Sørensen and S. Sørensen, *Pol. J. Food Nutr. Sci.*, 1997, **6**, 115.

[7] D.R. Gere, R. Boerd and D. McManigill, *Anal. Chem.*, 1982, **54**, 736.

[8] M. Novotny, W. Bertsch and A. Zlatkis, *J. Chromatogr.*, 1971, **61**, 17.

[9] P. Sandra, A. Medvedovici, A. Kot and F. David. Selectivity tuning in packed column supercritical fluid chromatography, in 'Supercritical Fluid Chromatography with Packed Columns' eds. K. Anton and C. Berger. Chromatography Science Series, 1997, **75**, 161.

[10] C.M. Hansen, Cohesion energy parameters applied to surface phenomena, in 'Handbook of Surface and Colloid Chemistry', ed. K.S. Birdi, CRC Press, Boca Raton, FL, 1997, ch. 10.

[11] J.C. Giddings, M.N. Myers, L. McLaren and R.A. Keller, *Science*, 1968, **162**, 67.

[12] E. Bender, Equation of State Exactly Representing the Phase Behavior of Pure Substances, in 'Proceedings of the 5th Symposium on Thermophysical Properties', ed. C.F. Bonilla, The American Society of Mechanical Engineers, Colombia, USA, 1970, 227.

[13] S. Angus, B. Armstrong and K.M. de Reuck, Carbon Dioxide. International Thermodynamic Tables of the Fluid State-3, IUPAC, Commision on Thermodynamics and Thermochemistry, Pergamon Press, New York, 1976.

[14] Anonymous, in K. von Baczko (ed.) 'Gmelins Handbuch der Anorganishen Chemie, Kohlenstoff', Verlag Chemie, GMBH, Weinheim, 1970, Teil C-Lîeferung 1. System Nummer 14, 373.

Subject Index

Absorbance, 98
ABTS, in immunoassay, 382
Acceleration, gravity, 2
ACES,
  buffer in HPCE, pK_a'-value, 253
  metal–buffer binding constant, 27
  structure, 24
  temperature dependency of pK_a'-value, 24
Acetate,
  buffer in HPCE, pK_a'-value, 253
  HPCE, *see* HPCE, application, FZCE
Acetic acid,
  buffer capacity, 22
  eluotrope series, 28
  influence on surface tension, 125
Acetone; *see also* HPCE, modifier, organic, and Protein, extraction
  as EOF marker, 210
  eluotrope series, 28
Acetonitrile; *see also* HPCE, modifier, organic
  eluotrope series, 28
3-Acetoxyindole, HPCE, *see* HPCE, application, MECC
Acetyl bromide method, *see* Lignin analysis
Acetylcholine, 280
N-Acetyl-5-methoxytryptamine, *see* Melatonin
Acid detergent fibre, *see* ADF
Acid phosphatase, *see* Phosphatase, acid
Acid, definition, 12
Actin, 318
Active site, protein/enzyme, 38
Activity coefficient, 14
Activity,
  biological, 206, 320, 333
  specific, 320
ADA,
  buffer in HPCE, pK_a'-value, 253
  metal–buffer binding constant, 27
  structure, 24
  temperature dependence of pK_a'-value, 24
Adenosine, UV spectrum, 101
ADF, 398
ADP, manganese–ligand binding constant, 46
ADP–glucose, HPLC, 303
Adrenalin, HPCE, *see* HPCE, application, MECC
A-eluate,
  alkaloids, 280
  anthocyanins, 289
  aromatic choline esters, 280
  basic amino acids, 284, 333
  basic heteroaromatics, 284
  biogenic amines, 280, 288
  lupine alkaloids, 281, 284
Affinity chromatography, 144–148, 326, 377, 391–393
  of antibodies, 377
  of antigens, 391–393
  of α-galactosidase, 351
  of myrosinase, 344
  protein separation strategy, 175–176, 321, 326
Affinity elution, 148
Agar gel, 186
Agarose gel, 186, 187
  Coomassie Brilliant Blue staining, 197, 205
  for gelfiltration, 144
  for ion exchange, 153–154, 158
  in crossed immunoelectrophoresis, 380–382
  in Ouchterlony technique, 379–380
  preparation, 204
    for IEF, 197

Agarose gel (*continued*)
  silver staining, 198
Agarose, activated, 391
Agarose, in CGE, 245
Agmatine, 74
Alanine, HPCE, *see* HPCE, application, MECC
β-Ala–Lys, HPCE, *see* HPCE, application, MECC
Ala–Tyr, HPCE, *see* HPCE, application, MECC
Albumin,
  contamination in antibody purification, 376
  distribution, 318
  emulsifying agent, 124
  pI-value, 84
  solubility, 79–80
Alcohol dehydrogenase, protein–ligand complex, $\Delta G^{\circ\prime}$, 37
Alcohol, $\lambda_{max}/\varepsilon$-values, 102
Aldehyde, $\lambda_{max}/\varepsilon$-values, 102
Alditol acetates derivative, in GLC, 401
Aldobiuronic acid, 400, 408
Aldonitrile acetate derivative, in GLC, 401
Alkaline phosphatase, *see* Phosphatase, alkaline
Alkaloid, 280–284
  colorimetric methods for determination, 112, 139
  group separation, 280
  HVE, 184
  lupine, 280–284
  structure, 283
Alkene, $\lambda_{max}/\varepsilon$-values, 102
Alkylsilane group, in RPC, 172
Allelochemical, 69
Allose,
  internal standard in GLC, 403
  structure, 403
Alphabet, Greek, 4
Amide, $\lambda_{max}/\varepsilon$-values, 102
Amido black, *see* Staining
Amine group, in fluorimetry, 112
Amine modifiers, *see* HPCE, modifier
Amine,
  biogenic, *see* Biogenic amine
  colorimetric methods for determination, 112, 333
  $\lambda_{max}/\varepsilon$-values, 102
Amine-group, conjugation to aromatic ring, 103
Amino acid,
  acidic, 292–293
  analysis from solid support blot of proteins, 328
  basic, 284
  colorimetric methods for determination, 112, 333
  derivatization,
    with dansyl chloride, 335
    with ninhydrin, 333
    with *o*-phthalaldehyde (OPA), 334
  group separation, 284, 292
  HPLC, 334
  HVE, 183–184
  in exchange chromatography, 163–164
  in protein quantification, 332
  ionization, 17
  MECC, 236, 251, 286, 288, 293, 302, 335
  molar extinction coefficient, 99, 104
  neutral, 292
  non-protein, 70, 291–294, 301–302
  PC/TLC, 138–140
  pI-value, 17–18
  pK_a-value, 42
  RPC purification, 149
  UV absorption,
    of aliphatic compounds, 103–104
    of aromatic compounds, 99, 101, 103, 104
2-Aminobenzimidazole, λ_{max}, ε and pK_a'-values, 254
4-Aminobenzoic acid, *see* Monosaccharide, derivatization
4-Aminobenzonitrile, *see* Monosaccharide, derivatization
p-Aminomethylphenylalanine, HPCE, *see* HPCE, application, MECC
m-Aminophenylalanine, HPCE, *see* HPCE, application, MECC
p-Aminophenylalanine, HPCE, *see* HPCE, application, MECC
2-Amino-2-phenylpropanol, HPCE, *see* HPCE, application, MECC
α-Aminopimelic acid, 74
9-Aminopyrene-1,4,6-trisulfonate, *see* Monosacchararide, derivatization

2-Aminopyridine, *see also* Monosaccharide, derivatization
  λ_{max}, ε and pK_a'-values, 254
4-Aminopyridine, λ_{max}, ε and pK_a'-values, 254
Ammonium molybdate, *see* Staining
Ammonium oxalate, for extraction of pectins, 409
Ammonium salt precipitation, ligand binding, 53
Ammonium sulfate, *see* Protein, precipitation
Ammonium sulfate, *see also* Staining
  density, 82
  precipitation of antibodies, 376
Ammonium, HPCE, *see* HPCE, application, FZCE
AMP, HPLC, 303
Amperometry, detection in HPCE, 226
Ampholyte, 16–18, 291–298
  chromatofocusing, 166–167
  CIEF, 245–250
  ion exchange chromatography, 164–166
  isoelectric focusing, 190–196
α-Amylase,
  in DF-analysis, 398
  $E^{1\%}_{1\,cm}$-value, 100
  inhibitor, 397
Amyloglucosidase, pI-value, 84
Analyte–electrolyte complexation, *see* Complexation
Angiotensin, HPCE, *see* HPCE, application, MECC
Angustifolin, structure, 283
3,6-Anhydrogalactose, in agar, 187
Aniline, *see* Staining
1-Anilinonaphthalene-8-sulfonate, HPCE, *see* ANS
Anion,
  general analysis, 280
  in Hofmeisters series, *see* Hofmeisters series
  HPCE, *see* HPCE, application, FZCE
ANS, 37
  as fluorescent, 206
  protein–ligand complex, ΔG°′, 37
  structure, 113
Anthocyanin, 104–106, 289–291
  analysis, 291
  extraction, 289–291
  group separation, 289
  HVE, 184
  stability, 289–290
  structure, 106
  UV absorption, 211
12-(9-Anthranoyl)stearic acid, *see* AS
Antibody, 374–379
  affinity, 377–379
  antibody–antigen binding, 49
  characterization, 378–379
  complex formation with antigen, HPCE, 390–391
  concentration, 378
  conjugation for immunoaffinity chromatography, 391–393
  ELISA, 384–387
  extinction coefficients, 378
  immunostaining, 387–388
  monoclonal, 375
  polyclonal, 374
  precipitation by ammonium sulfate, 376
  precipitin reactions, 379
  for protein characterization, 333
  purification, 375–378
  storage, 378
Antichaotrope compound, 81
Antigen determinant, 375
Antigen, 374–375
  antibody–antigen binding, 49
  hapten, 375
  precipitin reactions, 379–382
  purification by immunoaffinity chromatography, 391–393
Antinutrient, binding to protein/enzyme, 38
Apigenin glycoside, 104
Apparent mobility of analytes, *see* Mobility of analytes
Arabinans, DF component, 396
Arabinogalactans,
  I, DF component, 396
  II, DF component, 396
Arabinose,
  GLC, 402
  HPCE, *see* HPCE, application, FZCE
  HPLC, 406
  R_f-value, 139
  structure, 395
Arabinoxylans, DF component, 396

Arginine, HPCE, *see* HPCE, application, MECC
Aromatic carboxylic acids, HPCE, *see* HPCE, application, MECC
Aromatic choline ester, 280–282
  ^{13}C NMR, 116
  group separation, 280
  HPLC, 281
  HVE, 185
  MECC, 282
  PC, 141
AS, structure, 113
Ascorbic acid, HPCE, *see* HPCE, application, MECC
Ascorbic acid, R_f-value, 139
Ash, analysis, 279
Asparagine, HPCE, *see* HPCE, application, FZCE/MECC
Aspartate, structure, 109
Aspartic acid, HPCE, *see* HPCE, application, MECC
Asp–Lys, HPCE, *see* HPCE, application, MECC
Association, 35
Asymmetric peaks, 212
Atomic spectroscopy, 94
  analysis of cations, 279
  metal–ligand binding constant, determination of, 46
Aubergine, HPLC of anthocyanins, 291
Aurantinidin, structure, 290
Auxochrome group, 103
Avidin,
  ligand in affinity chromatography, 147
  protein–ligand complex, $\Delta G^{\circ\prime}$, 37
Avogadro's constant, 1, 95
Axial illumination, in HPCE, 227, 229
Azide, *see* Sodium azide
2,2′-Azinodi(ethylbenzthiazoline)-sulfonate, *see* ABTS
Azo-group, conjugation to aromatic ring, 103

Background electrolyte, indirect detection, 226–227, 273
Band broadening, in HPCE, 211–215
BAPA, *see* Trypsin assay
Barbarea intermedia, HVE of extracts from seeds, 185
Barbarea verna, HVE of extracts from seeds, 185
Barbarea vulgaris, HVE of extracts from seeds, 185
Barley, composition, 317–318
Base, definition, 12
Basic heteroaromatics, *see* Heteroaromatics, basic
Bathochromic shift, 103
BBI
  diffusion coefficient, 129
  inhibitor–enzyme binding, 49
  MALDI-TOF, 118
  partial specific volume, 129
  relative molecular mass, 129
BCA, *see* Bicinchoninin acid, and Staining
Bean, green, composition, 317
B-eluate,
  acidic amino acids, 292, 333
  acidic peptides, 292
  neutral amino acids, 292, 333
Benzene, $\lambda_{max}/\varepsilon$-values, 102, 254
Benzoate, pK_a'-values, 254
Benzoic acid derivatives,
  HPCE, *see* HPCE, application, MECC
  in DF, 410
  structure, 105, 282
  λ_{max}-values, 105
N-α-Benzoyl-L-arginine-4-nitroanilide, *see* BAPA
Benzylamine, λ_{max}, ε and pK_a'-values, 254
Betaine, zwitterion in HPCE, 259
Betalain,
  occurrence, 70, 104
  structure, 106
Bicinchoninin acid method, *see* Protein, quantification
Bicine,
  buffer in HPCE, pK_a'-value, 253
  metal–buffer binding constant, 27
  structure, 25
  temperature dependence of pK_a'-value, 25
Binding, theory of, 35–68
Bio-Gel A, *see* Gelfiltration
Bio-Gel P, *see* Gelfiltration
Biogenic amine,
  group separation, 280
  HPCE, *see* HPCE, application, MECC

Biogenic amine (*continued*)
  HVE, 184
  PC, 138
Biotin,
  HPCE, *see* HPCE, application, MECC
  labelling of proteins, 384
  protein–ligand complex, $\Delta G^{\circ\prime}$, 37
N,*N*-Bis(hydroxyethyl)glycine, *see* Bicine
Bis, in cross-linking of polyacrylamide gels, 187–188, 195
Bisubstrate reaction, *see* Kinetic, mechanism
Biuret method, *see* Protein, quantification
Bjerrum diagram, 19
Blocking, in immunoassay, 387–388
Blotting,
  from agarose gel, 387
  diffusional, 387
  dot, 384
  electrophoretic transfer, 387
  western, 206, 207
Boiling point, propanol, 259
Boltzmann's constant, 2
Borate,
  buffer in HPCE, pK_a'-value, 253
  complexation, 27, 230–231, 252, 406–407
Bouguer–Lambert–Beer's Law, *see* Lambert–Beer's law
Bovine serum albumine, *see* BSA
Bowman–Birk inhibitor, *see* BBI
Bradford method, *see* Protein, quantification in microtitre assay
Bradykinin, HPCE, *see* HPCE, application, MECC
Broccoli, composition, 317
Bromochloroindolyl, in immunoassay, *see* Immunoassay, staining
Bromophenol blue, *see* Staining
Brønsted theory, 12
BSA,
  for blocking in immunoassay, 384, 387
  as carrier, 386
  diffusion coefficient, 129
  $E^{1\%}_{1\ cm}$-value, 100
  HPCE, *see* HPCE, application, capillary SDS–PAGE
  macromolecule–ligand binding constant, determination of, 48
  partial specific volume, 129
  protein–ligand complex, $\Delta G^{\circ\prime}$, 37
  relative molecular mass, 129
  as standard, 73, 328
  UV absorption, 100
Bubble cell, in HPCE, 227–229
Buffer, 12–30
  capacity (β), 18, 20–22
  change of, 322
  composition, 20
  coordination complex, 27
  description, 26
  electrolytes, indirect UV detection, λ_{max}, ε and pK_a'-values, 254
  HPCE,
    evaporation, 217
    general, 253–264
    ion depletion, 217
    pK_a-values, 253
  HVE, 184
  maximum capacity (β_{max}), 20
  metal–buffer binding constant (K_m), 27
  preparation, 23
  range, 19
  selection, 26–28
  stock solution, 23
1,2-Butanediol, HPCE, *see* HPCE, modifier, organic
Butanol, *see* HPCE, modifier, organic
  eluotrope series, 28
Butyric acid, influence on surface tension, 125

Cabbage, composition, 317
Cadaverine, 259
Cadoxen, for extraction of cellulose, 409
CAE, 250
Caffeic acid,
  HPCE, *see* HPCE, application, MECC
  structure, 105
  λ_{max}-value, 105
cis-2-*O*-Caffeoyl-L-malate, HPCE, *see* HPCE, application, MECC
trans-2-*O*-(Caffeoyl)-L-malate, HPCE, *see* HPCE, application, MECC
Calcium,
  electrostatic interaction, 29

HPCE, *see* HPCE, application, FZCE
metal–ligand binding constant, 45
Canavanine, structure, 283
Capacity (k'), 134
Capacity, ion exchange material, 155–157
Capensinidin, structure, 290
Capillary,
active compounds, 124
affinity electrophoresis, *see* CAE
coating, 265, 269, 272
coil, 215
detection window, 225
effective length, 209, 221
electrochromatography, *see* CEC
electro-osmotic chromatography, *see* CEC
free zone electrophoresis, *see* CFZE
gel electrophoresis, *see* CGE
inactive compounds, 124
internal diameter, 221–222, 269, 272
isoelectric focusing, *see* CIEF
isotachophoresis, *see* CITP
preconditioning procedure, 217
quadratic, 227
rectangular, 227–228
silanol groups, 209
surface-to-volume ratio, 213–214
total length, 209, 221–222
tubing material, 221
wall interaction with analyte, 214–215, 220, 265
zone electrophoresis, *see* CZE
Capparales, 70, 235, 312, 342
CAPS, buffer in HPCE, pK_a'-value, 253
N-(Carbamoylmethyl)-2-aminoethanesulfonic acid, *see* ACES
N-(Carbamoylmethyl)iminodiacetic acid, *see* ADA
Carbazole reaction, in uronic acid analysis, 408
Carbinol pseudobase, structure, 290
Carbohydrate
analysis in general, 305–308
in animal products, 318
borate complexation, 27, 230–231, 252, 406–407
colorimetric methods for determination, 112
^{1}H NMR data, 117
HPCE, *see* HPCE, application, FZCE
PC/TLC, 138–139
in plant products, 317
R_f-value, 139
Carbon, analysis, 279
Carbonic anhydrase B (bovine), pI-value, 84, 191
Carbonic anhydrase B (human), pI-value, 84, 191
Carbonic anhydrase, HPCE, *see* HPCE, application, capillary SDS–PAGE
3-(4-Carboxybenzoyl)-2-quinolinecarboxaldehyde, *see* Monosaccharide, derivatization
3-Carboxycinnamic acid, HPCE, *see* HPCE, application, MECC
3-Carboxyglyoxylic acid, HPCE, *see* HPCE, application, MECC
m-Carboxy-4-hydroxy-φ-acetic acid, HPCE, *see* HPCE, application, MECC
m-Carboxy-4-hydroxy-cinnamic acid, HPCE, *see* HPCE, application, MECC
m-Carboxy-4-hydroxy-φ-glyoxylic acid, HPCE, *see* HPCE, application, MECC
3-(3-Carboxy-4-hydroxyphenyl)alanine,
HPCE, *see* HPCE, application, MECC
structure, 140, 288
3-Carboxy-4-hydroxyphenylglycine,
HPCE, *see* HPCE, application, MECC
structure, 140, 288
Carboxylate,
analysis in general, 298–301
C-eluate, 298–300
electrostatic interaction, 29–30
HPCE, *see* HPCE, application, MECC, carboxylic acid
HVE, 184–186
Carboxylic acid,
aliphatic, influence on surface tension, 124
colorimetric methods for determination, 112
ester, 308
HPCE, *see* HPCE, application, MECC
HVE, 184–186
PC/TLC, 138–141
$\lambda_{max}/\varepsilon$-values, 102

3-Carboxymandelic acid, HPCE, *see* HPCE, application, MECC
Carboxypeptidase A,
diffusion coefficient, 129
partial specific volume, 129
relative molecular mass, 129
3-Carboxyphenylacetic acid, HPCE, *see* HPCE, application, MECC
3-(3-Carboxyphenyl)alanine,
HPCE, *see* HPCE, application, MECC
structure, 140, 288
3-Carboxyphenylglycine,
HPCE, *see* HPCE, application, MECC
structure, 140, 288
3-Carboxyphenyllactic acid, HPCE, *see* HPCE, application, MECC
m-Carboxytyrosine, HPCE, *see* HPCE, application, MECC
β-Carotene,
structure, 107
UV spectrum, 107
Carotenoids, UV absorption, 105–107
Carrier, *see* Hapten
Casein, 317
HPCE, *see* HPCE, application, CGE
Catalase, 316
diffusion coefficient, 129
haem group, 107–108
partial specific volume, 129
relative molecular mass, 129
Catechine, 105
Catecholamine, 288–289
HPCE, *see* HPCE, application, MECC
Cation,
analysis in general, 279–280
crown ether complexation, 257
in Hofmeister's series, 81
HPCE, *see* HPCE, application, FZCE
tartaric acid complexation, 254
CDTA, for extraction of pectins, 409
CEC, 239–242, 271
applications, 243–244
Cellulose,
acetate, 182–186
derivatives; *see also* HPCE, capillary, permanent coating
in CGE, 244, 247
in CIEF, 248, 250
DF component, 395
for dialysis bags, 84
extraction, 409
for ion exchange material, 153–154
structure, 396
C-eluate,
carboxylates, 298–300
phosphates, 298, 302–303
sulfates, 298
sulfonates, 298
Centrifugation, 89–91
Centrospermae, 70, 104
Cereal, composition, 317
Cetyltrimethylammonium bromide, *see* CTAB
Cetyltrimethylammonium chloride, *see* CTAC
CFZE, *see* FZCE
CGE, 243–245, 270
applications, 245, 247
column materials, 243–244
Cgs system, 1
Chalcone, structure, 290
Chaotropic compound, 81
CHAPS, data of, 262
CHAPSO, data of, 262
Charge, *see* Electric charge
Chemical,
ionization, 118
potential, 14
potential gradient, 127
Chemotaxonomy, 69–70, 312
Chenopodiales, 70
CHES, buffer in HPCE, pK_a'-value, 253
Chick pea, composition, 317
Chiral chromatography, 174–175
Chiral selectors, 175
buffer additive, 256–257
Chiral surfactant, 30, 32, 260
Chloride, HPCE, *see* HPCE, application, FZCE
3-[3-(Chloroamidopropyl)dimethyl-ammonio]-1-propanesulfonate, *see* CHAPS
3-[3-(Chloroamidopropyl)dimethyl-ammonio]-2-hydroxy-1-propanesulfone, *see* CHAPSO
Chloroform,
eluotrope series, 28–29
extraction of lipids, 72
Chlorophyll,
structure, 107

UV absorption, 105–107
UV spectrum, 107
Cholic acid, structure, 32
Choline ester, 139, 174, 280, 308; *see also* Aromatic choline ester
separation by HPLC I-P RPC, 174
Chromate, λ_{max}, ε and pK_a'-values, 254
Chromatofocusing, 166–167
protein separation strategy, 321
Chromatographic DF analysis, 399–408
Chromatography,
affinity, *see* Affinity chromatography
immunoaffinity, *see* Immunoaffinity chromatography
of proteins, guide to boxes, 341
Chromophore, structure relation to λ_{max} and ε, 102–103
Chromoproteins, pI-value, 84
Chymotrypsin inhibitor assay, *see* Inhibitor, chymotrypsin, assay
Chymotrypsin,
assay, 367
disc gel electrophoresis of inhibitor, 198–199
$E_{1\,cm}^{1\%}$-value, 100
Chymotrypsinogen,
diffusion coefficient, 129
disc gel electrophoresis of inhibitor, 198–199
HPCE, *see* HPCE, application, CIEF
partial specific volume, 129
relative molecular mass, 129
Cibacron blue, ligand in affinity chromatography, 147
CIEF, 245–250, 270
applications, 250
Cinnamic acid derivatives,
in DF, 410
structure, 105, 282
λ_{max}-value, 105
Cinnamoyl derivatives, 300–301
CITP, 250
Citrate,
binding to protein, 27
buffer in HPCE, pK_a'-value, 253
HPCE, *see* HPCE, application, FZCE/MECC
metal complex, 27
C.m.c.-value, 30–34
selected surfactants, 33, 262
CMP, HPLC, 303
Cobalamins, 106
Cobalt, metal–ligand binding constant, 45
Cod, composition, 318
Codeine, 281
Colchicin, 281
Collagen, 318
diffusion coefficient, 129
partial specific volume, 129
relative molecular mass, 129
Colorimetry; *see also* DF analysis, Protein, quantification, Staining, 111–112
detection of low-M_r compounds, 138–139
Colour exhibition, plants, 104–106, 289–291
Combustion, 8
Compartmentalization, 69
Complexation,
antibody–antigen, *see* Precipitin methods
borate and carbohydrates, *see* Borate complexation
buffers, 27, 254
HPCE study, 389–390
Con A,
diffusion coefficient, 129
ligand in affinity chromatography, 147
myrosinase purification, 344
partial specific volume, 129
relative molecular mass, 129
Concanavalin A, *see* Con A
Concentration, of protein sample, 323
Conductance, *see* Electric conductance
Conductivity
detection in HPCE, 226
effect in sample stacking, 264–265
specific, in c.m.c.-determination, 32
thermal, 3
Convicine, 110
HPCE, *see* HPCE, application, MECC
structure, 109, 285–286
Coomassie Brilliant Blue,
protein quantification; *see also* Staining, 112, 205–206, 330–331
staining,
agarose gels, 197–198, 205, 381
polyacrylamide gels, 199
Cooperativity, ligand binding, 49–53

Copper, metal–ligand binding constant, 45
Correlated spectroscopy, *see* NMR, COSY
Correlation coefficient, 7
Corrosion, 8
Cosolvents, in protein extraction, 80
COSY, *see* NMR
Coulomb's law, 29
Coumaric acid,
  HPCE, *see* HPCE, application, MECC
  PC, 141
  structure, 105
  λ_{max}-value, 105
Coumaroylcholine,
  HPCE, *see* HPCE, application, MECC
  HVE, 185
  structure, 282
cis-2-*O*-(*p*-Coumaroyl)-L-malate, HPCE, *see* HPCE, application, MECC
trans-2-*O*-(*p*-Coumaroyl)-L-Malate, HPCE, *see* HPCE, application, MECC
Coumaroyllupinin, structure, 283
Counterion,
  buffer, 23
  detergent, 31–34, 254–255, 262, 299
  in ion exchange chromatography, 152
  in ion-pair reversed-phase chromatography, 173–174
Critical micellar concentration, *see* c.m.c.
Crown ether, 256–257
Cruciferous plants, composition, 317
Crude fibre method, *see* DF analysis, gravimetric
Crude protein determination, *see* Protein, quantification
C-S-lyases, transformation of β-thioalkyl amino acid, 71
CTAB,
  data for, 33, 262
  structure, 32
CTAC, data for, 33, 262
Curly kale, composition, 317
Current, *see* Electric current
Cut-off-values, UV-solvents, 97
Cyanidin, structure, 106, 290
Cyanocobalmin, HPCE, *see* HPCE, application, MECC
m-Cyanophenylalanine, HPCE, *see* HPCE, application, MECC
Cyclodextrin, 175, 256–257
  structure, 257–258
Cyclohexanediaminetetraacetic acid, *see* CDTA
Cysteine, 332
  microscopic constant, 42–44
Cystine, 332
Cytidine, UV-spectrum, 101
Cytisin, structure, 283
Cytochrome *c* (equine),
  diffusion coefficient, 129
  partial specific volume, 129
  relative molecular mass, 129
Cytochrome *c*, 107–108
  HPCE, *see* HPCE, application, CIEF
  pI-value, 84
Cytochrome oxidase, 106
Cytochrome *P*-450, 106–107
CZE, *see* FZCE

DAB, in immunoassay, 382
Dansyl chloride; *see also* Amino acid, derivatization
  as fluorescent, 206
  structure, 335
d-Camphor-10-sulfonate, 257
Debye unit, 28
Debye–Hückel equation, 15
Decapeptides, HPCE, *see* HPCE, application, MECC, peptides
Decyltrimethylammonium bromide, data for, 33
2,3-Dehydroflavones, 105
Decyltrimethylammonium chloride, data for, 33
Delphinidin, structure, 106, 290
Delphinidin-3-glycoside, HPLC, 291
Denaturation, proteins, *see* Protein, denaturation
3-Deoxyflavonols, 104
Deoxygalactose, HPLC, 406
2-Deoxy-D-ribose, HPCE, *see* HPCE, application, FZCE
Derivatization,
  amino acids, *see* Amino acid, derivatization
  with chromophore, *see* Chromophore group, derivatization
  monosaccharides, *see* Monosaccharides, derivatization

Desalting,
  by dialysis, 84–88, 322
  by gelfiltration, 142–144, 175, 322
Desulfoglucosinolates, HPCE, *see* HPCE, application, MECC
Desulfuration, 107
Detection,
  activity of various enzymes, 334–373
  in electrophoresis, 204–207
  in general, 92–119
  in HPCE, 225–229
Detergent methods, DF analysis, *see* DF analysis
Detergent, 30–34, 260–264, 268
  application,
    as chiral selector, 263
    as cosolvent, 80
    as emulsifying agent, 124–125
  c.m.c.-value, definition, 30
  data for, 33, 262
  in HICE, 237–238
  in MECC, 233–235
  micelle, *see* Micelle
  for protein extraction, *see* Protein extraction
  structure of selected, 32
Determinant, antigen, 375
Dextran,
  in CGE, 244
  as gelfiltration medium, 142–144, 146
  as ion exchange material, 153–154
DF,
  analysis, 394–411
    chromatographic, 399–408
      GLC, 400–403
      HPCE, 406–408
      HPLC, 403–406
    colorimetry, 400
    gravimetric, 397–399
      crude fibre method, 397
      detergent method, 397–398
      enzymatic, 398–399
  composition, 394–397
  dominant monosaccharides, 395
  fractionation, 408–410
  metal–ligand binding constant, 45
Dialysis, 84–88
  macromolecule–ligand binding constant, determination of, 47
  protein separation strategy, 321, 325
Diaminobenzidene, *see* DAB
1,4-Diaminobutane, HPCE, *see* HPCE, modifier, amine
1,5-Diaminopentane, HPCE, *see* HPCE, modifier, amine
1,3-Diaminopropane, HPCE, *see* HPCE, modifier, amine
2,3-Diaminopropionic acid, oxalyl amides, 294
Dielectric constant, 28–29, 81
Dietary fibres, *see* DF
Diethyl ether, extraction of lipids, 72
Diffuse layer, in electrophoresis, 179, 181
Diffusion coefficient (*D*), 3, 127–130
  proteins, 129
Digitonin, 262
2,3-Dihydroxybenzoylcholine, HPLC, 281
3,5-Dihydroxybenzoylcholine, HPLC, 281
o-Dihydroxy derivatives, 294
3,4-Diiodo-L-tyrosine, HPCE, *see* HPCE, application, MECC
3,5-Dimethoxybenzoylcholine, HPLC, 281
3,4-Dimethoxycinnamic acid, 74
3,4-Dimethoxycinnamoylcholine,
  HPCE, *see* HPCE, application, MECC
  HPLC, 281
  structure, 282
2,3-Dimethoxycinnamoylcholine, HPLC, 281
2,5-Dimethoxycinnamoylcholine, HPLC, 281
3,5-Dimethoxy-4-hydroxycinnamoylcholine, HPLC, 281
3,4-Dimethoxyphenylethylamine, 74
  HPCE, *see* HPCE, application, MECC
Dimethyl sulfoxide, *see* DMSO
Dimethylaminobenzaldehyde, *see* Staining
Dimethylphenol reaction, in uronic acid analysis, 408
2,4-Dinitrofluorobenzene, *see* Staining
2,4-Dinitrophenylhydrazine, *see* Staining
Dinitrosalicylate, *see* Staining
Diode array detection, 114
Dioxane,
  eluotrope series, 29
  as reference compound in NMR, 115
Dipeptides, HPCE, *see* HPCE, application, MECC, peptides

Diphenylamine, *see* Staining
Dipole binding, 35
Dipole moment, 28
Disc gel electrophoresis, *see* Electrophoresis
Dispersion,
  electromigration, 211–213
  medium, 124–125
Disrupting solvents, *see* Solvents, disrupting
Dissociation, 35–68
Dissociation constants,
  practical, 14
  thermodynamic, 14
Distribution constant, 121, 234
Disulfide, $\lambda_{max}/\varepsilon$-values, 102
Dithiothreitol,
  as cosolvent, 80
  reducing agent in SDS–PAGE, 199–200
DMSO
  hemicellulose extraction, 408
  organic modifier in HPCE, *see* HPCE, modifier
  RS dissolvation, 400
DNA,
  partial specific volume, 130
  protein–ligand complex, $\Delta G^{\circ\prime}$, 37
  sequencing reaction products, CGE, 245, 270
  UV absorption, 100–101
  UV spectrum, 100
N-Dodecyl betaine, data of, 262
Dodecyltrimethylammonium bromide, *see* DTAB
Dodecyltrimethylammonium chloride, *see* DTAC
Dodecyl-*β*-D-glucoside, data of, 262
Dolicholphosphates, extraction, 72
Donnan equilibrium, 87
Dopa,
  HPCE, *see* HPCE, application, MECC
  structure, 286
Dopa-glucoside,
  HPCE, *see* HPCE, application, MECC
  structure, 285–286
Dopamine, HPCE, *see* HPCE, application, MECC
Dot blotting, *see* Blotting, dot
Double layer, *see* Electric double layer
Dragendorff reagens, *see* Staining
Dry matter, determination of, 71
DTAB, data for, 33, 262
DTAC, data for, 33, 262
Dumas technique, *see* Protein, quantification
Dynamic coating, capillary, 265

$E^{1\%}_{1\,cm}$-values, 99–101, 329
Eadie–Hofstee plot, *see* Kinetics
EDTA, *see* Pectin, extraction
Efficiency, 135
Egg, hen, composition, 318
Electric charge, 3
Electric conductance, 3
Electric current, 3
Electric mobility, 3
Electric potential, 3
Electric resistance, 3
Electrochemical detection, HPLC/FPLC, 169
Electro-endo-osmosis, *see* EOF
Electrokinetic injection, *see* HPCE, injection
Electrolyte, background, *see* Background electrolyte
Electromagnetic radiation, 92–96
Electromigration dispersion, *see* Dispersion, electromigration
Electron sink, pyridine, 44
Electron volt, 1
Electronegativity, 28–29
Electron-impact mode (EI), 117
Electro-osmosis, *see* EOF
Electro-osmotic flow, *see* EOF
Electro-osmotic mobility; *see also* EOF, 181, 210
Electrophoresis; *see also* HPCE, 178–206
  blotting, 206–207
  convection current, 189
  density gradient, 189
  detection, 204–207
  diffuse layer, *see* Diffuse layer
  electric double layer, 181
  electroelution, 206
  electro-endo-osmosis, *see* EOF
  electro-osmotic flow, *see* EOF
  electrophoretic mobility, *see* Electrophoretic mobility

free solution; *see also* HPCE, 178, 188–190
gel,
disc, 196–200
native, 203–205
polyacrylamide, *see* PAGE
staining,
agarose gel, 197–198, 205
polyacrylamide gel, 195–196, 199, 202
two-dimensional, 186
immuno-, 203–204, 380–382
isotachophoresis, 189–190
moving boundary, 188–190
paper,
high voltage, *see* HVE
low voltage, 183
preparative, for protein purification, 328
Stern layer, 179–181
Stern plane, 179–181
support media, 182–188
temperature gradient, 189
zeta potential, 179
zone, 182–190
Electrophoretic mobility, 179, 209–211
electrolytes in HPCE, 254–255
micelle, determination of, 263
Electrostatic interaction, 29–30, 214
Elemental analyses, 279–280
ELISA,
blotting, 387–388
in microtitre plates, 384–387
Eluotrope series of solvents, 28
Emission spectroscopy, 92–94
Emulsion, 124–125
stabilization/destabilization, 125
water–oil, 124
Energy (units), 3
Energy,
exited state, 92–94
ground state, 92–94
level
rotational, 92–94
vibrational, 92–94
Enthalpy (unit), 3
Entropy (unit), 3
Enzymatic method, DF analysis, *see* DF analysis, gravimetric
Enzyme,
animal origin, 337
assay,
acid phosphatase, *see* Phosphatase, acid
alkaline phosphatase, *see* Phosphatase, alkaline
chymotrypsin, *see* Chymotrypsin
excess enzyme, 337–338
α-galactosidase, *see* α-Galactosidase
in general, 334–339
myrosinase, *see* Myrosinase
peroxidase, *see* Peroxidase
phenylalanine ammonia-lyase, *see* Phenylalanine ammonia-lyase
product inhibition, 337–338
substrate deficiency, 337–338
trypsin, *see* Trypsin
unstable product, 337–338
detection,
autochrome methods, 206
overlayered gel technique, 206
sandwich technique, 206
simultan–capture reaction, 206
inactivation, 53, 72–73
inhibitor assay,
ovomucoid, *see* Ovomucoid
proteintype proteinase inhibitor, *see* Trypsin inhibitor assay/ Chymotrypsin inhibitor assay
plant origin, 337
system for commercial food analysis, 338
unit, 2, 54, 334
Enzyme linked immunosorbent assay, *see* ELISA
EOF, 15, 180–182, 209–211
determination of size, 210
elimination, 245–246, 263, 265
influence,
of organic modifier, 258
of pH, 217, 255–256
marker, 210
regulation, 214, 217, 234, 251–266
Epiprogoitrin, HPCE, *see* HPCE, application, MECC
Epitope, 375
Epoxidation, 107
Equilibrium constant, 13, 36
between phases (K_d), 121

Erythritol,
internal standard in GLC, 403
structure, 403
ESR spectroscopy, 93
Essential nutrients, 69
Ester/amide, $\lambda_{max}/\varepsilon$-values, 102
Ethanol,
eluotrope series, 29
for homogenization, 73
for protein extraction, *see* Protein extraction
organic modifier in HPCE, *see* HPCE, modifier, organic
Ether, $\lambda_{max}/\varepsilon$-values, 102
Ethidium bromide, structure, 113
Ethylene glycol, organic modifier in HPCE, *see* HPCE, modifier, organic
Ethylenediaminetetraacetic acid, *see* EDTA
Europinidin, structure, 290
Evaporation from samples, in HPCE, *see* HPCE, buffer evaporation
Excitation, 92–94
External standards, 73–74
Extinction coefficient, 98
amino acids, 104
antibodies, 378
Extract, crude, preparation, 75–76
Extraction, 69–91
amphiphilic compounds, 72, 74–84
efficiency, 74
high-M_r compounds, 74–84
homogenization, 73, 75–76, 323
low-M_r compounds, 70–74
from liquid samples, 75
from solid samples, 76
protein, 79–80
repeated, 73
use of organic solvents, 72–73, 75–76, 84–85

FAB, 118
Faba bean, composition, 317
Faraday, 1
Fast atom bombardment ionization, *see* FAB
Fast Polymer Liquid Chromatography, *see* FPLC
Fatty acid,
extraction, 72–73
PC/TLC, 138–139
Favism, 110
FC, 166
protein separation strategy, 325
FDA, 10
Ferroprotoporphyrin IX, 107–108
Ferulic acid
HPCE, *see* HPCE, application, MECC
structure, 105
λ_{max}-value, 105
Feruloylcholine,
HPCE, *see* HPCE, application, MECC
structure, 282
cis-2-*O*-Feruloyl-L-malate, HPCE, *see* HPCE, application, MECC
trans-2-*O*-Feruloyl-L-malate, HPCE, *see* HPCE, application, MECC
Fibrinogen,
diffusion coefficient, 129
partial specific volume, 129
relative molecular mass, 129
Fick's first law, 128
Flame spectroscopy, 94, 96
Flash chromatography, *see* FC
Flavones, UV absorption, 104–105
Flavonoids, 70, 104–106, 308
adsorption, 142
HPCE, *see* HPCE, application, MECC
structure of selected, 106, 309
UV-absorption, 104–106
Flavonols, UV absorption, 104–105
Flavoproteins, as fluorescent, 113
Flavylium cation, structure, 290
Flow profile, *see* HPCE, flow profile
Flow rate, linear, 135–136
Fluorescamine,
as a fluorescent, 112, 206
structure, 113
Fluorescein, structure, 113
Fluorescence, 92–93, 111–114
detection
HPCE, 226
HPLC/FPLC, 169
Fluorimeter, 113–114
Flux (J), 127
Folic acid, HPCE, *see* HPCE, application, MECC

Folin–Ciocalteu reagent, Lowry, 112, 329
Food and Drug Administration, USA, *see* FDA
Formamide, influence on c.m.c.-value, 34
Formate, buffer in HPCE, pK_a'-value, 253
Formic acid, eluotrope series, 28
Fourier transform mode (FT), 114
FPLC, 168–177
  applications, 176–177
  column materials, 170–171
  detection systems, 169
  instrumentation, 169
  separation,
    affinity chromatography of proteins, 175–176
    chiral chromatography, 174–175
    chromatofocusing, 174
    gelfiltration of proteins, 174–175
    HIC of proteins, 174–175
    ion exchange liquid chromatography, 174
    ion-pair reversed-phase chromatography, 173–174
    normal-phase liquid chromatography, 172
    reversed-phase liquid chromatography, 172–173
    strategy for proteins; *see also* individual enzymes, isolation and characterization, 321, 326
Fractionation, DF, *see* DF
Free solution capillary electrophoresis, *see* FSCE
Free solution electrophoresis, *see* Electrophoresis
Free zone capillary electrophoresis, *see* FZCE
Freezing, 71
Freundlich equation, 41
Frictional coefficient, 126–127
Fronting of peaks, 212–213
FSCE, *see* FZCE
Fucose,
  GLC, 406
  HPLC, 406
  structure, 395
Fumarase, protein–ligand complex, $\Delta G^{\circ\prime}$, 37
Fumarate,
  HPCE, *see* HPCE, application, MECC
  protein–ligand complex, $\Delta G^{\circ\prime}$, 37
Functional groups of ion exchange materials, 156–157
FZCE, 230–232
  applications, *see* HPCE, applications, FZCE
  carbohydrates, *see* HPCE, applications, FZCE
  ions, *see* HPCE, applications, FZCE
  monosaccharides, *see* HPCE, applications, FZCE
  oligosaccharides, *see* HPCE, applications, FZCE
  organic acids, *see* HPCE, applications, FZCE

Galactans, DF component, 396
Galactoglucomannans, DF component, 396
Galactomannans, DF component, 396
α-Galactopyranosyl-*p*-nitrophenol, assay for α-galactosidase, 349
Galactose,
  in agar, 186–187
  GLC, 402, 406
  HPCE, *see* HPCE, applications, FZCE
  HPLC, 406
  R_f-value, 139
  structure, 395
α-Galactosidase, 345–347
  assay
    with α-galactopyranosyl-*p*-nitrophenol, 349
    in microtitre plates, 349
    with natural substrate, 350
  isolation and characterization, 351–352
  specific detection after electrophoresis, 352
β-Galactosidase, HPCE, *see* HPCE, application, capillary SDS–PAGE
Galacturonic acid,
  HPCE, *see* HPCE, applications, FZCE
  structure, 395
Galegin, structure, 283
Gallic acid
  structure, 105
  λ_{max}-value, 105
Gas constant, 2
Gas–liquid chromatography, *see* GLC

Gaussian distribution, 135, 211
GDP, manganese–ligand binding constant, 46
Gel electrophoresis, *see* Electrophoresis
Gelatin, emulsifying agent, 124
Gelfiltration, 142–144
binding constant determination, 46–47
FPLC, *see* FPLC, separation media, 142–146
protein separation strategy, 325
Δ^4-GlcUA-4,6-bis-*O*-sulfo-GalNAc, HPCE, *see* HPCE, application, MECC
Δ^4-GlcUA-GalNAc, HPCE, *see* HPCE, application, MECC
Δ^4-GlcUA-4-*O*-sulfo-GalNAc, HPCE, *see* HPCE, application, MECC
Δ^4-GlcUA-6-*O*-sulfo-GalNAc, HPCE, *see* HPCE, application, MECC
GLC, 120
in DF analysis, *see* DF analysis, chromatography
γ-Globulin, $E_{1\,\mathrm{cm}}^{1\%}$-value, 100
γ_2-Globulin, pI-value, 84
Globulin, 79–80, 316
distribution, 318
pI-value, 79–80, 316
GLP, 10
β-Glucans, mixed, DF component, 396
Glucoalyssin, HPCE, *see* HPCE, applications, MECC
Glucobarbarin,
HVE, 185–186
PC, 141
Glucobrassicanapin, HPCE, *see* HPCE, applications, MECC
Glucobrassicin, HPCE, *see* HPCE, applications, MECC
Glucoerucin, HPCE, *see* HPCE, applications, MECC
Glucoiberin, HPCE, *see* HPCE, applications, MECC
Glucoiberverin, HPCE, *see* HPCE, applications, MECC
Glucomannans, DF component, 396
Gluconapin, HPCE, *see* HPCE, applications, MECC
Gluconasturtiin,
HVE, 185–186
PC, 141
HPCE, *see* HPCE, applications, MECC
α-D-Glucopyranosyl, affinity chromatography, 147, 344–345
Glucoraphanin, HPCE, *see* HPCE, applications, MECC
Glucose,
GLC, 402, 406
HPCE, *see* HPCE, applications, FZCE
HPLC, 405–406
R_f-value, 139
structure, 395
Glucosibarin,
HVE, 185–186
PC, 141
Glucosinalbin,
HVE, 185–186
PC, 141
Glucosinolate; *see also* individual glucosinolates, 70, 303
HPLC I-P RPC, 174
HVE, 184–186
MECC, 218, 224, 239, 261, 304–305
on-column desulfatation, 303
PC/TLC, 138–139, 141
Glucotropaeolin, HPCE, *see* HPCE, applications, MECC
Glucuronic acid,
HPCE, *see* HPCE, applications, FZCE
structure, 395
Glucuronomannans, DF component, 396
Glucuronoxylans, DF component, 396
Glu–Lys, HPCE, *see* HPCE, applications, MECC
Gluphepa, *see* Chymotrypsin, assay
Glutamine, HPCE, *see* HPCE, applications, MECC
α-Glutamyl derivatives, HVE, 184
γ-Glutamyl derivatives, HVE, 184
γ-Glutamyl peptides, 291, 294–295
α-Glutaryl-L-phenylalanine-4-nitroanilide, *see* Gluphepa
Glutathione, *see* GSH
Glutathione dehydrogenase, 316
Glutathione, oxidized, *see* GSSG
Glutelin, 79, 316
distribution, 318
Gluthamic acid, HPCE, *see* HPCE, applications, MECC
Glycerol,
borate complex, 27

as cosolvent, 80
in dialysis tubings, 85
organic modifier in HPCE, *see* HPCE, modifier, organic
Glycine, HPCE, *see* HPCE, applications, MECC
Glycine amide, buffer in HPCE, pK_a'-value, 253
Glycinin, 316
Glycolipid, extraction, 72
Glycophosphoprotein, 318
Glycoprotein; *see also* Ovomucoid, Myrosinase and α-Galactosidase,
adsorption, 142
affinity chromatography, 147
borate complex, 27
detection, 206
DF component, extraction, 409
Glycosaminoglycans, HPCE, *see* HPCE, applications, MECC
Glycylglycine, *see* Gly–Gly
Gly–Gly,
buffer in HPCE, pK_a'-value, 253
metal–buffer binding constant, 27
structure, 25
temperature dependence of pK_a'-value, 25
GMP, HPLC, 303
Gold, *see* Staining
Good laboratory practice, *see* GLP
Grain legumes, composition, 316–317
Gramine,
HPCE, *see* HPCE, applications, MECC
structure, 283
Gravimetric DF analysis, *see* DF analysis, gravimetric
Gravity, specific (unit), 3
Group separation,
by ion exchange chromatography, 161–163
protein, 323–326
GSH, 292, 294
HPCE, *see* HPCE, applications, MECC
metal–ligand binding constant, 45
GSSG, 291, 294
HPCE, *see* HPCE, applications, MECC
Guajacol, *see* Peroxidase, assay
Guanidine, chaotropic compound, 81
Guanidinium, influence on c.m.c.-value, 34
Guanidinum thiocyanate, *see* Hemicellulose, extraction
Guanido compound,
analysis, in general, 280–281
colorimetric methods for determination, 112
PC/TLC, 138–139
Guanosine, UV spectrum, 101
Gums, DF component, 396

Haem group,
structure, 108
UV spectrum, 108
Haemoglobin, 106–107
binding of O_2, 51
diffusion coefficient, 129
partial specific volume, 129
pI-value, 84
relative molecular mass, 129
Haemoprotein; *see also* Peroxidase, 106–108
handling, 319
Haldane's relationship, 59
Halide-group, conjugation to aromatic ring, 103
Hanes plot, *see* Kinetics, delineation
Hapten, 375
characterization of conjugation by HPCE, 389–391
Hazardous chemicals, 8–9
Heat,
development, *see* HPCE, heat development
dissipation, *see* HPCE, heat dissipation
specific (unit), 3
Hemicellulose,
dominating polysaccharides, 395–396
extraction media, 409
Henry's law, 14
Heparin, ligand in affinity chromatography, 147
HEPES, buffer in HPCE, pK_a'-value, 253
HEPPS, buffer in HPCE, pK_a'-value, 253
Heptane, *see* Lipid, extraction
Hesperaline,
HPCE, *see* HPCE, application, MECC
structure, 282
Hesperis matronalis, 282

Heteroaromatics, 108–111, 284–287, 297–298
basic, group separation, 284
chromophore system, 110–111
HPCE, *see* HPCE, application, MECC
HPLC, 285
structure, 109
Heterotrope effector, 38
HETP, 135, 173
Hexadecanophenone, HPCE, *see* HPCE, application, HICE)
Hexametaphosphate, extraction of pectins, 409
Hexanes, extraction of lipids, 72
Hexapeptide, *see* HPCE, application, MECC, peptides
Hexosamines, colorimetric methods for determination, 112
HIC, 148–149, 175–176
ligand binding, 53
protein separation strategy, 321
HICE, 237–239
applications, 238–239, 242
High-M_r compounds
column chromatographic separation, 164–166
extraction, 69–91
ion exchange chromatography, *see* Ion exchange chromatography
occurrence, 278
High-performance capillary electrophoresis, *see* HPCE
High-performance liquid chromatography, *see* HPLC
High Voltage Electrophoresis, *see* HVE
Hill equation, 51–52
Hip–His–Leu, HPCE, *see* HPCE, application, MECC
Hippuric acid, HPCE, *see* HPCE, application, FZCE/MECC
Hirsutidin, structure, 290
Histamine,
HPCE, *see* HPCE, application, MECC
λ_{max}, ε and pK_a'-values, 254
Histones, pI-value, 84
Hofmeister's series, 81
HOHAHA/TOCSY (NMR), 116
Homoarginine, HPCE, *see* HPCE, application, FZCE
Homogenization, *see* Extraction
Homotrope effector, 38
Hordenine, HPCE, *see* HPCE, application, MECC
Hormones, analysis, 311
Horse heart myoglobin, *see* Myoglobin
Horseradish peroxidase, *see* Peroxidase
HPCE, 208–277
application
capillary SDS–PAGE
BSA, 246
carbonic anhydrase, 246
β-galactosidase, 246
α-lactalbumin, 246
myosin, 246
ovalbumin (chicken egg), 246
phosphorylase B, 246
protein, 246
soybean trypsin inhibitor, 246
CGE, 245
casein, 247
characterization of hapten conjugation, 389
CIEF, 250
chymotrypsinogen, 250
cytochrome *c*, 250
myoglobin (horse heart), 250
protein, 250
in DF analysis, 406–408
FZCE, 230–232
acetate, 232
ammonium, 232
anion, 232
arabinose, 231, 407
asparagine, 294
calcium, 232
carbohydrate, 230–231, 407
cation, 232
chloride, 232
citrate, 232
2-deoxy-2-ribose, 231
2-deoxy-D-ribose, 231
DMSO, 294
galactose, 231, 252, 407
galacturonic acid, 231
glucose, 231, 252, 407
glucuronic acid, 231
hippuric acid, 294–296
homoarginine, 294
ions, 232
lactate, 232

lactose, 230
Lathyrus sativus, crude extract, 293–294
magnesium, 232
mannose, 252
monosaccharides, 231, 252, 407
α-ODAP, 294
β-ODAP, 294
oligosaccharides, 230–307
organic acids, 232
oxalate, 232
phosphate, 232
potassium, 232
raffinose, 230, 307
rhamnose, 231
sodium, 232
stachyose, 230, 307
sucrose, 230, 307
sulfate, 232
verbascose, 230, 307
xylose, 252, 407
HICE
hexadecanophenone, 242
octadecanophenone, 242
tetradecanophenone, 242
MECC
3-acetoxyindole, 238
adrenalin, 289
β-Ala–Lys, 292–297
alanine, 236, 251
Ala–Tyr, 295–297
amino acids, 236, 251, 285–288, 293
p-aminomethylphenylalanine, 293
m-aminophenylalanine, 293
p-aminophenylalanine, 293
2-amino-2-phenylpropanol, 289
angiotensin, 295–297
arginine, 236
aromatic carboxylic acids, 301–302, 410
aromatic choline esters, 282
ascorbic acid, 310
asparagine, 236, 251
aspartic acid, 236, 251
Asp–Lys, 295–297
benzoic acid derivatives, 240
biogenic amines, 288–289
biotine, 310
bradykinin, 296–297
caffeic acid, 301
cis-2-*O*-caffeoyl-L-malate, 301
trans-2-*O*-caffeoyl-L-malate, 301
3-carboxycinnamic acid, 302
3-carboxyglyoxylic acid, 302
m-carboxy-4-hydroxy-φ-acetic acid, 302
m-carboxy-4-hydroxycinnamic acid, 302
m-carboxy-4-hydroxy-φ-glyoxylic acid, 302
3-(3-carboxy-4-hydroxyphenyl)alanine, 293
3-carboxy-4-hydroxyphenylglycine, 293
carboxylic acids, 240, 301–302, 410
3-carboxymandelic acid, 302
3-carboxyphenylacetic acid, 302
3-(3-carboxyphenyl)alanine, 293
3-carboxyphenylglycine, 293
3-carboxyphenyllactic acid, 302
m-carboxytyrosine, 293
catecholamine, 289
citrate, 299
convicine, 285–287
coumaric acid, 240, 301, 410
coumaroylcholine, 282
cis-2-*O*-(*p*-coumaroyl)-L-malate, 301
trans-2-*O*-(*p*-coumaroyl)-L-malate, 301
cyanocobalamin, 310
m-cyanophenylalanine, 293
desulfoglucosinolate, 304
3,4-diiodo-L-tyrosine, 312
3,4-dimethoxycinnamoylcholine, 282
3,4-dimethoxyphenylethylamine, 289
dopa, 286–287
dopa-glucoside, 286–287
dopamine, 288
epiprogoitrin, 239
ferulic acid, 240, 301, 410
feruloylcholine, 282
cis-2-*O*-feruloyl-L-malate, 301
trans-2-*O*-feruloyl-L-malate, 301
flavonoids, 309
folic acid, 310
fumarate, 299

HPCE (*continued*)
application (*continued*)
MECC (*continued*)
Δ^4-GlcUA-4,6-bis-*O*-sulfo-GalNAc, 307
Δ^4-GlcUA-GalNAc, 307
Δ^4-GlcUA-4-*O*-sulfo-GalNAc, 307
Δ^4-GlcUA-6-*O*-sulfo-GalNAc, 307
glucoalyssin, 239
glucobrassicanapin, 239, 304
glucobrassicin, 239, 304
glucoerucin, 239, 304
glucoiberin, 239, 304
glucoiberverin, 239, 304
gluconapin, 239, 304
gluconasturtiin, 239, 304
glucoraphanin, 239, 204
glucosinolates, 218, 224, 239, 261, 304–305
glucotropaeolin, 239, 304
Glu–Lys, 295–297
glutamine, 236, 251
glutamic acid, 251
glycine, 236, 251
glycosaminoglycans, 306–307
gramine, 288
GSH, 295–297
GSSG, 295–297
hesperaline, 282
heteroaromatics, 285–287
hippuric acid, 294–296
His–His–Leu, 296–297
histamine, 288–289
hordenine, 288
4-hydroxybenzoic acid, 236, 240, 410
4-hydroxyglucobrassicin, 239, 304
13α-hydroxylupanin, 284
3-hydroxymethylphenylalanine, 293
5-hydroxytryptophan, 238
indoles, 238
indol-3-ylacetic acid, 238
indol-3-ylacetonitrile, 238
indol-3-ylaldehyde, 238
indol-3-ylcarboxylic acid, 238
3-(indol-3-yl)-propionic acid, 237
3-iodo-L-tyrosine, 312
ions, 232
isoferulic acid, 240
isoferuloylcholine, 282
isoleucine, 236
isolupanin, 284
isorhamnetin-3-(6″-carboxyglucoside), 309
isovanillic acid, 240
isovanillylcholine, 282
isoxazoline, 286
kaempferol, 309
kaempferol-3-(6″-carboxyglucoside), 309
kaempferol-3,7-diglucoside, 309
kaempferol-3-glucoside, 309
kaempferol-3-sinapoylsophoroside-7-glucoside, 309
kaempferol-3-sophoroside-7-glucoside, 309
α-ketoglutarate, 299
KSTI, 295, 298, 389–391
leucine, 236
lupanin, 284
lupine alkaloids, 284
lysine, 236
malate ester, 301
maleinate, 299
malonate, 299
methionine, 236
4-methoxyglucobrassicin, 239, 304
5-methoxytryptamine, 289
3-methoxytyramine, 289
myoglobin, 250
myrosinase, 347
NAD^+, 310
$NADP^+$, 310
napoleiferin, 239, 304
neoglucobrassicin, 239
nicotinamide, 310
nitrate, 237
noradrenalin, 289
oxalate, 299
peptides, 295–297
phenethylamine, 289
phenylalanine, 236, 293
phosphatidic acid, 241
phosphatidylcholine, 241
phosphatidylethanolamine, 241
phosphatidylinositol, 241
phosphatidylserine, 241
phospholipids, 241
progoitrin, 239, 304

proline, 236
protein, 246–247, 250, 298, 340
proteinase inhibitor, 246, 298, 389–390, 392
pyridoxamine, 310
pyridoxamine phosphate, 310
pyridoxine, 310
quercetin, 309
quercetin-3-(6″-carboxyglucoside), 309
quercetin-3-glucoside, 309
riboflavin, 310
riboflavin-5′-phosphate, 310
RPPI, 298
rustocide, 309
rutin, 309
salicylic acid, 240
serine, 236, 251
serotonine, 289
sinapic acid, 240, 301, 410
sinapine, 282
cis-2-*O*-sinapoyl-L-malate, 301
trans-2-*O*-sinapoyl-L-malate, 301
sinigrin, 304
structural epimers, 235
Δ^4-2-*O*-sulfo-GlcUA-GalNAc, 307
Δ^4-2-*O*-sulfo-GlcUA-6-*O*-sulfo-GalNAc, 307
Δ^4-2-*O*-sulfo-GlcUA-sulfo-GalNAc, 307
T_3, 312
T_4, 312
tartrate, 299
thiamine, 310
thiocyanate ion, 237
threonine, 236, 251
trigonellinamide, 282, 286–287, 293, 295, 297, 302, 304, 310, 390
trigonelline, 286–287
tryptamine, 238, 288–289
tryptophan, 236, 238, 286–288, 293
tyramine, 288–289
tyrosine, 286–288, 293
tyrosine–glucoside, 288
valine, 236
vanillylcholine, 282
vicine, 285–287
water-soluble vitamin, 310
willardine, 286
basic principles, 208–220
buffer,
additive, 255–264, 267–268
crown ether, 256–257
cyclodextrin, 256–258
organic modifier, 257–259
zwitterions, 259
electrolyte, 253–255, 266–267
evaporation, 217–218
ion depletion, 217
capillary; *see also* Capillary, 221–222, 265–266, 269, 272
dynamic coating, 265
permanent coating, 265
preconditioning, 217
detection, 225–229
axial illumination, 227–228
bubble cell, 227–229
indirect, 226–227
laser-induced fluorescence, 226
limit, 227
multireflection cell, 227
radioactivity, 226
Z-cell, 227–229
EOF, 209–211
flow profile,
flat, 211
laminar, 126–127, 211
heat development, 213, 255
heat dissipation, 213–214, 228
influence of buffer additives,
amine modifiers (summary), 268
chiral selectors (summary), 267
ion-pairing reagents (summary), 268
organic modifiers (summary), 267
surfactants (summary), 268
zwitterions (summary), 268
influence of,
buffer electrolyte composition (summary), 266–267
buffer electrolyte concentration (summary), 267
capillary coating (summary), 269
capillary dimension/type (summary), 269
injection (summary), 269
pH (summary), 267
sample (summary), 269
separation parameters (summary), 266–269
temperature (summary), 266

HPCE (*continued*)
influence of (*continued*)
voltage (summary), 266
in immunochemical techniques, 389–391
injection, 222–225, 269
electrokinetic, 224–225
hydrodynamic, 222–224
plug length, 215
precision; *see also* HPCE, injection, various modes, 219–220
pressure, 222–224
siphoning action, 222–224
volume calculation, 223
instrumentation, 221–229
method development
optimization procedure, 271–276
validation, 275–276
method,
CAE, 250
CEC, 239–243, 271
CGE, 243–245, 270
choice of mode, 270
CIEF, 245–250, 270
CITP, 250
FZCE, 230–232
HICE, 237–239
MCE, 250
MECC, 233–237
micellar flow, 234
measurement by Sudan II, 263–264
measurement by Sudan IV, 263–264
migration time, relative, 218
migration time window, 234
modifier
amine, 259
organic, 257–259
parameters affecting separation, 251–266
peak area
normalized, 261
relative normalized, 220
pH, 255, 267
influence on,
electrophoretic mobility, 255
EOF, 255–256
hysteresis, 217
in protein analysis, 327
protein separation strategy, 321, 327
quantification, 219–220
sample, 264–265, 269
analyte–capillary wall interaction, 214–215
band broadening, 211–215
diffusion of analytes, 211–213
dispersion of analytes, 211–213
evaporation, 220
plug length, *see* HPCE, injection, plug length
protein adsorption, 327
stacking, 264–265
study of complex formation, antigen–antibody, 366, 389–391
surface-to-volume ratio in capillary, 213
temperature, 251–253, 266
temperature control system, 214
temperature gradient, 213–214
voltage, 251–253, 266
HPLC, 168–177,
affinity chromatography of proteins, 175–176
chiral chromatography, 174–175
column materials, 170–171
detection systems, 169
gelfiltration of proteins, 175–176
hydrophobic interaction chromatography, 175–176
in DF analysis, 403–406
instrumentation, 169
ion exchange liquid chromatography, 174
ion-pair reversed-phase chromatography, 173–174
normal-phase liquid chromatography, 172
protein separation strategy, 321, 326–327
reversed-phase liquid chromatography, 172–173
size exclusion chromatography, 175–176
Hughes–Klotz plot, *see* Kinetics, delineation
Humoral immune response, *see* Immune response
HVE, 183–186
Hydration,
gelfiltration media, 143, 146
ions, 13, 29

Hydrodynamic injection, *see* HPCE, injection
Hydrogen bonding, 12–13
Hydrogen, analysis, 279
Hydrogencarbonate, interactions, 27
Hydrolysis,
of polysaccharides, *see* Polysaccharide, hydrolysis
of protein, *see* Protein, hydrolysis
Hydrophobic interaction; *see also* HIC and HICE, 35
Hydrophobic interaction capillary electrophoresis, *see* HICE
Hydrophobic interaction chromatography, *see* HIC
p-Hydroxybenzoate, λ_{max}, ε and pK_a'-values, 254
4-Hydroxybenzoic acid
HPCE, *see* HPCE, application, MECC
structure, 105
λ_{max}-value, 105
4-Hydroxybenzoylcholine,
HPLC, 281
HVE, 185
structure, 282
4-Hydroxybenzylamine, HPLC, 281
Hydroxygalegin, structure, 283
4-Hydroxyglucobrassicin, HPCE, *see* HPCE, application, MECC
Hydroxy group, conjugation to aromatic ring, 103
4-Hydroxy-3-iodo-alkylbenzene, $\lambda_{max}/\varepsilon$-values, 102
13α-Hydroxylupanin,
HPCE, *see* HPCE, application, MECC
structure, 283
3-Hydroxymethylphenylalanine, HPCE, *see* HPCE, application, MECC
3-Hydroxy-4-methoxybenzoylcholine, HPLC, 281
3-Hydroxy-4-methoxycinnamoylcholine, HPLC, 281
4-Hydroxy-3-nitro-alkylbenzene, $\lambda_{max}/\varepsilon$-values, 102
5-Hydroxytryptophan, HPCE, *see* HPCE, application, MECC
Hyperchromic shift, 103
Hypochromic shift, 103
Hypsochromic shift, 103
Ice point, 2
IDF, 398
IEF, 190–196
in capillary, *see* CIEF
IsoPrime™, 204, 328
pI marker, 191
preparation of,
agarose gels, 197
polyacrylamide gels, 195
Imidazole, λ_{max}, ε and pK_a'-values, 102, 254
Immune response, 374
Immunization, 374
Immunoaffinity chromatography, 391–393
Immunoassay,
biotin labelling, 384
detection systems, 382–383
direct, 382
enzyme labelling, 382–383
indirect, 382
staining,
bromochloroindolyl, 382
nitro blue tetrazolium, 383
Immunochemical techniques, 374–393
Immunodiffusion, 379
Immunoelectrophoresis, 379–382
crossed, 380–381
double radial immunodiffusion (Ouchterlony), 379–380
rocket, 380
single radial immunodiffusion (Mancini), 379
Immunoglobulin, 374–375
Indirect detection, in HPCE, *see* HPCE, detection
Indolealkaloids, 281
Indoles, HPCE, *see* HPCE, application, MECC
Indol-3-ylacetic acid, HPCE, *see* HPCE, application, MECC
Indol-3-ylacetonitrile, HPCE, *see* HPCE, application, MECC
Indol-3-ylaldehyde, HPCE, *see* HPCE, application, MECC
Indol-3-ylcarboxylic acid, HPCE, *see* HPCE, application, MECC
Indolylglucosinolates, effects, 310
3-(Indol-3-yl)-propionic acid, HPCE, *see* HPCE, application, MECC

Indolyls, analysis, 308–310
Infrared spectroscopy, *see* IR spectroscopy
Inhibition,
  competitive, 64–67
  non-competitive, 64–67
  uncompetitive, 64–67
Inhibitor; *see also* Proteinase inhibitor
  binding, 49
  chymotrypsin, assay, 367–368
  disc gel electrophoresis, 198–199
  enzyme kinetic, 61–67
  isoinhibitor determination by MALDI-TOF, 118
  phenyl methyl sulfonyl fluoride (PMSF), 80, 369
  proteinase, 365–372
    isolation and characterization, 368–369
  staining, 199
  trypsin,
    assay, 365–366
    diffusion coefficient, 129
    partial specific volume, 129
    pI-value, 84
    relative molecular mass, 129
Ink, *see* Staining
Inorganic ion, analysis, 279
Inorganic phosphate, colorimetric methods for determination, 112
Inositol phosphate, 302
  metal–ligand binding constant, 45
Insoluble dietary fibres, *see* IDF
Internal standard, 73–74
3-Iodo-L-tyrosine, HPCE, *see* HPCE, application, MECC
Ion binding, 35
Ion exchange chromatography, 152–167
  application,
    amino acid analysis, 163–164
    high-M_r compounds, 155–159, 164–166
    low-M_r compounds, 155–159, 161–164
    proteins, 155–159, 164–166, 321
  column packing, 159–160
  FPLC, 159
  HPLC, 158
  material regeneration, 160
  materials, 153–154
  resolution, 159
Ion,
  HPCE, *see* HPCE, application, MECC
  hydration, *see* Hydration, ion
Ionic strength, 14–16
  calculation, 15
    influence on protein separation, 321
Ionization
  acid and base, 13–14
  microscopic constant, 43
Ion-pair,
  chromatography, 173–174
  reversed-phase chromatography, *see* I-P RPC
I-P RPC, 173–174
IR spectroscopy, 93
Iron, metal–ligand binding constant, 45
Isoelectric focusing, *see* IEF
Isoelectric point, *see* pI
Isoenzymes, IEF, 167, 190
Isoferulic acid
  HPCE, *see* HPCE, application, MECC
  structure, 105
  λ_{max}-value, 105
Isoferuloylcholine,
  HPCE, *see* HPCE, application, MECC
  HVE, 185
  structure, 282
Isoflavones, UV absorption, 104
Isoinhibitors, *see* Inhibitor, isoinhibitor determination by MALDI-TOF
Isoionic point, 17
Isoleucine, HPCE, *see* HPCE, application, MECC
Isolupanin,
  HPCE, *see* HPCE, application, MECC
  structure, 283
IsoPrime™, *see* IEF, IsoPrime™
Isopropanol, eluotrope series, 28
Isorhamnetin-3-(6″-carboxyglucoside), HPCE, *see* HPCE, application, MECC
Isorhamnetin, structure, 106, 309
Isotachophoresis, 188–190, 250
Isothiocyanate, 71
Isovanillic acid,
  HPCE, *see* HPCE, application, MECC
  structure, 105
  λ_{max}-value, 105
Isovanillylcholine,

HPCE, *see* HPCE, application, MECC
structure, 282
Isoxazoline,
derivatives, 100, 108, 294
HPCE, *see* HPCE, application, MECC
structure, 109, 286

Kaempferol,
HPCE, *see* HPCE, application, MECC
structure, 106, 309
Kaempferol-3-(6″-carboxyglucoside), HPCE, *see* HPCE, application, MECC
Kaempferol, 3,7-diglucoside, HPCE, *see* HPCE, application, MECC
Kaempferol-3-glucoside, HPCE, *see* HPCE, application, MECC
Kaempferol-3-sinapoylsophoroside-7-glucoside, HPCE, *see* HPCE, application, MECC)
Kaempferol-3-sophoroside-7-glucoside, HPCE, *see* HPCE, application, MECC
Katal, 2, 54, 334
α-Ketoglutarate, HPCE, *see* HPCE, application, MECC
Ketone,
colorimetric methods for determination, 112
$\lambda_{max}/\varepsilon$-values, 102
Kieselguhr, for TLE, 186
Kinetic energy, 37
Kinetics,
delineation,
Eadie–Hofstee plot, 39–41, 58, 66
Hanes plot, 39–41, 58
Hughes–Klotz plot, 39–41, 58
Lineweaver–Burk plot, 39–41, 58, 66
Scatchard plot, 39–41, 52, 58–59
Theorell–Chance mechanism, 61
mechanism,
bisubstrate reaction, 55, 59–63
monosubstrate reaction, 55–59, 337
multisubstrate reaction, 59–63
ping-pong mechanism, 60, 62
pseudomonosubstrate reaction, 55, 337
random mechanism, 60, 62
sequential ordered mechanism, 59, 62
pre-steady state, 57
reaction order,
first, 54, 56, 58, 61
pseudoreaction, 55
second, 54
zero, 56, 58
steady state, 57
Kjeldahl analysis, *see* Protein, quantification
Klason lignin, *see* Lignin, Klason
KOH, *see* Hemicellulose, extraction
KSTI,
HPCE, *see* HPCE, application, MECC
inhibitor–enzyme binding, 49
pI-value, 84
Kunitz soybean trypsin inhibitor, *see* KSTI

Laboratory practice, good, *see* GLP
Laboratory results,
recording, 8–11
reporting, 8–11
Laboratory safety, 7–8
Lac repressor, protein–ligand complex, $\Delta G^{\circ\prime}$, 37
α-Lactalbumin, HPCE, *see* HPCE, application, capillary SDS–PAGE
Lactate dehydrogenase,
diffusion coefficient, 129
partial specific volume, 129
protein–ligand complex, $\Delta G^{\circ\prime}$, 37
relative molecular mass, 129
Lactate, HPCE, *see* HPCE, application, FZCE
Lactic acid dehydrogenase, kinetics, 61
β-Lactoglobulin,
diffusion coefficient, 129
partial specific volume, 129
pI-value, 84
relative molecular mass, 129
Lactose,
HPCE, *see* HPCE, application, FZCE
R_f-value, 139
Lambert–Beer's law, 98
Laminar flow profile, *see* HPCE, flow profile
Langmuir, theory of adsorption, 41–42
Laser-induced fluorescence, *see* HPCE, detection
Lathyrism, 108, 293

Lathyrus sativus, HPCE, *see* HPCE, application, FZCE
LC; *see also* Ion-exchange chromatography, HPLC, FPLC, Electrophoresis, and HPCE, 120–151
Leading zeroes, 4
Lectin,
  adsorption, 142
  agglutinating activity, 334
  Helix pomatia, ligand in affinity chromatography, 147
  lentil,
    ligand in affinity chromatography, 147
    pI-value, 84, 191
  soybean, ligand in affinity chromatography, 147
Leek, composition, 317
Legumin, 316
Leguminosae, 281
Leguminous seeds, compounds in, 317, 347, 365, 386
Lepidium campestre, HVE of extracts from seeds, 185
Leucine, HPCE, *see* HPCE, application, MECC
Liebermann–Burchard, *see* Staining
Ligand
  affinity chromatography, 144–148
  binding,
    to macromolecules, 46–49
    to metal ions, 44–46
    to molecules with cooperative and non-identical sites, 49–53
    to molecules with *n* equivalent sites, 40–41
    to molecules with one binding site, 39
    to molecules with two equivalent sites, 40
    to two different sites, 42–44
Light petroleum, *see* Lipid extraction
Light,
  emission, *see* Emission
  intensity, 98
  monochromatic, 94
  plane polarized, 95
  speed, 2, 95
  unpolarized, 95
Lignin, 397
  analysis, 399
    acetyl bromide method, 399
    permanganate oxidation method, 399
  Klason, 399, 409
  structure, 397
Linear function,
  double reciprocal, 58–59
  single reciprocal, 58
Linear regression, 6–7
Lineweaver–Burk plot, *see* Kinetics, delineation
Lipase, 73, 147, 316
Lipid,
  determination,
    colorimetric methods, 112
    PC/TLC, 139
  extraction, 72
    light petroleum, 72–73
    perforation, 72
    SFE, *see* SFE
    Soxhlet, 72–73
Lipoprotein, 77, 147, 317–318, 324, 330
Lipoxygenase, 316
Liquid chromatography, *see* LC
Lissamine green, *see* Staining
Literature,
  chemical, 10–11
  search, 11
Low-M_r compounds,
  analytical determination; *see also* individual analysis methods, 278–314
  binding study, 48–49
  colorimetric methods for determination, 112
  column chromatographic separation, 163–164
  extraction, 69–91
  HVE, 184–186
  ion-exchange chromatography, *see* Ion-exchange chromatography
  occurrence, 69–70, 278
  UV–Vis absorption, 102–111
Low voltage paper electrophoresis, 183
Lowry, *see* Protein quantification
Lupanin,
  HPCE, *see* HPCE, application, MECC
  structure, 283
Lupine alkaloids, HPCE, *see* HPCE, application, MECC

Lupinin, structure, 283
Lupinus albus, MECC of alkaloids, 284
Lupinus angustifolius, MECC of alkaloids, 284
Luteolin glycoside, 104
Lyase, phenylalanine ammonia, *see* Phenylalanine ammonia lyase
Lyophilization, 71
  protein sample, 323
Lyotrope series, 81
Lyotropic salt, 81
Lysine, HPCE, *see* HPCE, application, MECC
Lysozyme (hen egg white),
  partial specific volume, 129
  diffusion coefficient, 129
  relative molecular mass, 129
  pI-value, 84

Mackerel, composition, 318
Macroscopic constant, 40, 42–44
Magnesium,
  HPCE, *see* HPCE, application, FZCE
  metal–ligand binding constant, 45
  phosphatase activator, 349, 359
Maize, composition, 317–318
Malate ester, 300
  HPCE, *see* HPCE, application, MECC
Malate, protein–ligand complex, $\Delta G^{\circ\prime}$, 37
MALDI-TOF, 118
Maleinate,
  buffer capacity, 22
  HPCE, *see* HPCE, application, MECC
Malonate, HPCE, *see* HPCE, application, MECC
Maltose,
  for permanent coating of capillary, 265
  R_f-value, 139
Malvidin, structure, 290
Mancini, *see* Immunoelectrophoresis, single radial immunodiffusion
Manganese,
  metal–ligand binding constant, 45
  phosphatase activator, 349, 359
Mannans, DF component, 396
α-D-Mannopyranosyl, affinity chromatography, 147, 344–345
Mannose,
  GLC, 402, 406
  HPCE, *see* HPCE, application, FZCE
  HPLC, 405–406
  R_f-value, 139
  structure, 395
Mannuronic acid, structure, 395
Margicassidin, structure, 290
Mass spectrometry, 117–118
  detection in HPCE, 226
Matrix volume (V_g), 121–122
MCE, 250
Mean deviation, 6
Mean value, 4
Meat,
  cow beef, composition, 318
  ham, composition, 318
  hen breast, composition, 318
MECC, 233–237
  amino acids, HPCE, *see* HPCE, application, MECC
  aromatic carboxylic acids, HPCE, *see* HPCE, application, MECC
  aromatic choline esters, HPCE, *see* HPCE, application, MECC
  biogenic amines, HPCE, *see* HPCE, application, MECC
  flavonoids, HPCE, *see* HPCE, application, MECC
  glucosinolates, HPCE, *see* HPCE, application, MECC
  heteroaromatics, HPCE, *see* HPCE, application, MECC
  ions, HPCE, *see* HPCE, application, MECC
  peptides, HPCE, *see* HPCE, application, MECC
  phenolics, HPCE, *see* HPCE, application, MECC
  phospholipids, HPCE, *see* HPCE, application, MECC
  proteins, HPCE, *see* HPCE, application, MECC
  separation principle, 233–235
  water-soluble vitamins, HPCE, *see* HPCE, application, MECC
  MEKC, *see* MECC
Melatonin, 311
Membrane,
  lipid, 31, 76–79
  protein, 31, 38, 76–79
MES,
  buffer in HPCE, pK_a'-value, 253

MES (*continued*)
metal–buffer binding constant, 27
structure, 24
temperature dependence of pK_a'-value, 24
Mesh, 135
Mesityl oxide, as EOF marker, 210
Metal ion, metal–ligand binding constant, 45
Methanol,
as buffer additive, *see* HPCE, buffer additive, modifier, organic
eluotrope series, 28–29
as EOF marker, 210
for homogenization, 73
Methionine, HPCE, *see* HPCE, application, MECC
1-Methoxyacetic acid, *see* Chiral selectors
4-Methoxyglucobrassicin, HPCE, *see* HPCE, application, MECC
5-Methoxytryptamine, HPCE, *see* HPCE, application, MECC
3-Methoxytyramine, HPCE, *see* HPCE, application, MECC
4-Methylbenzylamine, λ_{max}, ε and pK_a'-values, 254
Methyl cellulose, *see* EOF, regulation
N,*N*′-Methylenebisacrylamide, *see* Bis
4-*O*-Methyl-D-glucuronic acid, structure, 395
N-Methylimidazole, 401
1-Methylnicotinamide, *see* Trigonellinamide
N-Methylnicotinamide, structure, 109
N-Methylnicotinate, structure, 109
N-Methylpyrrolinium, structure, 109
4-Methylumbelliferone, structure, 113
Micellar Electrokinetic Capillary Chromatography, *see* MECC
Micellar Electrokinetic Chromatography, *see* MEKC
Micellar flow, 233–235
Micelle,
aggregation number, 262
measurement of electrophoretic mobility, 263–264
mixed, 262
structure, 31–33
type of counterion, 31, 33, 254–255, 262, 299
Michaelis–Menten constant, 57, 337
Michaelis–Menten equation, 57–58
Michaelis–Menten kinetics, inhibition, 64–67
Microemulsion Capillary Electrophoresis, *see* MCE
Microscopic constant, 40, 42–44
Microtitre assay, immunosorbent, *see* ELISA
Migration time, *see* MT
Milk, composition, 318
Mobility; *see also* Electric mobility
analytes, electrophoretic, *see* Electrophoreic mobility
electro-osmotic, *see* EOF
Modifier,
amine, in HPCE, *see* HPCE, buffer additives
organic, in HPCE, *see* HPCE, buffer additives
organic, in SPE, 150
Molecular mass, relative (M_r) for proteins, 129
Molecular sieving, 182
Molecular weight determination,
FPLC gelfiltration, 339
HPCE, 339
MALDI-TOF, 339
SDS–PAGE, 199–203, 339
Monochromatic light, *see* Light, monochromatic
Monoclonal antibody, 375
Monosaccharide,
complexation with borate, *see* Borate, complexation
derivatization, 407
4-aminobenzoic acid, 407
4-aminobenzonitrile, 407
9-aminopyrene-1,4,6-trisulfonate, 407
2-aminopyridine, 407
3-(4-carboxybenzoyl)-2-quinolinecarboxaldehyde, 407
in DF, 395
HPCE, *see* HPCE, application, FZCE
Monosubstrate reaction, *see* Kinetics, mechanism
MOPS, buffer in HPCE, pK_a'-value, 253
MOPSO, buffer in HPCE, pK_a'-value, 253
Morphine, 281

Morpholine, buffer in HPCE, pK_a'-value, 253
2-Morpholinoethanesulfonic acid, *see* MES
Moving boundary electrophoresis, 188–190
MS, *see* Mass spectrometry
MT,
  relative, *see* HPCE, relative migration time
  window, *see* HPCE, migration time window
Mucilage, DF component, 396
Mucoproteins, pI-values, 84
Multidimensional NMR, *see* NMR
Multireflection cell, in HPCE, *see* HPCE, detection
Myoglobin, 106–107, 318
Myoglobin (horse),
  HPCE, *see* HPCE, application, CIEF
  pI-value, 84, 191
Myoglobin (sperm whale),
  diffusion coefficient, 129
  partial specific volume, 129
  relative molecular mass, 129
Myoinositol,
  internal standard in GLC, 403
  structure, 403
Myosin, 318
  diffusion coefficient, 129
  HPCE, *see* HPCE, application, capillary SDS–PAGE
  partial specific volume, 129
  relative molecular mass, 129
Myrecetin, structure, 106
Myrosinase,
  adsorption, 142
  assay,
    in microtitre plate, 343
    sinigrin substrate, 342–343
  detection in gels, 345
  FPLC chromatofocusing, 167, 345–346
  HPCE, *see* HPCE, application, MECC
  isoforms, 343, 346
  isolation and characterization, 344–347
  purification, 342–347
  transformation of glucosinolates, 71, 342

NA, equation, 219
NAD^+,
  HPCE, *see* HPCE, application, MECC
  UV spectrum, 110–111
NADH,
  as fluorescent, 113
  protein–ligand complex, $\Delta G^{\circ\prime}$, 37
  UV spectrum, 110–111
$NADP^+$, HPCE, *see* HPCE, application, MECC
NADPH, as fluorescent, 113
NaOH, *see* Hemicellulose extraction
Napoleiferin, MECC, HPCE, *see* HPCE, application, MECC
Naringin, 105
Native gel electrophoresis, *see* Electrophoresis
Natural products,
  amphiphilic, TLC separation, 137
  lipophilic, TLC separation, 137
NC blot,
  gold staining, 388
  immunostaining, 388
NC, blotting to NC from agarose gel, 387
NDF, 398
Neoglucobrassicin, HPCE, *see* HPCE, application, MECC
Neutral detergent method, *see* NDF
Nickel, metal–ligand binding constant, 45
Nicotinamide,
  HPCE, *see* HPCE, application, MECC
  structure, 109
Nicotinate, structure, 109
Nicotine, structure, 109
Nigrosine, *see* Staining
Ninhydrin, *see* Amino acid, derivatization, and Staining
Nitrate, HPCE, *see* HPCE, application, MECC
Nitrile, $\lambda_{max}/\varepsilon$-values, 102
Nitro blue tetrazolium, *see* Immunoassay, staining
Nitrocellulose, *see* NC
Nitrogen, analysis, Kjeldahl/Dumas, *see* Protein, quantification
p-Nitrophenylphosphate,
  in immunoassay, 382
  phosphatase assay, 357, 360
NMR, 114–117
  COSY, 116
  NOE, 116

NMR (*continued*)
  NOESY, 116
  TOCSY/HOHAHA, 116
NOE, *see* NMR
NOESY, *see* NMR
Non-covalent interaction, 29
Non-Newtonian solution, 127
Non-starch polysaccharides, *see* NSP
Noradrenalin, HPCE, *see* HPCE, application, MECC
Normal distribution, 6
Normalized peak area, *see NA*
NSP, 394
Nuclear Magnetic Resonance, *see* NMR
Nuclear Overhauser effect, *see* NMR, NOE
Nucleic acid,
  borate complex, *see* Borate, complexation
  detection, 112
  ionization, 18
  precipitation, 324
  UV absorption, 100
Nucleoside, UV spectra, 101
Nucleotide, 302–303
Nucleotide phosphate, metal–ligand binding constant, 45–46
Nudiflorine, structure, 109

Oat, composition, 317–318
Octadecanophenone, HPCE, *see* HPCE, application, HICE
Octylglucoside,
  data for, 33, 262
  structure, 32
Octyl-β-D-glucopyranoside, *see* Octylglucoside
α-ODAP,
  HPCE, *see* HPCE, application, FZCE
  lathyrism, 293
β-ODAP,
  HPCE, *see* HPCE, application, FZCE
  lathyrism, 293
Oil body, 77
Oligonucleotides, HPCE, *see* HPCE, application, CGE
Oligosaccharide, 347
  borate complexation, *see* Borate, complexation
  DF associated, 397, 410
  extraction, 410
  HPCE, *see* HPCE, application, FZCE/CGE
OPA, *see* Amino acid, derivatization
OPD, *see* Immunoassay, detection systems
Orcinol, *see* Staining
Organic acids, HPCE, *see* HPCE, application, FZCE
Organic modifier; *see also* HPCE, buffer additive, modifier,
  influence on c.m.c.-value, 33, 260
  influence on EOF, 258
  influence on pK_a'-value, 30
Organic solvent,
  extraction of lipids, *see* Lipid, extraction
  polarity, 28
Osmotic pressure, 87, 125–126
Ouchterlony, *see* Immunoelectrophoresis, double radial immunodiffusion
Ovalbumin, 317
  diffusion coefficient, 129
  $E^{1\%}_{1\,cm}$-value, 100
  HPCE, *see* HPCE, application, capillary SDS–PAGE
  partial specific volume, 129
  pI-value, 84
  relative molecular mass, 129
Ovomucoid, 317, 372
  enzyme inhibitor assay, 365–366, 372
  isolation and characterization, 372
  pI-value, 84
Ovotransferrin, 317
Oxalate, HPCE, *see* HPCE, application, FZCE/MECC
α-Oxalylamino-β-aminopropionic acid, *see* α-ODAP
β-Oxalylamino-α-aminopropionic acid, *see* β-ODAP
Oxidoreductase cytochrome *P*-450, 106–107
Oxidoreductase, kinetics, 59
Oxygen, analysis, 279

PAGE, 187–188
Papaverin, 281
Paper chromatography, *see* PC
Paper electrophoresis, *see* Electrophoresis, paper

Partial specific volume,
for proteins, 129–130
unit, 3
Particle diameter, 135–136
Partition chromatography, 130
PAS, *see* Staining
Pauly reaction, 112
PC, 130–142
R_f-values for carbohydrates, 139
solvents and detection, 138–139
Pea,
composition, 317–318
FZCE of oligosaccharides, 230, 307
α-galactosidase, 345–352
inhibitor, 365–369
Peak,
area, *see* HPCE, peak area
fronting, *see* Fronting of peaks
shape, *see* Shape of peaks
symmetry, *see* Symmetry of peaks
tailing, *see* Tailing of peaks
Peanut, composition, 317–318
Pectin,
dominating polysaccharides, 396
extraction media, 409
gel, 30
PEG, *see* HPCE, buffer additive, modifier, organic
Pelargonidin, structure, 106, 290
Pentanes, *see* Lipid, extraction
Pentanol, organic modifier in HPCE, *see* HPCE, buffer additive, modifier, organic
Peonidin, structure, 106, 290
Pepsin, pI-value, 84
Peptide, 291–292, 294–297
adsorption to capillary in HPCE, 214
detection by ninhydrin, 333
general analysis, 294–297
group separation, 161–163, 292
HPCE, *see* HPCE, application, MECC/CIEF
HVE, 183–184
hydrolysis pattern, 'fingerprint', 340
PC/TLC, 138–139
RPC purification, 149
Perchlorate, *see* Hemicellulose, extraction
chaotropic compound, 81
Perforation, *see* Lipid, extraction
Performance,
analytical equipment, 10
method, 10
Periodate, *see* Staining
Periodic acid-Schiff, *see* PAS
Periodic table, 5
Permanent coating, capillary, *see* HPCE, instrumentation, capillary
Permanganate oxidation method, *see* Lignin analysis
Permeation chromatography, 142–144
Peroxidase, 316
assay,
guajacol substrate, 353
TMB substrate in microtitre plates, 353–354
haem group, 106, 108
horseradish, in immunoassay, 382
isolation and characterization, 354–356
kinetics, 61
purification,
milk, 354
plant material, 355
Peroxygenation, 107
Petunidin, structure, 290
Petunidin-3,5-diglucoside, HPLC, 291
pH, 15–16
influence on electrophoretic mobility, *see* HPCE, pH
influence on EOF, *see* HPCE, pH
hysteresis, *see* HPCE, pH hysteresis
pH meter,
cleaning, 26
general handling, 26
Pharmacia Phast Gel Electrophoretic System™, 191–195
Phenethylamine, HPCE, *see* HPCE, application, MECC
Phenol; *see also* Staining
$\lambda_{max}/\varepsilon$-values, 102
Phenolate, $\lambda_{max}/\varepsilon$-values, 102
Phenolic compounds, 235, 300
colorimetric methods for determination, 112
DF associated, 300, 394
general analysis, 300–302
group separation, 161–163
ionization, 43–44
UV absorption, 104–108

Phenylalanine,
HPCE, *see* HPCE, application, MECC
UV absorption, 103–104
UV spectrum, 103
Phenylalanine ammonia-lyase, 369–372
assay, 370
isolation and characterization, 370–372
Phenylboronate, ligand in affinity chromatography, 147
Phenylenediamine, ortho-, *see* OPD
Phenylmethylsulfonyl flouride, *see* PMSF
pH_i, *see* pI
Phosphatase,
acid, 356–362
assay, 360
isolation and characterization, 361–362
activators, 349, 359
alkaline, 341, 347–349
assay, 357
in immunoassay, 382
isolation and characterization, 357–359
Phosphate,
analysis, 302
buffer in HPCE, pK_a'-value, 253
C-eluate, 298, 302–303
HPCE, *see* HPCE, application, FZCE
interactions, 27
metal–ligand binding constant, 45
Phosphate anhydride, metal–ligand binding constant, 45
Phosphate ester, 302
HPLC, 303
HVE, 184
metal–ligand binding constant, 45–46
Phosphatidic acid, HPCE, *see* HPCE, application, MECC
Phosphatidylcholine, HPCE, *see* HPCE, application, MECC
Phosphatidylethanolamine, HPCE, *see* HPCE, application, MECC
Phosphatidylinositol, HPCE, *see* HPCE, application, MECC
Phosphatidylserine, HPCE, *see* HPCE, application, MECC
Phospholipid,
amphiphatic property, 30
extraction, 72
general analysis, 302
HPCE, *see* HPCE, application, MECC
in membranes, 77
PC/TLC, 138–139
Phosphorescence, 92
Phosphoribosyl pyrophosphate, *see* PRPP
Phosphorylase B, HPCE, *see* HPCE, application, capillary SDS–PAGE
o-Phthalaldehyde, *see* OPA
Phthalate, λ_{max}, ε and pK_a'-values, 254
Phytase, 316
Phytate, metal–ligand binding constant, 45
pI, 17–18
determination; *see also* IEF/CIEF, 339–340
marker, 191
selected proteins, 84
Piperazine-*N*,*N'*-bis(ethanesulfonic acid), *see* PIPES
PIPES,
buffer in HPCE, pK_a'-value, 253
metal–buffer binding constant, 27
structure, 24
temperature dependence of pK_a'-value, 24
Pisum sativum; *see also* Pea
α-galactosidase, 348, 351–352
HPLC of low-M_r phosphate esters, 303
MECC of heteroaromatics, 286–287
monosaccharides in IDF, 405
pK_a'-value,
buffer in HPCE, 253
change of, 24–25, 27, 30
practical, 14, 16
thermodynamic, 14, 16
Planck constant, 1, 95
Plane polarized light, *see* Light, plane polarized
Plant cell wall, 76, 394
PMSF, 80, 369
Poiseuille's law, 222–223
Polarity, eluotrope series, 28–29
Polyacrylamide,
in CGE, 243–245
for permanent coating of capillary, 265
Polyacrylamide gel, 187–188
Coomassie Brilliant Blue staining, 199
C-value, 188
electrophoresis, *see* PAGE
formation of, 187–188

preparation,
for IEF, 195
for SDS–PAGE, 201
silver staining, 195–196, 202
T-value, 188
Polyampholyte, 16–17
Polyclonal antibody, 374
Polyelectrolyte, 16–17, 87, 126
Polyethylene glycol,
in CGE, 244
for dialysis, 88
for permanent coating of capillary, 265
Polyethyleneimine, for permanent coating of capillary, 265
Polymer, polyprotic, 16
Polynucleotides, HPCE, *see* HPCE, application, CGE
Polyoxyethylene(9.5)*p-t*-octylphenol, *see* Triton X-100
Polyprotic polymer, *see* Polymer, polyprotic
Polysaccharide; *see also* DF
HPCE, *see* HPCE, application, CGE
hydrolysis, 399–400
ionization,
carboxymethylated, 18
phosphorylated, 18
Polyvinyl alcohol, for permanent coating of capillary, 265
Pore size,
in CGE, 245
in dialysis tubing, 84
in gels, 186–188
in ion-exchange chromatography, 158–159
Porphyrin, metal ion cofactors, 45
Post-column derivatization, 227
Potassium,
HPCE, *see* HPCE, application, FZCE
metal–ligand binding constant, 45
Potato,
composition, 317
inhibitor, 316, 365
Potential, *see* Electric potential
Potentiometry, detection in HPCE, 226
Power function, 7
Power, unit, 3
Precipitation,
protamin sulfate, 324
of protein,
with ammonium sulfate, 81–82, 324
at isoelectric pH, 82–83, 324
in organic solution, 75, 84, 324–325
protein separation strategy, 321, 324–325
Precipitin methods, antibody analyses, 379–382
Pre-column derivatization, 227
Prefix, 4
Preparative Isoelectric Membrane Electrophoresis, *see* IsoPrime™
Pressure injection, in HPCE, *see* HPCE, injection
Pressure, unit, 3
Pressure–volume product, 2
Pretreatment of samples,
general, 71
SDS–PAGE, 201
Procion blue, *see* Staining
Procion S, *see* Staining
Proenzyme, 362–363
Progoitrin, HPCE, *see* HPCE, application, MECC
Prolamin, 79
distribution, 316, 318
Proline, HPCE, *see* HPCE, application, MECC
Propanol, *see* HPCE, buffer additive, modifier, organic
eluotrope series, 28
Protamin, 79
Protamin sulfate, precipitation of DNA, 324
Protease inhibitors, as cosolvent, 80
Protein,
activity,
biological, 320
specific, 320, 322
adsorption to capillary wall, 214–215, 327
binding properties; *see also* Antigen–antibody binding, 37–38, 48, 53, 79, 149
denaturation,
acid induced, 75
organic solvent induced, 75
precautions, 79, 84
detection,
fluorescence, 112, 206
staining, 112, 205–206
UV, 99–101, 103–104, 206

Protein (*continued*)
DF associated, 394, 397
diffusion coefficient, 129
$E^{1\%}_{1\,cm}$-values, 100
extraction,
acetone, 84–85
detergent, 78–80
ethanol, 84–85
organic solvents, 28, 75, 319, 324
water, 79
FPCL,
affinity chromatography, 175–176, 326–327
chromatofocusing, 174, 321
gelfiltration, 175–176, 325
HIC; *see also* HIC, 175–176, 321
ion-exchange liquid chromatography, 174–175
gel electrophoresis,
disc gel, 196, 198, 200
IEF, 190–196, 339
immunoelectrophoresis, 379–382
SDS–PAGE, 199–203
group separation, 321, 323–326
guide to boxes, separation, 341
handling, 319–320
solubilization, 80, 315, 319
storage, 319, 378
HIC, 148–149
homogenization, 323
HPCE, *see* HPCE, application, capillary SDS–PAGE
HPCE, *see* HPCE, application, CGE
HPCE, *see* HPCE, application, CIEF
HVE, 184–185
hydrolysate, soy bean, 340
hydrolysis, 332
integral, 31, 77, 315
ion-exchange chromatography, 155–159, 164–166, 321
isoelectric points, *see* Protein, pI-value
lipoprotein, precipitation, 324
natural conformation, 79
partial specific volume, 129–130
peripheral, 77–78, 315
pI-value, 17–18, 84
precipitation,
fractional precipitation by ammonium sulfate, 81–83, 324
isoelectric, 82–83, 324
organic solvent, 84–85, 324–325
protamine sulfate, 324
present in,
animals, 317–318
cereals, 316–318
crucifers, 317
fruit, 316
grain legumes, 316–318
membranes, *see* Membrane, protein
vegetables, 316–318
purification,
efficiency, 320
FPLC/HPLC, 175–176
in general, 315–373
HIC, 148–149
of hydrolysed protein, 333
low-M_r RPC, 149–150
strategy, 320–321
quantification,
Bicinchoninin acid method, 112, 330
Biuret method, 112, 330
Bradford method, in microtitre assay, 331
Coomassie Brilliant Blue, 112, 206, 330–331
crude protein, 279
Dumas technique, 279, 330–332
interference, 27, 100–101, 324
Kjeldahl analysis, 279, 330–332
Lowry, 112
silver, 112, 206
UV, 99–101, 206, 328–329
relative molecular mass, 129
sample,
concentration, 323
desalting, 88–89, 175–176, 322
solubility, 79–84, 319
effect of salt, 80–82
as a function of pH, 82–85
subclasses, 79
UV spectrum, 99
Protein A,
affinity purification of antibodies, 326, 377–378
ligand in affinity chromatography, 147
Proteinase inhibitor; *see also* Inhibitor, 365–372
binding, 49

disc gel electrophoresis, 198–199
HPCE, *see* HPCE, application, MECC
in protein extraction, 79
Protein G, ligand in affinity chromatography, 147
Protolytic activity, 12–14, 16–18
Protonization,
degree of, 18–19
macroscopic constant, *see* Macroscopic constant
microscopic constant, *see* Microscopic constant
Pseudomonosubstrate reaction, *see* Kinetics, mechanism
Pseudoreaction order, *see* Kinetics, reaction order
Pulchellidin, structure, 290
Purine, 100–101
Putrescine, 290
Pyridine, 100, 110–111
electron sink, 44
spectroscopic properties, 111
λ_{max}, ε and pK_a'-values, 254
Pyridoxalphosphate, microscopic constant, 44
Pyridoxamine, HPCE, *see* HPCE, application, MECC
Pyridoxamine phosphate, HPCE, *see* HPCE, application, MECC
Pyridoxine,
HPCE, *see* HPCE, application, MECC
microscopic constant, 42–44
Pyrimidine, 100–101, 110
derivatives, structure, 109
glycoside, adsorption, 142
Pyromellitic acid, λ_{max}, ε and pK_a'-values, 254

Quartz, capillary tubing material, 221
Quaternary amines, *see* HPCE, instrumentation, capillary, dynamic coating
Quercetin,
HPCE, *see* HPCE, application, MECC
structure, 106, 309
Quercetin-3-(6″-carboxyglucoside), HPCE, *see* HPCE, application, MECC
Quercetin-3-glucoside, HPCE, *see* HPCE, application, MECC
Quinoidal base, structure, 290
Quinolate, structure, 109
Quinolizidine alkaloids, 281

Radiation,
reflection, 98
scattering, 98
Radioactivity, detection in HPCE, 226
Radix althaeae, FZCE of monosaccharides, 231
Raffinose,
^{1}H NMR data, 117
HPCE, *see* HPCE, application, FZCE
HPLC, 307
natural substrate, 350
Raoult's law, 126
Rapeseed,
cinnamic/benzoic acid derivatives, IDF, 410
composition, 317
inhibitor, *see* RPPI
MECC of glucosinolates, 239, 261, 304–305
monosaccharides,
IDF, 407
SDF, 402
protein-type proteinase inhibitors, *see* RPPI
myrosinase, *see* Myrosinase
Redox cofactors, 108–111
Reflection of radiation, *see* Radiation
Refractive index, *see* RI
Reinheits Zahl, *see* RZ
Relative migration time, *see* RMT
Relative normalized peak area, *see* *RNA*
Repeatability, 6, 217, 219–220, 275
Reproducibility, 6, 219–220, 275
Reseda odorato, MECC of amino acids, 293
Resistance, *see* Electric resistance
Resistant starch, *see* RS
Resolution, 135, 215
Response factor, 220
Restriction digests, HPCE, *see* HPCE, application, CGE
Retention,
coefficient (R), 133
factor (k'), 134
time (t_R), 132–133

Reversed-phase liquid chromatography, *see* RPC
Reversed-phase chromatography, *see* RPC
R_f-value, 132, 139
Rhamnogalacturonans, DF component, 396
Rhamnose,
  GLC, 406
  HPCE, *see* HPCE, application, FZCE
  HPLC, 406
  R_f-value, 139
  structure, 395
Rhombifolin, structure, 283
Riboflavin,
  as fluorescent, 113
  HPCE, *see* HPCE, application, MECC
Riboflavin-5′-phosphate, HPCE, *see* HPCE, application, MECC
Ribonuclease,
  diffusion coefficient, 129
  partial specific volume, 129
  relative molecular mass, 129
Ribonuclease A, $E_{1\,\mathrm{cm}}^{1\%}$-value, 100
Ricinine, 281
  structure, 109
RI detection,
  HPLC/FPLC, 169
  HPCE, 226
RMT, equation, 218
RNA,
  equation, 220
  partial specific volume, 130
  UV absorption, 100–101
  UV spectrum, 100
Rosinidin, structure, 290
RPC, 149–150, 172–173
RPPI, 295, 365–372
  HPCE, *see* HPCE, application, MECC
RS,
  determination by ELISA reader, 410–411
  DF component, 396–397, 400
Rustocide,
  HPCE, *see* HPCE, application, MECC
  structure, 309
Rutin,
  HPCE, *see* HPCE, application, MECC
  structure, 309
Rye, composition, 317–318
RZ, 108, 347

Safety rules, 7–8
Sakaguchi reaction, 112, 139
Salicylic acid,
  HPCE, *see* HPCE, application, MECC
  structure, 105
  λ_{max}-value, 105
Salmon, composition, 318
Salt, effect on protein solubility, *see* Protein, solubility
Salting in, *see* Protein, solubility
Salting out, *see* Protein, solubility
Sample; *see also* HPCE, sample
  pretreatment, 71
  stacking, *see* HPCE, sample
Sampling, 71
Saponin,
  analysis, 299–300
  antibody detection, 385–386
  conjugation to KSTI, 389
  PC/TLC, 138–139
Savoy cabbage, MECC of desulfoglucosinolates, 304
Scaling down, 71
Scanning optics (Schlieren), 188
Scatchard plot, *see* Kinetics, delineation
Scattering of radiation, *see* Radiation
SCS, data for, 33
SDF, 398, 402
SDS,
  data for, 33, 262
  denaturation, 199
  structure, 32
SDS–PAGE, 199–203
  capillary, 246
  preparation of polyacrylamide gels, 201
  procedure, 201–202
Selectivity, 135, 176
Separation efficiency, 215–216
Separation factor (α), 134
Sephacryl, *see* Gelfiltration
Sephadex, *see* Gelfiltration
Sephadex material, for dialysis, 88
Sepharose, *see* Gelfiltration
Serine, 332
  HPCE, *see* HPCE, application, MECC
Serotonine, HPCE, *see* HPCE, application, MECC

Serum albumin, *see* BSA
SFC, Appendix
SFE, 74, Appendix
Shape of peaks, 211–213
Silanol groups, capillary surface, *see* Capillary
Silica,
  capillary tubing material, 221
  for electrophoresis, 186
  for HPLC/FPLC, 171
  for ion-exchange chromatography material, 158
  for TLE, 136
Silver staining, *see* Electrophoresis, gel, staining, agarose gel, Electrophoresis, gel, staining, polyacrylamide gel, and Staining
Sinapic acid,
  HPCE, *see* HPCE, application, MECC
  HVE, 185
  structure, 105
  λ_{max}-value, 105
Sinapic chromophore, PC, 141
Sinapine,
  HPCE, *see* HPCE, application, MECC
  HPLC, 281
  HVE, 185
  structure, 282
Sinapis alba, 281–282, 347
cis-2-*O*-Sinapoyl-L-malate, HPCE, *see* HPCE, application, MECC
trans-2-*O*-Sinapoyl-L-malate, HPCE, *see* HPCE, application, MECC
Sinigrin, *see* Myrosinase assay
  HPCE, *see* HPCE, application, MECC
Siphoning action injection, *see* HPCE, injection
Size-exclusion chromatography, 133, 142–144, 175–176
Slab gel electrophoresis, *see* Electrophoresis, gel
Sodium azide, 23, 320
Sodium cetyl sulfate, *see* SCS
Sodium chloride, as cosolvent, 80
Sodium cholate,
  data for, 33, 262
  structure, 32
Sodium deoxycholate,
  data for, 33, 262
  structure, 32
Sodium-*N*-dodecanoyl-L-valinate, enantiomeric detergent, 260
Sodium dodecyl sulfate, *see* SDS
Sodium dodecyl sulfate–polyacrylamide gel electrophoresis, *see* SDS–PAGE
Sodium hexadecyl sulfate, data for, 33
Sodium, HPCE, *see* HPCE, application, FZCE
Sodium octadecyl sulfate, *see* SOS
Sodium taurodeoxycholate, data of, 262
Sodium tetradecyl sulfate, *see* STS
Solid phase extraction, *see* SPE
Soluble dietary fibres, *see* SDF
Solvation, 14
Solvents, disrupting, 70
Soret peak, 108
SOS, data for, 33
Southern blotting, 206
Soxhlet extraction, *see* Lipid, extraction
Soyasaponin I, *see* Saponin
Soybean,
  composition, 317–318
  inhibitor; *see also* BBI and KSTI, 49, 365
  MECC of phospholipids, 241
  protein hydrolysate, 340
Soybean trypsin inhibitor, HPCE, *see* HPCE, application, capillary SDS–PAGE
Spacer arm, affinity chromatography, 146–147
Spartein, structure, 283
SPE, 74, 149–150
Specific gravity, *see* Gravity, specific
Specific heat, *see* Heat, specific
Spectroscopy, 92–119
  absorption; *see also* UV–Vis spectroscopy, 92
  correlated, *see* COSY
  emission, 92
  mass, 117–118
  NOE, *see* NOESY
  totally correlated, *see* TOCSY/HOHAHA
Speed of light, *see* Light
Spiking, 74, 218–219
Spin number, 114

Stachyose,
^{1}H NMR data, 117
HPCE, *see* HPCE, application, FZCE
HPLC, 307
Stacking, *see* Sample, stacking
Staining,
amido black, 205–206, 388
ammonium molybdate, 112
ammonium sulfate, 112, 139
aniline, 112, 138–139
BCA, *see* Bicinchoninin acid method
biuret, *see* Biuret method
bromophenol blue, 205
Coomassie Blue, *see* Coomassie Brilliant Blue
dimethylaminobenzaldehyde, 112
2,4-dinitrofluorobenzene, 112
2,4-dinitrophenylhydrazine, 112
dinitrosalicylate, 112
diphenylamine, 112, 138
Dragendorff reagens, 112, 139
Folin, *see* Folin–Ciocalteu
gold, of NC blot, 388
immuno, of NC blot, 388
ink, 388
Liebermann–Burchard, 112
lissamine green, 205
Lowry, *see* Folin–Ciocalteu
nigrosine, 205
ninhydrin; *see also* Amino acid, derivatization, 112
orcinol, 112
PAS, 205–206
periodate, 112, 139
phenol, 112
procedures; *see also* specific subjects, 111–112, 205–206, 382
Procion blue, 205
Procion S, 205
silver, 112, 195–196, 198, 202, 205–206
thymol, 112, 139
Standard,
external, *see* External standard
internal, *see* Internal standard
Standard deviation (of mean value), 6
Starch,
determination by ELISA reader, 410–411
DF component, 396
Starch gel, pore size, *see* Pore size
Statistics, significant figure, 4
Stern layer, 179–181
Stern plane, 179–181
Steroid glycoside, TLC separation, 137–139
Steroids, colorimetric methods for determination, 112, 139
Stokes' equation, 179
Stokes' law, 90, 126
Stoldt fat, 72
Structural epimers, HPCE, *see* HPCE, application, MECC
STS, data for, 33, 262
Succinate, buffer in HPCE, pK_a'-value, 253
Sucrose,
as cosolvent, 80
HPCE, *see* HPCE, application, FZCE
HPLC, 307
Sudan II, *see* HPCE, micellar flow
Sudan IV, *see* HPCE, micellar flow
Sulfanylethanol,
antioxidant, 369
reducing agent in SDS–PAGE, 199–200
Sulfate,
C-eluate, 298–299
general analysis, 303
HPCE, *see* HPCE, application, FZCE
Sulfate ester,
HVE, 184
separation by HPLC I-P RPC, 174
Δ^4-2-*O*-Sulfo-GlcUA-GalNAc, HPCE, *see* HPCE, application, MECC
Δ^4-2-*O*-Sulfo-GlcUA-6-*O*-sulfo-GalNAc, HPCE, *see* HPCE, application, MECC
Δ^4-2-*O*-Sulfo-GlcUA-sulfo-GalNAc, HPCE, *see* HPCE, application, MECC
Sulfonate,
C-eluate, 298–299
general analysis, 303
Sulfur, analysis, 279
Sulfuric acid,
for extraction of cellulose, 409
for polysaccharide hydrolysis, 399–400
Supercritical fluid extraction, *see* SFE
Supercritical fluid chromatography, *see* SFC
Superdex, *see* Gelfiltration

Superose, *see* Gelfiltration
Support media, *see* Electrophoresis
Surface area of matrix, 122, 124, 135–136, 168
Surface shear, 179–181
Surface tension, 124–125
  capillary active compounds, 124
  capillary inactive compounds, 124
  influence of butyric acid, *see* Butyric acid
  influence of valeric acid, *see* Valeric acid
Surfactant, *see* Detergent
Swelling, 142–144, 155–158
Symbol(s), 1–2
Symmetry of peaks, 212–213

T_3, 311
  HPCE, *see* HPCE, application, MECC
T_4, 311
  HPCE, *see* HPCE, application, MECC
Tailing of peaks, 212–213
Tartaric acid, complexation with cations, 216
Tartrate, HPCE, *see* HPCE, application, MECC
Taurine, zwitterion in HPCE, 259
TDF, 398, 409
Teflon, capillary tubing material, 221
TEMED, in polymerisation of polyacrylamide gels, 187–188
Temperature,
  absolute, 2
  gradient, in HPCE, 213–214
TES,
  buffer in HPCE, pK_a'-value, 253
  metal–buffer binding constant, 27
  structure, 24
  temperature dependence of pK_a'-value, 25
Tetradecanophenone, HICE, HPCE, *see* HPCE, application, HICE
Tetradecyltrimethylammonium bromide, *see* TTAB
Tetradecyltrimethylammonium chloride, *see* TTAC
Tetrahydrorhombifolin, structure, 283
Tetrahydroxyborate ion, 406
Tetramethylbenzidine (3,3′,5,5′-), *see* TMB
N,*N*,*N*′,*N*′-Tetramethylenediamide, *see* TEMED
Tetramethylsilane, *see* TMS
Tetrapyrrole haem structure, 106
Theoretical plates,
  FPLC/HPLC, 173
  HPCE, 215–216
  LC, 134
Thermal conductivity, *see* Conductivity, thermal
Thermal lens, detection in HPCE, 226
Thiamine, HPCE, *see* HPCE, application, MECC
Thin-layer chromatography, *see* TLC
Thin-layer electrophoresis, *see* TLE
Thiocyanate, 71, 235
  chaotropic compounds, 81
  HPCE, *see* HPCE, application, MECC
β-Thioglucoside glucohydrolase, *see* Myrosinase
Thiol,
  compound, as cosolvent, 80
  derivatives, 294
  in protein extraction, 79–80
  microscopic constant, 42–44
  $\lambda_{max}/\varepsilon$-values, 102
Thiolate, $\lambda_{max}/\varepsilon$-values, 102
Thiophil adsorption, of antibody, 377
Threonine, HPCE, *see* HPCE, application, MECC
Thymidine, UV spectrum, 101
Thymohistone, pI-value, 84
Thymol, *see* Staining
L-Thyroxine, *see* T_4
Time,
  in mobile phase (t_m), 132
  in stationary phase (t_s), 132
Tiselius apparatus, 188
Titration curve, 19
TLC, 130–142
  solvents and detection, 138–139
TLE, 186
TMB,
  in immunoassay, 382, 385
  in peroxidase assay, 353–354
TMP, HPLC, 303
TMS, as reference compound in NMR, 115
TOCSY/HOHAHA, *see* NMR
Toluene, eluotrope series, 28–29

Tomato, composition, 317
Total dietary fibres, *see* TDF
Totally correlated spectroscopy, *see* NMR, TOCSY/HOHAHA
Transferase, kinetics, 61
Transferrin, contamination in antibody purification, 376
Transmittance, 98–99
Trichloromethane, *see* Chloroform
Tricine,
  buffer in HPCE, pK_a'-value, 253
  metal–buffer binding constant, 27
  structure, 24
  temperature dependence of pK_a'-value, 25
Trifluoroacetate, chaotropic compounds, 81
Trifluoroacetic acid, for polysaccharide hydrolysis, 399–400
Trigonellinamide,
  HPCE, *see* HPCE, application, MECC
  as standard, 73
  structure, 109, 186
  UV spectrum, 110
Trigonelline,
  HPCE, *see* HPCE, application, MECC
  structure, 109, 286
3,4,5-Trihydroxybenzoylcholine, HPLC, 281
L-Triiodothyronine, *see* T_3
Trimellitic acid, λ_{max}, ε and pK_a'-values, 254
Trimethylsilyl derivatives, in GLC, 401
Tripeptide, HPCE, *see* HPCE, application, MECC, peptides
N-Tris(hydroxymethyl)-2-aminoethanesulfonic acid, *see* TES
Tris(hydroxymethyl)methylamine, *see* TRIS
N-Tris(hydroxymethyl)methylglycine, *see* Tricine
TRIS,
  buffer capacity, 22
  buffer in HPCE, pK_a'-value, 253
  metal–buffer binding constant, 27
  structure, 24
  temperature dependence of pK_a'-value, 25
Triton X-100
  data for, 33, 262
  structure, 32
Tropomyocin,
  diffusion coefficient, 129
  partial specific volume, 129
  relative molecular mass, 129
Tropomysin, 318
Troponin, 318
Trypsin, 362–365
  assay, 362–363
  $E_{1\,cm}^{1\%}$-value, 100
  inhibitor assay, *see* Inhibitor, trypsin, assay
  inhibitor; *see also* Inhibitor, and Proteinase inhibitor
  inhibitor (KSTI), pI-value, 84
  isolation and characterization, 363–365
Trypsin (bovine),
  diffusion coefficient, 129
  partial specific volume, 129
  relative molecular mass, 129
Trypsinogen, pI-value, 84
Tryptamine, HPCE, *see* HPCE, application, MECC
Tryptophan,
  as fluorescent, 113
  HPCE, *see* HPCE, application, MECC
  macromolecule–ligand binding constant, determination, 48
  structure, 109, 286
  UV absorption, 103–104
  UV spectrum, 103
TTAB, data on, 33
TTAC, data on, 33
Two-dimensional electrophoresis, *see* Electrophoresis, gel
Tyramine, HPCE, *see* HPCE, application, MECC
Tyrosine,
  as fluorescent, 113
  HPCE, *see* HPCE, application, MECC
structure, 286
  UV absorption, 103–104
  UV spectrum, 103
Tyrosine–glucoside, HPCE, *see* HPCE, application, MECC

UDP, manganese–ligand binding constant, 46

UDP–glucose, HPLC, 303
Ultrafiltration, 88–89
  protein separation strategy, 321, 323, 325–326
UMP, HPLC, 303
Units, 1–4
  SI, 1–2
Unpolarized light, *see* Light, unpolarized
Urea; *see also* Hemicellulose, extraction
  additive in MECC, 263–264, 269
  chaotropic compounds, 81
  influence on c.m.c.-value, 34
Urease,
  diffusion coefficient, 129
  partial specific volume, 129
  relative molecular mass, 129
Uridine, UV spectrum, 101
Uronic acid, analysis, 408
UV–Vis spectrophotometer,
  double-beam, 97
  multi-beam, 98
  single-beam, 98
UV–Vis spectroscopy; *see also* Protein, quantification, 93–111
  cut-off values for solvents, 97
  in HPCE, 226
  in HPLC/FPLC, 169
  instrumentation, 97–98

Vacuum injection, *see* HPCE, injection
Valeric acid, influence on surface tension, 125
Valine, HPCE, *see* HPCE, application, MECC
Van der Waals forces, 35
Vanillic acid,
  structure, 105
  λ_{max}-value, 105
Vanillylcholine,
  HPCE, *see* HPCE, application, MECC
  structure, 282
Vegetables, composition, 317–318
Verbascose,
  HPCE, *see* HPCE, application, FZCE
  HPLC, 307
Vicia faba,
  HPLC of heteroaromatics, 285
  MECC of heteroaromatics, 285–287
Vicilin, 316
Vicine,
  HPCE, *see* HPCE, application, MECC
  structure, 110
Virial coefficient, 87
Viscosity, 124, 126–127
  unit, 3
Vitamin,
  analysis, 310–311
  HPCE, *see* HPCE, application, MECC
  lipid soluble, TLC separation, 137
  water-soluble, HPCE, *see* HPCE, application, MECC
Vitamin B_6, microscopic constant, 42
Volume, partial specific for proteins, 129–130
Volumetric flow, velocity, 136

Warburg–Christian method, 100
Water,
  acid–base properties, 13
  buffer capacity, 21–22
  eluotrope series, 28–29
  for extraction of pectins, 409
  ion product, 13
  organized structure, 13
Wavelength, 95
Wavenumber, 95
Western blotting, 206–207
Wheat, composition, 317–318
Whey, 317
Willardine,
  HPCE, *see* HPCE, application, MECC
  structure, 286

Xenobiotica, 69
  hydroxylation, 106
X-ray spectroscopy, 93–94
Xylans, DF component, 395, 409
Xyloglucans, DF component, 395
Xylose,
  GLC, 402, 406
  HPCE, *see* HPCE, application, FZCE
  HPLC, 406
  R_f-value, 139
  structure, 395

Youden experimental design, 275–276

Z-cell, *see* HPCE, detection
Zeta potential, 180–182
Zinc ion, phosphatase activator, 349, 359
Zinc, metal–ligand binding constant, 45
Zone electrophoresis, *see* Electrophoresis, zone
Zwitterion; *see also* HPCE, capillary, dynamic coating, 17